IUTAM Symposium on Smart Structures and Structronic Systems

SOLID MECHANICS AND ITS APPLICATIONS
Volume 89

Series Editor: G.M.L. GLADWELL
Department of Civil Engineering
University of Waterloo
Waterloo, Ontario, Canada N2L 3GI

Aims and Scope of the Series

The fundamental questions arising in mechanics are: *Why?*, *How?*, and *How much?* The aim of this series is to provide lucid accounts written by authoritative researchers giving vision and insight in answering these questions on the subject of mechanics as it relates to solids.

The scope of the series covers the entire spectrum of solid mechanics. Thus it includes the foundation of mechanics; variational formulations; computational mechanics; statics, kinematics and dynamics of rigid and elastic bodies: vibrations of solids and structures; dynamical systems and chaos; the theories of elasticity, plasticity and viscoelasticity; composite materials; rods, beams, shells and membranes; structural control and stability; soils, rocks and geomechanics; fracture; tribology; experimental mechanics; biomechanics and machine design.

The median level of presentation is the first year graduate student. Some texts are monographs defining the current state of the field; others are accessible to final year undergraduates; but essentially the emphasis is on readability and clarity.

For a list of related mechanics titles, see final pages.

IUTAM Symposium on

Smart Structures and Structronic Systems

Proceedings of the IUTAM Symposium
held in Magdeburg, Germany,
26–29 September 2000

Edited by

U. GABBERT
Otto-Von-Guericke-Universität Magdeburg, Germany

and

H.S. TZOU
University of Kentucky,
Lexington, U.S.A.

SPRINGER-SCIENCE+BUSINESS MEDIA, B.V.

A C.I.P. Catalogue record for this book is available from the Library of Congress.

ISBN 978-94-010-3836-2 ISBN 978-94-010-0724-5 (eBook)
DOI 10.1007/978-94-010-0724-5

Cover illustration: Active Camber Rotar 9ACR0 wit smart tab actuated in bending.

Printed on acid-free paper

CONTENTS

Participants of the IUTAM Symposium on *Smart Structures and Structronic Systems* in front of the conference hotel in Magdeburg on 29 September 2000

Preface

Synergistic integration of smart materials, structures, sensors, actuators and control electronics has redefined the concept of "structures" from a conventional passive elastic system to an active controllable structronic (structure + electronic) system with inherent self-sensing, diagnosis, and control capabilities. Such structronic systems can be used as components of high performance systems or can be an integrated structure itself performing designated functions and tasks. Due to the multidisciplinary nature of structronic systems their development has attracted researchers and scientists from theoretical and applied mechanics and many other disciplines, such as structures, materials, control, electronics, computers, mathematics, manufacturing, electromechanics, etc., see Figure 1. This field was first introduced about mid-80 and it is quickly becoming a new emerging field recognized as one of the key technologies of the 21st century. This new field focuses on not only multi-field and multi-discipline integrations, but has also enormous practical applications impacting many industries and enriching human living qualities.

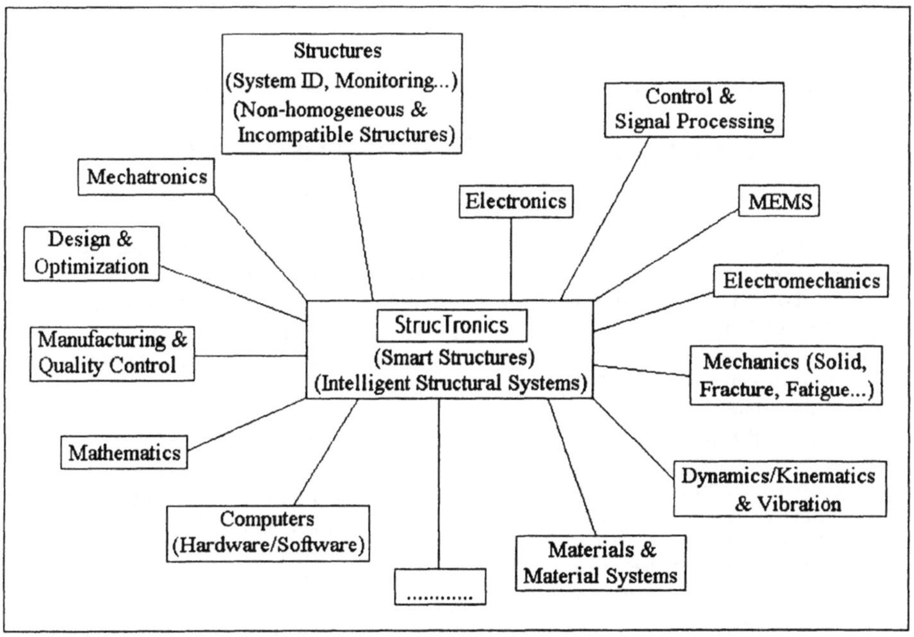

Figure 1 Multi-disciplinary integration of structronic systems.

To reflect the rapid development in smart structures and structronic systems, the objective of the IUTAM 2000 Symposium on Smart Structures and Structronic Systems, the first IUTAM symposium in this new emerging area, is to provide a forum to discuss recent research advances and future directions or trends in this field. The IUTAM Symposium took place from September 26th to 29th, 2000, at the Otto-von-

Guericke-University of Magdeburg, Germany, with 79 participants coming from 16 countries around the world.

The Symposium focussed on the fundamental mechanics and electromechanics of structures and structronic systems, consisting of smart materials, sensors, actuators, and control electronics. Multi-field phenomena, such as coupled elastic, electric, temperature and light phenomena related to structronic systems, as well as control effectiveness and other related topics were also discussed at the meeting. It was the intention of the Scientific Committee to invite leading scientists and researchers with different expertise to present their research findings at the Symposium. The open and friendly environment during the Symposium provided an excellent opportunity for intensive discussions and exchanging ideas among all participants. At the end of the Symposium a very challenging and exciting panel on "Prospects of Smart Structures and Structronic Systems" took place. Distinguished researchers or program managers (Prof. U. Gabbert, University of Magdeburg; Dr. E. Garcia, DARPA; Prof. I. Hagiwara, Tokyo Institute of Technology; Prof. D.J. Inman, Virginia Tech; Prof. H. Irschik, University of Linz; Prof. Y.P. Shen, Xian Jiaotong University; Prof. J. Tani, Tohoku University; Prof. V.V. Varadan, Penn State University) first reported their research activities, visions, etc. to the symposium. Open challenging research issues, current needs, unsolved problems, and future directions were also discussed and they are summarized in Figure 2.

Issues of Smart Structures and Structronic Systems

- Nonlinear modeling, simulation and **design tools/ / control-structure interaction/ /benchmark problems**
- System integration and system **design criteria, standards,** tools, limits, ..
- **Material** processing, new material with enhanced engineering properties and temperature stability, restrictions, **material library/database, diffusion-driven smart materials**
- New material **evaluation** techniques, tools,
- Product demos, new **mechanisms,** new materials applied to micro-electromechanical systems (**MEMS**)
- **Applications to manufacturing**
- Material incompatibility, material/structure integration, ...
- **Micro-mechanics**: bonding, fracture, fatigue, nonlinear behavior, etc.
- Health monitoring & diagnosis, NDE/NDT (new modeling tools)
- Real-time system identification and control, implementation, electronics (power efficiency)
- Biological inspired structures, self-growth/repair
- **Education** (industry, public, government, students,...), **network, web info,** "system" approach,
- **Distributed control** of continua (PDE) via structronics technology, demos,...

Figure 2 Summary of research issues and future directions.

Furthermore, assembling presented papers from distinguished invited speakers and reporting to the technical community is an important mission of the symposium. This symposium proceedings represents the symposium highlight, which consists of 43 papers presented by distinguished scientists from 15 countries. The editors sincerely hope that the symposium proceedings will serve as a milestone of this new emerging field and further promote the technology in both scientific research and practical applications of structronic systems. The 43 papers collected in this proceedings are presented according to the subsections of the IUTAM Symposium.

Finally the editors would like to thank all speakers and participants of this Symposium for their invaluable contributions to the field of smart structures and structronic systems. The editors also wish to acknowledge the active supports of the Scientific Committee in preparing the Symposium and the excellent organization by Dr. Harald Berger, Dr. Karl Fuchs and Dr. Friedemann Laugwitz as well as the great administrative assistance from Mrs. Ilona Hesse. In addition, many thanks are due to Kluwer Academic Publishers for its support and cooperation.

Ulrich Gabbert
Horn-Sen Tzou

Magdeburg, Germany
October 2000

Opening Address by the Chairman of the Symposium

Dear Colleagues, Ladies and Gentlemen:

On behalf of the International Union of Theoretical and Applied Mechanics I open the Symposium on "Smart Structures and Structronic Systems". I welcome you very warmly here in Magdeburg as our guests, as guests of the Otto-von-Guericke-University and the city of Magdeburg, and I wish you pleasant and successful days here in the Capital of the Federal State of Saxony-Anhalt. Especially, I would like to greet the participants in the symposium who are about 80 and come from all parts of Europe, from America and from Asia but also from Australia, the continent which gave us much pleasure by hosting the Olympic Summer Games. It is a special honor for me to welcome the President of IUTAM, Professor Werner Schiehlen from Stuttgart University, and I look forward to his opening address.

Yesterday I received a letter from Dr. Willi Polte, the Lord Mayor of the city of Magdeburg. He also warmly welcomes you in Magdeburg and extends his best wishes for a successful meeting to you. He recommends us not only to follow the scientific schedule of the symposium but also to have a look at the interesting historical attractions of the city, such as the cathedral - the first Gothic Cathedral in Germany - a must for visitors, the Monastery of Our Lady which is the oldest building of the city and dates back to the 11th century, the city hall with the oldest free-standing horseman north of the Alps, a monument which shows the first German Emperor Otto the Great, or the monument of the most important mayor and scientist of our city, Otto von Guericke who is well known all over the world due to his famous experiments with the Magdeburg hemispheres. Our University is named after Otto von Guericke and you will see three of his experiments with vacuum during our banquet on Wednesday at St. John's Church. But several other well known historical persons came from our city. Our colleagues from the United States should visit the monument of General von Steuben, a son of this city, and one of the most important Generals of George Washington during the War of Independence. But I have to stop here, as we are of course not at a meeting of historians.

You are here because you are interested in smart structures research and you have been invited due to your important research results in this field - an emerging scientific topic with a great potential to stimulate the development of new industrial products in many fields of engineering, such as aerospace, transportation, medicine, civil engineering, etc. This new area is a highly interdisciplinary one and successful research requires a close cooperation of scientists from different engineering disciplines such as mechanics, material sciences, control, electronics, process engineering, informatics, mathematics, etc. I believe we all are fascinated by the new engineering challenges and possible applications which we may have in mind talking about intelligent structures, smart structures, adaptronics or structronic systems.

A scientist should have visions, he needs visions which give him the motivation to create something new and to follow new ideas. But we all know that a lot of hard work is required to obtain new interesting results and to move forward step by step.

Nature provides us with an infinite amount of surprising ideas to solve engineering problems. It is fascinating to see how biological systems are able to adapt to changing environmental conditions, such as plants moving the leaves to the light or dolphins changing the shape of their skin in order to reduce flow resistance. Also the behavior of birds and insects is so fascinating and has attracted many scientists in aeronautics since the time of Leonardo da Vince and Otto Lilienthal. We should try to understand nature and learn from nature in order to create new ideas and solution concepts which are then transformed into a mathematical language and technical solutions. I hope that our symposium will contribute to these interesting developments and I also hope that at the end of our meeting during the panel discussion we will be able to draw some conclusions about questions which are still unanswered, steps required in future, priority research activities and so on. Of course, we should have dreams and we need visions, but we also know that we can be only successful when we have a good team with excellent knowledge, interdisciplinary cooperation, patience and the financial support we need.

In 1993 this university established the research network *ADAMES - Adaptive Mechanical Structures*, which has been financially supported by the German Research Foundation (DFG) since 1996. Based on this support, which we gratefully acknowledge here, several new research activities and research projects as well as industrial cooperation have been initiated by this interdisciplinary college. But in Germany there are also several other research centers which have focused on smart structures, such as the German Aerospace Research Center, the Fraunhofer Society and several other universities, e.g. in Hamburg, in Saarbrücken, in Munich, and especially in Stuttgart, where a research center for *Adaptive Aerospace and Lightweight Structures* was founded. I am glad that all these German centers have sent invited papers to our symposium.

This IUTAM-Symposium has been organized by members of the Adaptive Structures Research Center and the Conference Office of Magdeburg University. I would like to thank Ilona Hesse from the Conference Office and Harald Berger, Friedemann Laugwitz and Karl Fuchs as members of the Local Organization Committee for their help and excellent contribution to organizing this symposium. The staff of Ratswaage Hotel - our conference site - as well as the members of the Local Organizing Committee are prepared to help you in all of your questions and problems. Please, don't hesitate to contact us if you need any support or help.

The quality of a conference and the benefit the participants may get from attending a conference decisively depend on the rules, guidelines and objectives in organizing such a meeting, the active role of the Scientific Committee and of course on the participants and their contributions to the meeting. I would like to thank the Co-Chairman, H.S. Tzou, for his excellent contribution to organizing the symposium, but also Dan Inman, Vasundara Varadan, Wang Dajun, Shen Yapeng and Junji Tani have done a great job in actively supporting the symposium.

We have tried to do our best to follow the rules of IUTAM in preparing a successful meeting and to meet the high scientific standards of IUTAM. In particular, I'd like to ask you for a open atmosphere and frank discussions during the symposium as I think that this a precondition for a successful meeting and for learning from each other. Therefore, I cordially ask all lecturers to leave enough time for discussions. Indeed, the program is tough and there might be too many papers, but all contributions were very valuable and each lecture will contribute to a successful meeting.

As you know the conference papers will be published in a book series by Kluwer Academic Publishers. It was my intention to publish the book as soon as possible and therefore I asked you to send us the manuscripts before the meeting. Most of the manuscripts we have received are of an excellent quality. Some information to fulfil the publishing criteria of KLUWER Academic Publisher will be given to the authors during the conference. It is our objective and the objective of KLUWER to produce a very valuable book in an excellent printing quality. I would like to ask you for your support.

Finally, I would like to take the opportunity to thank IUTAM, the German Research Foundation (DFG), and Kluwer Academic Publishers for their financial support of the symposium.

Dear participants, dear colleagues and friends, I whish you a very successful meeting, interesting new impressions and a good time in Magdeburg.

Ulrich Gabbert
Chairman of the Symposium

Magdeburg, September 26, 2000

Welcome Address by the President of IUTAM

Mr. Chairman, Mr. Co-Chairman, Dear Colleagues from all over the world,
Ladies and Gentlemen,

True science does not recognize economical or social systems. Cooperation between scientists from different countries and parts of the world has a long tradition in mechanics.

Organized meetings between scientists in the field of mechanics were initiated 78 years ago, namely in 1922, when Prof. Theodore von Kármán and Prof. Tullio Levi-Civita organized the world's first conference in hydro- and aero-mechanics. Two years later, in 1924, the First International Congress was held in Delft, The Netherlands encompassing all fields of mechanics, that means analytical, solid and fluid mechanics, including their applications. From then on, with exception of the year 1942, International Congresses in Mechanics have been held every four years.

The disruption of international scientific cooperation caused by the Second World War was deeper than that caused by the First World War, and the need for reknotting ties seemed stronger than ever before, when the mechanics community reassembled in Paris for the Sixth Congress in 1946. Under these circumstances, at the Sixth Congress in Paris, it seemed an obvious step to strengthen bonds by forming an international union, and as a result IUTAM was created and statutes were adopted. Then, the next year, in 1947, the Union was admitted to ICSU, the International Council for Science. This council coordinates activities among various other scientific unions to form a tie between them and the United Nations Educational, Scientific and Cultural Organization, well known as UNESCO.

Today, IUTAM forms the international umbrella organization of nearly 50 national Adhering Organizations of mechanics from nations all over the world. Furthermore, a large number of international scientific organizations of general or more specialized branches of mechanics are connected with IUTAM as Affiliated Organizations. As a few examples, let me mention: the European Mechanics Society (EUROMECH), the International Association of Computational Mechanics (IACM), and the International Association for Vehicle System Dynamics (IAVSD).

Within IUTAM the only division used so far is related to solid and fluid mechanics as indicated by our two Symposia Panels. But more recently six Working Groups have been established by the General Assembly of IUTAM devoted to specific areas of mechanics.

These areas are:

- Mechanics of Non-Newtonian Fluids,
- Dynamical Systems,
- Fracture Mechanics and Damage,
- Mechanics of Materials,
- Electromagnetic Processing,
- Computational Mechanics.

On recommendation of the IUTAM Assessment Panel the Working Groups will be developed into Standing Committees. In addition, other specific areas of mechanics may be identified to support the international cooperation in more branches of mechanics.

IUTAM carries out an exceptionally important task of scientific cooperation in mechanics on the international scene. Each national Adhering Organization of IUTAM, like the German Committee for Mechanics (DEKOMECH), is represented by a number of scientists in IUTAM's General Assembly. In particular, the German delegates with IUTAM are

Professor Ulrich Gabbert, the Chairman of this Symposium
Professor Egon Krause, Rheinisch-Westfälische Technische Universität Aachen
Professor Günter Kuhn, Universität of Nürnberg-Erlangen
Professor Siegfried Wagner, Universität Stuttgart.

Mechanics is a very well developed science in Germany represented at most universities and some national laboratories. Since 1949 there have been held more than 250 IUTAM symposia worldwide. Out of them 28 IUTAM symposia where organized in Germany. And this IUTAM Symposium is the first one hosted by the Otto von Guericke Universität in Magdeburg.

As I mentioned before, IUTAM organizes international congresses and symposia all over the world. Just four weeks ago the 20th International Congress of Theoretical and Applied Mechanics was held in Chicago, Illinois, USA. This quadrennial congress is also considered as the Olympics of Mechanics. With 1500 participants the Chicago Congress was the central millenium event in mechanics to celebrate the turn of the century, too. The Twentyfirst International Congress of Theoretical and Applied Mechanics will be held in Warsaw, Poland, from 15th to 21st August 2004, what means in four years from now.

Announcements of this forthcoming congress will be widely distributed and published in many scientific journals.

The present Symposium is exceptionally interesting because it deals with new developments in mechanics. The Symposium covers important approaches: Technologies, Materials, Controls and their Application to Smart Structures. IUTAM found that the proposal of Professor Gabbert for such a symposium was not only very timely, but also very well founded in the outstanding research carried out in this field at the Otto von Guericke University. Thus, the proposal for the Symposium was readily accepted and granted by the General Assembly of IUTAM. There is no doubt that IUTAM considers Structronics as an important field of mechanics.

On behalf of IUTAM, I wish to express my sincere thanks to the Otto von Guericke University, in particular to the Chairman, Professor Gabbert, for the invitation to host this significant scientific event. I welcome all the invited participants for their readiness to come and to contribute to the success of the Symposium by very active participation in the lectures and the scientific discussions, as well as in the social program.

Finally, I would like to mention that to sponsor a scientific meeting is one thing, to organize one is another. A heavy burden is placed on the shoulders of the Chairman, the Co-Chairman and their associates who are in charge of the scientific programme and the practical local arrangements. All who have tried this before know perfectly well how much work has to be done in organizing a meeting like this one.

Thus, we should be thankful, not only to the International Scientific Committee, but also to the Chairman, Professor Ulrich Gabbert, and the associates who assisted him in carrying the heaviest load and responsibility.

It is up to us now, Ladies and Gentlemen, to harvest the fruits of the Organizer's work. Let us contribute our share to make this IUTAM Symposium a meeting that will be long remembered as a very successful one!

On behalf of IUTAM, I greet you all and wish you great success!

Werner Schiehlen
President of IUTAM

Magdeburg, Germany
26 September 2000

International Scientific Committee

U. Gabbert (Germany, Chairman)
I. Hagiwara (Japan)
D.J. Inman (USA)
H. Irschik (Austria)
R.S.W. Lee (Hong Kong, P. R. China)
W. Schiehlen (Germany, President of IUTAM)
Y.P. Shen (P. R. China)
B.F. Spencer (USA)
J. Tani (Japan)
G.R. Tomlinson (UK)
H.S. Tzou (USA, Co-Chairman)
V.V. Varadan (USA)
D. Wang (P. R. China)

Local Organizing Committee

H. Berger
K. Fuchs
I. Hesse
F. Laugwitz (Chairman)
C.-T. Weber

Sponsors of the IUTAM Symposium on Smart Structures and Structronic Systems

International Union of Theoretical and Applied Mechanics (IUTAM)
Deutsche Forschungsgemeinschaft (DFG)
Kluwer Academic Publishers

SIMULTANEOUS ACTIVE DAMPING AND HEALTH MONITORING OF AIRCRAFT PANELS

DANIEL J. INMAN
Center for Intelligent Material Systems and Structures
Department of Mechanical Engineering

MEHDI AHMADIAN
Advanced Vehicle Dynamics Laboratory
Department of Mechanical Engineering

RICHARD O. CLAUS
Fiber and Electro-Optics Research Center
Departments of Electrical Engineering and of Materials Science

Virginia Polytechnic Institute and State University
Blacksburg, VA 24061

1. Introduction

This paper reports on an experimental implementation of active control methods using smart structures for the purpose of performing simultaneous health monitoring and active damping of panels characteristic of aircraft components. Here smart structures refers to the integrated use of piezoelectric actuators and fiber optic sensors as the measurement transducers and control actuators required to actively damp a panel and to provide health monitoring of its structural integrity. The experiments focus on a flat aluminum panel with fundamental structural frequency of range. The active damping performance will be designed to be work across a range of temperatures from 70° to 120° F. The health monitoring aspects are impedance based using frequencies in the kilohertz range to avoid interacting with the structural control that is targeted to include the first six or seven modes. Both theoretical considerations and experimental verification is given. The major goal here is to show experimentally that simultaneous control and monitoring is possible.

U. Gabbert and H.S. Tzou (eds.), IUTAM Symposium on Smart Structures and Structronic Systems, 1–8.

Current aircraft pay a large weight penalty for damping (usually constrained layer) used to mitigate fatigue. In addition, there is a great deal of interest in monitoring the structural health of aircraft panels. Each of these problems may be addressed by using smart materials. However, if treated separately, the amount of additional hardware becomes overwhelming. Thus we are motivated to try and address both problems with the same set of hardware. A sample test panel was used in the form of a clamped plate mounted in a small acoustic chamber. Modeling and testing where performed. The active control consists of a positive-position-feedback compensation circuit built around measured frequencies served as the control law. The health monitoring is accomplished by using an impedance-based method in the high frequency range. The vibration suppression system is low frequency. Thus the two systems will not interfere with each other. The result is a proof of concept experiment that illustrates that simultaneous control and health monitoring is feasible across a range of temperatures.

2. Test Apparatus

Although motivated by aircraft panels, the work was done in that laboratory on a test stand, allowing a variety of different panels with a variety of treatments. A test stand was designed and fabricated for testing and evaluating the effectiveness of piezoelectric damping materials for reducing vibration and structure-borne noise. The test stand enables vibration and acoustic measurements and analysis on a plate with clamped-clamped boundary conditions. The plate, simulating an aircraft panel, is clamped rigidly around its edges and excited over a frequency range of 0-400 Hz. Various passive damping materials, and smart damping materials, can be added to the panel in order to evaluate their effect on reducing the panel vibration and the noise that it causes.

The test stand, shown in Figure 1, includes an electromechanical shaker, a panel excitation frame, a sound-insulating enclosure, and data acquisition equipment. Measurements are taken with two accelerometers, located on the plate and excitation frame, and a microphone positioned in the upper reception chamber. The reception chamber and bottom enclosure is designed to eliminate background noise and isolate the noise generated by the vibrating panel.

To begin validation testing, a standard test plate was clamped into place with 14 bolts tightened to a torque of 25 N-m. The standard plate was a 500mm X 600mm, 20-guage, galvanized steel plate. The plate was bolted as in Figure 2 such that the outside 10 cm along the edges were clamped and the remaining test plate area was 400mm X 500mm. The bolts were always tightened in a criss-crossing pattern, similar to that for lug nuts on a car, to improve the repeatability of the boundary conditions for the plate.

Accelerometers were attached in the center underneath the top beam of the excitation frame and on the test plate underneath its center. A microphone was hung in the reception chamber such that it was 0.5m above the center of the test plate when the test stand was fully assembled. The data acquisition was set up according to the test schematic shown in Figure 3. A Hewlett Packard dynamic signal analyzer served as the data recorder, the fast Fourier transformer, the band pass filter, and the signal generator for controlling the shaker. Initial tests and experiments were performed with a number of different excitation functions and sampling techniques. A Hewlett Packard Impedance Analyzer was used to make impedance measurements of the structure for the diagnostic aspects of this study.

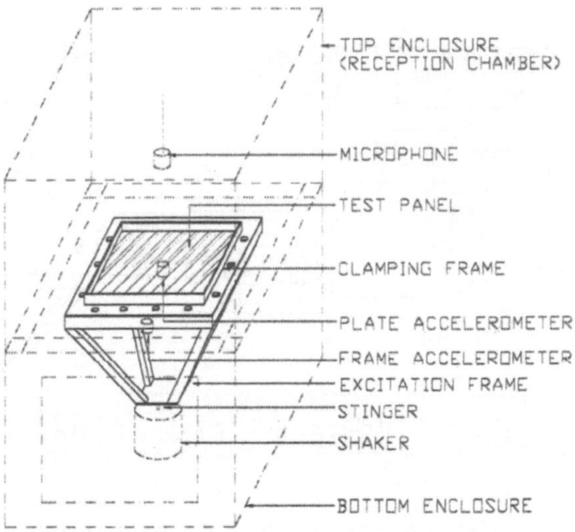

Figure 1. Vibration and Acoustics Test Stand Schematic

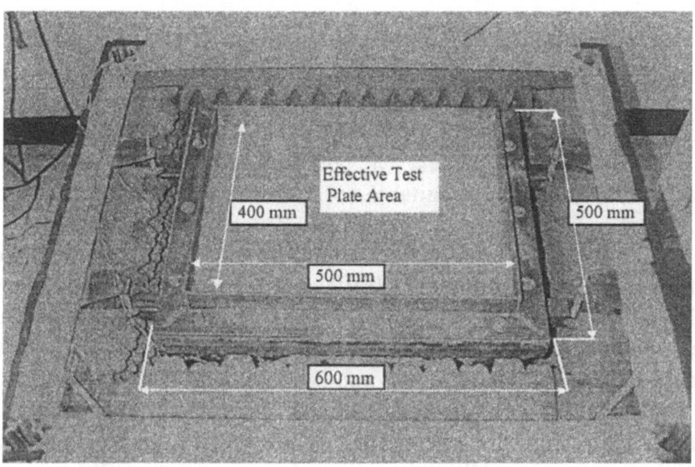

Figure 2. Standard Test Plate in Testing Position

Although the generated input signal is an ideal signal for testing the frequency response for a plate, the direct input excitation for the plate is from the frame not from the HP Analyzer. The desired excitation range for the plate is between 0 and 400 Hz, and poor data will result if there are any resonant frequencies of the frame within this range. The frequency response of the frame was then analyzed to ensure that this was not the case. Data was first collected for the excitation frame and clamping frame without the plate in place. This test clearly shows that the major frame structural resonant frequencies occur above 500 Hz. The frequency spectrum of the frame within

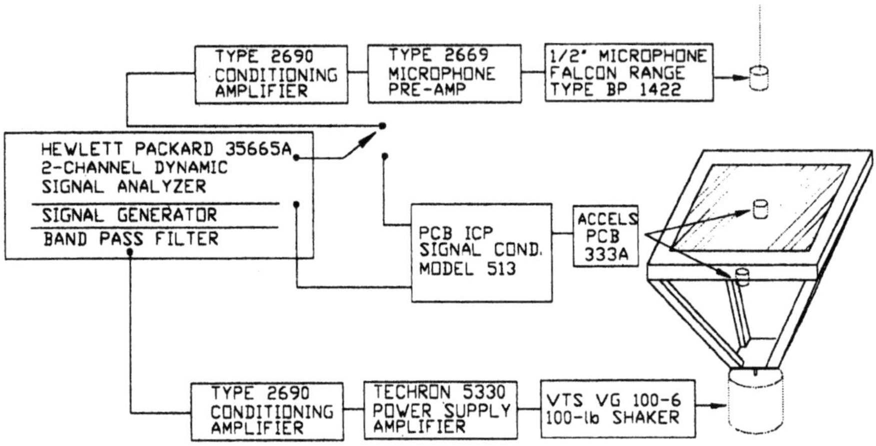

Figure 3. Shaker Table Test Stand Data Acquisition Schematic

3 Active Control and Monitoring Hardware

A piezoceramic patch was used as an actuator and a fiber optic was used for the sensor. Fiber optic sensors are widely used as physical parameter gauges in various structural applications, such as strain and vibration sensing and damage detection. After examining the various types of sensors, we concluded that the Fabry-Perot fiber sensors demonstrated the a response suitable for use in an active feedback control system.

The Fabry-Perot (FP) interferometric strain sensors can be classified into two main types: intrinsic FP interferometers (IFPI) and Extrinsic FP interferometers (EFPI). The EFPI sensor is constructed by fusion splicing a glass capillary tube with two optical fibers. Compared to the IFPI sensor, the EFPI-based sensor is relatively simple to construct and the FP cavity length can be accurately controlled. The EFPI sensor can also be easily configured to suit different applications with desired strain range and sensitivity by altering the type of fibers, the capillary tube, air-gap distance and the length of the sensor. In addition, a major advantage of the EFPI sensor is its low temperature sensitivity, which makes it possible to interrogate the EFPI sensor with simple signal processing techniques. An EFPI sensor can be constructed using either single-mode or multi-mode fibers. The single-mode design offers higher accuracy and low insensitivity to unwanted disturbance while the multi-mode design offers higher power coupling efficiency. In our experiment, single-mode fibers are used to deliver and collect the light, and are used as an internal reflector as well. Two EFPI sensors were fabricated and measured under different circumstances, both on a bench and the vibration test stand.

4. Control Results

In this section we show an example of our test results. Figure 4 illustrates the strong reduction of a single tone excitation controlled by a piezoceramic patch actuator and fiber optic sensor. Similar results were obtained for chirp random excitation and for acoustic excitation. A high frequency, impedance based health monitoring system is superimposed over the active control signal and results are shown that indicate that small changes in the bolt torque on the boundary (simulating weld cracking) can be detected. It is interesting to note that the active control system reduces the time response, as it should, but that health monitoring in general needs a long time response. Thus it is not obvious that the two can be implemented simultaneously.

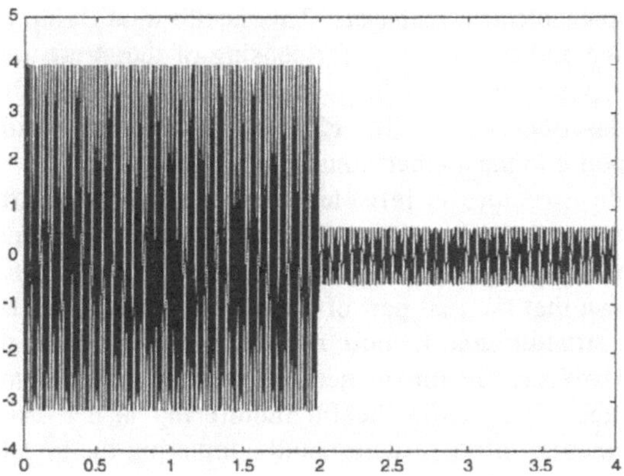

Figure 4 Sample active control results as measured from the optical sensor.

The specific control law used for the active control tests is called Positive Position Feedback [1,2]. This control law uses a generalized displacement measurement from the test article to accomplish the feedback and uses a displacement measurement to accomplish control. Positive position feedback control is a stable and relatively simple control method for vibration suppression. The control law for a positive position feedback controller consists of two equations, one describing the structure and one describing the compensator:

$$\text{Structure: } \ddot{\xi}(t)+2\gamma\omega\dot{\xi}(t)+\omega^2\xi(t)=g\omega^2\eta(t) \tag{1}$$

$$\text{Compensator: } \ddot{\eta}(t)+2\gamma_f\omega\dot{\eta}(t)+\omega_f^2\eta(t)=g\omega_f^2\xi(t) \tag{2}$$

where g a positive scalar gain, ξ is the modal coordinate, η is the filter coordinate, ω and ω_f are the structural and filter frequencies, respectively, and γ and γ_f are the structural and filter damping ratios, respectively. The positive position terminology in the name positive position feedback comes from the fact that the position coordinate of the structure equation is positively fed to the filter, and the position coordinate of the compensator equation is positively fed back to the structure. In effect, a positive position feedback controller behaves much like an electronic vibration absorber.

The system's parameters and the overall control system's performance were measured using a two-channel HP Dynamic Signal Analyzer. This signal analyzer was used to get the frequency response function between the plate and the clamping frame using two accelerometers, one on the bottom center of the plate and one on the clamping frame. This yields equation (1) for each mode. Then, the filter defined by equation (2) was designed to provide the required damping.

5. Health Monitoring Results

The basic concept of this impedance-based structural health monitoring technique is to monitor the variations in the structural mechanical impedance caused by the presence of damage. Since structural mechanical impedance measurements are difficult to obtain, this non-destructive evaluation technique utilizes the electromechanical coupling property of piezoelectric materials. This health monitoring method uses one PZT patch for actuating and a fiber optic for sensing of the structural response. The interaction of a PZT patch with the host structure can be described by a simple impedance model as described in [3]. The PZT is considered as a thin bar undergoing axial vibrations in response to the applied sinusoidal voltage.

Damage in the structure is reflected in changes of the parameters such as mass, stiffness, or damping. Assumed that the PZT's parameters remain constant any changes in the mechanical impedance Z_s change the overall admittance. Previous experiments have shown that the real part of the overall impedance contains sufficient information about the structure and is more reactive to damage than the magnitude or the imaginary part. Therefore, all impedance analyses are confined to the real part of the complex impedance. The actual health monitoring is performed by saving a healthy impedance signature of the structure and comparing the signatures taken over the structure's service life. The impedance measurements were taken with an HP 4194A Impedance Analyzer. A frequency range from 45 kHz to 55 kHz proved to be an optimum for this structure. To simulate damage on the plate one or two bolts of the clamping frame were loosened from 25 ft-lb. to 10 ft-lb. Figure 2 shows a photograph of the frame and names the bolts to be loosened. Adding mass to the test specimen could not be used to simulate damage, because the mechanical impedance is mostly defined by the boundary conditions of the clamped plate.

Since the task consisted of simultaneous health monitoring and active control with the same actuators for both the control and the health monitoring system needed to be de-coupled. The impedance method is very sensitive to disturbing voltages in the measuring circuit. The controller however creates exactly those disturbances by generating the control signal. A simple capacitor of 390 nF in series with the impedance analyzer blocked efficiently the control signal from the impedance analyzer. All health monitoring data presented in this report was taken while the shaker was exiting the plate with a periodic chirp signal from 0 to 200 Hz. The active controller was also switched on and increased the damping of the first three modes of the plate significantly. For comparing impedance signatures, a qualitative damage assessment has been developed. The assessment is made by computing a scalar damage metric, defined as the sum of the squared differences of the real impedance at every frequency step. Equation (3) gives the damage metric M in a mathematical form. The used variables include: $Y_{i,1}$ the healthy impedance at the frequency step i, $Y_{i,2}$ the impedance of the structure after the structure has been altered, and n the number of frequency steps

$$M = \sum_{i=1}^{n} \left[\mathrm{Re}(Y_{i,1}) - \mathrm{Re}(Y_{i,2}) \right]^2 \tag{3}$$

The damage metric simplifies the interpretation of the impedance variations and summarizes the information obtained by the impedance curves. Different damage metric values of the plate are presented in Figure 5. Note the difference in the metric between one bolt loosened and two bolts loosened. It supports the idea of a damage threshold value to warn an operator when this threshold value has been reached.

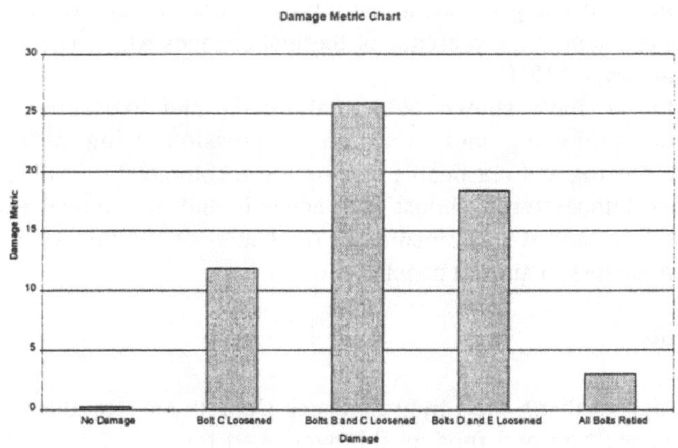

Figure 5. Damage Metric Chart of Different Impacts to the Plate.

The chart also illustrates that the closer the damage to the sensor happens the earlier the damage can be detected. Loosening bolts D and E surely represent a similar damage then loosening bolts B and C. However, the damage metric of bolts D and C loosened is smaller than the damage metric of bolts B and C loosened. This is due to the relative distance of the damage impact to the PZT sensor. Figure 5 also shows a problem of the impedance technique. After tightening all bolts back to 25 foot-pounds torque, another set of impedance values was taken and the damage metric was computed. The metric value on the right demonstrates that the impedance signatures from the undamaged plate and the "repaired" plate were noticeable different. The impedance technique is so sensitive that it is almost impossible to achieve the same impedance signature as before altering after the structure was altered and re-altered.

To prove the reliability of the NDE technique several measurements in intervals of several hours were taken. During all measurements, the shaker applied a periodic chirp vibration signal to the plate and the active controller was in operation. No damage was induced to the test specimen for this experiment. The results are presented in Figure 7.6. The hours in the graph specify the time passed after the first data set was taken. Figure 7.6 also contrasts the repeatability metrics to an actual impact on the structure.

6. Conclusions

Several fiber optic sensors were examined and an Extrinsic Fabry-Perot Interferometer sensor type was chosen to be the most compatible with vibration control and sensing. This sensor type was constructed and we verify that it functions effectively as a vibration sensor. We compared PZT and Fiber optic sensing and showed them to be comparable. We demonstrated active vibration suppression with PZT and this was successfully performed with both acoustic and mechanical excitations. Positive position feedback control and impedance based health monitoring were selected as being compatible for simultaneous health monitoring and vibration suppression. The successful implementation of simultaneous health monitoring and vibration suppression was completed using the same hardware for both tasks. In addition we showed vibration suppression in the presence of thermal changes which was performed from room temperature up to 118°F.

In summary we have shown both analytically and experimentally that simultaneous health monitoring and vibration suppression using piezoceramic actuation, fiber optic sensing and reasonable electronics is completely feasible and can do so over a range of temperatures against both acoustic and mechanical excitation. This result is encouraging and if implemented in large aircraft has the potential for a great reduction in fatigue loss in aircraft panels.

7. Acknowledgements

The authors would like to thank the Flight Sciences Department, Raytheon Systems Company for both suggesting and funding this work and for the excellent technical monitoring of Dr. Richard A. Ely. In addition, we thank the National Science Foundation (CMS-9713453-001) and the Airforce of Office of Scientific Research (F49620-99-1-0231) for funding work in control, smart structures and health monitoring used in this effort. The detailed work for this effort was performed by our students: Thomas Hegewald, Andrew Deguilo and Zhaoju Luo.

8. References

1. Fanson, J.L. and Caughey, T.K., 1990, "Positive Position Feedback Control for Large Space Structures", *AIAA Journal*, **28**(4), 717-724.
2. Inman, D.J., 1995, "Programmable Structures for Vibration Suppression", *International Conference on Structural Dynamics, Vibration, Noise and Control*, Hong Kong, December 1995, 100-107.
3. Park, G., Kabeya, K., Cudney, H. H. and Inman, D. J., 1999. "Impedance-Based Structural Health Monitoring for Temperature Varying Applications," *JSME International Journal*, Series A, Vol. 42, No. 2, pp. 249-258.

DECENTRALIZED VIBRATION CONTROL AND COUPLED AEROSERVOELASTIC SIMULATION OF HELICOPTER ROTOR BLADES WITH ADAPTIVE AIRFOILS

BORIS A. GROHMANN
present affiliation: EADS Deutschland GmbH, Industrial Research and Technology, Postfach 800465, 81663 München

PETER KONSTANZER
present affiliation: DaimlerChrysler AG, Research and Technology FTK/A, Epplestraße 225, 70546 Stuttgart

BERND KRÖPLIN
Institut für Statik und Dynamik der Luft- und Raumfahrtkonstruktionen, Universität Stuttgart Pfaffenwaldring 27, 70569 Stuttgart

1 Introduction

In helicopters, a high vibration level of the airframe occurs due to higher harmonic aerodynamic loads acting on the rotor blades. However, when the airfoil shape is adaptive, the aerodynamic loads can be affected to reduce vibration and moreover, the airfoil shape can be adjusted to the periodically changing flow conditions to increase aerodynamic efficiency. Adaptation of the airfoil shape may be achieved by discrete trailing edge flaps or continuously by a variation of the airfoil camber and bending of smart tabs. For vibration reduction applying feedback [4], stability augmentation of the lagging modes becomes indispensible. This is discussed first for rotors with discrete trailing edge flaps. The disadvantages of a discrete flap, such as kinks of the airfoil contour and concentrated masses, may be overcome by continuous shape adaptation, investigated in the second part.

2 Decentralized Vibration Control

A method for lagging mode stability augmentation for helicopter rotors based on decentralized control is proposed. Since helicopter rotors consist of N rotor blades, the application of decentralized controllers to each rotor blade intuitively

U. Gabbert and H.S. Tzou (eds.), IUTAM Symposium on Smart Structures and Structronic Systems, 9–16.

appears to be a natural approach, see Fig. 1. Furthermore, since all blades may be assumed to be identical, there is no reason for the individual blade controllers to be non-identical. This introduces a further conceptual simplification into the control system. Second, the interactions of the rotor blades with the vortices generated by the preceding blades – the blade-vortex interactions – introduce uncertainties in the interconnection of the rotor blade subsystems. Decentralized control [9] inherently provides robustness with respect to this type of uncertainty.

The aeroservoelastic behavior of active helicopter rotors about the trim state can be described by a linear time-periodic LTP system [3]

$$\dot{\tilde{x}} = A(\psi)\tilde{x} + B(\psi)\tilde{u}, \qquad \tilde{y} = C(\psi)\tilde{x} \tag{1}$$

where $A(\psi) = A(\psi + 2\pi), B(\psi) = B(\psi + 2\pi), C(\psi) = C(\psi + 2\pi)$ and $\tilde{x} = [\tilde{x}_1, \ldots, \tilde{x}_N]^T, \tilde{u} = [\tilde{u}_1, \ldots, \tilde{u}_N]^T, \tilde{y} = [\tilde{y}_1, \ldots, \tilde{y}_N]^T$ denote state, input and output vector of the N individual blades. The LTP system can be approximated by a linear time-invariant LTI system applying the multiblade coordinate MBC transformation [3], subsequent expansion of the system matrices in Fourier series, and neglecting the periodic coefficients. This yields an approximation in terms of an LTI system in MBC

$$\dot{x} = Ax + Bu, \qquad y = Cx \tag{2}$$

which represents the periodic system characteristics about the trim state. The state vector x consists of the collective x_0, cyclic x_{1c}, x_{1s} and differential form x_2 of the rotor state. Similar for the input vector u and the output vector y. Figure 1 compares an LTI system in multiblade coordinates MBC and an LTI system in individual blade coordinates IBC with the exact LTP system. It is observed that the response of the LTP system due to a collective impulse input $u_0(t) = \delta(t - T)$ is approximated with sufficient accuracy for the collective output y_0 by both LTI systems. The response of the cyclic outputs y_{1c}, y_{1s} is captured by the LTI system in MBC only whereas in IBC this interconnection does not appear at all. The interconnection of non-differential (e.g. collective) and differential form

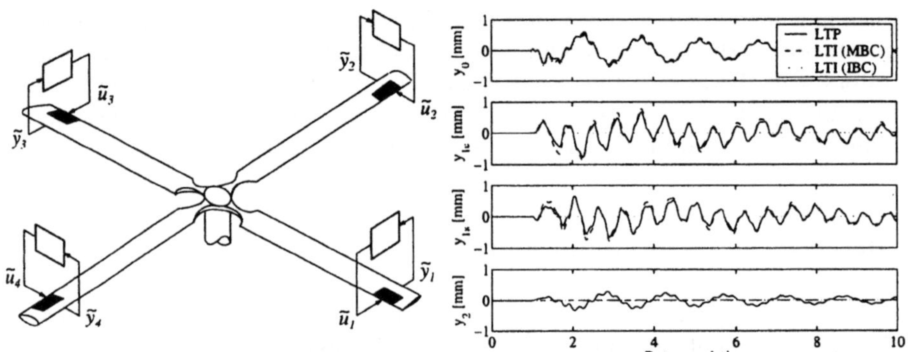

Figure 1. Decentralized control and response due to a collective impulse input $u_0(t) = \delta(t - T)$

is neither represented by the LTI system in IBC nor by the LTI system in MBC. This can be observed from the response of the differential output y_2. Thus, for a constant coefficient approximation of an LTP system, the MBC transformation recovers some of the periodic system characteristics, but a considerable error in the interconnection of differential and non-differential form is inevitable. Therefore, when designing a controller based on an LTI system in MBC, robustness with respect to uncertainties in this interconnection is essential. The subsequently proposed design method will inherently provide this robustness.

Let the plant (rotor) consist of N interconnected subsystems (blades) and consider N individual subsystem (blade) controllers. These controllers may be constructed from a generalized second-order filter [13]

$$F(s) = k\frac{s^2/\omega_z^2 + 2\zeta_z s/\omega_z + 1}{s^2/\omega_p^2 + 2\zeta_p s/\omega_p + 1} \tag{3}$$

where the coefficients $k, \zeta_z, \zeta_p, \omega_z, \omega_p$ specify various filters with different gain-phase characteristics such as lowpass, highpass, bandpass, or nonminimum-phase allpass filter. Based on these basic filters, a frequency-shaped subsystem controller can be constructed by simple series connection, e.g. a bandpass and a nonminimum-phase allpass filter may be combined to target a certain mode and provide the proper phase characteristics. Since the subsystems are identical, there is no reason for the subsystem controllers to be non-identical. After transforming the individual subsystem controllers into MBC, the overall stabilizing controller for $N = 4$ rotor blades is given by

$$\dot{x}_s = \underbrace{\begin{bmatrix} \hat{A}_s & 0 & 0 & 0 \\ 0 & \hat{A}_s & -\Omega I & 0 \\ 0 & \Omega I & \hat{A}_s & 0 \\ 0 & 0 & 0 & \hat{A}_s \end{bmatrix}}_{A_s} x_s + \underbrace{\begin{bmatrix} \hat{B}_s & 0 & 0 & 0 \\ 0 & \hat{B}_s & 0 & 0 \\ 0 & 0 & \hat{B}_s & 0 \\ 0 & 0 & 0 & \hat{B}_s \end{bmatrix}}_{B_s} y \tag{4}$$

$$u = \underbrace{\begin{bmatrix} \hat{C}_s & 0 & 0 & 0 \\ 0 & \hat{C}_s & 0 & 0 \\ 0 & 0 & \hat{C}_s & 0 \\ 0 & 0 & 0 & \hat{C}_s \end{bmatrix}}_{C_s} x_s + \underbrace{\begin{bmatrix} \hat{D}_s & 0 & 0 & 0 \\ 0 & \hat{D}_s & 0 & 0 \\ 0 & 0 & \hat{D}_s & 0 \\ 0 & 0 & 0 & \hat{D}_s \end{bmatrix}}_{D_s} y \tag{5}$$

where Ω denotes the rotor speed and $\hat{A}_s, \hat{B}_s, \hat{C}_s, \hat{D}_s$ the system matrices of the 4 identical subsystem controllers constructed from the generalized second-order filter. Defining $u_s := \dot{x}_s, y_s := x_s$, we obtain an expanded system consisting of plant and stabilizing controller

$$\underbrace{\begin{bmatrix} \dot{x} \\ \dot{x}_s \end{bmatrix}}_{\dot{x}_e} = \underbrace{\begin{bmatrix} A & 0 \\ 0 & 0 \end{bmatrix}}_{A_e} \underbrace{\begin{bmatrix} x \\ x_s \end{bmatrix}}_{x_e} + \underbrace{\begin{bmatrix} 0 & B \\ I & 0 \end{bmatrix}}_{B_e} \underbrace{\begin{bmatrix} u_s \\ u \end{bmatrix}}_{u_e} \tag{6}$$

$$\underbrace{\begin{bmatrix} y_s \\ y \end{bmatrix}}_{y_e} = \underbrace{\begin{bmatrix} 0 & I \\ C & 0 \end{bmatrix}}_{C_e} \underbrace{\begin{bmatrix} x \\ x_s \end{bmatrix}}_{x_e} \quad \text{and} \quad \underbrace{\begin{bmatrix} u_s \\ u \end{bmatrix}}_{u_e} = \underbrace{\begin{bmatrix} A_s & B_s \\ C_s & D_s \end{bmatrix}}_{K} \underbrace{\begin{bmatrix} y_s \\ y \end{bmatrix}}_{y_e} \tag{7}$$

where all unknown parameters are contained in the static output feedback matrix K of which the structure depends on the choice of the subsystem controllers. The structure is given by the arbitrary feedback structure formulation defined as

$$K = K_p + \sum_{i=1}^{m} t_i k^T U_i \tag{8}$$

where K_p denotes the matrix of prescribed coefficients, k the vector of feedback parameters to be designed, and t_i and U_i are vectors and matrices, respectively, which determine the entry of the feedback parameters k into the gain matrix K. The remaining static output feedback problem [10] is to find a static output feedback matrix K of arbitrarily specified structure. Decentralized optimal output feedback DOOF may be applied to find a matrix K which minimizes an infinite horizon quadratic performance index of the type

$$J = \frac{1}{2}\int_0^\infty (x_e^T Q x_e + u_e^T R u_e)\, dt \tag{9}$$

where Q and R are symmetric positive semidefinite weighting matrices.

A lagging mode stability augmentation system is designed for a Bo105 helicopter rotor in forward flight with trailing edge flaps located at 65% radial station [4]. The lagwise displacement signals at the radial station of the flaps are assumed to be amenable for measurement. The subsystem controllers consist of a fixed bandpass and two nonminimum-phase allpass filters of which the parameters are calculated by DOOF with an additional constraint on the allowable pole region as indicated in Fig. 2. As shown in Tab. 1, active damping enhancement of about 600% is achieved for the 1st as well as the 2nd lagging mode. Robustness with respect to uncertainties in the interconnections is shown by the closed-loop system response due to a differential impulse input $u_2(t) = \delta(t - T)$ shown in Fig. 2. Although the LTI design model does not generate any response in the non-differential outputs y_0, y_{1c}, y_{1s} the controller applied to the time-periodic plant provides excellent regulation in the non-differential as well as the differential outputs.

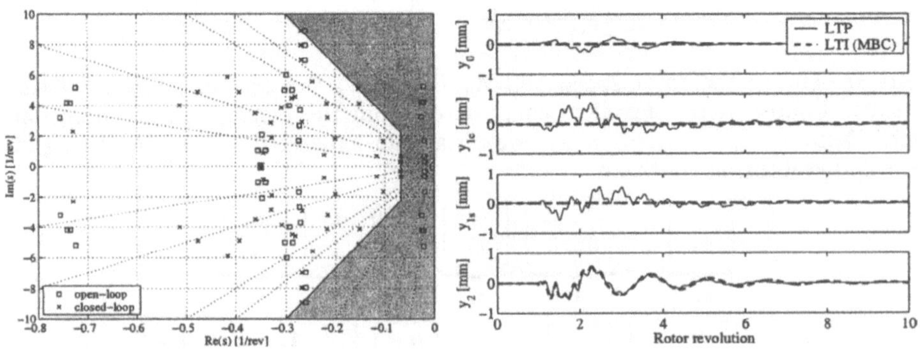

Figure 2. Pole map and closed-loop system response due to a differential impulse input

TABLE 1. Controller parameters and critical damping

	controller parameters					critical damping	
	ζ_1	ζ_2	$\omega_1[1/s]$	$\omega_2[1/s]$	k	1st lag	2nd lag
uncontrolled	–	–	–	–	–	1 - 5 %	0.5 - 1 %
controlled	0.92	1.17	90.4	112.2	-9695	6 - 20%	3 - 6 %

3 Aeroservoelastic Design of Active Camber Rotor (ACR)

The investigation on decentralized vibration control employs the concept of a smart trailing edge flap with discrete piezo actuators which has been investigated in the framework of RACT (Rotor Active Control Technology), see Geissler *et al* [1] and Schimke *et al* [6]. In the following, the structural concept for a smart helicopter rotor with active material components integrated in and distributed across the host structure resulting in continuous shape adaptation is discussed.

Aeroelastically Unbalanced Concept. Due to the integration of the active material into the host structure, a balanced design between the conflicting basic functionalities *dynamic shape adaptation* and *load carrying capability* is necessary to guarantee the efficiency of actuation. This is attempted by the concept depicted in Figure 3. The stiff, passive nose structure, reaching up to approximately 30-40% of the chord, supplies bending and torsion stiffness and carries the centrifugal loads. The flexible trailing edge area is responsible for dynamic shape adaptation by opposite chord-wise strain actuation of the upper and lower active composite layers. These layers may consist of thin piezoceramic plates actuated in 31-mode via standard electrodes, in 33-mode via interdigitated electrodes or piezoceramic fibers. The foam core must be very flexible to avoid losses or it may even be replaced by a different type of supporting skeleton that prevents the active composite layers from buckling. Displacement amplification increases with decreasing airfoil thickness ratio. Note that the trailing edge structure still has to carry local (aerodynamic) loads. Since the smart airfoil concept depicted in Figure 3 has limited authority, it may be combined with a smart tab actuated in bending, see Figure 4.

Coupled Multiphysics Simulation. The contradictory design paradigms "*as stiff as necessary to carry loads*" and "*as flexible as possible for shape adaptation*" of smart aeroservoelastic structures are the basic motivation for the development of

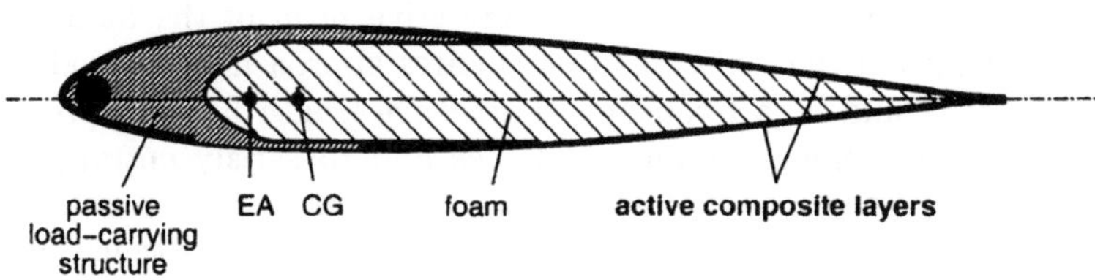

Figure 3. Smart Airfoil

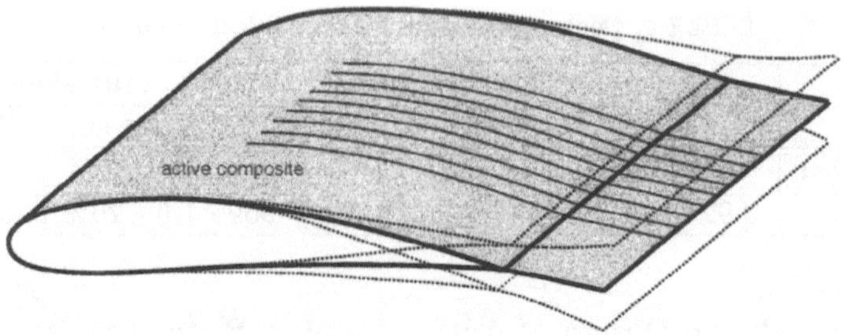

Figure 4. Active Camber Rotor (ACR) with smart tab actuated in bending

the coupled aeroservoelastic simulation described in the following. See Grohmann [2] for details on the formulation.

Electromechanics. The smart composite structure including piezoceramic layers is modeled by curved shell elements employing classical laminate theory. Two different models of the active material are used. First, its effect is modeled by equivalent actuator loads. Second, a coupled electromechanical model for actuation in 31-mode based on linear piezoelectricity is employed. Due to the different characteristic time scales, the electric field is treated as quasi-steady.

Aeroelasticity. The two-dimensional Euler equations are employed for compressible and inviscid aerodynamics. A representative blade segment at 85% of the rotor diameter is investigated. The varying flow conditions during one revolution of the rotor blades in forward flight are simulated by oscillatory mesh motion in streamwise direction. Concerning fluid-structure interaction, the displacements of the structure are considered as deformation of the fluid domain and the aerodynamic pressure acts as load on the structure. This leads to momentum and energy transfer at the fluid-structure interface. For the correct coupled response, conservation of mass, momentum and energy at the interface is essential.

Stabilized Space-Time Finite Element Discretization. The time-discontinuous Galerkin method yields implicit, unconditionally stable time discretization, see Wallmersperger *et al* [12]. The deformation of the fluid domain is modeled by a space-time isoparametric approach for the geometry of the fluid elements, see Tezduyar *et al* [11]. Stabilization of the convective term of the fluid equations is attained by means of the Galerkin/least squares approach which yields the required "upwinding" and high frequency/short wave length filtering property. In order to obtain monotonic solutions at shocks and to satisfy entropy stability, i.e. the second law of thermodynamics, nonlinear higher order discontinuity capturing operators are employed, see Shakib [8]. The stabilized space-time finite element discretization yields accurate, locally conservative and stable discretization for electromechanics, transonic aerodynamics and fluid-structure interaction.

Solution procedure. In the case of the linear model for piezoelectricity, it is possible to eliminate the electric field of the smart composite by static condensation and the structure is written in terms of mechanical displacement unknowns only. In aeroelasticity, staggered solution procedures for fluid-structure interaction are very popular. However, potential deficiencies concerning conservation and stability are well known, see Piperno *et al* [5]. For this reason, an efficient and reliable block iterative solver based on a coupled Newton-Raphson formulation for fluid-structure interaction including deformation of the fluid mesh is formulated. It does not degrade conservation, stability and accuracy of the underlying finite element discretization.

Aerodynamic Effectiveness of Smart Airfoil. Results of quasi-steady numerical simulations are presented for transonic flow around a NACA23010 airfoil at an angle of attack $\alpha = 0°$ and a Mach number $M = 0.74$. This corresponds to forward flight conditions of the advancing blade at a radial station of 85%. Consider the smart airfoil according to Fig. 3 with the upper and lower active composite layers contracting and expanding, respectively. Table 2 shows a comparison of the change of the angle of attack $\Delta\alpha$, lift C_l and moment coefficient C_m for different values of equivalent free actuator strain ϵ. It is seen that significant actuation is required for aerodynamic effectiveness.

Comparison of Smart Airfoil, Smart Tab and Discrete Trailing Edge Flap. For a discrete trailing edge flap $\lambda_F = c_F/c$ where c_F and c are the chord of the flap and the total chord of the wing section, respectively. According to Schlichting *et al* [7], a large $\lambda_F \to 1$ and small flap $\lambda_F \to 0$ behave like a direct lift and servo flap, respectively. In Figure 5 it is also seen, that the variable camber of the smart airfoil according to Figure 3 behaves like a direct lift flap where $\lambda_C = c_C/c$ and c_C is the chord length of the flexible part of the airfoil. For the smart tab according to Figure 4 we employ $\lambda_T = c_T/c$ where c_T and c are the chord of the tab itself and the chord of the plain airfoil without tab, respectively. The tab bending actuation leads to a servo effect, slightly depending on λ_T. Both concepts may be combined to tailor the aerodynamic characteristics for specific needs.

TABLE 2. Influence of equivalent free actuator strain ϵ on angle of attack $\Delta\alpha$, lift C_l and moment C_m coefficient

ϵ	$\Delta\alpha$	C_l	Cm
0.00%	0.00	0.18	−0.016
0.15%	−0.58°	−0.01	0.003
0.25%	−0.96°	−0.14	0.015
0.35%	−1.35°	−0.27	0.028

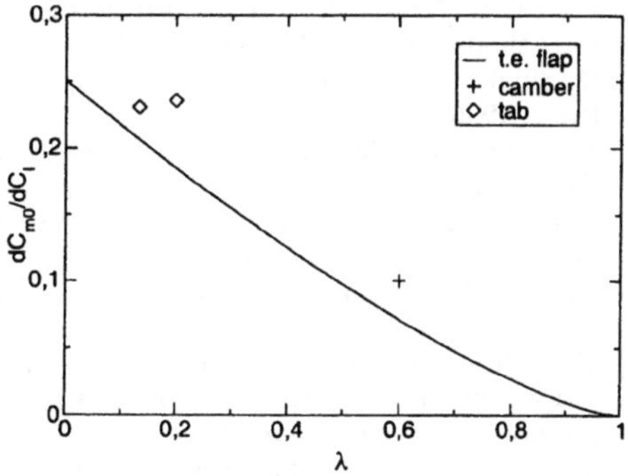

Figure 5. Comparison of trailing edge flap, variable camber of smart airfoil and smart tab

4 Conclusions

A method for lagging mode stability augmentation for active helicopter rotors based on decentralized control has been proposed. Active damping enhancement of about 600% has been achieved while providing robust performance when applying the controller to the time-periodic plant. Moreover, the derived decentralized controller consisting of N identical subsystem controllers is conceptually simple, permitting a certain degree of tuning during experiments.

Furthermore, an Active Camber Rotor (ACR) with distributed active material components and continuous shape adaptation has been proposed. Because it is an aeroelastically unbalanced concept, aeroservoelastic design turns out to be essential. For this reason, multiphysics simulation for coupled electromechanics and aeroelasticity of the smart composite structure including piezoceramic layers in inviscid, transonic flow has been developed based on stabilized space-time finite elements. Numerical simulations have been presented demonstrating the aerodynamic effectiveness of shape adaptation and comparing the ACR to a discrete trailing edge flap.

References

[1] W. Geissler, H. Sobieczky, and H. Vollmers. Numerical study of the unsteady flow on a pitching airfoil with oscillating flap. In *24th European Rotorcraft Forum*, Marseilles, France, 15th-17th September 1998.

[2] B. A. Grohmann. *Stabilized Space-Time Finite Elements for Transonic Aeroelasticity.* PhD thesis, Universität Stuttgart, June 2000. submitted.

[3] W. Johnson. *Helicopter Theory.* Dover Publications Inc., 1980.

[4] P. Konstanzer and B.-H. Kröplin. Performance considerations in the control of helicopter vibration. In *SPIE Proceedings of Smart Structures and Materials 1999: Mathematics and Control in Smart Structures*, volume 3667, 1999.

[5] S. Piperno and C. Farhat. Energy based design and analysis of staggered solvers for nonlinear transient aeroelastic problems. In *Structures, Structural Dynamics and Materials Conference.* AIAA/ASME/ASCE/AHS/ASC, April 2000. Atlanta, Georgia, AIAA-2000-1447.

[6] D. Schimke, P. Jänker, V. Wendt, and B. Junker. Wind tunnel evaluation of a full scale piezoelectric flap control unit. In *24th European Rotorcraft Forum*, 1998.

[7] H. Schlichting and E. Truckenbrodt. *Aerodynamik des Flugzeuges.* Springer Verlag, 1969.

[8] F. Shakib. *Finite Element Analysis of the compressible Euler and Navier-Stokes equations.* PhD thesis, Stanford University, 1988.

[9] D. Siljak. *The Control Handbook*, chapter Decentralized Control, pages 779 – 793. CRC Press, 1996.

[10] V. Syrmos, C. Abdallah, P. Dorato, and K. Grigoriadis. Static output feedback – a survey. *Automatica*, 33(2):125 – 137, 1997.

[11] T. E. Tezduyar, M. Behr, and J. Liou. A new strategy for finite element computations involving moving boundaries and interfaces – The deforming-spatial-domain/space-time procedure: I. The concept and the preliminary numerical tests. *Comput. Methods Appl. Mech. Engrg.*, 94:339–351, 1992.

[12] T. Wallmersperger, B. A. Grohmann, and B.-H. Kröplin. Time-discontinuous stabilized space-time finite elements for PDEs of first- and second-order in time. In *European Conference on Computational Mechanics ECCM '99*, München, Germany, 1999. GACM.

[13] B. Wie and K.-W. Byun. New generalized structural filtering concept for active vibration control synthesis. *Journal of Guidance, Control, and Dynamics*, 12:147 – 154, 1989.

DESIGN OF REDUCED-ORDER CONTROLLERS ON A REPRESENTATIVE AIRCRAFT FUSELAGE

M. J. ATALLA*, M. L. FRIPP, J. H. YUNG AND N. W. HAGOOD
Massachusetts Institute of Technology
77 Massachusetts Avenue, 37-331
Cambridge, MA 02139 USA

1. Introduction

The traditional approach to minimizing structural vibration implements either passive dampers or global model-based active controllers. Unfortunately, passive control becomes massive for control of low frequency disturbances. Model-based active controllers are difficult to apply to complex structures, because the model needs to be of roughly the same order as the system that it describes in order to achieve robust performance. Modeling errors due to mismodeled dynamics, missed dynamics, or time-varying dynamics can be performance degrading and potentially destabilizing [6]. This paper focuses on the lightly damped and modally dense systems where modeling errors are more significant.

Proper actuator and sensor choices can simplify the plant dynamics and, hence, simplify the feedback control system design. Ideally, actuators and sensors should be chosen to emphasize the important modes in the loop transfer function while de-emphasizing the modes that may contribute to instability and to controller effort. The reduction in modal complexity and the enhanced separation of the modes improves stability margins.

Previous studies have focused upon the use of modal transducers to isolate individual modes and to simplify the loop transfer function [5]. Modal transducers can isolate dynamically important modes and allow the design of reduced-order controllers focusing on a narrow frequency band. However, most implementation of modal transducers are limited to simple structures where an accurate model is available, or where modal information can be accurately measured.

Active structural-acoustic control due to a broadband disturbance has not been successfully implemented on a complex structure with high modal density. This study is motivated by a desire to minimize the interior acoustics of aircraft arising from an unknown broadband acoustic disturbance. Such structural-acoustic systems have very high modal density due to the three-dimensional nature of the

*Author for correspondence: Email: MJ@atallas.com, Phone +1 (617) 258-5920.

U. Gabbert and H.S. Tzou (eds.), IUTAM Symposium on Smart Structures and Structronic Systems, 17–24.

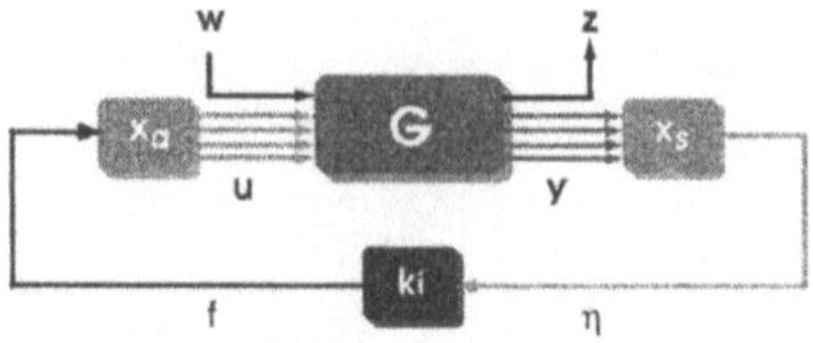

Figure 1: The weights on the sensor array, x_s, serve as a filter to the sensors signals y, while the weights on the actuator array, x_a, serves as a filter to the actuator signals u. The weighted signals, $\eta$and f, can be designed to be modal signals, which allows for the design of simpler controllers, k_i.

acoustic problem. Adding to the control difficulty, the coupled structural-acoustic dynamic behavior is sensitive to temperature, humidity, and other environmental factors. Attempts by the authors to numerically model the structural-acoustic dynamics of an aircraft fuselage test-bed have not been successful.

Arrays of transducers designed to provide robust feedback control with high performance and limited modeling are the subject of this paper. A reconfigurable array technique recently developed [3] enables the design of reduced-order controllers for complex structures and offers the potential to improve closed-loop robustness and to broaden the region of good performance even as the plant changes. The weighted summation of sensor signals senses the modes that are relevant to performance while rejecting the remaining modes; therefore reducing the required complexity of the controller.

This paper presents improvements over the weighting selection developed in earlier work and realizes feedback control of vibrations due to a broadband acoustic disturbance on a representative aircraft fuselage. The next section describes the approach while section 3 presents early experimental results along with new results obtained with the new proposed cost function for the computation of the weighting vectors.

2. Proposed Approach: Reconfigurable Arrays of Transducers

Reconfigurable arrays are arrays of discrete actuator or sensor elements with a weighting that is tuned to target individual modes. Multiple different weighting vectors can be used so that multiple modes can be isolated from the same array elements. The weighting can be updated to track the modes as the system changes.

The goal of reconfigurable arrays is to enable the design of reduced-order controllers for complex structures, improve closed-loop robustness and broaden the region of good performance in the presence of plant uncertainty. The key to reconfigurable arrays lies in determining the optimal weighting for the sensor and actuator elements. The optimal weights are those that map a single mode from forcing to measurement acting as pre- and post- filters to the control algorithm, allowing the implementation of reduced-order controllers (fig. 1).

There are various ways to compute the weighting vectors. On structures with simple boundary conditions and low damping the weights can be the real part of the residues of the desired modes evaluated at the sensor locations (in this case, $\mathbf{x}_s = \mathbf{x}_a$). Another alternative, also valid when $\mathbf{x}_s = \mathbf{x}_a$, is to invert the transfer function matrix, so that the weighting vectors are given by:

$$x_s = (\mathbf{G}_{yu}^T \mathbf{G}_{yu})^{-1} \mathbf{G}_{yu}^T f_d, \tag{1}$$

where $\mathbf{G}_{yu}$ is the matrix of measured transfer functions from actuators, $\mathbf{u}$, to sensors, $\mathbf{y}$, and $\mathbf{f}_d$ is the desired frequency response of the weighted signal. In this case the weights will be complex in systems with non-zero damping and, therefore, difficult to implement experimentally.

A more general approach is to minimize a cost function that represents the inverse of a performance metric. Consider the transfer function matrix $\mathbf{G}_{yu}$, given by

$$\mathbf{y} = \mathbf{G}_{yu}\mathbf{u}. \tag{2}$$

The array weights on the actuators and sensors create a weighted sensor response, η, and a weighted actuator input, f, according to:

$$\eta = \mathbf{x}_s^T \mathbf{y} \quad \text{and} \quad \mathbf{u} = \mathbf{x}_a f. \tag{3}$$

Note that the weighting vectors are not necessarily identical. Substituting equation (3) into equation (2) yields the transfer function from the weighted actuators to the weighted sensors,

$$\eta = \mathbf{x}_s^T \mathbf{G}_{yu} \mathbf{x}_a f. \tag{4}$$

When targeting multiple modes for control, $\mathbf{x}_s$ and $\mathbf{x}_a$ becomes matrices that, when properly chosen, render the weighted transfer function matrix diagonal, mapping modal excitation to modal response. The authors [3] have previously used cost functions designed to minimize the modal response of unwanted modes, R_n, while maximizing the residue of the desired mode, R_m, in the weighted transfer function $\mathbf{x}_s^T \mathbf{G}_{yu} \mathbf{x}_a$. The H_2 and H_∞ cost functions used in [3] were

$$J_2 = \frac{\sum_n R_n^* R_n}{R_m^* R_m} \quad \text{or} \quad J_\infty = \frac{\max_n R_n^* R_n}{R_m^* R_m}, \tag{5}$$

where $()^*$ indicates the complex conjugate. These cost functions can also include penalty terms if the weighted plant, $\mathbf{x}_s^T \mathbf{G}_{yu} \mathbf{x}_a$, does not have collocated behavior or does not roll-off with increasing frequency.

The approach described above and investigated in detail in [2] has shown to be viable, however problems with performance and robustness of the closed-loop system were observed. In order to address these issues a MIMO analysis can be performed. The reason for the poor performance and low robustness can be seen by analyzing the closed-loop MIMO transfer function, as follows. Given a control input vector $\mathbf{u}$ and a disturbance vector $\mathbf{w}$, the response of the sensors is given by

$$\mathbf{y} = \mathbf{G}_{yu}\mathbf{u} + \mathbf{G}_{yw}\mathbf{w}. \tag{6}$$

Introducing equation (3) leads to

$$\mathbf{y} = \mathbf{G}_{yu}\mathbf{x}_a f + \mathbf{G}_{yw}\mathbf{w}, \tag{7}$$

and using the fact that $f = -k_i\eta$ for the reduced-order controllers of interest in this study, yields:

$$\mathbf{y} = -\mathbf{G}_{yu}\mathbf{x}_a k_i \mathbf{x}_s^T \mathbf{y} + \mathbf{G}_{yw}\mathbf{w} \Longrightarrow \mathbf{y} = (\mathbf{I} + \mathbf{G}_{yu}\mathbf{x}_a k_i \mathbf{x}_s^T)^{-1}\mathbf{G}_{yw}\mathbf{w} \tag{8}$$

It is clear from the equation above that the behavior of the MIMO system is determined by the matrix $\mathbf{G}_{yu}\mathbf{x}_a k_i \mathbf{x}_s^T$. It turns out that choosing the weighting vectors so that $\mathbf{x}_s^T\mathbf{G}_{yu}\mathbf{x}_a$ features only the modes of interest does not guarantee that $\mathbf{G}_{yu}\mathbf{x}_a\mathbf{x}_s^T$ will be diagonalized by the eigenvectors of the system. In fact, this has not been observed in this study.

A new cost function is proposed to compute the weighting vectors. Its form is given by

$$J = \frac{\|\mathbf{G}_{yu}\mathbf{x}_a\mathbf{x}_s^T\|_{\omega \in n}}{\|\mathbf{G}_{yu}\mathbf{x}_a\mathbf{x}_s^T\|_{\omega \in m}}, \tag{9}$$

where $\omega \in n$ represents the frequency points where the response should be minimized and $\omega \in m$ represents the frequencies of the modes of interest. This function is equivalent to minimizing the maximum singular value of the matrix $\mathbf{G}_{yu}\mathbf{x}_a\mathbf{x}_s^T$ over the range of frequencies to be suppressed while maximizing the maximum singular value at the frequencies of interest. The controller k_i is left out of the cost function since its form is such that its magnitude is large only at $\omega \in m$, therefore only significantly effecting the denominator.

Another important factor to be considered is spatial aliasing, which is a limiting factor in the performance of discrete element arrays. Spatial aliasing is behaviorally similar to aliasing in temporal signals in that shorter wavelength modes are indistinguishable from longer wavelength modes. Spatial aliasing sets an upper limit to the number of modes that can be minimized in the weighted response. The effect of spatial aliasing can be mitigated by using distributed transducers and taking advantage of the spatial filtering effect, a feature addressed in the next section.

3. Modal Isolation and Feedback Control

A representative aircraft fuselage test-bed has been built to focus the control efforts and to serve to compare different control approaches. The fuselage test-bed was designed to capture the essential dynamic characteristics of rotorcraft and fixed-wing aircraft, preserving the important structural-acoustic characteristics. Hybrid scaling parameters were used to replicate not only the geometric dimensions of the aircraft but also the salient dynamics of the structure. The final design of the test-bed features a thin aluminum skin riveted to a frame of ribs and stringers.

The fuselage test-bed holds the same panel dynamics, global dynamics, acoustic dynamics, and coupled structural-acoustic dynamics as a representative sample of six aircraft structures and models [4]. Figure 2(a) shows a picture of the fuselage test-bed and figure 2(b) shows representative experimental transfer functions. The very high modal density of the system is apparent by the number of modal resonances that are observed; there are 47 distinct modal resonances between 380 Hz and 580 Hz. Clearly, model-based control would be very difficult to implement on the fuselage test-bed.

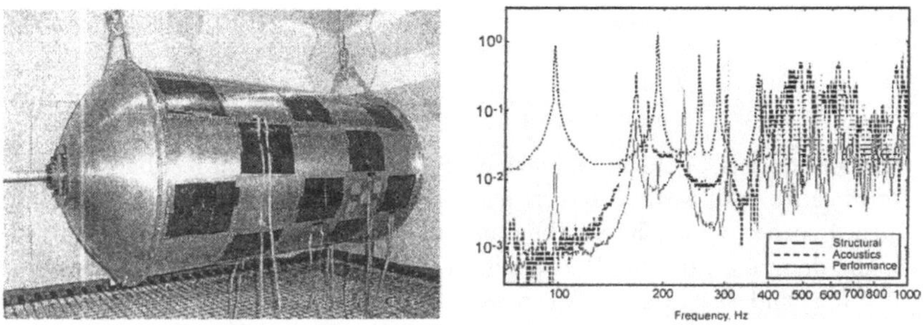

(a) (b)

Figure 2: (a) Fuselage test-bed designed to represent a typical aircraft. The picture shows the sensor-actuator plies developed to control the vibration of the fuselage's skin. (b) Typical transfer functions (structural, acoustic and performance) measured on the fuselage test-bed.

A sketch of an active ply developed for this study is shown in figure 3. Six panels were instrumented with 5 sensor-actuator pairs per panel. The size of the transducers was chosen to simplify the control design, since the size of the transducers influences the spatial filtering of the measured strain. Spatial filtering initiates when the transducer spans multiple structural wavelengths. Thus, larger sized transducers will tend to have reduced coupling to higher frequency modes. The size of the transducers on the fuselage test-bed was chosen so that there would be good modal observability up to 2 kHz. Experimental iteration with different sized patches indicated that the chosen transducer size was of the appropriate size. Although square transducers were used in this study, circular transducers would improve the roll-off behavior of the loop transfer function [1].

Previous experiments [2, 3] performed using the cost functions shown in equation (5) and the PVDF-PZT sensor-actuator pairs showed promising performance. Two strain-rate feedback controllers, implemented according to equation (10), were used to reduce the vibration of the modes at 925 Hz and 975 Hz due to a structural disturbance source (a piezoceramic wafer).

Figure 3: Lay-up of the active ply that is composed of collocated piezoceramics actuators and piezopolymer sensors. Electrodes for actuators and sensors are etched into copper-covered Kapton polyimide sheets.

$$k_i(s) = \left(\frac{s - \omega_c}{s^2 + 2\zeta_c\omega_c s + \omega_c^2}\right)\left(\frac{g_i\ \omega_i^2}{s^2 + 2\zeta_i\omega_i s + \omega_i^2}\right) \tag{10}$$

The controller consists of the combination of a positive-position feedback (PPF) controller in series with a stabilized integrator. The controller variables were the low frequency corner, $\omega_c = 150$ Hz, the damping ratio of the low frequency corner, $\zeta_c = 0.5$, the damping ratio for the target frequencies, $\zeta_1 = \zeta_2 = 0.05$, the gains, $g_1 = g_2 = 0.5$, and the target frequencies for each mode, $\omega_1 = 950$ Hz for the first mode and $\omega_2 = 1000$ Hz for the second mode. Figure 4(a) shows that the dual-mode controller reduced the weighted sensor response by 15.4 dB (RMS value computed from 850 Hz to 1000 Hz).

However, the acceleration of the shell as measured by an array of surface mounted accelerometers (**z** in figure 1) revealed that the closed-loop radial motion of the shell had increased. This is attributed to the fact that the input signal to the controller was the weighted in-plane strain, and the reduction of this quantity does not necessarily lead to the reduction of the radial motion of the shell. Another problem was that the disturbance was provided by a piezoceramic wafer instead of a loudspeaker, simplifying the problem.

In order to address these issues two actions were taken. The first was to mount accelerometers at the center of each actuator element in order to obtain a direct measurement of the surface's radial motion. New weighting vectors were computed using equation (5) and new controllers were designed and implemented with no significant improvement. The drawback of this approach can be seen in figure 4(b), which shows a comparison of the norms of the loop transfer function matrices for three cases due to a 50-1000 Hz acoustic disturbance. It is clear that the previous approach does not simply the MIMO behavior of the system, leaving room for controller spillover.

A genetic algorithm implemented in MATLAB was used in order to compute the weighting vectors. Genetic algorithms were chosen due to the high non-linearity of the objective function and the very high number of local minima. Weighting vectors were computed according to the proposed objective function to target the mode at 426 Hz (fig. 5(a)), with convergence being achieved after about 100 generations.

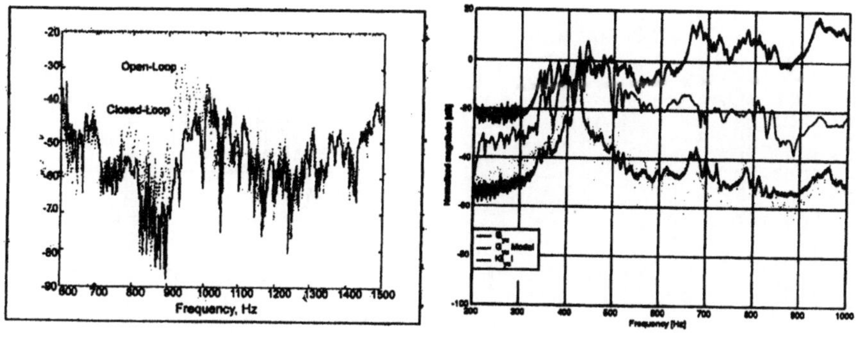

(a) (b)

Figure 4: (a) Weighted-strain open-loop and closed-loop responses using the PVDF sensor array with controllers targeting the modes at 925 Hz and 975 Hz. (b) Comparison among the maximum singular values of the open-loop G_{yu}, weighted open-loop according to eq. (5) and weighted open-loop according to eq. (9).

A controller with the form shown in equation (10) was experimentally implemented and the comparison between open-loop and closed-loop is shown in figure 5(b). The experimental results were obtained with a dSPACE real-time controller sampling at 10 kHz and using a 50-1000 Hz acoustic disturbance. The controller design variables are $\omega_c = 50$ Hz, $\zeta_c = 0.707$, $g = 1e9$, $\zeta_1 = 7e-3$, and $\omega_1 = 426$ Hz. The measured closed-loop responses show a consistent 5 to 9 dB acceleration reduction throughout the sensor array.

4. Conclusions

A reconfigurable array of piezoceramic actuators and accelerometer sensors was used to isolate individual modes and to minimize the vibrations of a representative aircraft fuselage test-bed. Most of the modes that were observable to the actuator and sensor arrays could be isolated. A new objective function is proposed to compute the weighting vectors of the sensor and actuator arrays so that the resulting filtered transfer function appears to be modal. This enables the design of reduced-order controllers for complex structures and is demonstrated experimentally. Further work is being carried out to control the interior acoustic response of the test-bed fuselage.

5. Acknowledgements

The authors are grateful for funding by the Army Research Office MURI DAAH 4-95-1-0104 monitored by Dr. Gary Anderson. Significant technical assistance was

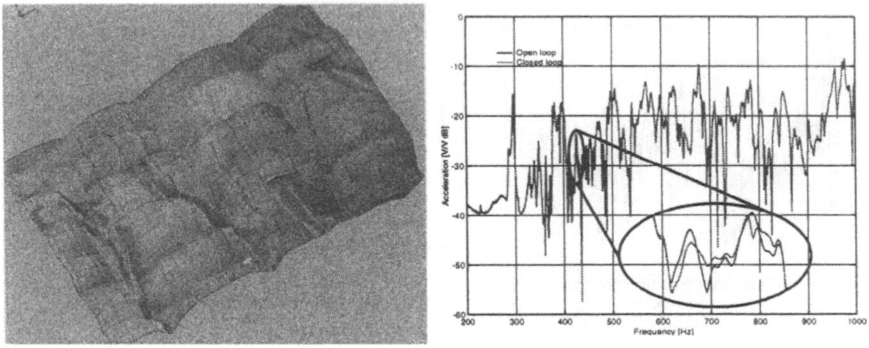

(a) (b)

Figure 5: (a) Mode shape of the mode at 426 Hz, target for control. (b) Comparison between open-loop and closed-loop responses obtained using the accelerometer array and the weighting vectors computed according to the new objective function (eq. 9). The closed-loop response is identical to the open-loop outside the band shown above.

provided by Steven Hall and significant manufacturing assistance was provided by Stephen Tistaert, Daniel Kwon, Christian D. Garcia and Eric Coulter.

6. References

[1] M. S. Andersson and E. F. Crawley. Structural shape estimation using shaped sensors. In *AIAA/ASME/ASCE/AHS Structures, Structural Dynamics & Materials Conference*, volume 5, pages 3368–3378, 1995.

[2] M. L. Fripp. *Weighted Arrays for Modal Isolation and Active Control of Complex Structures*. PhD thesis, Massachusetts Institute of Technology, Department of Aeronautical and Astronautical Engineering, 2000.

[3] M. L. Fripp, M. J. Atalla, N. W. Hagood, C. Savran, and S. Tistaert. Reconfigurable arrays for broadband feedback control of aircraft fuselage vibrations. In *10th International Conference on Adaptive Structures and Technologies*, volume 1, pages 447–456, 1999.

[4] M. L. Fripp, D. Q. O'Sullivan, S. R. Hall, N. W. Hagood, and K. Lilienkamp. Test-bed design and modeling for aircraft interior acoustic control. In *SPIE Conference on Smart Structures and Integrated Systems*, volume 3041, pages 88–99, 1997.

[5] Michael L. Fripp and Mauro J. Atalla. A review of modal sensing and actuation techniques. *Shock and Vibration Digest Journal*, 33(1), 2001.

[6] A. H. von Flotow. The acoustic limit of control of structural dynamics. In S. N. Atluri and A. K. Amos, editors, *Large Space Structures: Dynamics and Control*, pages 213–237. Springer Verlag, New York, 1988.

NUMERICAL ANALYSIS OF NONLINEAR AND CONTROLLED ELECTRO-MECHANICAL TRANSDUCERS

R. LERCH, H. LANDES, R. SIMKOVICS, M. KALTENBACHER
University of Erlangen, Department of Sensor Technology
Paul-Gordan Str. 3/5, D-91052 Erlangen, Germany

1. Introduction

Short product lifetime cycles, fast time to market and cost reduction as well as an increasing technical complexity are only some of the challenges developers of electromechanical transducers are faced with. Since the fabrication of prototypes and experimental based design is a lengthy and costly process, the need for appropriate numerical simulation tools arises. While linear numerical simulation used for sensor and actuator design is nowadays state of the art, high power actuators are still developed in manner of experimental trial and error. At low frequencies, as occurring in smart structure applications or positioning systems, controlled nonlinear amplifiers are used to overcome the power limitations of the transducers due to nonlinearity. This method is not applicable for piezoelectric transducers used for generating high intensity ultrasound due to the high frequencies used. Nonlinear effects have to be considered during the design process and therefore, simulation tools capable to analyze the complex interactions of different nonlinearities are necessary.

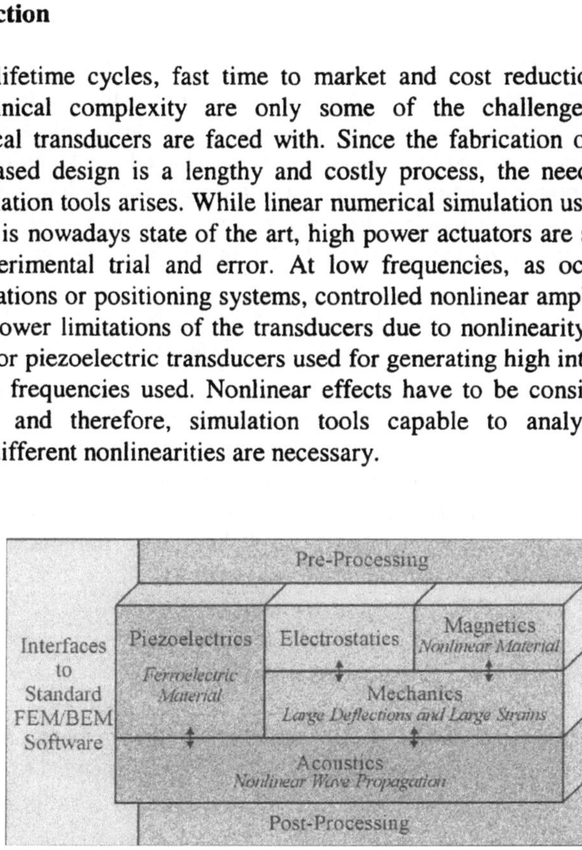

Figure 1. Linear and nonlinear features of the software system CAPA

In fig. (1) the basic building blocks of our software system *CAPA* are shown [1]. In this modeling environment various combinations of nonlinearities including geometric, electrostatic, magnetic, piezoelectric and acoustic nonlinearities can be analyzed. Therewith, large deformation and deflections as well as different kinds of material nonlinearities of magnetic and piezoelectric materials and nonlinear wave propagation can be dealt with this numerical analysis software.

U. Gabbert and H.S. Tzou (eds.), IUTAM Symposium on Smart Structures and Structronic Systems, 25–32.

In this paper we will focus on some applications of CAPA in the analysis of nonlinear effects in piezoelectric transducers. The rest of this paper is organized as follows: first we will give a short presentation of the underlying theory. Several practical examples will then demonstrate the applicability of our simulation system to real-world devices. These include the analysis of piezoelectric transducers under electrical and mechanical prestressing, the influence of hysteresis effects in a piezoelectric disc transducer and the optimization of controlled piezoelectric bimorphs.

2. Theory

The piezoelectric ceramic materials as used for ultrasonic power transducers show strong nonlinear dependencies of the material parameter on the electric field strength and the mechanical stresses applied. Therefore, changes in the modulus of elasticity, the modulus of piezoelectricity and the dielectric constants have to be taken into account for the design of high power actuators. In contrast to the standard nonlinear formulation for piezoelectric crystals which requires higher order material parameters, our method is based on a generalized piezoelectric material relation which can be stated as

$$\begin{aligned} \boldsymbol{T} &= \boldsymbol{c}^{\mathrm{E}}(T)\,\boldsymbol{S} - \boldsymbol{e}^{t}(T,E)\,\boldsymbol{E} \\ \boldsymbol{D} &= \boldsymbol{e}(T,E)\,\boldsymbol{S} + \boldsymbol{\varepsilon}^{\mathrm{S}}(E)\,\boldsymbol{E}\,. \end{aligned}$$

In large signal analysis the elements of the elastic tensor $\boldsymbol{c}^{\mathrm{E}}$ depend on the mechanical stresses $\boldsymbol{T}$ and the dielectric constants $\boldsymbol{\varepsilon}^{S}$ depend on the electric field strength $\boldsymbol{E}$, while the piezoelectric tensor $\boldsymbol{e}$ is influenced by both, mechanical stresses and electric field strength. Therefore, the material relation becomes nonlinear. In order to include ferroelectric effects the dielectric polarization $\boldsymbol{P}$ has to be introduced in the above equation, leading to

$$\begin{aligned} \boldsymbol{T} &= \boldsymbol{c}^{\mathrm{E}}\boldsymbol{S} - \boldsymbol{e}^{t}\boldsymbol{E} \\ \boldsymbol{D} &= \boldsymbol{e}\;\boldsymbol{S} + \boldsymbol{\varepsilon}_0\boldsymbol{E} + \boldsymbol{P}\,. \end{aligned}$$

Hysteresis effects can be modeled by considering the changes of ferroelectric polarization due to the load history the transducer had been suspended to.

2.1 FINITE ELEMENT FORMULATION

The application of the finite element discretization process to the above equations leads to the nonlinear finite element formulation

$$\begin{pmatrix} \boldsymbol{K}_{uu} & \boldsymbol{K}_{u\Phi} \\ \boldsymbol{K}_{u\Phi}^{t} & \boldsymbol{K}_{\Phi\Phi} \end{pmatrix} \begin{pmatrix} \{\Delta U\} \\ \{\Delta\Phi\} \end{pmatrix} = \begin{pmatrix} \{R\} - \{F\} \\ \{Q\} - \{q\} \end{pmatrix}$$

which is similar to the one used for linear piezoelectricity [2,3]. In contrast to the stiffness matrices of the linear formulation the mechanical, electrical and piezoelectric stiffness matrices in this equation now vary according to the state of stress and electric field. On the right hand side a residual vector of the externally applied nodal forces R and charges Q and their corresponding internal reactive nodal loads F and charges q is formed. Applying standard finite element procedures, the static formulation given above can be extended to a general transient calculation scheme.

2.2 CONSTITUTIVE MODEL

For the large signal analysis of piezoceramic materials a highly efficient mathematical model has to be used in the numerical calculation scheme. The parameter of the model should be easy to derive from measurements on specimens.

2.2.1 *Material Nonlinearities*

The implemented nonlinear material model is based on the specification of the material tensor as a function of the current states of stress and electric field. So, a functional dependency of the modulus of elasticity $\boldsymbol{c}$ on the mechanical stresses $\boldsymbol{T}$, as well as dependencies of the modulus of piezoelectricity $\boldsymbol{e}$ and dielectric constants $\boldsymbol{\varepsilon}$ on the electric field strength $\boldsymbol{E}$ is considered. This is advantageous since the influence of a single parameter on the entire nonlinear behavior of a transducer can be examined. In fig. (2) nonlinear dependencies of the piezoelectric modulus e_{31} and e_{33} and dielectric constants ε_{33} for different materials are shown, as derived from data published by the manufacturers [4-7]. Since these nonlinear dependencies can be measured directly, in contrast to higher order material parameters, this constitutive model offers a more practical approach to nonlinearity.

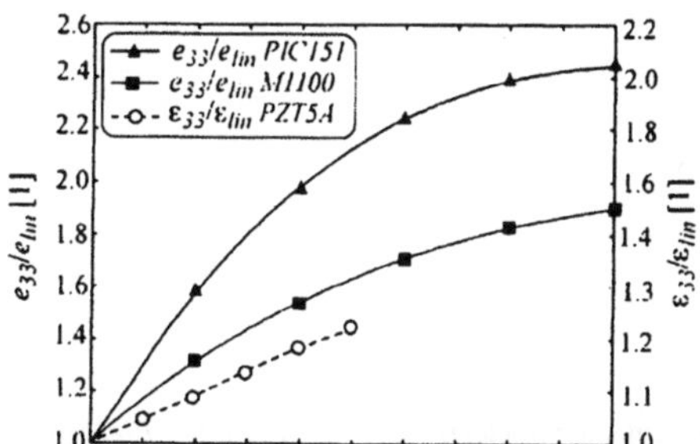

Figure 2. Dependencies of piezoelectric and dielectric constants on electric field strength

2.2.2 *Hysteresis Model*

In piezoceramic materials strong electric fields and large mechanical stresses lead to a change in the state of polarization $\boldsymbol{P}$ based on a spontaneous switching of the electric dipole orientation of ferroelectric grains. The energy needed for this change in orientation causes losses, which lead to a splitting of the positive and negative load path and resulting in a ferroelectric hysteresis [8]. Therefore, the knowledge about the current internal state is insufficient for the characterization of the behavior of the piezoelectric ceramic. Instead, the whole load history the transducer has been suspended to has to be taken into account. The effects of ferroelectric hysteresis show numerous similarities with ferromagnetic material behavior, which is well researched and for which a variety of mathematical models of hysteresis have been developed [9]. In our case a Preisach model [10] is used, which results from averaging the functional value of the Preisach operator over the region of interest. Details of the formulation can be found in [11] and will not be recalled here.

3. Applications

3.1 PRESTRESSED PIEZOELECTRIC DISK

3.1.1 *Experimental Setup*

As a first application, piezoelectric disk resonators subject to electric and mechanical prestressing have been analyzed. In the case of electrical prestressing, a disk resonator has been exposed to a bias voltage producing electric field strength throughout the resonator up to 400 V/mm. The constant electric field has been superimposed by a low amplitude sinusoidal signal with sweeping frequency, as shown in fig. (3). Due to the nonlinear change of the piezoelectric modulus $\boldsymbol{e}$ and the dielectric constant ε with increasing electric field strength, as shown in fig. (2), a change of resonance frequency can be expected. In the case of mechanical prestressing, the piezoelectric disk was mounted between two plates, which have been used to compress the disk and, therewith, apply a pressure surface load to the disk (fig. (3)). Due to the compression and the piezoelectric effect, a strong electric field is generated inside the disk and consequently, the material tensor is modified.

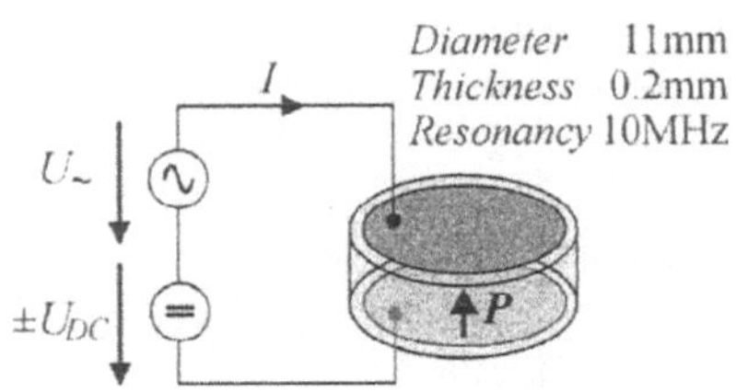

Figure 3. Piezoelectric disk subjected to electric prestressing

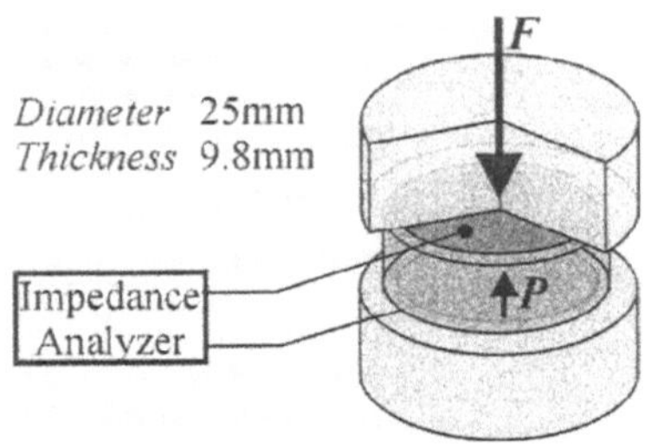

Figure 4. Piezoelectric disk resonator under mechanical prestressing

3.1.2 *Simulated and Measured Results*

In a first step, the electric input impedance of the electrical prestressed disk has been considered, when an electric field of 400 V/mm is applied. The disk consisted of *PIC151* ceramics and the impedance of both the prestressed and the free disk have been calculated. Furthermore, corresponding measurements have been performed. In measurements and simulations a frequency shift of approx. 0.7 % has been observed. In order to gain more insight into the dependency of the frequency shift on the applied electrical field, a sequence of simulation runs and measurements have been performed. The results of these investigations for two material data sets are shown in fig. (5). As can be seen, good agreement between measurement and simulations is observed.

It should be noted, that the electric field strength applied is rather large in comparison to the one typically used for polarization, which is about 2 kV/mm. For that reason, partial depolarization effects can be observed during measurement. Thus, the results shown in fig. (5) can be obtained only when a strictly increasing electric load is applied.

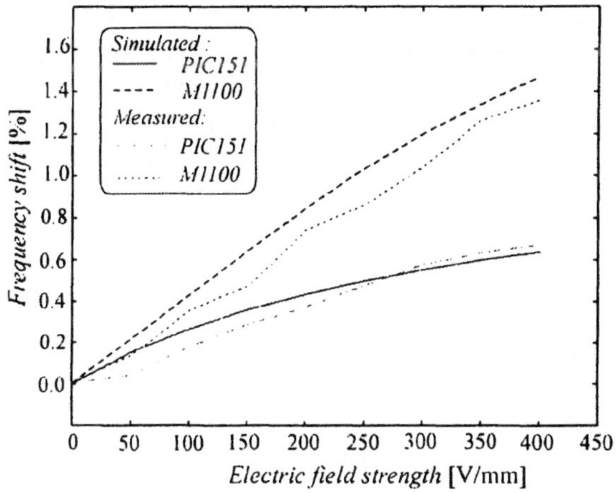

Figure 5. Dependency of the resonance frequency on the electric field strength for the ceramic materials PIC151 and M1100

In the next step we considered the case of mechanical prestressing. Increasing the pressure loads results in a strong decrease in intensity of the thickness mode and furthermore, in a strong frequency shift of the resonance itself. The frequency shift, as observed during simulations, is mainly due to nonlinear changes of the dielectric constant according to the applied external stresses. In fig. (6) the results of measurements and calculations for the case of a mechanically prestressed disk are compared with those for the free disk. Again, good agreement between measurements and simulations is found.

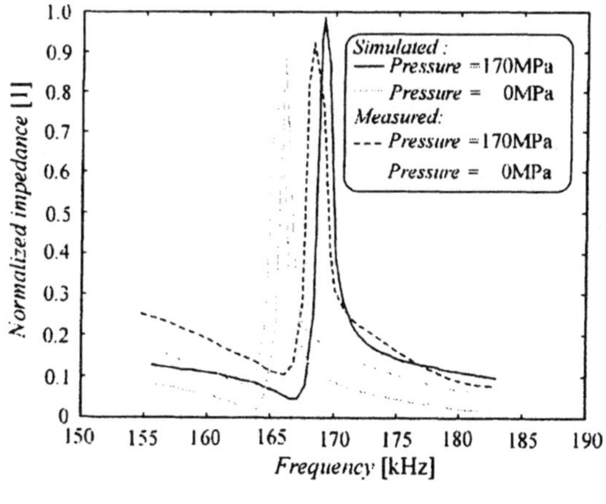

Figure 6. Resonance frequency shift of a piezoelectric disk under mechanical load

3.2 DYNAMIC LOADCASE

While the preceeding experiments are primary quasi-static load cases this section will deal with a dynamic large signal analysis. In contrast to the setup shown in fig. (3) the amplitude of the high frequency signal is of about the same magnitude as the bias voltage applied. Thus, with an increasing high frequency voltage distortions of the sinusoidal form of the input current of the transducer can be detected. Changes of the state of ferroelectric polarization are mainly responsible for these characteristics of the piezoceramic disk. In order to overcome strong thermal influences due to the heat generation caused by internal friction in the piezoceramic material a sine burst signal with a small pulse-pause-ratio is used for excitation. In fig. (7) the time signal of the input current and voltage can be observed. The occurrence of higher harmonics and the unsymmetrical response signal are caused by ferroelectric hysteresis.

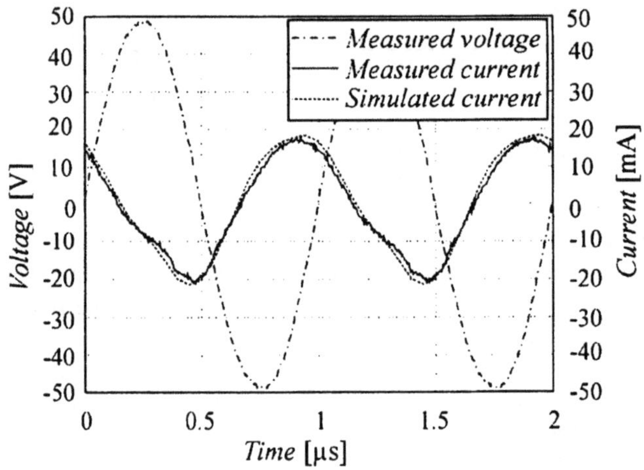

Figure 7. Input voltage and current

3.3 PIEZOELECTRIC BIMORPH ACTUATOR

As a second application we consider here the study of a controlled actuator consisting of a piezoelectric bimorph, which is fixed on one side. In a first simulation, the mechanical displacement of the bimorph due to a discrete force F applied to the free end of the bimorph was analyzed. In addition to the force F, an electrical control voltage U was applied to the electrodes in order to minimize the deflection. To measure the deflection of the structure, the square weighted average normal displacement $\overline{y}$ defined by

$$\overline{y} = \left(\frac{1}{L} \int_0^{L_b} \eta(x)^2 \, dx \right)^{\frac{1}{2}}$$

was used. Besides the standard, fully electroded configuration of the bimorph, partially electroded modifications have been considered. In fig. (8) the dependency of $\overline{y}$ on the control voltage U for bimorph structures with electrode/bimorph length ratios L_e/L_b of $L_e/L_b = 1.0$, i.e. fully electroded, $L_e/L_b = 0.8$ and, $L_e/L_b = 0.6$ is shown.

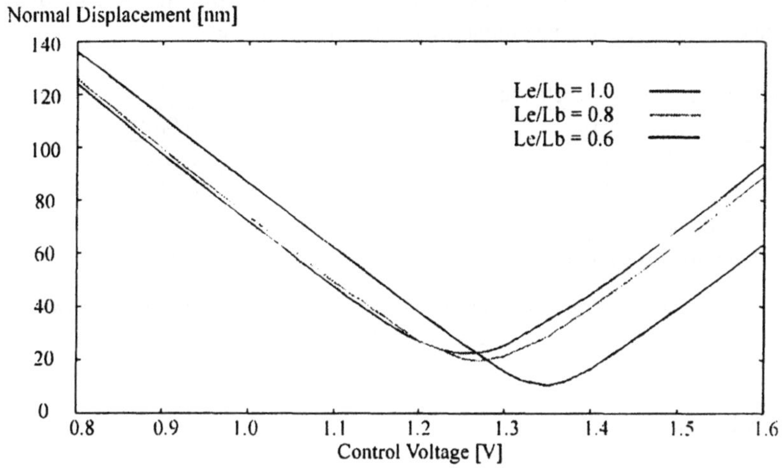

Figure 8. Average normal displacements of piezoelectric bimorph beam

As can be seen, partially electroded bimorphs lead to better results, i.e. minor average deflection, when compared with a bimorph which is fully covered by electrodes.
In order to gain an even better control, bimorphs equipped with an array electrode consisting of 4 electrically separated elements have been considered. The corresponding electrode patterns are shown in figs. (9) and (10).

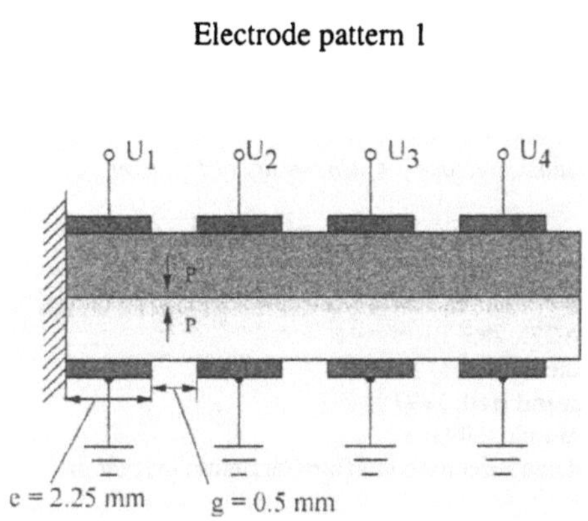

Figure 9. Bimorph with array electrodes (pattern 1)

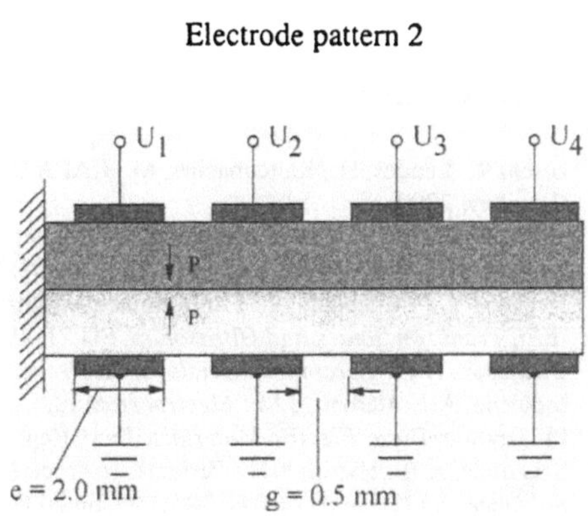

Figure 10. Bimorph with array electrodes (pattern 2)

Since these electrodes can be loaded by 4 independent voltages $U_1,..,U_4$ the seek for an optimum control results in an optimization problem with a 4-dimensional parameter space. The optimum values for the control voltages have been determined for both configurations using an optimization module available in CAPA.

As above, the optimization criterion used was the minimization of the square weighted average deflection $\bar{y}$ of the bimorph. These average displacements have been normalized with respect to those of the fully electroded bimorph with no control applied. The obtained results have been summarized in the following table.

Uncontrolled, fully electroded	1.00
$L_e/L_b = 1.0$	0.082
$L_e/L_b = 0.8$	0.070
$L_e/L_b = 0.6$	0.037
4 electrodes (pattern 1)	0.0016
4 electrodes (pattern 2)	0.0023

TABLE 1. Average displacement of bimorph for different electrode configurations and optimal control voltages

4. Conclusion

We have presented a numerical calculation scheme for the analysis of nonlinear effects in piezoelectric transducers. This scheme has been implemented into the FE/BE package CAPA. Piezoceramic transducers have been analyzed taking nonlinearities of the material properties of the piezoceramic material and changes of the state of ferroelectric polarization into account. For these actuators measurements and numerical simulations have been performed subjecting the ceramic to various cases of prestressing. The coupling with controller and optimization modules further expands the applicability of CAPA in the design process of sensor and actuator devices.

5. References

1. Lerch, R.; Landes, H.; Kaltenbacher, M.: CAPA User Manual, *Release 3.4, University of Erlangen, Germany*, 2000.
2. Lerch, R.: Simulation of piezoelectric devices by two- and three-dimensional finite elements, *IEEE Trans. on Ultras., Ferroel. and Freq. Control*, 37, No. 3, 1990, pp 233-247
3. Kagawa, Y.; Yamabuchi, T.: Finite element simulation of two-dimensional electromechanical resonators, *IEEE Trans. on Sonics and Ultrasonics*, SU 21, 1974, pp 275-283
4. Buchanan, R.C.: *Ceramic materials for electronics*, Dekker, 1991
5. Moulson, A.J.; Herbert, J.M.: *Electroceramics*, Chapman and Hall, 1997
6. PI-Ceramic: *Piezo-Electric Materials*, Tech. Rept., PI-Ceramic, 1998
7. Schäufele, A.B.; Härdtl, K.H.: Ferroelastic properties of Lead Zirconate Titanate Ceramics, *Journal of the American Ceramic Society*, 78, 1996, pp 2637
8. Boser, O.: Statistical theory of hysteresis in ferroelectric materials, *Journal of Applied Physics*, Vol. 62, 1987, pp 1344-1348
9. Visintin, A.: Mathematical models of Hysteresis, *Topics in Nonsmooth Mechanics*, ed. J.J. Moreau, P.D. Panagiatopoulos and G. Strang, Birkhauser, VIII, pp 295-323
10. Mayergoyz, I.D.; *Mathematical Models of Hysteresis*, Springer Verlag, 1991
11. Simkovics, R.; Landes, H.; Kaltenbacher, M.; Hoffelner J.; Lerch R.: Finite Element Analysis of Hysteresis Effects in Piezoelectric Transducers; March, 2000; *Proceedings of SPIE*, Vol. 3984; pp. 33-44

SMART STRUCTURES IN ROBOTICS

F. DIGNATH, M. HERMLE, W. SCHIEHLEN
Institute B of Mechanics
University of Stuttgart
70550 Stuttgart
Germany

1. Introduction

The control of flexible manipulators requires control strategies considering the elastic deformations of the robot. In this paper a hierarchical control concept is presented for serial–chain manipulators equipped with electrical drives in the joints and, additionally, actuating and sensing devices located on the elastic links between the joints, see Figure 1. With this additional actuators and sensors, Static Dissipative Controllers (SDC) can be used resulting in a low–authority control strategy counterbalancing elastic disturbances. The decentralized linear control allows also a static correction of the elastic deformations and can be combined with any given joint level control for the gross motion of the manipulator. In this paper, for a SCARA type robot an inverse dynamics based controller and Static Dissipative Controllers are combined and adapted by parameter optimization.

2. Modeling

Figure 1 depicts a portion of the serial chain for a flexible manipulator. In order to describe the elastic deformations, the slender links are adequately modeled as Euler–Bernoulli beams with clamped–free boundary conditions.

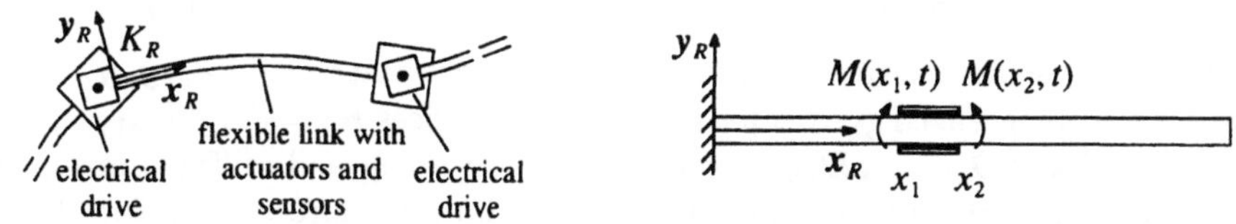

Figure 1. Flexible manipulator serial chain and actuated beam

We assume that sensor and actuator strips are mounted on each elastic link of the robot. The actuation of the beam is equivalent to two torques of equal size and opposite direction, $M(x_2,t) = M(x_1,t) = u(t)$, where $u(t)$ denotes the control input of the system. The output signals are the differences of the angular

U. Gabbert and H.S. Tzou (eds.), IUTAM Symposium on Smart Structures and Structronic Systems, 33–40.

displacements and angular velocities at the sensor endpoints x_i, $i = 1, 2$.

$$s_a(t) = \gamma(x_2, t) - \gamma(x_1, t) \ , \quad s_r(t) = \dot{\gamma}(x_2, t) - \dot{\gamma}(x_1, t) \ . \tag{1}$$

Such collocated sensors and torque actuators can be realized e.g. using piezoelectric materials, see Schiehlen and Schönerstedt [8].

The mathematical model of the j–th link can be obtained by finite element modeling. For analysis and design purposes the order of the system is reduced using a modal truncation approach. The system equations in modal coordinates for the r_j remaining modes are written as

$$\boldsymbol{M}_{ej}\ddot{\boldsymbol{q}}_j + \boldsymbol{D}_{ej}\dot{\boldsymbol{q}}_j + \boldsymbol{K}_{ej}\boldsymbol{q}_j = \boldsymbol{\Gamma}_{ej}^T \boldsymbol{u}_{ej} \ , \tag{2}$$

$$\boldsymbol{s}_{aj} = \boldsymbol{\Gamma}_{ej}\boldsymbol{q}_j \ , \qquad \boldsymbol{s}_{rj} = \boldsymbol{\Gamma}_{ej}\dot{\boldsymbol{q}}_j \tag{3}$$

where $\boldsymbol{q}_j \in I\!R^{r_j}$ is the vector of modal amplitudes, and $\boldsymbol{M}_{ej}$, $\boldsymbol{D}_{ej}$, $\boldsymbol{K}_{ej}$ are the generalized mass, damping and stiffness matrices. Due to the collocation of sensors and actuators, $\boldsymbol{\Gamma}_{ej}^T \in I\!R^{r_j \times m_j}$ and $\boldsymbol{\Gamma}_{ej}$ are the corresponding control input and output matrices, respectively. The vector $\boldsymbol{u}_{ej} \in I\!R^{m_j}$ represents the control inputs, and $\boldsymbol{s}_{aj}$, $\boldsymbol{s}_{rj} \in I\!R^{m_j}$ are the angular displacement and angular velocity measurements.

Now, the equations of motion for the complete robot with flexible and rigid links can be derived by the method of flexible multibody systems, see Shabana [9] and Schiehlen [7]. The vector of generalized coordinates $\boldsymbol{y} = [\boldsymbol{y}_r^T \ \boldsymbol{y}_e^T]^T$ consists of the n joint coordinates $\boldsymbol{y}_r$, describing the rigid body motion and the n_e modal amplitudes of equation (2) as generalized elastic coordinates $\boldsymbol{y}_e$. Then, the equations of motion read as follows.

$$\underbrace{\begin{bmatrix} \boldsymbol{M}_{rr}(\boldsymbol{y}) & \boldsymbol{M}_{re}(\boldsymbol{y}) \\ \boldsymbol{M}_{re}^T(\boldsymbol{y}) & \boldsymbol{M}_{ee}(\boldsymbol{y}) \end{bmatrix}}_{\boldsymbol{M}(\boldsymbol{y})} \begin{bmatrix} \ddot{\boldsymbol{y}}_r(t) \\ \ddot{\boldsymbol{y}}_e(t) \end{bmatrix} + \underbrace{\begin{bmatrix} \boldsymbol{0} & \boldsymbol{0} \\ \boldsymbol{0} & \boldsymbol{D}_e \end{bmatrix} \begin{bmatrix} \dot{\boldsymbol{y}}_r(t) \\ \dot{\boldsymbol{y}}_e(t) \end{bmatrix} + \begin{bmatrix} \boldsymbol{0} & \boldsymbol{0} \\ \boldsymbol{0} & \boldsymbol{K}_e \end{bmatrix} \begin{bmatrix} \boldsymbol{y}_r(t) \\ \boldsymbol{y}_e(t) \end{bmatrix}}_{\boldsymbol{k}_i(\boldsymbol{y}, \dot{\boldsymbol{y}})}$$

$$+ \underbrace{\begin{bmatrix} \boldsymbol{k}_{cr}(\boldsymbol{y}, \dot{\boldsymbol{y}}) \\ \boldsymbol{k}_{ce}(\boldsymbol{y}, \dot{\boldsymbol{y}}) \end{bmatrix}}_{\boldsymbol{k}_c(\boldsymbol{y}, \dot{\boldsymbol{y}})} = \begin{bmatrix} \boldsymbol{I} & \boldsymbol{0} \\ \boldsymbol{0} & \boldsymbol{\Gamma}_e^T \end{bmatrix} \begin{bmatrix} \boldsymbol{u}_r(t) \\ \boldsymbol{u}_e(t) \end{bmatrix} - \underbrace{\begin{bmatrix} \boldsymbol{g}_r(\boldsymbol{y}) \\ \boldsymbol{g}_e(\boldsymbol{y}) \end{bmatrix}}_{\boldsymbol{g}(\boldsymbol{y})} \tag{4}$$

where $\boldsymbol{M}(\boldsymbol{y})$ is the positive definite symmetric mass matrix and $\boldsymbol{k}_c(\boldsymbol{y}, \dot{\boldsymbol{y}})$ denotes the vector of generalized gyroscopic forces. The matrices $\boldsymbol{K}_e = \text{diag}\{\boldsymbol{K}_{e1}, ..., \boldsymbol{K}_{en_e}\}$ and $\boldsymbol{D}_e = \text{diag}\{\boldsymbol{D}_{e1}, \ldots, \boldsymbol{D}_{en_e}\}$ consist of the damping and stiffness matrices of the n_e flexible links and define the vector of internal forces $\boldsymbol{k}_i(\boldsymbol{y}, \dot{\boldsymbol{y}})$. On the right hand side of equations (4) the identity matrix is denoted by $\boldsymbol{I} \in I\!R^{n \times n}$, $\boldsymbol{\Gamma}_e^T = \text{diag}\{\boldsymbol{\Gamma}_{e1}^T, \ldots, \boldsymbol{\Gamma}_{en_e}^T\}$ is composed of the control input matrices of equations (2), and $\boldsymbol{g}(\boldsymbol{y})$ represents the vector of gravitational forces.

3. Control of Flexible Robots

In this section, a hierarchical control for flexible manipulators is presented combining a controller for the rigid body motion with Static Dissipative Controllers for the flexible links. The deviation of the end–effector's path from its desired trajectory $\boldsymbol{y}^d$ due to the elasticity of the links is compensated by a quasi–static correction of the desired trajectory resulting in the input $\boldsymbol{y}^{d*}$.

3.1. STATIC DISSIPATIVE CONTROLLERS (SDC)

As shown by Joshi [4], dissipative controllers using collocated actuators and sensors offer an attractive strategy for active control of flexible systems. The simplest controller of this type for the flexible beam described by equations (2) and (3) is the Static Dissipative Controller (SDC)

$$\boldsymbol{u}_{ej} = -\boldsymbol{G}_{Pj}\boldsymbol{s}_{aj} - \boldsymbol{G}_{Dj}\boldsymbol{s}_{rj} \; . \tag{5}$$

The gain matrices $\boldsymbol{G}_{Pj}, \boldsymbol{G}_{Dj} \in I\!R^{m_j \times m_j}$ are chosen to be symmetric, constant, and positive definite, using the piezo actuators introduced in section 2. For simplicity, we omit the subscript j in the following. Applying the control law (5) to the flexible beam described by equations (2) and (3) the equations of motion read as

$$\boldsymbol{M}_e\ddot{\boldsymbol{q}} + \left(\boldsymbol{D}_e + \boldsymbol{\Gamma}_e^T\boldsymbol{G}_D\boldsymbol{\Gamma}_e\right)\dot{\boldsymbol{q}} + \left(\boldsymbol{K}_e + \boldsymbol{\Gamma}_e^T\boldsymbol{G}_P\boldsymbol{\Gamma}_e\right)\boldsymbol{q} = \boldsymbol{0} \; . \tag{6}$$

The two quadratic forms in this equation are obtained due to the collocation of actuators and sensors following the assumption in section 2. Since $\boldsymbol{G}_D$ is positive definite, the quadratic form $\boldsymbol{\Gamma}_e^T\boldsymbol{G}_D\boldsymbol{\Gamma}_e$ is at least positive semi–definite and, therefore, the velocity feedback of the SDC enhances structural damping. Through the position feedback signal in equation (5) with the matrix $\boldsymbol{G}_P$, the additional stiffness term $\boldsymbol{\Gamma}_e^T\boldsymbol{G}_P\boldsymbol{\Gamma}_e \geq 0$ appearing in equation (6) reduces the deviation from the equilibrium $\boldsymbol{q} = \boldsymbol{0}$ of the flexible beam.

3.2. CONTROL OF RIGID BODY MOTION

For the design of the controller the robot is assumed to be ideal rigid, i.e. $\boldsymbol{y}_e = \dot{\boldsymbol{y}}_e = \ddot{\boldsymbol{y}}_e = \boldsymbol{0}$. Then, from equation (4) it follows

$$\boldsymbol{M}_{rr}(\boldsymbol{y}_r)\ddot{\boldsymbol{y}}_r(t) + \boldsymbol{k}_{cr}(\boldsymbol{y}_r, \dot{\boldsymbol{y}}_r) + \boldsymbol{g}_r(\boldsymbol{y}_r) = \boldsymbol{u}_r \; . \tag{7}$$

If the tip location $\boldsymbol{r}_t = \boldsymbol{f}(\boldsymbol{y}_r)$ of the manipulator is of interest, the vector of desired joint coordinates $\boldsymbol{y}_r^d$ has to be computed using the inverse kinematic equation $\boldsymbol{y}_r^d = \boldsymbol{f}^{inv}(\boldsymbol{r}_t^d)$. For the control of the gross motion any conventional control algorithm may be used, for a discussion see Spong and Vidyasagar [10], and be combined with the SDC for the flexible links.

3.3. HIERARCHICAL CONTROL

The hierarchical control consists of three parts: (1) Damping of the vibrations in the elastic links by the velocity feedback of the SDC. (2) Rigid body control including a quasi–static correction of the tracking error, resulting from the elastic deformations and (3) Adaptive stiffening about the quasi–static elastic deformation by the position feedback of the SDC.

3.3.1. *Damping*

In order to damp the elastic vibrations of the flexible links, we use the velocity feedback of the SDC from equation (5)

$$\boldsymbol{u}_{ej} = -\boldsymbol{G}_{Dj}\boldsymbol{s}_{rj} \ , \quad j = 1(1)n_e \ , \tag{8}$$

enhancing the structural damping of all flexible links of the manipulator. Applying the control laws (8), the equations of motion (4) read as

$$\boldsymbol{M}(\boldsymbol{y})\ddot{\boldsymbol{y}}(t) + \boldsymbol{k}_c(\boldsymbol{y},\dot{\boldsymbol{y}}) + \boldsymbol{g}(\boldsymbol{y}) + \begin{bmatrix} \boldsymbol{0} & \boldsymbol{0} \\ \boldsymbol{0} & \boldsymbol{D}_e + \boldsymbol{\Gamma}_e^T\boldsymbol{G}_D\boldsymbol{\Gamma}_e \end{bmatrix} \begin{bmatrix} \dot{\boldsymbol{y}}_r(t) \\ \dot{\boldsymbol{y}}_e(t) \end{bmatrix} + \begin{bmatrix} \boldsymbol{0} & \boldsymbol{0} \\ \boldsymbol{0} & \boldsymbol{K}_e \end{bmatrix} \begin{bmatrix} \boldsymbol{y}_r(t) \\ \boldsymbol{y}_e(t) \end{bmatrix} = \begin{bmatrix} \boldsymbol{I} & \boldsymbol{0} \\ \boldsymbol{0} & \boldsymbol{\Gamma}_e^T \end{bmatrix} \begin{bmatrix} \boldsymbol{u}_r(t) \\ \boldsymbol{0} \end{bmatrix} \tag{9}$$

where $\boldsymbol{G}_D = \text{diag}\{\boldsymbol{G}_{D1},\ldots,\boldsymbol{G}_{Dn_e}\}$ and $\boldsymbol{\Gamma}_e = \text{diag}\{\boldsymbol{\Gamma}_{e1},\ldots,\boldsymbol{\Gamma}_{en_e}\}$. The active damping of the elastic vibrations is achieved by the additional positive semi-definite term $\boldsymbol{\Gamma}_e^T\boldsymbol{G}_D\boldsymbol{\Gamma}_e$ in the damping matrix of the generalized elastic coordinates.

3.3.2. *Quasi–Static Correction*

Having achieved a damping of elastic vibrations, a quasi–static analysis of the flexible manipulator yields a trajectory of quasi–static elastic displacements $\boldsymbol{y}_e^d(t)$, see Kleemann [5]. From these displacements the disturbance of the end–effector position can be calculated and the rigid joint coordinates $\boldsymbol{y}_r^d$ from section 3.2. can be corrected by the corresponding variables $\boldsymbol{z}_r^c$.

This yields the corrected reference motion $\boldsymbol{y}^{d*} = [\boldsymbol{y}_r^{d*T}, \boldsymbol{y}_e^{dT}]^T$ consisting of the corrected joint coordinates $\boldsymbol{y}_r^{d*} = \boldsymbol{y}_r^d + \boldsymbol{z}_r^c$ and the corresponding quasi–static elastic displacements $\boldsymbol{y}_e^d$, see Hermle and Eberhard [2]. For the control of the gross motion of the manipulator any conventional control algorithm may be used, see section 3.2.

3.3.3. *Stiffening*

In order to achieve stiffening about the reference attitude $\boldsymbol{y}_e^d$ we modify the attitude feedback signal of the SDC

$$\boldsymbol{u}_e^* = -\boldsymbol{G}_P\boldsymbol{\Gamma}_e\left(\boldsymbol{y}_e - \boldsymbol{y}_e^d\right) = -\boldsymbol{G}_P\boldsymbol{\Gamma}_e\boldsymbol{z}_e^* \tag{10}$$

where $\boldsymbol{G}_P = \text{diag}\{\boldsymbol{G}_{P1},\ldots,\boldsymbol{G}_{Pn_e}\}$. This term is added to the control law (8). Using the linearization of the equations of motion (9) with respect to the corrected reference motion $\boldsymbol{y}^{d*}$, Hermle and Eberhard [2] show that this modified position feedback of the SDC yields the desired adaptive stiffening.

4. Point–to–Point Motion and Trajectory Tracking of a SCARA Type Robot with Elastic Link

In this section firstly, the model of a SCARA robot is presented and secondly a time optimal PTP–motion is calculated. Thirdly, the task of following a given reference trajectory is investigated and solved by means of spline interpolation.

The hierarchical control has been tested in simulations of the SCARA type robot "Adept One", see [2], which is shown in Figure 2. In the simulations the rotations Φ_1 in the shoulder joint and the rotation Φ_2 in the elbow joint are considered.

For the studies of this paper, the gripper of the robot has been replaced by a slender tetragonal link, which is modelled as a flexible beam, with a sphere at its end representing the end–effector. The vector of generalized elastic coordinates is $\boldsymbol{y}_e = [q_{y1}q_{y2}q_{z1}q_{z2}]^T$, consisting of two modal coordinates in the y and the z–direction. Two collocated sensor/actuator pairs are mounted on each side of the flexible link.

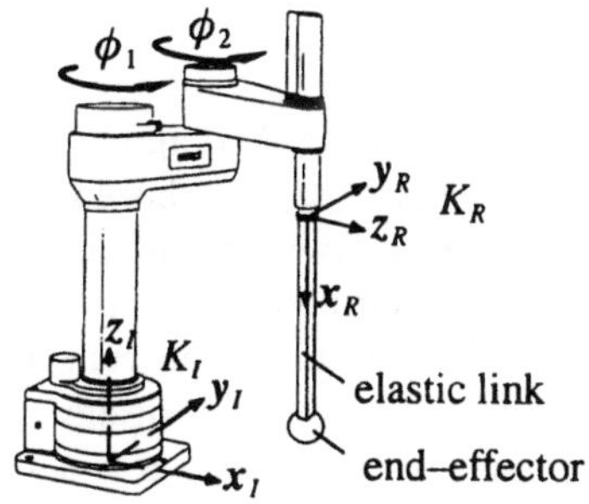

Figure 2. SCARA type robot

4.1. POINT–TO–POINT MOTION

For the task of moving from a given initial position to a desired end–location the time optimal problem is investigated, i.e. the time t_e after which the end–position is reached with acceptable accuracy is minimized.

Following the procedure described in the previous sections, at first the desired joint angles in the initial position $\boldsymbol{r}_t^i = [800\,\text{mm}\ \ 0\,\text{mm}\ \ 395\,\text{mm}]^T$ and the end position $\boldsymbol{r}_t^e = [-265\,\text{mm}\ \ 690\,\text{mm}\ \ 395\,\text{mm}]^T$ are calculated, which corresponds to the solution of a simple inverse kinematics problem, since the purely static deformations are zero for the chosen example. Next, a time optimal Bang–Bang trajectory is imposed for the PTP–motion. Inverse dynamics control, see Spong and Vidyasagar [10] is chosen for the gross motion, and SDC are applied in order to control the elastic deformations excited during the PTP–motion. This leads to the hierarchical control structure in Figure 3.

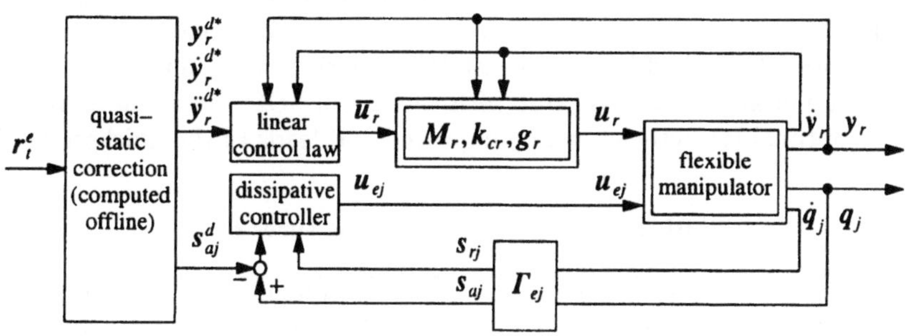

Figure 3. Hierarchical Control Structure

For the linear controller supplemented by the corrected reference motion

$$\bar{\boldsymbol{u}}_r = \ddot{\boldsymbol{y}}_r^{d*} + \boldsymbol{K}_D\left(\dot{\boldsymbol{y}}_r^{d*} - \dot{\boldsymbol{y}}_r\right) + \boldsymbol{K}_P\left(\boldsymbol{y}_r^{d*} - \boldsymbol{y}_r\right) , \tag{11}$$

with the constant diagonal gain matrices $\boldsymbol{K}_P = \text{diag}\{\omega_1^2, \omega_2^2\}$ and $\boldsymbol{K}_D = \text{diag}\{2\delta_1\omega_1, 2\delta_2\omega_2\}$ is chosen, leading to a PT_2 dynamics of each link. In order to achieve an advantageous behavior with respect to disturbances critical damping values are applied for each axis, $\delta_1 = \delta_2 = 1.0$.

The hierarchical control structure is optimized with respect to a minimal end time t_e. The vector of design variables consists of the natural frequencies ω_i and the gains of the SDC, $\boldsymbol{p} = [\omega_1\, \omega_2\, g_{P1} \ldots g_{P4}\, g_{D1} \ldots g_{D4}]^T$, and the optimization criterion is $\psi = t_e$. The optimization problem was solved by the software package NEWOPT/AIMS [1] using a stochastic Simulated Annealing Method, see Ingber [3].

The resulting PTP–motion is shown in Figure 4 in comparison to the resulting dynamics when using the concept of a smooth sinusoidal reference motion, see Kleemann [5], which mostly avoids the excitation of vibrations of the elastic links (so SDC cannot improve this result). Table 1 compares the optimal time t_e to

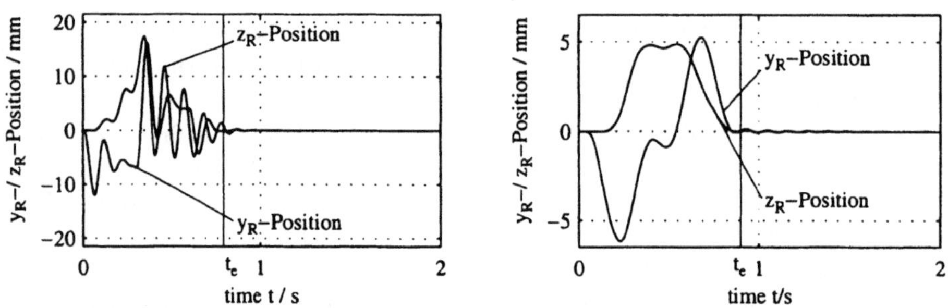

Figure 4. Optimal Point-to-Point motion with SDC and time optimal trajectory (left) and sinusoidal trajectory (right)

times achieved by the use of the smooth sinusoidal reference motion and a PD–controller with compensation of gravity forces. It can be seen that the smooth sinusoidal trajectories lead to the fastest PTP–motion when no active damping of the elastic vibrations is used. But with the use of SDC the dynamic behavior using time optimal trajectories can be improved and leads to the quickest motion.

TABLE 1. Comparison of position times t_e

Control of rigid body motion	Inverse dynamics sinusoidal trajectories	Inverse dynamics time–opt. trajectories	PD–Control with g–compensation
without SDC	0.9 s	2.6 s	2.0 s
with SDC	—	0.79 s	1.23 s

4.2. TRAJECTORY TRACKING

If the end–effector is to follow a given path, e.g. for applying a tool to a workpiece, the described quasi–static deformation compensation, see section 3.3., needs to be applied in order to calculate corrected reference values for the joint coordinates of the rigid model Φ_1 and Φ_2.

Usually, a precise realization of the desired path can only be achieved by control methods that use next to prescribed positions additionally prescribed velocity and acceleration references. Therefore, a derivation of the corrected reference trajectories $\boldsymbol{y}_r^{d*}$ is needed, what cannot in general be done analytically.

From $\boldsymbol{y}_r^{d*} = \boldsymbol{y}_r^d + \boldsymbol{z}_r^c$ and $\boldsymbol{z}_r^c = f(\boldsymbol{y}_r^d, \dot{\boldsymbol{y}}_r^d, \ddot{\boldsymbol{y}}_r^d)$, follows $\ddot{\boldsymbol{y}}_r^{d*} = f(\boldsymbol{y}_r^{d^{(4)}})$, requiring the planned path to be four times differentiable in order to avoid discontinuous accelerations that would strongly excite vibrations in the elastic links.

As an alternative way of implementing the trajectory tracking, the use of spline interpolation is proposed. Instead of the continuous correction of elastic deformations, key points of the trajectory are corrected only and new trajectories $\boldsymbol{y}_r^{d*}, \dot{\boldsymbol{y}}_r^{d*}, \ddot{\boldsymbol{y}}_r^{d*}, \boldsymbol{s}_a^{d*}$, see Figure 3 are calculated by spline interpolation of these points, see Kolender [6]. I.e. we use a time discrete version of the quasi–static correction from section 3.3., see also Hermle and Eberhard [2].

Figure 5 shows a circular reference motion to be accomplished by the robot's end–effector and the corresponding key points. The control gains for this case study are again found by optimization using the simulated annealing algorithm, but the vector of design variables $\boldsymbol{p}$ is augmented by the damping ratios δ_1 and δ_2. As the optimization criterion the quadratic tracking error of the end–effector $\psi = \int_0^{t^e} f^2 dt$ is chosen.

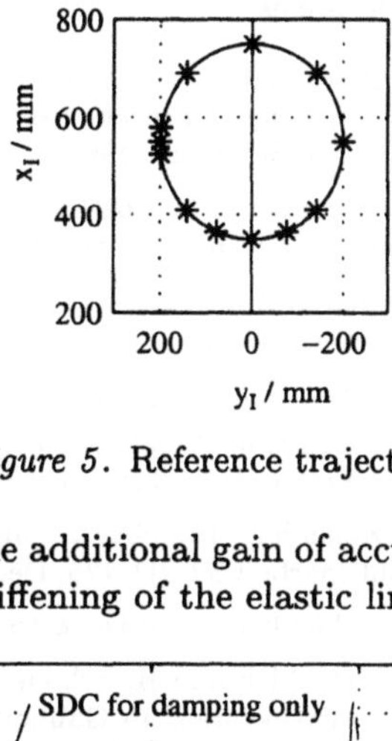

Figure 5. Reference trajectory

A simulation of this task is presented in Figure 6, which shows the advantage of the discrete compensation of the elastic deformations and the additional gain of accuracy by the use of optimized SDC for the damping and stiffening of the elastic link.

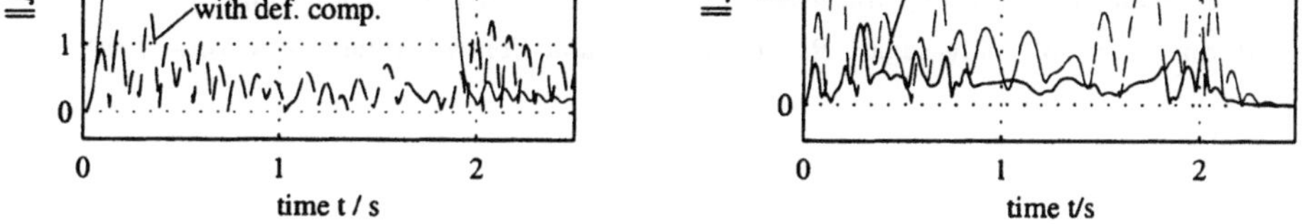

Figure 6. Simulation of tracking the circle reference motion without SDC (left) and with full hierarchical control (right)

5. Conclusions

The hierarchical control concept presented in this paper offers the combination of any given joint level control for the gross motion of a flexible robot manipulator with a decentralized linear control of the elastic deformation of each flexible link. For the joint level control, we propose the design of an inversion based control law under the assumption of infinitely stiff links. Adding a quasi–static correction to SDCs realizing a low–authority control of the elastic deformations, the hierarchical control achieves an excellent tracking performance.

In the numerical case studies using a SCARA type robot with one flexible link, the hierarchical control has proven its efficiency in simulations of a Point–to–Point motion and a trajectory tracking task applying spline interpolation. Using optimized parameters for the controllers several PTP–trajectories and joint level controllers are compared. It is shown that the PTP–motion can be accomplished in the shortest time when applying a time–optimal trajectory, since the SDC accomplish a quick decay of the excited structural vibrations. The hierarchical control also proves well suited for trajectory tracking tasks. It is shown that the elastic deformations related to a circular reference trajectory can be reduced by an order of magnitude using optimized control parameters.

6. References

[1] D. Bestle and P. Eberhard. *NEWOPT/AIMS 2.2. Ein Programmsystem zur Analyse und Optimierung von mechanischen Systemen.* Manual AN–35. Institute B of Mechanics, University of Stuttgart, 1994.

[2] M. Hermle and P. Eberhard. Control, and parameter optimization of flexible robots. *Mechanics of Structures and Machines*, accepted for publication.

[3] L. Ingber. Very fast simulated reannealing. *Mathematical Computational Modeling*, **12**, 8, 967–973, 1989.

[4] S. Joshi. *Control of Large Flexible Space Structures.* Springer Verlag; Berlin, 1989.

[5] U. Kleemann. *Regelung elastischer Roboter, VDI–Fortschritt–Berichte.* Reihe 8, No. 191, VDI Verlag; Düsseldorf, 1989.

[6] L. Kolender. Bahnberechnung für elastische Roboter mit Hilfe der Spline–Interpolation. Studienarbeit STUD–175, December 1999. Institute B of Mechanics, University of Stuttgart.

[7] W. Schiehlen. *Technische Dynamik.* B.G. Teubner; Stuttgart, 1986.

[8] W. Schiehlen and H. Schönerstedt. Reglerentwurf zur aktiven Schwingungsdämpfung von Balkenstrukturen. In Gabbert, U. (Ed.), *Smart Mechanical Systems - Adaptronics, VDI–Fortschritt–Berichte.* Reihe 11, Nr. 244, VDI-Fortschritt–Berichte, VDI Verlag; Düsseldorf, 1997.

[9] A. Shabana. *Dynamics of Multibody Systems.* University Press; Cambridge, 1998.

[10] M. Spong and M. Vidyasagar. *Robot Dynamics and Control.* Wiley; New York, 1989.

AN APPROACH FOR CONCEPTUAL DESIGN OF PIEZOACTUATED MICROMANIPULATORS

K D. HRISTOV, FL. IONESCU*, K. GR. KOSTADINOV
Mechatronic Systems Lab.-Institute of Mechanics & Biomechanics
**University of Applied Sciences - Konstanz, Germany*
Acad. G. Bonchev St, Block 4, 1113 Sofia, Bulgaria

1. Introduction

The most challenging task to be solved in the field of scientific research and engineering is the development of a full operational system with all its components, sub-systems and functions. A design process generally consists of iterative procedures between reference task function analysis, conceptual design and embodiment design. In this paper we deal with the first two stages of the design process. So, a conceptual design approach for piezoactuated micromanipulators is developed here. The reference task function is considered as an area in the manifold of the possible functions, which describes the state of the micromanipulation system to be designed, including the motions, qualitative parameters, constraints (e.g. force, space, etc.).

The previous experience of the designer and all possible data should be taken into consideration by developing a "state-of-the-art" task definition in the frame of the investigated field of problems.

The aim of the proposed paper is the modelling and simulation of piezo actuated micromanipulators performed in SDS, Matlab and ANSYS programs. The specific spectrum of micromanipulation tasks is the main point which should be taken into account in the process of pre-design of the kinematic structures of the micromanipulators. The study of the specific behaviour of piezo actuators (both stack and bimorph), is also a very important feature applying them as driving elements for the micromanipulators.

2. Formulation of the reference task

As a first phase of pre-design of micromanipulation systems with piezo actuators we consider the statement of the need. The reference task function that has to be achieved by the micromanipulator systems to be designed is considered in dimensional, energetic (impedance) and working spaces. The micromanipulators discussed are used for precise product orientation in the work space and successive manipulation or feeding during given operations, such as cell penetration, micro-assembling, investigation of thin films, in atomic force microscopes and scanning tunnelling microscopes, etc. Here, the main

U. Gabbert and H.S. Tzou (eds.), IUTAM Symposium on Smart Structures and Structronic Systems, 41–48.

goal of the design engineer is to provide the appropriate accuracy and the functionality of the micromanipulation systems. An example of the formulation the task reference for cell manipulations in biology is examined in [1]

3. Modelling of Piezoactuated Micromanipulators

There are several ways to obtain a mathematical description of the behaviour of a physical system by the modelling of dynamic systems. Mostly, we use energy and equilibrium methods.

In our case the dynamic modelling of piezo actuated micromanipulation systems is based on graph theory and the Orthogonality Principle [2]. Such systems can be treated as components of a more complex system which consists of a mechanical subsystem, an electrical subsystem and a servo-controller. The substance of the method lies in the decomposition of the complex system into relatively simple components, and the derivation of characteristic terminal equations for each component. That is convenient in the design process, because the hybrid systems are described mathematically from one point of view. To the system which has to be designed, we assign a general graph which reflects the structure of the interrelations between individual components. The system is discussed jointly with its general energy space, which consists of energy subspaces, depending on the types of energy in the particular problem. The energy is considered as a general measure of the motion of the material objects and is also used as a quantitative estimate of the material interactions. The dynamic equations are used to analyse how the required task function would be performed [3,4].

3.1.MODEL DEVELOPMENT

3.1.1.Theoretical model of a bimorph piezoelement.

In order to describe the dynamical behaviour of a bimorph piezoelement a theoretical model is introduced. Hence, the achieved theoretical results can be later compared with the results from FEM simulations and experiments.

The bimorph piezoelement is considered as a cantilever beam with the corresponding coordinate system (fig.1).

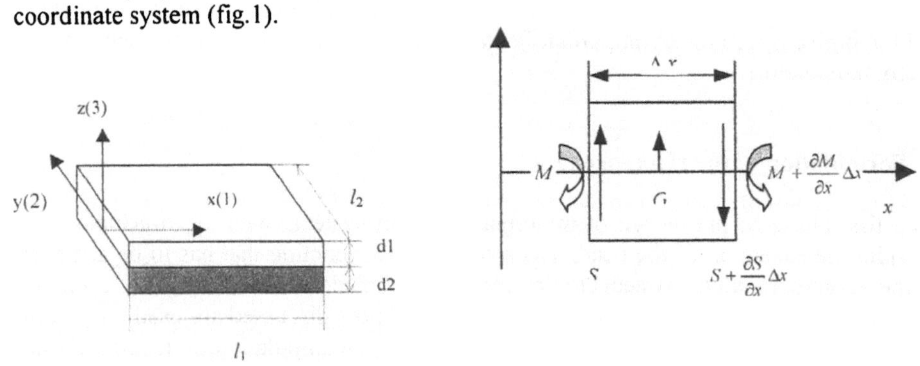

a. Co-ordinate system of the element b. A segment of the element

Figure 1. Bimorph piezoelement

We consider the case in which the element only bends; we neglect expansion along the x-axis. According to the chosen coordinate system, the *neutral plane* (the element is in initial unbent condition) can be associated with plane $z= 0$. It is assumed that the bending takes place only in the xz-plane, but not in y. Now, let $\omega(x,t)$ is the deviation of a point x, where ($x\in[0,l]$, l–the length of the element) along the Z-axis in a moment of time t. Further, the following notations are chosen: – $S(x,t)$ the force along the element, and $M(x,t)$ – the bending moment (fig.1b.). The movement of the center of mass of a segment Δx of the piezoelement is described by the following equation [8]:

$$\rho A \Delta x \frac{\partial^2 w}{\partial t^2} = \frac{\partial S}{\partial x} \Delta x\,, \tag{1}$$

where A is the aria of the cross-section, ρ is the density of the element.
The equation of motion of the segment around its center of mass can be written as follows:

$$\rho A \frac{\partial^2 w}{\partial t^2} = \frac{\partial^2 M}{\partial x^2}\,, \tag{2}$$

and finally

$$\frac{\partial^2 w}{\partial t^2} = -\frac{EI}{\rho A}\frac{\partial^4 w}{\partial x^4} \tag{3}$$

So, the motion of the bimorph element is described by the function $w(x, t)$, with the partial differential equation (3), and the following boundary conditions:

$$w(0,t)=0,\quad \frac{\partial w}{\partial x}(0,t) = 0\,, \qquad (4),$$

which refers to a fixed end (at x=0)

$$\frac{\partial^2 w(l,t)}{\partial x^2} = 0\,,\quad \frac{\partial^3 w(l,t)}{\partial x^3} = 0 \qquad (5),$$

which refers to a free end at $x=l$.
For the characteristic frequency we have:

$$\nu_n = \frac{\omega_n}{2\pi} = \frac{1}{2\pi}\frac{a_n^2 \pi^2}{l^2}\sqrt{\frac{EI}{A\rho}} \qquad (6),$$

where the first 3 values for $\beta_1 l$ are: $\beta_1 l=0{,}60\,\pi$, $\beta_2 l=1{,}49\pi$, $\beta_3 l=2{,}5\pi$.
The element type and the properties of the material used are presented in Tables1 and 2.

TABLE 1 Dimensions of LVPZT element type P803.40 (PI, Germany)

Dimensions	Length	width	height
	39mm	12mm	0.65mm

TABLE 2 Material Properties of LVPZT element model P803.40 (PI, Germany)

Type	Young's Module	Density	Moment of Inertia	Poisson Ratio
P-803.40	2,2e11	7500	$3{,}3.10^{-12}$	0,3

3.2. ANSYS MODEL

3.2.1. Modal Analysis

Modal Analysis in ANSYS is done in 4 steps:

- Building the model (Pre-processing)
- Applying loads and obtaining solution (Solution)
- Expanding the modes (Solution)
- Reviewing the results (Post Processing)

The first 5 modes are determined (Table3).

TABLE 3. The first 5 modes of the element

Mode	Time/Freq., Hz
1	1294.4
2	8080.8
3	22493
4	34754
5	43719

3.2.2. Harmonic Analysis

For this problem, we will conduct a harmonic forced response test on a bimorph element. We will do this by applying a cyclic load (harmonic) at the end of the beam and observing the response at that location. The frequency of the load will be varied from 1 - 10000 Hz. ANSYS provides 3 methods for conducting a harmonic analysis - Full, Reduced and Modal Superposition methods. In our case the Full method is used. The results are presented in Fig.2.

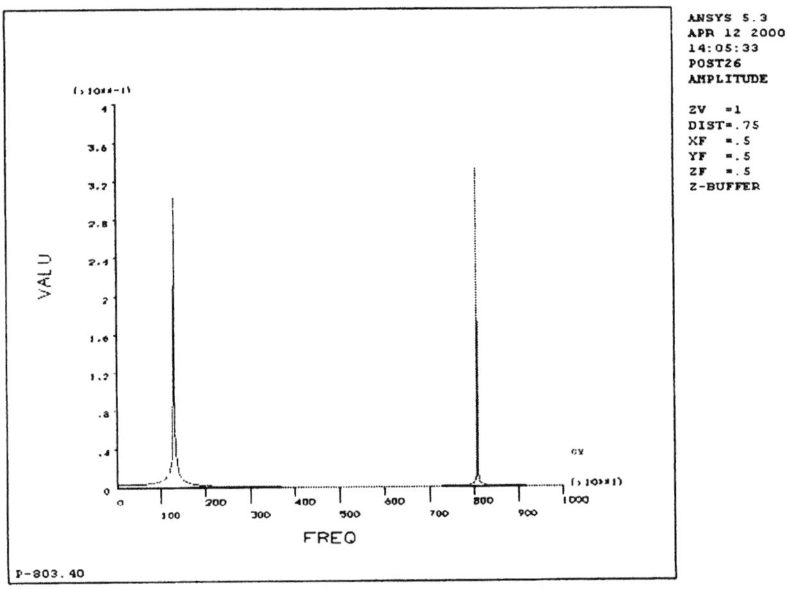

Figure 2. Harmonic response of the element in the range of 1 – 10000Hz

4. Conceptual Design

In order to represent the design process of a piezo actuated system in general, the following basic block-scheme is introduced (fig.3). The programs mentioned in the algorithm have the following features.

SDS programme [5] is a software package designed to process dynamic mechanisms in 3D environment. It allows us to implement dynamic calculations of mechanisms composed of rigid bodies, kinematic calculations, or complex calculations (law of forced motion on the articulations of the mechanisms, free motion on the other joints and possible closing of the remaining joints) and the calculations of flexible bodies using modelling of multi-jointed parts.

The procedure of the simulation goes through several steps:

- construction of the micromanipulator model;
- definition of the dynamic environment in which the micromanipulator model is functioned – e.g. cell manipulations in biological research;
- simulation and analysis of the micromanipulator model.

The conception of the micromanipulator mechanical system occurs in the Constructor unit. Here the development of the mechanical model's geometric structure is made. By positioning different articulations with up to 6 d.o.f. a tree-like structure presenting the micromanipulator model is obtained. The design of the different parts building up the model with all dimensions and material properties is made in the Designer unit.
In order to find the appropriate model for the proposed manipulation task, the following criteria for comparison are used: D.o.F. of the micromanipulator; Type of the kinematic structure; Number and type of the actuators; Optimum of coincidence of working space image with the configuration one; Dimensions of the micromanipulator; Mounting and sensing possibilities; Minimum of task function deviation; Desired kinematic and dynamic accuracy.
MATLAB/Simulink software package [6] is in our case for modelling, simulation and analysis of dynamical systems. Simulink™ program has two phases of use: model definition and model analysis.

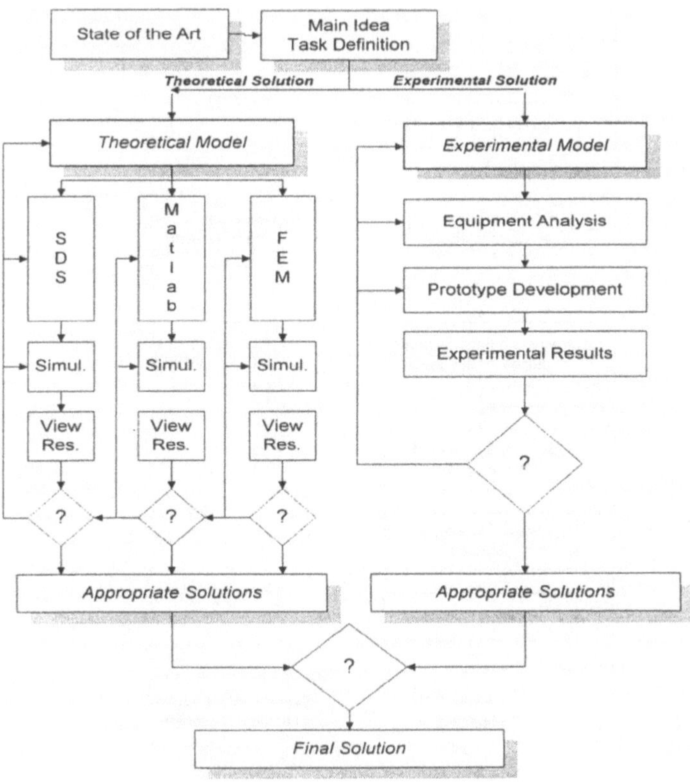

Figure 3. Design process algorithm

ANSYS is a general-purpose finite element modelling (FEM) package for numerically solving a wide variety of mechanical problems [7]. These problems include static/ dynamic structural analysis, heat transfer and fluid problems, as well as acoustic and electro-magnetic problems. Preprocessing: defining the problem; is consisted of the following major steps:

- Define keypoints/lines/areas/volumes
- Define element type and material/geometric properties
- Mesh lines/areas/volumes as required

The amount of detail required will depend on the dimensionality of the analysis (i.e. 1D, 2D, axi-symmetric, 3D). Solution: assigning loads, constraints and solving; here we specify the loads, constraints and finally solve the resulting set of equations.
Postprocessing: further processing and viewing of the results - in this stage one may wish to see: Lists of nodal displacements; Element forces and moments; Deflection plots; Stress contour diagrams.

An algorithm for FEM analysis of piezo actuators and systems is presented in fig.4.

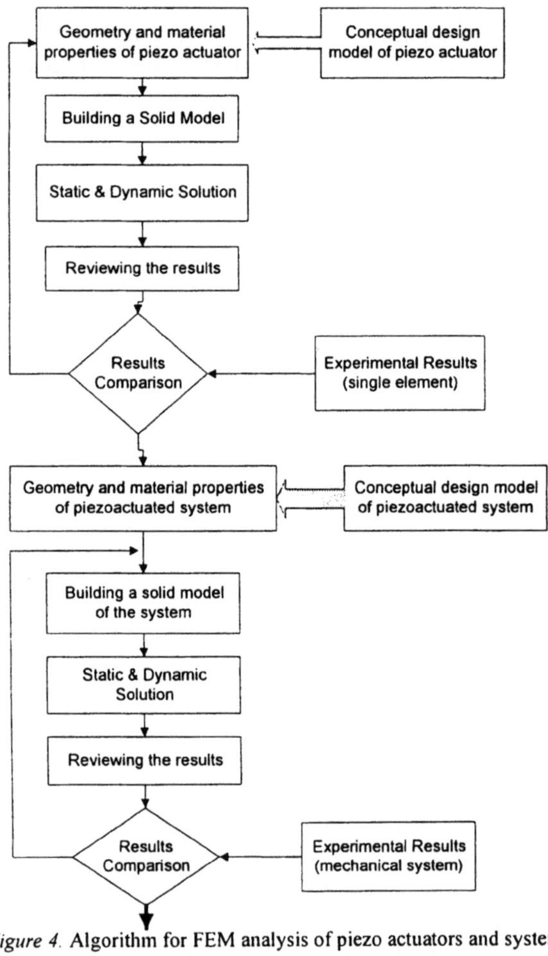

Figure 4. Algorithm for FEM analysis of piezo actuators and systems

In our case, dynamic analysis of bimorph piezoelements is performed in ANSYS. The following steps are considered:

- Modal Analysis: Determining the mode frequencies and mode shapes of the elements.
- Harmonic Analysis: Analysing the steady-state behaviour of the elements subject to cyclic loads.
- Transient Analysis: Determining the dynamic response of the elements under more general time-dependant loads.
- Spectral Analysis: Studying the response of the elements under the action of loads with known "spectra" (e.g. random loading conditions).

As an example, modal analysis (§3.2.1) and harmonic analysis (§3.2.2) of a bimorph piezoelement are performed (this corresponds to the upper half of fig. 4 as well).

As two examples, of the discussed pre-design, a microrobot with 3 d.o.f. and 3 bimorph piezo actuators, and a microrobot with 3 d.o.f. and 5 stack piezo actuators (antagonistic impedance control) are shown in fig.5 and fig.6.

For the bimorph piezoelements used (fig.5), the robot can produce displacements up to 450μm, but the forces cannot exceed 1N. Hence, such a robot can be used for applications where no great interaction forces are needed.

For the stack piezoelements (fig.6), the robot can produce displacement of only 45μm, but forces up to 3000N in the "push" direction. Here the actuators respectively in x and y direction work as antagonistic couples, not only to increase the stiffness of the micromanipulator in the xy- plane, but to work as an actuator-sensor couple as well. Hence, such a robot can be used for applications where no large displacements are necessary, but the big forces are expected and the high stiffness is necessary.

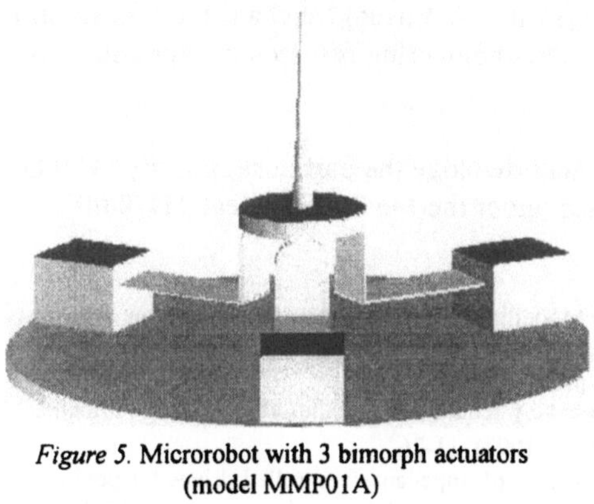

Figure 5. **Microrobot with 3 bimorph actuators (model MMP01A)**

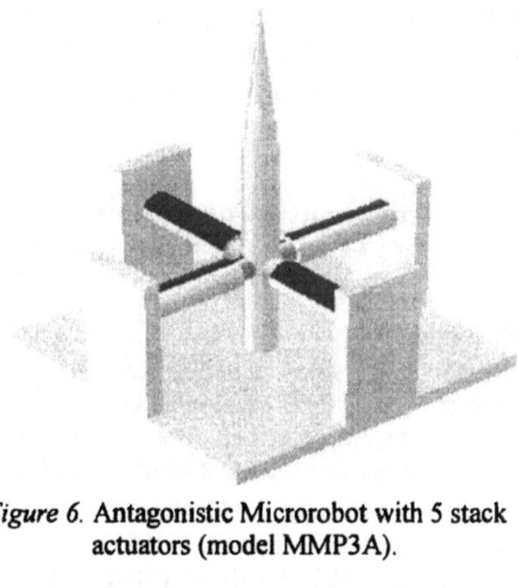

Figure 6. **Antagonistic Microrobot with 5 stack actuators (model MMP3A).**

Some simulation results for the two examples are shown in fig.7 and fig.8.

Fig.7 shows the simulation results of a position control. For a certain position of the platform the positions of the 3 actuators are obtained, where: q[2a], q[3a], q[4a], the rotations of the 3 actuators.

In Fig.8 concerning the model in fig.6, a low of motion is introduced, where: *esp3*-position, *vit3*-velocity of the translation in Z axis, *q(6a)* & *q(6b)* rotation around axis X & Y respectively, a3 acceleration in Z axis

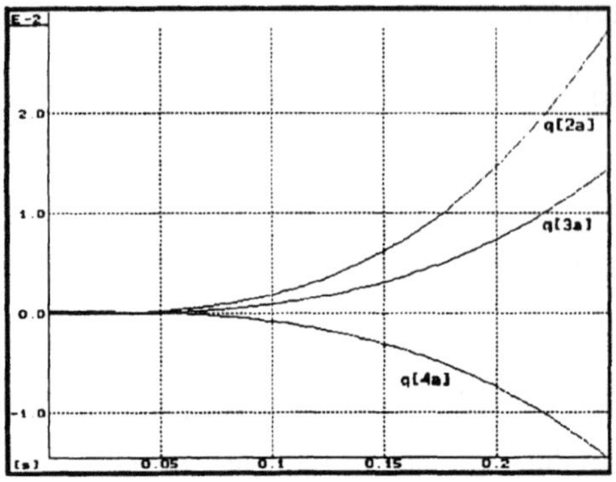

Figure 7. Simulation results (model MMP01A)

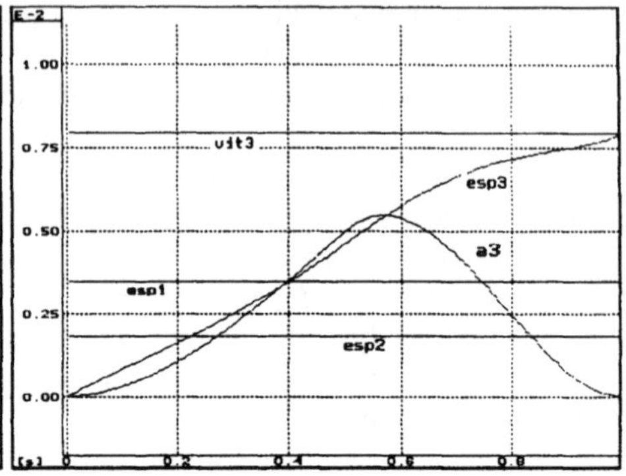

Figure 8. Simulation results (model MMP03A

5. Conclusion

The proposed approach for conceptual design of piezo actuated micromanipulators is a part of the development of a Virtual Mechatronic Lab. Based on the theoretical basics and the specific features of piezo actuators combined with the well known modelling and simulation programs, the approach presents an alternative (faster) way to solve a certain task in the field of microrobot development. The paper shows only a few possibilities in the mainstream of solutions that may occur according to the different research approaches.

The team will greatly appreciate any remarks and ideas towards the development of a general approach for design of mechatronic systems. A Virtual Mechatronic Lab should be considered as a tool not only for scientific and engineering research but for education as well.

Acknowledgements: The authors gratefully acknowledge the partial support by DAAD, DFG and BG Ministry of Education & Science under the *InMechS* Project TH708/97.

6. References:

1. Hristov K., Fl. Ionescu, and K. Kostadinov (1999) Modelling and Simulation of A Microrobot with Piezo Actuators for Cell Manipulations, in R. Kasper...(eds.) *Entwicklungsmethoden und Entwicklungsprozesse im Maschinenbau*, Berlin: Logos-Verlag, pp.201-208.
2. Boiadjiev G., Lilov L., Dynamics of Multicomponent Systems Based on the Orthogonality Principle, J. *Theoretical and Applied Mechanics*, Year XXIV, N 1, (1993), 11-26.
3. Kostadinov K., G. Boiadjiev (1997) Dynamic modelling of impedance controlled drives for positioning robots, in G.M.L. Gladwell (Series Editor), *Solid mechanics and its applications* - vol.52, Kluwer Academic Publishers, Dordrecht, , pp.183-190.
4. Kostadinov K., G. Boiadjiev (1998) Impedance Control Method of robots and Mechatronic Systems, in Gabbert, U. (Ed.) *Modelling and Control of Adaptive Mechanical Structures*, Fortschr.-Ber. VDI Reihe 11 Nr.268, Duesseldorf: VDI Verlag, pp.257-266.
5. SDS Manual, Version 3-50, 1993-1997
6. Matlab/Simulink, Version 5.3.
7. ANSYS Finite Element Analysis, Version 5.5
8. SASHIDA T. and T. KENJO An Introduction to Ultrasonic Motors, Shinsei Industries, Japan, University of Industrial Technology, Kanagawa, Japan

MODELLING AND OPTIMISATION OF PASSIVE DAMPING FOR BONDED REPAIR TO ACOUSTIC FATIGUE CRACKING

L. R. F. ROSE AND C. H. WANG
Aeronautical and Maritime Research Laboratory,
Defence Science and Technology Organisation,
506 Lorimer Street, Fishermans Bend VIC 3207, Australia

1. Introduction

High-performance aircraft experience severe acoustic loads and consequently suffer from acoustic fatigue cracking in service. Although adhesively bonded repairs have been implemented quite successfully in a variety of aircraft applications for the past twenty years, a recent application to repair acoustic-fatigue cracking on the engine nacelle of the F/A-18 has proved ineffective [1], contrary to expectations from current design practice for such repairs. Accordingly, an extensive research program is in progress, aimed at (i) an improved repair-design analysis that addresses the limitations of conventional repair design in the present context; (ii) modelling and optimisation of passive and active damping treatments in conjunction with bonded repair; (iii) laboratory and flight-test evaluation of optimised repair and damping schemes.

Research undertaken to date has addressed theoretical modelling issues for two types of passive damping treatment. The first is constrained-layer damping (CLD), which is a well-established approach [2] relying on shear dissipation in a visco-elastic layer. Commercially available CLD generally have a limited temperature range of effectiveness, so that a stack of several layers is often used to provide effective damping over a broad temperature range required in practice [2]. It is recognised that this form of damping involves shear load-transfer through the visco-elastic layer to the covering constraining-layer, and that design optimisation can be viewed as tuning this load-transfer length to the length of the cover [3]. However, there is an important feature of this load transfer that has not been fully appreciated in previous work, *viz.* the maximum shear stress does not occur at the ends of the cover plate, as implied by simplified models [3] where the vibrating panel (substrate) is considered to be much stiffer than the constraining layer. These simplified models lead to a second-order differential equation for the shear stress, whereas the simplest level of modelling that accounts correctly for the structural impact of CLD leads to a sixth-order equation for the deflection [2]. This feature of CLD, and its implications for optimal design of both passive and active CLD [4,5] will be presented in detail elsewhere.

U. Gabbert and H.S. Tzou (eds.), IUTAM Symposium on Smart Structures and Structronic Systems, 49–56.

This paper addresses the modelling of a novel form of passive damping that relies on a shunted piezoelectric to dissipate vibrational energy. An insightful analysis of this damping mechanism has been given by Hagood and von Flotow [6]. The emphasis in the present work is to provide a more detailed and explicit model of the structural impact of the piezoelectric elements (or layers) attached to the vibrating panel. Much of the recent literature involving piezoelectric elements as sensors or actuators has tended to ignore the structural impact of these elements on the dynamic response of the host panel (or structure). While this simplification is usually justified for sensors, it is more questionable for actuators, and it is inappropriate for the type of passive damping treatment being considered here. It will be seen that shunted piezoelectric damping (SPD) provides an electro-mechanical analogue of a dynamic vibration absorber [2]. It therefore constitutes an excellent example of a *structronic system*, in line with the theme of this symposium.

2. Problem Statement and Preliminary Analysis

Aircraft panels subjected to acoustic loading can often be idealised as flat rectangular plates, clamped along the edges, for the purposes of a simplified design analysis [1]. In the present context, a further simplification is appropriate: the panel can be assumed to be of infinite extent in one dimension, corresponding to the y-axis in Fig.1. The resulting one-dimensional configuration can be regarded as a beam of length ℓ, clamped at both ends. The resonant frequencies and mode shapes for this simple structure are well known. A piezoelectric patch (of uniform thickness) is assumed to be rigidly bonded to this structure, with its faces fully electroded, and the electrodes connected through a shunting circuit consisting of a resistor R and an inductor L in series, as indicated by the impedance Z_{ext} in Fig.1. *The objective is to estimate theoretically the (non-dimensional) structural loss factor that can be achieved with this configuration, for comparison with CLD.*

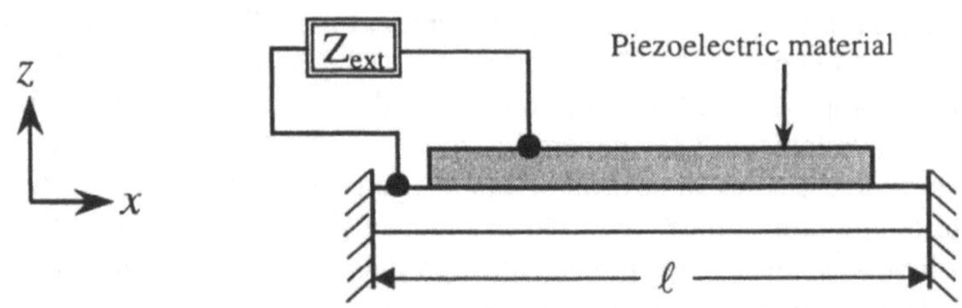

Fig.1 Clamped beam with a shunted piezoelectric patch bonded to one surface

Even for the simplified configuration of Fig.1, an analysis that accounts properly for the structural vibrational impact of the added piezoelectric layer turns out to be quite complicated, due to (i) the step discontinuity in mass (per unit length) and stiffness (extensional and flexural) at the ends of the patch; (ii) the coupling of extensional and flexural deformations resulting from asymmetry relative to the neutral surface of the composite structure, when a piezoelectric patch is bonded to only one face of the original structure. Accordingly, it is desirable to further simplify the problem formulation, and to proceed in stages, adding one complicating factor at a time. Thus, consider first the case where (i) identical piezoelectric patches are bonded to both the top and bottom surfaces of the clamped plate, so that extensional and flexural vibrations are now uncoupled; (ii) these patches (and the electrodes) extend over the full length (and width) of the vibrating plate; (iii) the poling direction is aligned with the z-axis for the top patch, and is in the opposite direction for the bottom patch.

The present paper outlines the main steps of the analysis for the case of forced extensional vibration, as illustrated in Figs.2 and 3. A more detailed presentation of the governing equations and of the intermediate steps is given in [8].

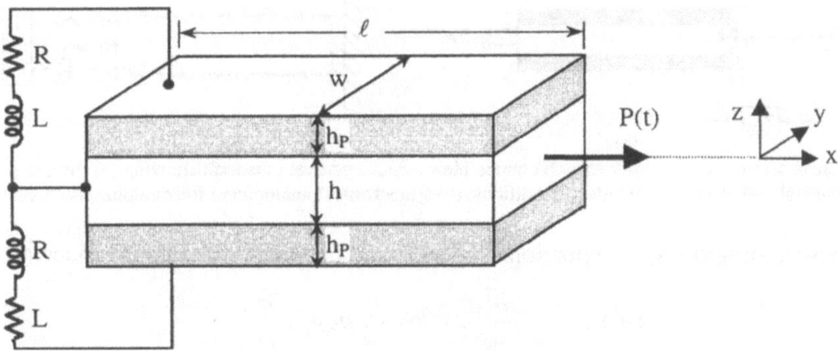

Fig.2 A rod with two full-length PZT patches (shaded) subjected to a longitudinal force.

3. Governing Equations: Forced Extensional Vibration

Theoretical analysis of piezoelectric sensors and actuators that have appeared in the literature involve varying levels of rigour in the formulation of the governing equations. The present work follows the formulation given by Tiersten [7] for linear piezoelectric materials. The general equations of motion can be greatly simplified for particular cases by the use of appropriate kinematic assumptions. For the case shown in Fig.2, the key assumptions are:

(i) The elastic displacement $\boldsymbol{u}$ involves only one non-zero component u_x, which is uniform across the thickness,

$$\boldsymbol{u}(\boldsymbol{x},t) = [u(x,t), 0, 0], \quad 0 < |z| < \tfrac{h}{2} + h_p \tag{1}$$

(ii) The electric displacement $\boldsymbol{D}$ involves only one non-zero component D_3, which is uniform across the thickness, *ie.*

$$\boldsymbol{D}(\boldsymbol{x},t) = [D_3(x,t), 0, 0], \quad \tfrac{h}{2} < |z| < \tfrac{h}{2} + h_p \tag{2}$$

(iii) The electric potential is constant on each electrode, so that the in-plane components of the electric field $\boldsymbol{E}$ vanish (to a first order approximation, for small h_p / ℓ) and

$$E(x,t)=[0,0,E_3(t)]\,,\quad 0<x<\ell\,,\quad \tfrac{h}{2}<|z|<\tfrac{h}{2}+h_p \tag{3a}$$

$$E_3(t)=-\frac{V(t)}{h_p} \tag{3b}$$

where $V(t)$ denotes the voltage across the electrodes of each piezoelectric patch, which is not specified in the present context, but must be determined as part of the solution process.

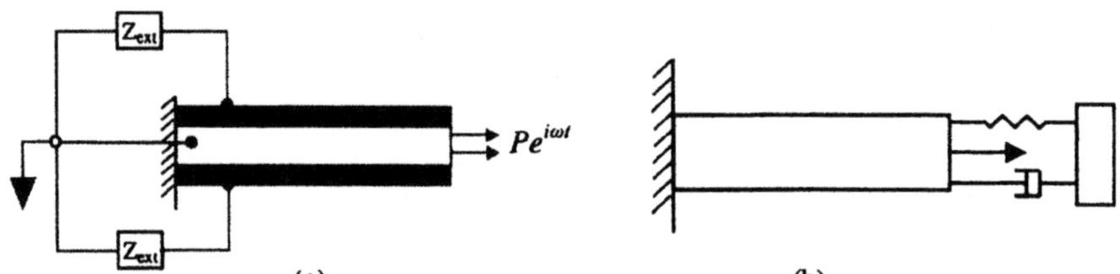

Fig.3 Cross-sectional view of a rod with bonded piezoelectric patches (shaded), showing (a) the geometry, mechanical and electrical boundary conditions; (b) a mechanical analogue of the damping mechanism.

With these assumptions, the equation of motion and boundary conditions reduce to

$$(Yh+2Y_ph_p)\frac{\partial^2 u}{\partial x^2}=(\rho h+2\rho_p h_p)\frac{\partial^2 u}{\partial t^2} \tag{4a}$$

$$u(x=0,t)=0 \tag{4b}$$

$$hw\sigma(x=\ell,t)=P(t)\,,\quad 0<|z|<\tfrac{h}{2} \tag{4c}$$

$$\sigma_p(x=\ell,t)=0\,,\quad \tfrac{h}{2}<|z|<\tfrac{h}{2}+h_p \tag{4d}$$

where the dimensions (ℓ=length, h=height, w=width) are shown in Fig.2; Y and ρ denote the Young's modulus and density; σ denotes the only non-zero component of the stress tensor ($\sigma\equiv\sigma_{xx}$); and the subscript p is used to distinguish properties pertaining to the piezoelectric layers. It is noted that the equation of motion (4a) does not directly involve the electro-mechanical coupling; this coupling enters *via* the boundary conditions (4c, 4d), the kinematic assumptions (1-3) and the following constitutive equations

$$\sigma=Y\frac{\partial u}{\partial x} \tag{5a}$$

$$\sigma_p=Y_p\frac{\partial u}{\partial x}-e_{31}E_3 \tag{5b}$$

$$D_3=e_{31}\frac{\partial u}{\partial x}+\epsilon_{33}\,E_3 \tag{5c}$$

which can be shown [8] to lead to the following boundary condition

$$(Yh + 2Y_p h_p)\frac{\partial u(x \to \ell_-, t)}{\partial x} = \frac{P(t)}{w} - 2e_{31}V(t) \tag{6}$$

As noted above, the voltage $V(t)$ is not specified independently in the present context. However, it is related to the current $I(t)$ flowing through the shunting circuit, and to the total charge $Q(t)$ collected by the electrodes, by

$$V(t) = Z_{ext}I(t) \tag{7a}$$

$$I(t) = \dot{Q}(t) \tag{7b}$$

$$Q(t) = \int D_3 dA = w\int_0^{\ell} D_3(x,t)dx = w\left[e_{31}u(x=\ell,t) - \epsilon_{33}\, V(t)\ell / h_p\right] \tag{7c}$$

where equation (7c) follows from using the constitutive equation (5c), together with the boundary condition (4b) and the kinematic assumption (3a, b).

Thus the governing equation is given by (4a), with the boundary conditions (4b) and (6), where the unknown voltage $V(t)$ in (6) has to be determined so as to be consistent with (7). In general, these equations must be supplemented by a statement of the initial conditions at time $t = 0$. However, we shall focus here on the steady-state response to a harmonic excitation

$$P(t) = Pe^{i\omega t} \tag{8a}$$

The response is given by

$$u(x,t) = \hat{u}(x,\omega)e^{i\omega t} \quad , \; V(t) = \hat{V}(\omega)e^{i\omega t} \tag{8b}$$

where

$$\frac{d^2\hat{u}}{dx^2} + k^2\hat{u} = 0 \tag{9a}$$

$$\hat{u}(x=0,\omega) = 0 \tag{9b}$$

$$(Yh + 2Y_p h_p)\frac{d\hat{u}(x=\ell,\omega)}{dx} = \frac{P}{w} - 2e_{21}\hat{V}(\omega) \tag{9c}$$

$$\hat{V}(\omega)\left[1 + i\omega Z_{ext}C_p\right] = i\omega Z_{ext} w e_{31}\hat{u}(x=\ell,\omega) \tag{9d}$$

Here C_p denotes the intrinsic capacitance of each piezoelectric patch,

$$C_p = \frac{\epsilon_{33}\, \ell w}{h_p} \tag{9e}$$

and k denotes the wave number, which is related to the wave speed c as follows,

$$k = \omega / c \tag{9f}$$

$$c = \sqrt{\frac{Yh + 2Y_p h_p}{\rho h + 2\rho_p h_p}} \tag{9g}$$

The solution of (9a-d) is readily derived as

$$\frac{\hat{u}(x,\omega)}{u_{st}} = \frac{1}{k\ell} \frac{(1-\Omega^2 + i\eta)\sin k\ell \sin kx}{(1-\Omega^2 + i\eta)\cos k\ell + \mu_0(i\eta - \Omega^2)\sin k\ell} \tag{10a}$$

where Ω and η denote the non-dimensional frequency and dissipation coefficient,

$$\Omega = \omega / \omega_e\,, \quad \omega_e = 1/\sqrt{LC_p} \tag{10b}$$

$$\eta = \omega R C_p \tag{10c}$$

The parameters L and R denote the inductance and resistance of the shunting circuit, *ie.*

$$Z_{ext} = R + i\omega L \tag{10d}$$

The parameter u_{st} denotes the static extension due to an applied load P,

$$u_{st} = P\ell / (Yh + 2Y_p h_p) w \tag{10e}$$

and

$$\mu_0 = k_{31}^2 \frac{2Y_p h_p}{Yh + 2Y_p h_p} \tag{10f}$$

$$k_{31}^2 = \frac{e_{31}^2}{\in_{33} Y_p} \tag{10g}$$

where k_{31} denotes the electromechanical coupling coefficient [9]. It can be seen from (9c,d) that the shunted piezoelectric acts as a boundary damping mechanism. A mechanical analogue is shown in Fig.3b.

4. Design Optimisation

The optimal design of a mechanical vibration absorber attached to a continuous structure, as in Fig.3(b), was first discussed by Dana Young [2], as an extension of the classical analysis where the structure is represented by a one-degree-of-freedom (DOF) mass-spring system. The approach is to represent the structural response in terms of the normal modes of vibration. The spring-damper can then be tuned to a particular natural frequency and mode of vibration. A more convenient approach in practice is to replace the continuous structure by a one-DOF mass-spring system, using the appropriate modal mess and stiffness for a selected vibration mode, and to adopt the classical approach for a one-DOF system. This provides an excellent first approximation, which can be further refined, if desired, by recourse to the exact representation for the structural response. The detailed analysis following this approach for the case of shunted piezoelectric damper is given in [8]. The main results can be summarised as follows.

A key non-dimensional parameter for the mechanical vibration absorber is the mass ratio. The corresponding parameter for the shunted piezoelectric damper is μ, which is related to the electro-mechanical coupling coefficient k_{31},

$$\mu = \frac{8}{\pi^2} k_{31}^2 \frac{2Y_p h_p}{Yh + 2Y_p h_p} \tag{11a}$$

For a given value of μ, the optimisation process involves two stages. First, the parameters of the external shunting circuit are adjusted so that the electric resonant frequency ω_e, given by (10b), is approximately tuned to the structural frequency, which is selected here to be ω_1, corresponding to the fundamental mode. The tuning condition gives

$$\omega_e = \sqrt{1+\mu}\,\omega_1 \tag{11b}$$

Secondly the dissipation coefficient η is adjusted to maximise the damping. This leads to

$$RC_p\omega_1 = \sqrt{\frac{3\mu}{(1+\mu)(2+\mu)}} \tag{11c}$$

as a first approximation for the optimum value of the external resistance R. The frequency response is shown in Fig.4. A refined estimate can be obtained iteratively by recourse to the exact structural response [8].

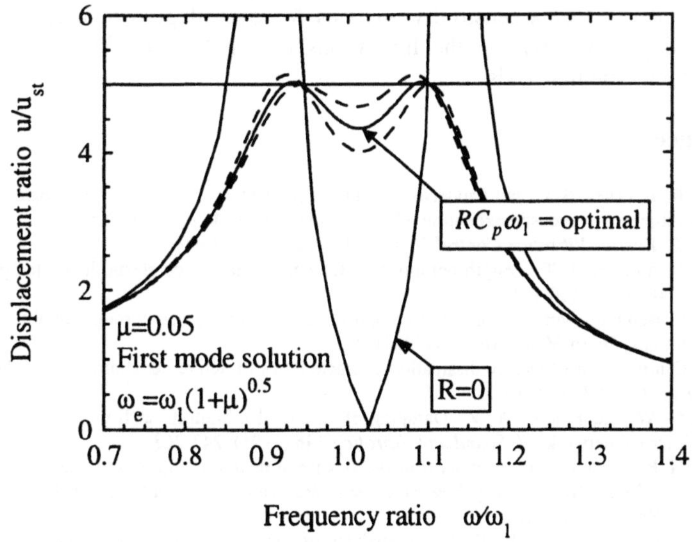

Fig. 4 First mode transfer function at optimal tuning.

The structural loss factor η_s, defined by

$$\frac{u_{max}}{u_{st}} = \frac{8}{\pi^2}\frac{1}{\eta_s} \tag{12a}$$

provides an important *performance index*. For the optimal design (11b, c) corresponding to a given μ, an explicit representation can be derived for η_s, *viz.*

$$\eta_s = \sqrt{\frac{\mu(1+\mu)}{2}} \tag{12b}$$

5. Discussion and Conclusion

Analytical results have been derived that show clearly the structural impact of added piezoelectric patches and the optimal design parameters for shunted piezoelectric damping, for the case of forced extensional vibration. The structural impact appears through the static extension u_{st} defined in (10c), and through the wave speed c in (9g), or the wave number k for a given frequency ω (9f). For the case of balanced stiffness (i.e. $2Y_p h_p = Yh$), and for a typical value of the electro-mechanical coupling constant ($k_{31} = 0.37$, corresponding to PZT5500), optimal design leads to a structural loss factor $\eta_s = 0.17$. This represents a significant level of damping, quite competitive with CLD.

Further work currently in progress aims to (i) extend the analysis to account for extensional-flexural coupling and geometrical optimisation for one-sided patch application; (ii) conduct a sensitivity analysis for overall performance under random acoustic loading; (iii) examine the limitations imposed by the strain allowables for ceramic piezoelectric materials.

6. References

1. Callinan, R. J., Galea, S. C. and Sanderson, S.: Finite element analysis of bonded repairs to edge cracks in panels subjected to acoustic excitation, *Composite Structures* **38** (1997) 649-660.
2. Mead, D. J.: *Passive Vibration Control*, Wiley, Chichester, UK (1999).
3. Plunkett, R. and Lee, C.T.: Length optimisation for constrained visco-elastic layer damping, *J. Acous. Soc. Amer.* **48** (1970) 150-161.
4. Shen, I. Y.: Bending-vibration control of composite and isotropic plates through intelligent constrained-layer treatments, *Smart Mater. Struc.* **3** (1994) 59-70.
5. Baz, A.: Optimisation of energy dissipation characteristics of active constrained layer damping, *Smart Mater. Struc.* **6** (1997) 360-368.
6. Hagood, N. W. and Flotow, A. V.: Damping of structural vibrations with piezoelectric materials and passive electrical networks, *J. Sound and Vibration* **146** (1991), 243-268.
7. Tiersten, H. F.: Equations for the extension and flexure of relatively thin electroelastic plates undergoing large electric fields, *Mechanics of Electromagnetic Materials and Structures*, AMD Vol. 161/MD Vol. 42, ASME, New York (1993).
8. Wang, C. H. and Rose, L. R. F.: Stress analysis of the impact of shunted piezoelectric patches on structural vibration; Part 1: extensional vibration, *DSTO Research Report*, Aeronautical and Maritime Research Laboratory, Melbourne, Australia (2000).
9. IEEE Std 176-1987: *IEEE Standard on Piezoelectricity*, The Institute of Electrical and Electronics Engineers, New York (1988).

A LOCALIZATION CONCEPT FOR DELAMINATION DAMAGES IN CFRP

S. KEYE, M. ROSE, D. SACHAU
Institut für Strukturmechanik
Deutsches Zentrum für Luft- und Raumfahrt (DLR)

1. Introduction

The problem of surveying the integrity of industrial structures becomes an increasingly important issue as operational safety requirements turn more stringent and maintenance costs need to be reduced. Using fiber reinforced materials in modern aerospace structures not only involves new design criteria but also requires advanced safety inspection techniques to account for the nature of damages associated with the new materials. For online or in-flight applications, damage detection and localization methods are required to work with electronic devices which are readily available on the structure, e.g. to monitor and control critical operating conditions [1].

With these settings given, any concept for detecting and localizing structural damage from vibration data must be capable to work with information from a small number of, generally inappropriately placed, excitation points and response sensors. The available mode shape data in this case suffers a high degree of spatial and modal incompleteness and is therefore considered to be of no value for damage localization. It is well known however, that eigenfrequencies alone do not provide sufficient information to detect and locate structural damage in complex 3D-structures. For modal damping values though, a high sensitivity, especially with respect to internal delaminations of CFRP, was found in earlier experimental investigations at DLR [2].

Based on these observations, it appears promising to develop a method which is capable of relating modal damping deviations caused by strucutral damage to the damage location on the structure.

2. Damage Localization Method

2.1. GENERAL METHODOLOGY

The method proposed here is based on the observation, that the influence of a structural damage on the structure's modal quantities varies from mode to mode. Comparing the dynamic response before and after a damage has occurred, one will note distinct changes in the individual eigenfrequencies, mode shapes, damping values and generalized masses/stiffnesses. Depending on its location, each damage causes a unique pattern of deviations to the modal parameters. The idea now is to solve the inverse

U. Gabbert and H.S. Tzou (eds.), IUTAM Symposium on Smart Structures and Structronic Systems, 57–64.

problem, i.e. to use given sets of modal data from an undamaged and a damaged structure to identify and localize the damage.

Simple as it appears, no working method currently exists for structures exceeding the complexity of simple beams, especially when the experimental data is incomplete and contains measurement errors [3-5]. As in most inverse problems, ill-conditioning is a major concern here. To avoid this, it is essential to limit the damage localization problem to its basic aspects [6,7], i.e. to reduce the unknown quantities such that only a few parameters need to be determined. Therefore, the investigations will be focussed on a single concentrated damage and linear behavior is assumed before and after damage has occurred. The parameters left to be identified are location and magnitude.

When only a single or a small number of sensors is available, the experimental data provides no spatial information on the dynamic response, i.e. on the mode shapes, and the process of localizing structural damage has to cope with eigenfrequencies and modal damping values, which can be extracted from the measured frequency response(s). In this case, a finite element model is used to locate the damage.

The localization process is based on the changes of modal properties rather than on the modal properties themselves. This involves measuring the structural response before and after damage has occurred. Since in practice the damage location is primarily unknown, a whole set of damage locations has to be simulated in the numerical model and the modal data must be computed for each damage case. The 'damaged' model which best fits the experimental data now yields the information on the most probable damage location on the test structure. The success of the method not only depends on the quality of the experimental data, but also on a realistic damage simulation in the mathematical model.

2.2. LOCALIZATION BASED ON MODAL DAMPING

Despite some promising results [7,8], the sensitivity of natural frequencies with respect to structural damage generally is very low. Particularly for delamination damages in CFRP it appears more encouraging to focus on changes in the damping factors than in the eigenfrequencies. In this case however, the experimental data has to be related to modal parameters of a finite element model that is needed to localize the damage but usually is designed to compute real eigenvalues and eigenvectors and therefore does not contain a damping distribution.

2.2.1. *Analytical Damping Model*

A widely used solution for this problem is the *proportional damping approach*

$$[d] = \gamma_m \cdot [m] + \gamma_k \cdot [k] \tag{1a}$$

where γ_m and γ_k are initially unknown. There is no physical justification for this approximation, but there seems to be no significantly better linear model available right now [9]. Rewriting eq. (1a) using generalized (modal) properties for each mode i gives

$$D_i = \gamma_m \cdot M_i + \gamma_k \cdot K_i \quad i = 1,\ldots,n \ . \tag{1b}$$

The *modal damping factor* is defined

$$\beta_{\mathrm{ana},i} = \frac{\mathrm{D}_i}{2 \cdot \sqrt{\mathrm{K}_i \cdot \mathrm{M}_i}} , \tag{2a}$$

or, using eq. (1b)

$$\beta_{\mathrm{ana},i} = \frac{\gamma_m}{2 \cdot \omega_i} + \frac{\gamma_k \cdot \omega_i}{2} . \tag{2b}$$

The proportionality constants γ_m and γ_k are tuned such, that the $\beta_{\mathrm{ana},i}$ provide the best approximation to the measured modal damping values of the undamaged structure in a least squares sense.

2.2.2. *Computing Modal Damping from Frequency Response Data*

Eqs. (1)-(2) allow to derive modal damping factors from the system matrices of an analytical model. To obtain the corresponding experimental damping values, measured frequency response functions will be analyzed.

Let us first consider a proportionally damped one-degree-of-freedom system with mass m and resonance frequency ω_0. The response $u(\omega)$ to a stationary harmonic excitation force $F(\omega)$ is given by

$$u(\omega) = \frac{F(\omega)}{m\left(-\omega^2 + 2j\beta_{\mathrm{exp}}\omega_0\omega + \omega_0^2\right)} . \tag{3}$$

For $F(\omega)$ = const., i.e. the excitation force does not vary with ω, the inverse of eq. (3)

$$\frac{1}{u(\omega)} = \underbrace{-\frac{m}{F}}_{a} \cdot \omega^2 + \underbrace{2j\beta_{\mathrm{exp}}\omega_0 \frac{m}{F}}_{b} \cdot \omega + \underbrace{\frac{m}{F}\omega_0^2}_{c} \tag{4}$$

is a quadratic polynomial in ω with coefficients a, b and c. Assuming that $u(\omega)$ is a measured response and the resonance frequency ω_0 is given by

$$\mathrm{im}(u(\omega))\Big|_{\omega=\omega_0} \rightarrow \max. , \tag{5}$$

a common polynomial curve fit, available as a standard tool within commercial computer software, around the resonance frequency ω_0 yields the coefficients in eq. (4) and the unknown damping factor β_{exp} is easily computed from

$$\beta_{\mathrm{exp}} = \frac{-b}{2j\omega_0 \cdot a} \quad \text{or} \quad \beta_{\mathrm{exp}} = \frac{\omega_0 \cdot b}{2j \cdot c} . \tag{6}$$

An example using a measured response from the test structure to be introduced in Chapter 3 is plotted in Figure 1. Five frequency steps around the resonance frequency are used for the curve fit and the damping factor computed from eq. (6) is 0.124%.

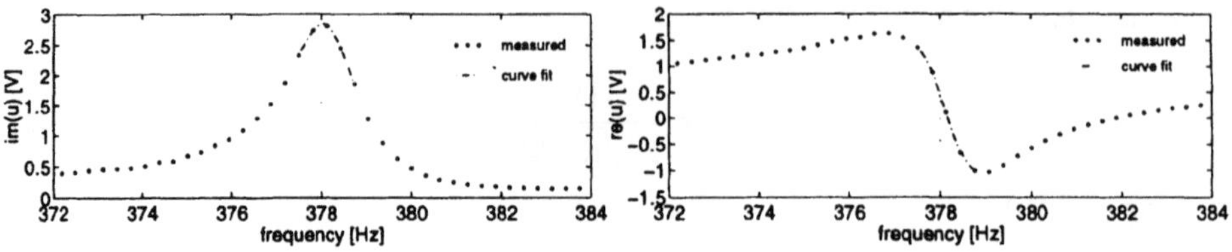

Figure 1. Curve Fit for Modal Damping Calculation.

The method may be extended to multi-degree-of-freedom systems when the damping is sufficiently low and the eigenfrequencies $\omega_{0,i}$ are well spaced.

2.2.3. *Correlation of Experimental and Analytical Modal Damping Deviations*

The damage localization criterion will be based on the deviations of modal damping factors between the healthy (subscript 'H') and the damaged (subscript 'D') structure

$$\delta\beta_i = 2\frac{(\beta_{D,i} - \beta_{H,i})}{(\beta_{D,i} + \beta_{H,i})} \ . \tag{7}$$

A vector of modal damping deviations is defined for the experimental modes

$$\{\delta\beta\}_{\text{exp}} = \{\delta\beta_{\text{exp},1}, \ldots, \delta\beta_{\text{exp},n}\}^{\text{T}} \tag{8a}$$

and for the appropriate analytical modes of each 'damaged' model k

$$\{\delta\beta\}_{\text{ana},k} = \{\delta\beta_{\text{ana},1}, \ldots, \delta\beta_{\text{ana},n}\}_k^{\text{T}} \tag{8b}$$

respectively. The experimental and analytical damping deviations are correlated through a modified *modal assurance criterion (MAC)* [10]

$$C_k = \frac{(\{\delta\beta\}_{\text{ana},k}^{\text{T}} \cdot \{\delta\beta\}_{\text{exp}})^2}{(\{\delta\beta\}_{\text{ana},k}^{\text{T}} \cdot \{\delta\beta\}_{\text{ana},k}) \cdot (\{\delta\beta\}_{\text{exp}}^{\text{T}} \cdot \{\delta\beta\}_{\text{exp}})} \ . \tag{9}$$

The best correlation between a numerical model k and the experimental data produces the highest value of C_k and indicates that the damage in model k lies closest to the test structure's actual damage.

Eq. (9) may also be used to correlate eigenfrequency deviations, combinations of both damping and eigenfrequency deviations, or frequency response functions.

3. Test Structure

3.1. EXPERIMENTAL SETUP

The test structure, Figure 2, consists of a plate with two stiffening stringers on the upper side. It is made of bi-directional carbon fiber mesh and is intended to represent a characteristic aircraft component. The dimensions are 500×400×44mm^3 (L×W×H) and the total weight is approximately 1.45kg. Measurements are taken with the plate clamped at one side. A piezoelectric actuator, bonded to the top surface between the stringers, is used for excitation, and a similar device, mounted in the same position but on the bottom surface, serves as a response sensor. The square areas labeled 1 to 4 represent the locations of artificial delaminations which will later be introduced into the mathematical model to perform a damage simulation study (see Chapter 3.4).

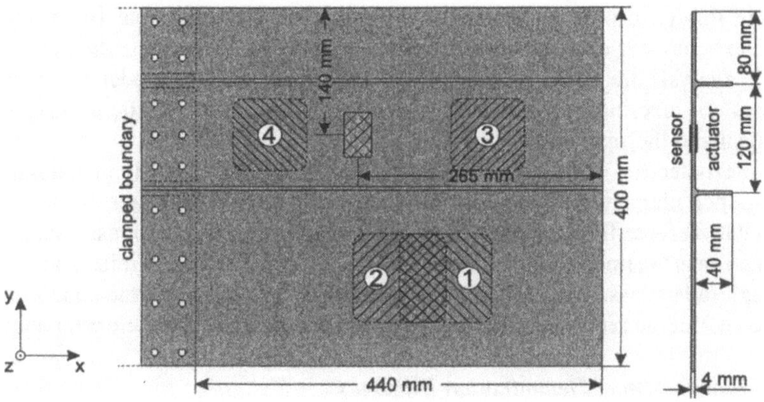

Figure 2. Test Structure.

3.1.1. *Mode Shapes and Frequency Response*

The experimental mode shapes are needed to enable an initial allocation of corresponding measured and analytical eigenfrequencies. This information is used to correctly assign experimental and analytical frequency shifts and damping variations in the succeeding damage detection process. A single-channel laser scanning vibrometer measures the out-of-plane deflections of the base plate (z-components). A harmonic excitation is applied through the integrated piezoelectric actuator. The setup assures that no additional masses from shakers or sensors can affect the results.

The frequency response is obtained from the output signal of the piezoelectric sensor. From the frequency response functions of the undamaged and damaged test structures, natural frequencies are determined and experimental modal damping factors are computed. The frequency range is 0-2000 Hz.

3.1.2. *Experimental Detection of Delamination Damages*

Delamination damages are caused by surface impacts which have sufficient energy to destroy the internal bonding between neighboring plies but generally do not produce visible exterior damage. They affect the transmission of sound waves and can therefore be detected by ultrasonic wave scans. This method will be used to identify the exact size, shape, and position of the delamination damages for comparison to the modal-based results after the test specimens have been damaged.

3.2. FINITE ELEMENT MODEL

The finite element model consists of 2400 shell elements with isotropic material properties. A resolution this high is certainly not necessary for a standard dynamic analysis, but is needed for a good spatial resolution in the modeling of damage. At the clamped boundary a rigid suspension is assumed.

3.2.1. *Piezoelectric Components*

Keeping in mind the purpose of these investigations, the main goal is not to exactly describe the behaviour of a piezoelectric element in detail. Only the features relevant for a dynamic analysis are considered here. Therefore, a simplified model has been set up which does not take into account any interactions between the elastic deformation and the electrical and thermal fields.

The piezoelectric actuator is modeled as a homogeneously contracting and expanding shell element with constant in-plane strain across its surface. This results in a radial excitation force field, which will be used in the frequency response calculations.

For the sensor element, a linear relation between the in-plane strains and the output signal magnitude is assumed. Again, only qualitative statements on the suitability of the sensors to collect the required vibration data are needed in the finite element analysis.

3.2.2. *Modelization of Delamination Damages*

As outlined in Chapter 2.1, the localization problem shall be restricted to a single concentrated damage with two remaining parameters, i.e. damage magnitude and damage location. This requires to predefine a characteristic delamination size and shape, which will be obtained from the ultrasonic wave scan images. The damage magnitude is described as a reduction of the in-plane shear stiffnesses G_{13} and G_{23}.

3.3. SELECTION OF ACTUATOR & SENSOR POSITION

3.3.1. *Introduction*

This section describes the criterions upon which the selection of an optimized position for the placement of the piezoelectric actuator and sensor are based. From the physical point of view the in-plane surface strain components ε_x and ε_y at the chosen location must be high enough to allow for the introduction of a sufficient amount of energy from the actuator into the structure, and to generate a sufficiently high output voltage in the sensor element.

In an earlier investigation [11] an optimized piezo location has been determined from the maximum sum of the in-plane strain components ε_x and ε_y for modes no. 1 to 20. This, however, only yields the position with the highest overall deflection magnitude. A better approach is to search for the location where a maximum number of mode shapes in a given frequency range can be excited and measured. This ensures that the largest possible amount of information is acquired from the experiments.

3.3.2. *Modal Analysis and Frequency Response*

To find the best actuator/sensor position in the sense mentioned, the piezoelectric elements have to be placed in an area where for as many modes as possible the surface strain components are different from zero. With 20 modes and due to the finite dimensions of the piezoelectric elements though, it can not always be avoided, that a nodal line comes close to or crosses the actuator/sensor area. The most suitable position, i.e. the position at which an actuator can excite the largest possible number of mode shapes in the given range from mode no. 1 to no. 20, is shown in Figure 2. Since, due to the stringer stiffness, the ε_y are generally higher than the ε_x, the piezoelectric elements are oriented in y-direction. A numerical frequency response for the chosen actuator/sensor location at 0.5% critical damping is plotted in Figure 3.

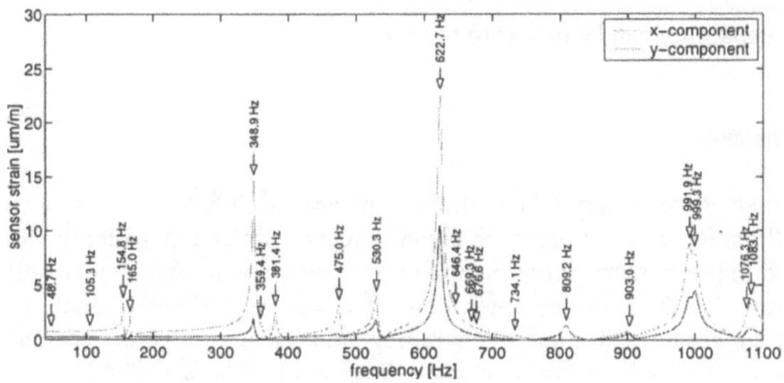

Figure 3. Analytical Frequency Response at Sensor Location.

3.4. DAMAGE SIMULATION

To assess the suitability of the damage modelization and the proportional damping approach for damage localization, the influence of various simulated delamination damages, modeled according to Chapter 3.2.2., is investigated. The damage locations are sketched in Figure 2 and the generalized damping value deviations for modes 1 to 20 with respect to the 'undamaged' model are plotted in Figure 4.

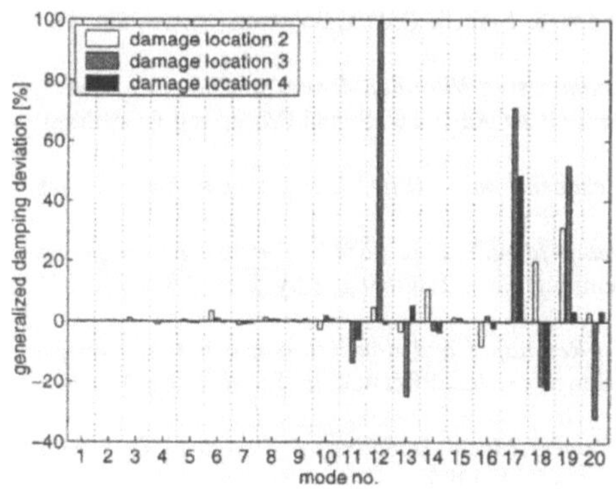

Figure 4a). Damping Deviation for Damage Locations 2 – 4.

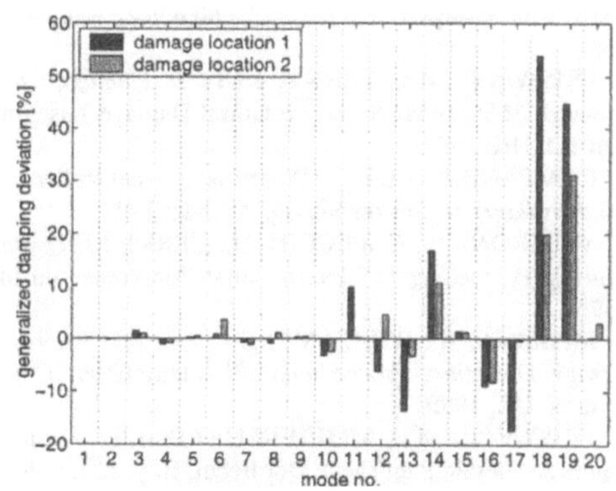

Figure 4b). Damping Deviation for Damage Locations 1 & 2.

Comparing damage locations 2, 3 and 4, Figure 4a), the damping deviations display unique and easily distinguishable patterns. This, conversely, permits a clear separation of the various damage locations in the finite element model, since some modes show a considerable response to a given damage, whereas others are not significantly influenced. Additionally, the sensitivity generally increases for higher mode numbers.

For the more critical case of two overlapping damages, Figure 4b), the spectrum of damping value deviations still contains some noticeable differences (e.g. modes no. 8, 11, 12, 17, 20) which are sufficient to distinguish between the two locations.

The simulation study indicates, that the concept of processing modal damping for damage localization is capable of yielding a high spatial resolution and that a good localization capability can be expected in practical use.

4. Conclusions

A new method of localizing delamination damages in CFRP has been introduced. So far, an identification concept based on eigenfrequency shifts and damping variations has been verified in preliminary experimental investigations with one- and two-dimensional test structures. For the more complex case of a stringer-stiffened panel, an advanced approach is suggested and some encouraging numerical results are presented. Future work will be related to the experimental verification of the new method.

5. References

[1] BOLLER, C.,BIEMANS, C.: *Structural Health Monitoring in Aircraft – State-of-the-Art, Perspectives and Benefits*, Proceedings International Workshop on Structural Health Monitoring, Stanford, CA, 1997.

[2] KAISER, S., MELCHER, J., BREITBACH, E., SACHAU, D.: *Structural Dynamic Health Monitoring of Adaptive CFRP-Structures*, Proceedings 6th Annual International Symposium on Smart Structures and Materials, Newport Beach, CA, 1999.

[3] CAWLEY, P., ADAMS, R.D.: *The Location of Defects in Structures from Measurements of Natural Frequencies*, Journal of Strain Analysis, Vol. 14 (3), pp. 49-57, 1979.

[4] VESTRONI, F., CERRI, M.N., ANTONACCI, E.: *The Problem of Damage Detection in Vibrating Beams*, Proceedings Eurodyn '96 Conference, Augusti, Borri & Spinelli (Eds.), Balkema, Rotterdam, NL, 1996.

[5] FRISWELL, M.I., PENNY, J.I.T.: *Is Damage Location Using Vibration Measurements Practical ?*, Euromech 365: DAMAS '97, Structural Damage Assessment Using Advanced Signal Processing Procedures, Sheffield, UK, 1997.

[6] CAMPANILE, L.F.: *Diagnose struktureller Schäden mit Hilfe der strukturdynamischen Schadensdiagnose*, Shaker Verlag, Aachen, 1997.

[7] VESTRONI, F., CAPECCHI, D., CERRI, M.N.: *Damage Identification Based on Measured Frequency Changes*, Proceedings 2nd International Conference on Identification in Engineering Systems, Swansea, UK, 1999.

[8] MESSINA, A., WILLIAMS, E.J.: *Use of Changes in Resonance and Anti-Resonance Frequencies for Damage Detection*, Proceedings 2nd International Conference on Identification in Engineering Systems, Swansea, UK, 1999.

[9] FRISWELL, M.I., MOTTERSHEAD, J.E.: *Finite Element Model Updating in Structural Dynamics*, Kluwer Academic Publishers, Dordrecht, Boston, London, pp. 15-18, 1995.

[10] ALLEMANG, R.J., BROWN, D.L.: *A Correlation Coefficient for Modal Vector Analysis*, Proceedings 1st International Modal Analysis Conference, Orlando, FL, pp. 110-116, 1982.

[11] KESSELS, J.F.A., NOACK, J.: *Optimal Sensor Positioning for Health Monitoring*, Internal Report, Deutsches Zentrum für Luft- und Raumfahrt e.V. (DLR), IB 131-99/30, 1999.

STRUCTURES WITH HIGHEST ABILITY OF ADAPTATION TO OVERLOADING

J. HOLNICKI-SZULC, T. BIELECKI
Institute of Fundamental Technological Research,
Swietokrzyska 21, 00-049 Warsaw, Poland
E-mail: holnicki@ippt.gov.pl

1. Introduction

Adaptive structures (structures equipped with controllable semi-active dissipaters, so called *structural fuses*) with highest ability of adaptation to extremal overloading are discussed. The quasistatic formulation of this problem allows developing effective numerical tools necessary for farther considerations concerning dynamic problem of optimal design for the best structural crash-worthiness (see [2]). The structures with the highest impact absorption properties can be designed in this way. The proposed optimal design method combines sensitivity analysis with remodelling process, allowing approach (with material distribution as well as stress limits controlled) to an optimally redesigned structure. So called Virtual Distortion Method (see [1]), leading to analytical formulas for gradient calculations, has bee used in numerically efficient algorithm.

2. VDM Based Structural Remodelling and Sensitivity Analysis

Let us concentrate on the sensitivity analysis for the truss structure under progressive collapse process due to extremal load applied. The superposition of virtual, plastic-like distortions β^0_i, simulating non-linear member behaviour, with distortions ε^0_i, modelling modifications of design variables (e.g. A_i), turns out to be productive in this case.

The strains and stresses, calculated with respect to initial cross-sections, can be expressed as follows (see [1], [4]):

$$\begin{aligned} \sigma'_i &= E_i\left(\varepsilon_i - \varepsilon^0_i - \beta^0_i\right) = E_i\left(\varepsilon^L_i + \sum_j \left(D_{ij} - \delta_{ij}\right)\varepsilon^0_j + \sum_k \left(D_{ik} - \delta_{ik}\right)\beta^0_k\right) \\ \varepsilon_i &= \varepsilon^L_i + \sum_j D_{ij}\varepsilon^0_j + \sum_k D_{ik}\beta^0_k \end{aligned} \tag{1}$$

where D_{ij} denote deformations caused in the members i by the unit virtual distortions ε^0_j generated in members j. The corresponding derivatives take the following form:

U. Gabbert and H.S. Tzou (eds.), IUTAM Symposium on Smart Structures and Structronic Systems, 65–72.

$$\frac{d\sigma_i'}{d\varepsilon_j^o} = E_i(D_{ij} - \delta_{ij})\frac{d\sigma_i'}{d\beta_k^o} = E_i(D_{ik} - \delta_{ik})$$
$$\frac{d\varepsilon_i}{d\varepsilon_j^o} = D_{ij}\ \frac{d\varepsilon_i}{d\beta_k^o} = D_{ik} \tag{2}$$

The subscripts j and k in the above formulas run through all modified and plastified members, respectively. Taking advantage of two expressions for the internal forces applied to so called *distorted* (with modification of material distribution modelled through virtual distortions) and *modified* (with redesigned cross-sections from A'_i to A_i, without imposing virtual distortions) structure:

$$P_i = E_i A_i'(\varepsilon_i - \varepsilon_i^o - \beta_i^o)$$
$$P_i = E_i A_i(\varepsilon_i - \beta_i^o) \tag{3}$$

(where components of ε''_i, β''_i are non-zero only in distorted or plastified members, respectively), the following formula can be derived:

$$A_i\left(\varepsilon_i^L + \sum_j D_{ij}\varepsilon_j^o + \sum_k (D_{ik} - \delta_{ik})\beta_k^o\right) = A_i'\left(\varepsilon_i^L + \sum_j (D_{ij} - \delta_{ij})\varepsilon_j^o + \sum_k (D_{ik} - \delta_{ik})\beta_k^o\right) \tag{4}$$

what can be expressed alternatively:

$$\sum_j \left[A_i'(D_{ij} - \delta_{ij}) - A_i D_{ij}\right]\varepsilon_j^o + \sum_k \left[(A_i' - A_i)(D_{ik} - \delta_{ik})\right]\beta_k^o = (A_i - A_i^o)\varepsilon_i^L \tag{5}$$

Calculation of the derivative with respect to A_m leads to:

$$\delta_{im}\left[\ \varepsilon_i^L + \sum_j D_{ij}\varepsilon_j^o + \sum_k (D_{ik} - \delta_{ik})\beta_k^o\ \right] + A_i\left(\sum_j D_{ij}\frac{\partial\varepsilon_j^o}{\partial A_m} + \sum_k (D_{ik} - \delta_{ik})\frac{\partial\beta_k^o}{\partial A_m}\right) =$$
$$= A_i'\left(\sum_j (D_{ij} - \delta_{ij})\frac{\partial\varepsilon_j^o}{\partial A_m} + \sum_k (D_{ik} - \delta_{ik})\frac{\partial\beta_k^o}{\partial A_m}\right) \tag{6}$$

After rearranging the above formula, we have:

$$\sum_j \left[A_i'(D_{ij} - \delta_{ij}) - A_i D_{ij}\right]\frac{\partial\varepsilon_j^o}{\partial A_m} + \sum_k \left[(A_i' - A_i)(D_{ik} - \delta_{ik})\right]\frac{\partial\beta_k^o}{\partial A_m} = (\varepsilon_i - \beta_i^o)\delta_{im} \tag{7}$$

The associated conditions for derivatives $\partial\beta_i^o/\partial A_i$ and $\partial\varepsilon_i^o/\partial A_m$ can be determined from the yield criterion (cf. Fig.1), written for the *modified structure* (with modified cross-sections A_i):

$$\sigma_i - \sigma^* = \gamma_i E_i \left(\varepsilon_i - \varepsilon^*\right) \tag{8}$$

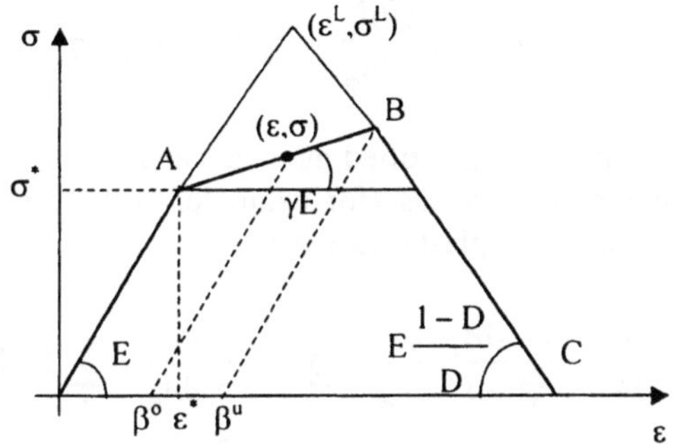

Figure 1. Yield criterion for the modified structure

For the *modified structure*, where ε^o affects the stress formula in an implicit way through modified deformations (cf. Eqs. (1) for *distorted structure*), we get the following strains and stresses, with respect to remodelled cross-sections A_i:

$$\begin{aligned} \sigma_i &= E_i\left(\varepsilon_i - \beta_i^o\right) = E_i\left(\varepsilon_i^L + \sum_j D_{ij}\varepsilon_j^o + \sum_k \left(D_{ik} - \delta_{ik}\right)\beta_k^o\right) \\ \varepsilon_i &= \varepsilon_i^L + \sum_j D_{ij}\varepsilon_j^o + \sum_k D_{ik}\beta_k^o \end{aligned} \tag{9}$$

Substituting (9) to (8) we obtain:

$$\begin{aligned} &\sum_k B_{lk}\beta_k^o + \sum_j \left(1-\gamma_l\right) D_{lj}\varepsilon_j^o = -\left(1-\gamma_l\right)\left(\varepsilon_l^L - \varepsilon^*\right) \\ &\textit{where} \quad B_{lk} = \left(1-\gamma_l\right)D_{lk} - \delta_{lk} \end{aligned} \tag{10}$$

Indices l and k run through plastified members and j runs through distorted members. The matrix B (so-called simulation matrix in collapse analysis) is non-positive definite. The mechanical interpretation of VDM simulation requires that all diagonal elements of B are non-positive. Therefore the following constraint imposed on the softening parameters:

$$\gamma_k \geq -\frac{1-D_{kk}}{D_{kk}} \tag{11}$$

for all members k has to be satisfied, to get correct solution through the VDM approach. If a member does not satisfy the above constraint, its contribution to the stress redistribution drops to zero and we have to apply the equation of line BC (Fig. 1) to model the corresponding local stress-strain characteristic. Now, calculating derivatives with respect to A_m we can get the following set of l' equations:

$$\sum_k [(1-\gamma_l)D_{lk} - \delta_{lk}]\frac{\partial \beta_k^o}{\partial A_m} + \sum_j (1-\gamma_l)D_{lj}\frac{\partial \varepsilon_j^o}{\partial A_m} = 0 \tag{12}$$

(where l' denotes the number of plastified members and $l, k=1, 2... l'$).

Finally, to calculate the sensitivities (for example, with respect to modifications of material distribution) for elasto-plastic structure:

$$\begin{aligned} \frac{\partial \sigma_i'}{\partial A_m} &= \sum_j \frac{\partial \sigma_i'}{\partial \varepsilon_j^0}\frac{\partial \varepsilon_j^o}{\partial A_m} + \sum_k \frac{\partial \sigma_i'}{\partial \beta_k^0}\frac{\partial \beta_k^o}{\partial A_m} = \sum_j E_i D_{ij}\frac{\partial \varepsilon_j^o}{\partial A_m} + \sum_k E_i(D_{ik} - \delta_{ik})\frac{\partial \beta_k^o}{\partial A_m} \\ \frac{\partial \varepsilon_i}{\partial A_m} &= \sum_j \frac{\partial \varepsilon_i}{\partial \varepsilon_j^o}\frac{\partial \varepsilon_j^o}{\partial A_m} + \sum_k \frac{\partial \varepsilon_i}{\partial \beta_k^o}\frac{\partial \beta_k^o}{\partial A_m} = \sum_j D_{ij}\frac{\partial \varepsilon_j^o}{\partial A_m} + \sum_k D_{ik}\frac{\partial \beta_k^o}{\partial A_m} \end{aligned} \tag{13}$$

the partial derivatives determined by l' equations (12) and m equations (7) (for each chosen design variable $\mu_m=A_m/A'_m$) have to be determined from the following set (15) of equations:

$$\begin{bmatrix} m\{ \overbrace{(1-\mu_i)D_{ij} - \delta_{ij}}^{m} & \vdots & \overbrace{(1-\mu_i)D_{ij}}^{l} \\ \cdots & & \cdots \\ l\{(1-\gamma_i)D_{ij} & \vdots & (1-\gamma_i)D_{ij} - \delta_{ij} \end{bmatrix} \begin{bmatrix} \varepsilon_j^o \\ \cdots \\ \beta_j^o \end{bmatrix} = \begin{bmatrix} -(1-\mu_i)\varepsilon_i^L \\ \cdots \\ -(1-\gamma_i)(\varepsilon_i^L - \varepsilon_i^*) \end{bmatrix} \tag{14}$$

$$\begin{bmatrix} m\{ \overbrace{(1-\mu_i)D_{ij} - \delta_{ij}}^{m} & \vdots & \overbrace{(1-\mu_i)D_{ij}}^{l} \\ \cdots & & \cdots \\ l\{(1-\gamma_i)D_{ij} & \vdots & (1-\gamma_i)D_{ij} - \delta_{ij} \end{bmatrix} \begin{bmatrix} \frac{d\varepsilon_j^o}{d\mu_k} \\ \cdots \\ \frac{d\beta_j^o}{d\mu_k} \end{bmatrix} = \begin{bmatrix} \frac{(\varepsilon_i - \beta_i^o)}{A_i'}\delta_{ik} \\ \cdots \\ 0 \end{bmatrix} \tag{15}$$

On the other hand, the set (14), with the same main matrix, describes the virtual distortion fields simulating modified structure.

The above formulas can be, for example, applied to the gradient based optimal remodelling processes of adaptive structures. The plastic-like behaviour (simulated through β^o) corresponds to the performance of actuators, while the material redistribution modified during the redesign process is modelled through virtual distortions ε^o. The gradients computed from Eqs.15 allow calculation of gradients (13) and finally, the gradient of an objective function. Then, corrections for the material distribution leading to reduction of the objective function can be performed, the corresponding modifications of virtual distortions can be determined from (14) and again new gradients can be computed from (15). Following this algorithm we can approach step by step the minimum of the objective function.

If stress limits $\sigma_k^* \leq \sigma''$ will be considered as design variables, rather than material redistribution, the gradient formulas (15) will take the following form:

$$\begin{bmatrix} m\{\overbrace{(1-\mu_i)D_{ij}-\delta_{ij}}^{m} & \vdots & \overbrace{(1-\mu_i)D_{ij}}^{l} \\ \cdots & & \cdots \\ l\{(1-\gamma_i)D_{ij} & \vdots & (1-\gamma_i)D_{ij}-\delta_{ij} \end{bmatrix} \begin{bmatrix} \dfrac{d\varepsilon_j''}{d\sigma_k^*} \\ \cdots \\ \dfrac{d\beta_j''}{d\sigma_k^*} \end{bmatrix} = \begin{bmatrix} 0 \\ \cdots \\ \dfrac{(1-\gamma_i)}{E_k}\delta_{ik} \end{bmatrix} \tag{15a}$$

3. Design of Adaptive Structures for Maximal Energy Dissipation.

The optimal design problem leading to maximal structural ability of adaptation to overloading can be now formulated (e.s. for the ideal elasto-plastic case, without hardening) as requirement of maximisation of the global energy f dissipation:

$$\text{where } f = \begin{cases} \sigma_i^* \beta_i'' \mu_i A_i' l_i & \text{if} \quad \beta_i'' \leq \beta'' \\ \sigma_i^* \beta_i'' \mu_i A_i' l_i & \text{if} \quad \beta_i'' > \beta'' \end{cases} \tag{16}$$

subject to:

$$\begin{gathered} \mu_i = \frac{\varepsilon_i - \varepsilon_i''}{\varepsilon_i} \geq 0 \\ \sum_i \mu_i A_i' l_i = V_o \\ |\sigma'_i| \leq \sigma_i^* \leq \sigma'' \\ \sigma_i \leq \frac{\sigma_i^L}{D_{ii}} - \frac{1-D_{ii}}{D_{ii}} E\varepsilon_i \\ \sigma_i \beta_i'' \geq 0 \end{gathered} \tag{17}$$

where: σ_i and ε_i are expressed by the formula (1), V_o denotes the constant material volume and σ^*_i, μ_i are control parameters (together with the associated virtual distortions ε_i'', β_i''). The parameters μ_i are responsible for material redistribution, while σ^*_i control the best adaptation of the yield stress limits of dissipative devices. β'' denotes the maximal stroke of structural fuses. The condition (17)[4] contains σ-ε response to points below the line BC in Fig. 1. The gradients of the objective function and the side constraints with respect to the control parameters μ_i and σ^*_i can be calculated (through virtual distortions) making use of the formulas analogous to (15). In the consequence, an efficient numerical algorithm for the gradient based optimal redesign process can be proposed. Heaving control parameters modified (due to gradient calculations) in the iterative process, the virtual distortions ε_i^o, β_i^o can be updated solving Eqs.(14). Let us now discuss particular cases of the structural redesign problem.

3.1 STRUCTURAL REMODELLING

Assuming constant plastic-like properties of dissipaters ($\sigma_i^*=\sigma^u=const.$) the above *adaptive structure design* problem leads to the *optimally remodelled* elasto-plastic, isostatic substructure with active constraints $(17)^3$ and $(17)^5$ (see [4]). This fully loaded ($|\sigma_i|=\sigma^u$) and fully distorted ($|\beta_i^o|=\beta^u$) substructure can be determined using numerically efficient *Tracing Active Constraints* method.

3.2 GENERAL PROBLEM OF ADAPTIVE STRUCTURE DESIGN

It can be demonstrated through numerical tests that the above, particular, *optimal remodelling* solution (case 3.1) will be reached also for the general *adaptive structure redesign* problem (16), (17) (for both: σ_i^* and μ_i control parameters) when only one loading state is considered. However, normally, we have to take into account several possible extremal loading scenarios, and that is why more complex results are normally expected. The gradient-based approach can be applied also to the above, general redesign problem. However, this formulation can be substituted, in the first approximation, by the following, simpler, decomposed, two-steps problem: i) *Structural remodelling –Multiload Case* and ii) *Structural Multiload Adaptation.*

3.3 DECOMPOSED PROBLEM OF ADAPTIVE STRUCTURE DESIGN

The *Tracing Active Constraints* algorithm (mentioned above, section 3.1) can be generalised for several load states (describing all possible extremal loading states). Assuming $\sigma_i^*=\sigma^u=const.$ the material redistribution μ_i can be determined. Each structural element is fully loaded ($|\sigma_i|=\sigma^u$) at least in one load state, but the structure is no more isostatic. The structure with fixed geometry (determined in the first step (i)) can be now optimally adapted (σ_i^* controlled) to particular (detected in real time) load state using the following approach of structural multi-load adaptation.

Assuming constant material distribution (μ_i=1) the adaptive structure design problem leads to the control of plastic-like stress limits σ_i^* in order to maximise the energy dissipation. The iterative, gradient based method to solve this problem can be described by the algorithm shown in Table 1., where gradient calculations are performed accordingly to formulas $(15a)^2$, while plastic distortions (β_i^o) are updated to modify σ_i^* using formulas $(14)^2$.

Finally, gradient of the objective function (16) can be determined:

$$\frac{\partial f}{\partial \sigma_k^*} = \sum_j \frac{\partial f}{\partial \beta_j^o} \frac{\partial \beta_j^o}{\partial \sigma_k^*} \tag{18}$$

where the first component can be calculated differentiating formula (16):

$$\frac{\partial f}{\partial \beta''_i} = \begin{cases} 2\mu_i A'_i l_i E_i \sum_j \left(D_{ij} - \delta_{ij}\right)\beta''_i + \mu_i A'_i l_i \sigma^L_i \\ \mu_i A'_i l_i E_i \left(D_{ij} - \delta_{ij}\right) \end{cases} \tag{19}$$

while the second component can be determined solving Eqs.(15a)[2].

TABLE 1. Flow Chard for the Structural Adaptation Algorithm

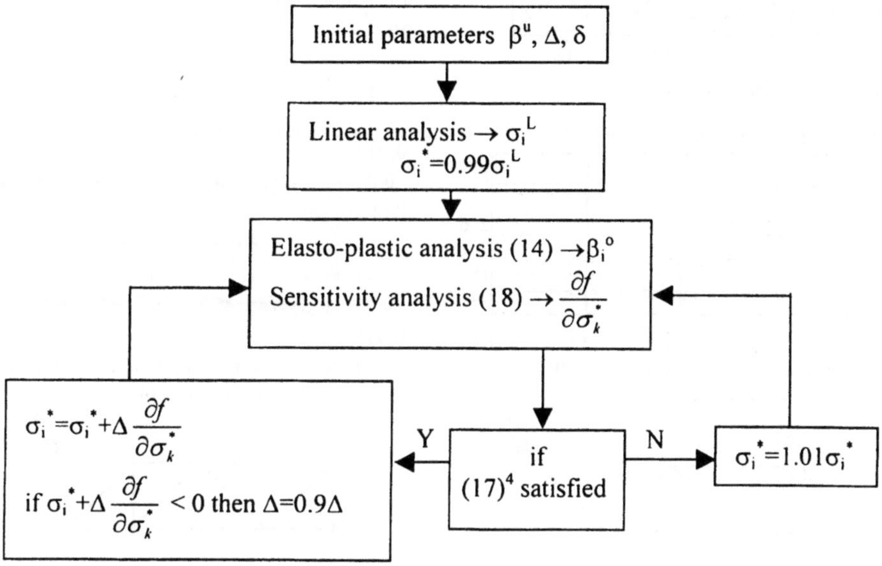

4. Numerical Examples

The hyper-static truss structure shown in Fig.2 has been used to demonstrate the structural adaptation procedure. Assuming fixed structural geometry (μ_i=1=const., A=0.0201, E=10^8, β^u=0.01) the optimal stress limits σ_i^* for each member have been determined (for "detected" extreme load P) using the algorithm described in section 3.3. To make the problem differentiable for all structural elements (plastified as well as still elastic), elasto-plastic behaviour with small hardening (small γ=0.01) has been applied.

The results of the optimisation process have been shown in Figs 3-6. The stress limits (Fig.4) and plastic-like distortions' (Fig.5) iteration show that the final result can be realised as the iso-static substructure demonstrated in Fig.6a. However, this result describes a local maximum of the objective function (see Fig.3 for its iteration, f=85kJ). To find the improved solution (f=104kJ, cf.Fig.6b) some special treatment (on-line corrections of the optimisation procedure) has to be applied. The solution demonstrated in Fig.6b is close to the result (obtained through another, numerically expensive approach) reported in [3].

5. Conclusion

Realisation of the best iso-static solutions described above can be done with help of the controllable structural fuses allowing tuning of the stress limits σ_i^* according to the pre-computed distribution (Fig.4) and identified (in real time) loading, what means that a group of members has to be disconnected from the structure.

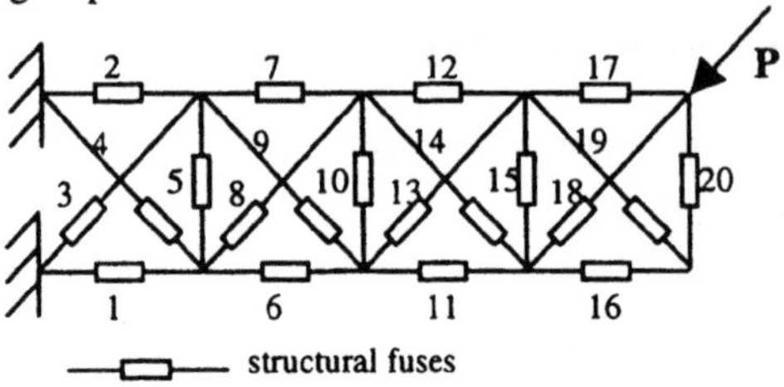

Figure 2. Adaptive truss structure example

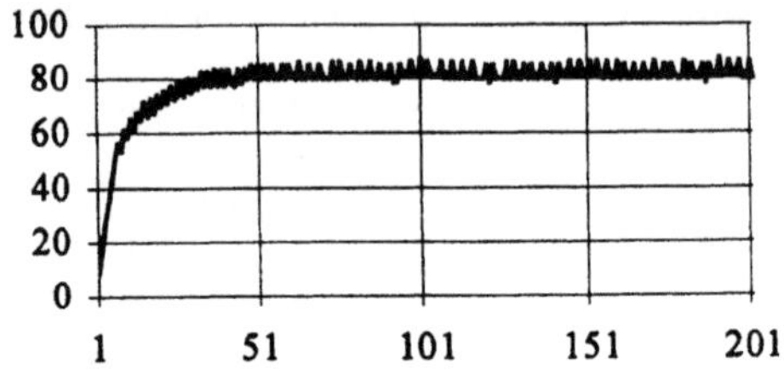

Figure 3. The objective function iteration [kJ]

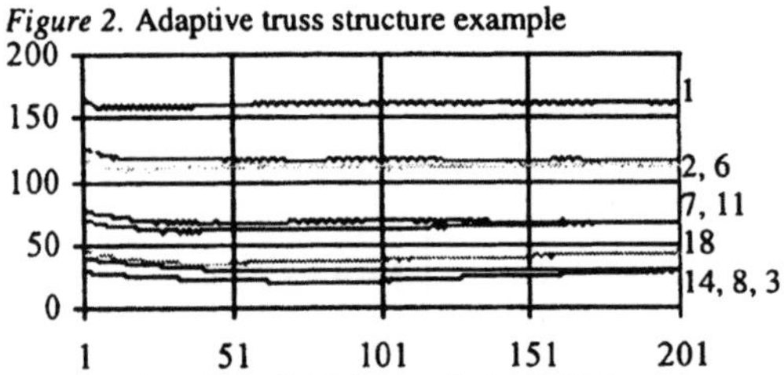

Figure 4. Iteration of yield stress limits [MPa] in remaining members

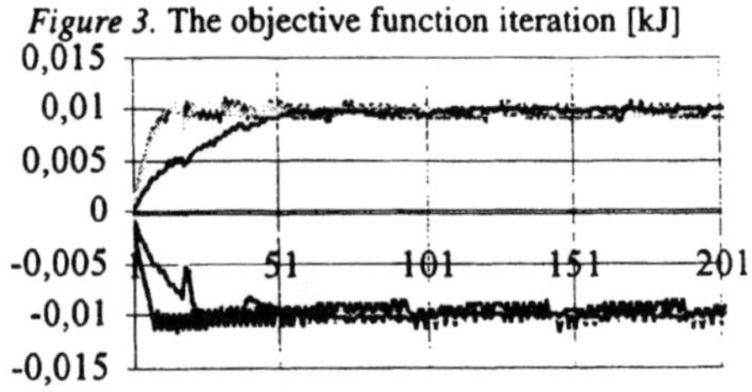

Figure 5. Iteration of plastic-like distortions in remaining members

Figure 6. The best iso-static substructure: a) local maximum, b) improved solution.

6. Acknowledgement

This work was supported by the grant No. KBN7T07A02516 from the Institute of Fundamental Technological Research, funded by the National Research Committee and presents a part of the Ph.D. thesis of the second author, supervised by the first author.

7. References

1. Holnicki-Szulc, J., Gierliński, J.T.: *Structural Analysis, Design and Control by the VDM Method*, J.Wiley & Sons, Chichester, 1995.
2. Knap, L., Holnicki-Szulc, J.: *Optimal Design of Adaptive Structures for the Best Crash-Worthiness*, Proc. 3rd World Congress on Structural and Multidisciplinary Optimisation, Buffalo, May 1999.
3 Holnicki-Szulc, J., Mackiewicz, A., Kołakowski, P.: *Design of Adaptive Structures for Improved Load Capacity*, AIAA Journal vol.36, No.3 March 1998.
4 Holnicki-Szulc, J., Bielecki, T.: *Gradient vs. Tracing Active Constraints Approach in Structural Remodelling*, Proc. 3rd World Congress on Structural and Multidisciplinary Optimization, Buffalo,May 99

BIO-INSPIRED STUDY ON THE STRUCTURE AND PROCESS OF SMART MATERIALS AND STRUCTURES

B.L. ZHOU, G.H. HE, J.D. GUO
Institute of Metal Research, Chinese Academy of Sciences
International Center for Materials Physics, Chinese Academy of Sciences
72 Wenhua Road, Shenyang, 110015, China

1. Introduction

The trend of development of structural materials in the new century can be expected to be composite, intelligent, multifunctional and ecological. Compared with other materials, the most conspicuous feature of biomaterials is the function of self-adjustment, i.e. being living organisms, biomaterials can adjust their physical and mechanical properties in some way to fit the environmental conditions. One of the outstanding features of an organism is its ability for regeneration. The organism itself can repair a fracture after the injury occurs. The purpose of our work is to investigate their structural and functional features and apply them to the design and manufacturing of advanced smart materials[1][2].

Some progress in bio-inspired study on the structure and process of smart materials and structures are summarized in this paper. These are: (1) The fractal-tree-like metallic materials, taken seaweeds as mould plate and analyzed on the fractal model, these materials can be used as special fiber-reinforcement for composite materials and catalyst carrier with much higher reactional surface than that of porous media. (2) Self-adaptive flux valve, mimicking the feed-back system of biomaterials, which is designed and composed on the principle of osmotic pressure. The response time of it is sensitive enough for the water supply of plants in water-lacking areas. (3) Nano-composites, simulating nacre, composed of alternately distributed hard and soft layers of nano-structural metallic ribbons with covered plating to make laminate composites.

Several examples of damage reversal and crack healing, such as (1) the microcrack healing on 1045 steels, (2) the persistent slip band reduction on copper single crystals and (3) the globular and fine grain of solidified casting on zinc-aluminum alloy, all treating by the electropulsing of high current density are briefly discussed.

2. Bio-inspired Thinking on Materials and Structures

2.1. FRACTAL-TREE-LIKE METALLIC MATERIALS

The foamed metal has the application for filtering and catalyzing purpose in the petro-chemical industry. Several properties of foamed metals, such as the porosity, average

U. Gabbert and H.S. Tzou (eds.), IUTAM Symposium on Smart Structures and Structronic Systems, 73–80.

pore diameter, effective surface areas, homogeneity and connexion of pores etc., must be considered. As we know, the fractal structure reveals some distinguish features including good connexion and larger specific surface with the fractal dimension increase.

Seaweeds, one kind of sea plants, has the natural fine fractal-tree-like structure. We use seaweeds as the plate mould and calculate their fractal dimension, and treat them by the procedures as shown in Figure 1.

Pretreatment → Chemodeposition → Electrodeposition → Erosion → Deoxidization

Figure 1. Flow chart of material preparation

The calculation of fractal dimension for two kinds of seaweeds, (a) gelidium amensii and (b) gelidium pacificum, by the box-counting dimension method:

$$D = (\ln N_L)/(-\ln L) \tag{1}$$

where D is the fractal dimension, L is the factor of proportionality, N_L is the number of boxes. The results of fractal dimension, $D_a = 1.69 \pm 0.06$, $D_b = 1.62 \pm 0.11$, show that these seaweeds has fine fractal structures.

After plating metal (nickel with thickness of 30 μm) on the surface of seaweeds, the followed important step is to resolve the seaweeds matrix and the hollowed fractal-tree-like structure of nickel has been obtained as shown in Figure 2. The further works will focus on the properties determination and bulk materials preparation.

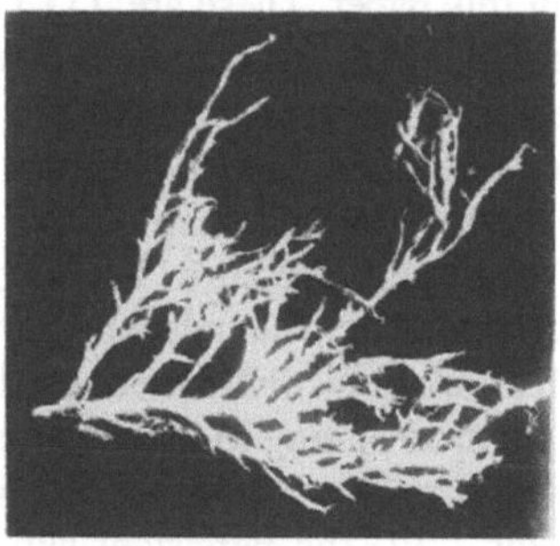

Figure 2. Hollowed fractal-tree-like structure of nickel

2.2. SELF ADAPTIVE POLYMER FLUX VALVE

The irrigation of saving water is one of the most important problems in the water-lacking areas, but the traditional irrigating methods such as drip irrigation and spray irrigation waste inevitably lots of the water.

By the inspiration of adaptability of being things to environmental change, we have gotten a scheme for controlling the water flow automatically according to the humidity of soil. The smart system is composed of the hydrogel, one kind of smart polymer, which can be synthesized by special process, the main performance of hydrogel is its volume change ability from swelling in water to shrinking in dry conditions. The principle scheme of the smart switch is shown in Figure 3.

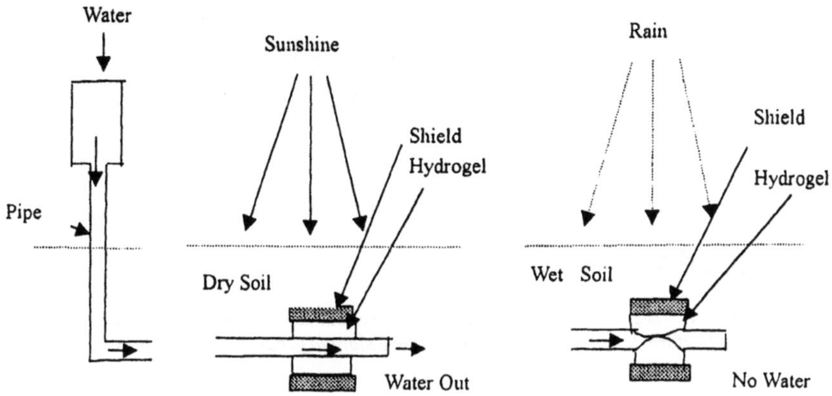

Figure 3. Principle Scheme of Smart Water Switch..

At present, we have proved the feasibility basically, the water flow in pipe can be stopped within three to five seconds when put the system into water and the water flow continues in dry conditions. The following works, such as improving the sensitivity of hydrogel polymers, miniaturizing the system size and adapting them to practical circumstances have been carried out already.

2.3. ALTERNATE HARD AND SOFT LAMINATE COMPOSITES

Nanocrystalline alloys exhibit high hardness and yield strength but low ductility. The ductility of nanocrystallines synthesized from amorphous precursor is greatly limited to a low level far below that of singe-phase nanocrystalline. People hope to have some methods for improving their ductility.

The nacre, one kind of shells, with the combination of hard inorganic crystals and soft biological matters showed the excellent mechanical properties and better ductility. By the inspiration of these attractive structures, a hopeful approach is to produce laminate composites that take advantage of high yield strength of nanocrystallines and high ductility caused by ductile laminates.

Experimentally the FeMoSiB amorphous alloy ribbons with the thickness of about 25 μm are cleaned by acetone first. Then the Ni layers of about 10 - 20 μm thick are electrodeposited onto the surface of the ribbons. After cutting the ribbons into the pieces of 5×8 mm^2, 25 pieces are stacked one by one and pressed tightly in a special designed clip and electropulsing of high current density is used to bond them together. The samples are then isothermally annealed at 873K in a vacuum furnace to make the amorphous alloy transform completely into nanocrystallines alloys. The sectional view of the bonded samples is shown in Figure 4.

The average microhardness of the nanocrystalline matrix is 1100 Hv, while the average hardness of the laminates is about 650 Hv. The ductility of the laminates may be improved owing to the deflection effect and the plastic deformation of the soft Ni layers. In the future, study should be carried out to optimize the processes and properties of the laminates.

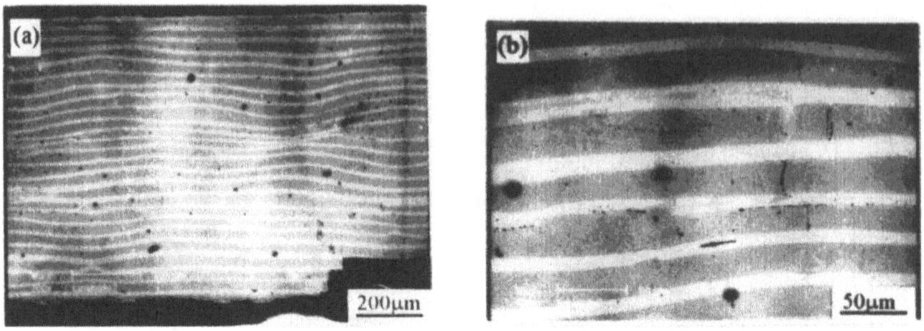

Figure 4. Sectional view of FeMoSiB nanocrystalline/Ni laminate composites.
a. Uniform bonding of the laminates b. Enlarged graph of the laminates

3. Bio-inspired Healing and Recovering on Metals

3.1. MICRO-CRACK HEALING ON 1045 STEELS

The damage is always discovered in the process and service of materials. Engineering materials cannot feel and cure the damage themselves. According to the viewpoint of open systems, the transient input of energy, for example electropulsing (EP), can be supplied to the materials with micro-crack and may result in healing effect and/or preventing the damage from growing [3].

The specimens of 1045 steels with the size width × thickness of 2.5 × 0.2 mm^2 were quenched in water after heating at 840°C. The micro-crack along the direction of width and through the thickness has been observed as shown in Figure 5a. Prior to electropulsing treatment, the specimens were polished. Figure 5b shows the SEM micrograph just after electropulsing without cleaning.

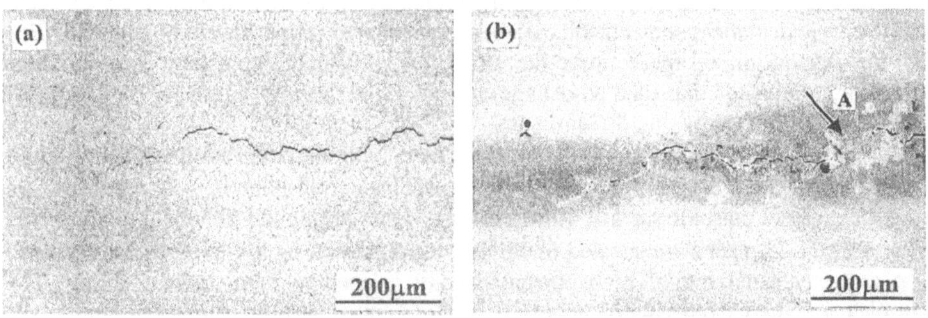

Figure 5. SEM micrographs of crack before (a) and after (b) EP.

Figure 6 shows the SEM micrograph of healing area A of crack. The average temperature rise of specimen has been estimated at about 200°C by discharging parameters. In the Figure 5b indistinct margin area darker than other part of specimen shows the temperature higher than 200°C. The microstructure of healing area A (see Figure 6) shows the uniform and fine structure without melting.

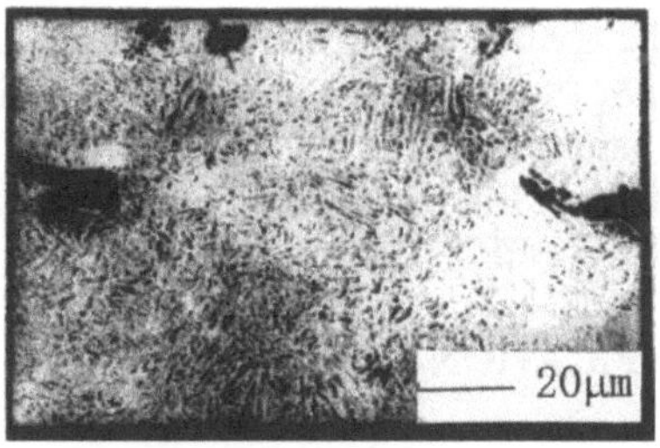

Figure 6. SEM micrograph of healing part A of the crack.

3.2. PERSISTANT SLIP BAND REDUCTION ON SINGLE-CRYSTAL COPPER

The persistent slip bands (PSBs) of metallic materials during fatigue tests has caused scientists much interesting in it because PSBs relates to the mechanism research of fatigue fractures closely. Some reports showed that the electropulsing (EP) on metals could promote the homogenization of slip band and raise the fatigue life for Cu and 316 stainless steel [4][5]. The present work will focus on the fatigued metal crystal to improve its fatigue life [6].

After making the pure copper single crystal with loading axis [$\bar{1}$23] (purity 99.999%) by Bridgmen method and cutting the specimens with the square cross-section of 4.3 × 2.0 mm^2, the push-pull fatigue tests were carried out under a constant plastic strain amplitude (5×10^{-4}) at frequency 0.3 Hz.

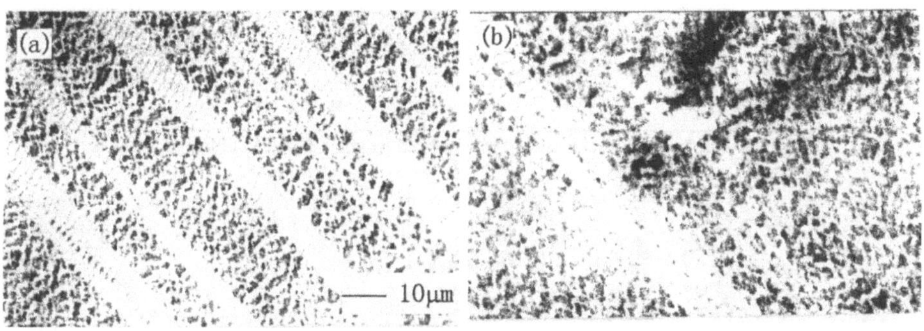

Figure 7. SEM micrograph of PSBs Characteristic.
a. without electropulsing b. PSBs vanish locally with electropulsing

At the end of 5000 cycles fatigued, the specimens were unloaded and treated by electropulsing of high current density (maximum current of 8.95×10^4 A, 110μs duration). Figure 7 shows the typical SEM micrograph of copper single crystal specimens after 5000 cycles fatigue without electropulsing (a) and with electropulsing (b).

It can be seen from Figure 7 (a and b) that the width of PSBs is reduced. The fatigue life N_f can be expressed as [5]:

$$N_f = C\gamma_P^{-2}\left(\frac{1}{X}\right)^2 \tag{2}$$

where c is a constant, γ_p is the applied plastic strain amplitude, X is the average spacing of the PSBs that are the values from 10.2 μm to 6.0 μm determined on a large number of PSBs for the conditions before and after EP respectively. The results calculated by Equation (2) show that the fatigue life of copper single crystal after EP could be raised three times higher than that before EP.

3.3. IMPROVING SOLIDIFICATION STRUCTURE ON Zn-Al ALLOYS

Casting of sand mould is one of the most popular metal processes because of its high efficiency and low costs. But some defects in castings, for example the shrink holes, segregation, porosity and dendritic crystal etc., will affect their qualities and applications. One can improve the solidification structure of Sn-Pb alloys by electropulsing (EP)[7]. At present, we give an example to improve the casting quality of Zn-Al27 alloy, which possesses higher melting point (≈ 460 °C) than that (≈ 280 °C) of Sn-Pb alloy, both in mechanical properties and structures by EP. The results of tensile strength σ_b, elongation δ and density ρ of Zn-Al 27 alloys have been listed in Table 1. And the typical solidification structures have been shown in Figure 8.

It can be seen from Table 1 and Figure 8 that the mechanical properties of Zn-Al 27 alloys have been increased significantly by EP and their structure reveals finer globular grains with EP instead of dendritic grains treatment without EP.

TABLE 1. Tensile strength, elongation and density of Zn-Al 27 casting alloys[a]

	Zn-Al 27 without EP	Zn-Al 27 with EP
σ_b (MPA)[b]	228	343
δ (%)[c]	2.3	3.7
ρ (g . cm^{-3})[b]	4.71	4.82

a Samples with 5 mm diameter and 30 mm marked length, b average for five samples
c average for four samples

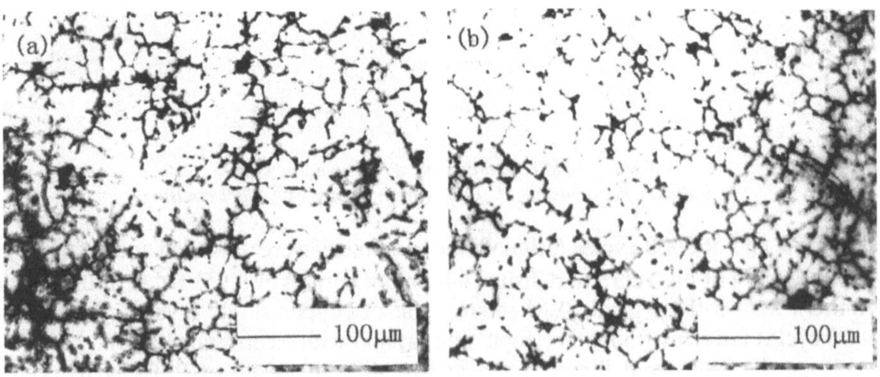

Figure 8. Solidification structure of Zn-Al 27 alloys. a. without EP b. with EP

4. Probable Mechanism of Healing and Recovering Effect on Metals

The electropulsing on the metals results in the effects of micro-crack healing, fatigue prolongating and grain size globing etc. The electroplastic model indicated that the significant EP effect may be attributed to electron wind effect and Joule heating is ignored [8]. But in our experiments the temperature increases (ΔT) calculated by using the EP parameters can not be neglected, for example in section 3.1 ΔT of 1045 steel is about 200°C and in section 3.2 ΔT of copper single crystal is about 69°C. What is the reason to introduce the change of properties and structures during EP treatment?

We noted the dynamic process of thermal expansion gives us some key points [9]. The anharmonic vibration of an one-dimensional lattice can be described as:

$$m\frac{du_n^2}{dt^2} = 2c[(u_{n+1} - u_n) - (u_n - u_{n-1})] - 3g[(u_{n+1} - u_n)^2 - (u_n - u_{n-1})^2] \quad (3)$$

where m is the mass of the particle, t is time, u_n is the displacement of the nth particle, c is the stiffness of the first order, and g is the stiffness of the second order. By analytical treatment, the exact solution of the nonliner differential equation set is:

$$u_n(t) = p\left[i\left(\frac{18gu_o}{m}\right)^{1/2}\left(\frac{T}{T_o}\right)^{1/6} a^{3/4}t + \phi\right]u_o\left(\frac{T}{T_o}\right)^{1/3}\left(n^3 + \frac{n}{2}\right)a^3 + \frac{nc}{3g} \quad (4)$$

where a is the lattice constant, T_o and T are temperatures, u_o and ϕ are constants, and $p[\lambda t + \phi]$ is the Weierstrass elliptical function. Taking statistical average, this solution yields

$$\tau_N = \frac{3Ng}{4c^2}\sqrt{\frac{\pi mKT}{2}} \quad (5)$$

where K is the Boltzmann constant. Equation (5) implies that the characteristic time τ_N for the thermal expansion of the whole lattice is proportional to the length of the N-particle lattice.

The theoretical prediction of the dynamic process of thermal expansion had been verified by experimental observation [10][11]. Nonsynchronous change of temperature and thermal expansion under transient heating was observed as shown in Figure 9.

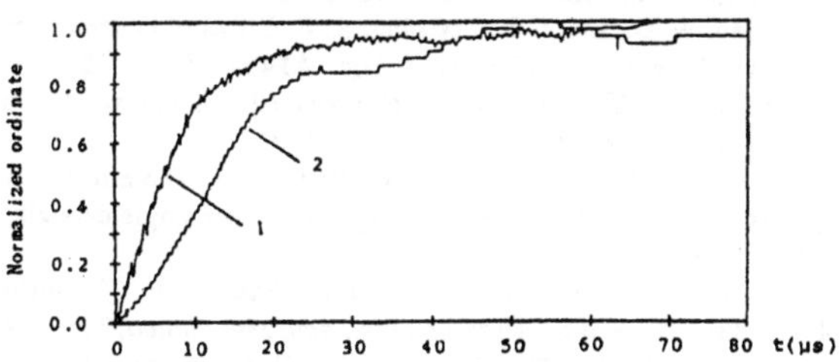

Figure 9. Comparision of temperature rise and thermal expansion curve [10].
1-temperature rise 2-thermal expansion

The nonsynchronous effect of temperature and thermal expansion will result in huge thermal pressure in the sample under transient heating and it can be expressed as :

$$\sigma(t) = E\alpha\Delta T_{max}[\Theta(t) - l(t)] \quad (6)$$

where E is the elastic modulus, α the thermal expansion, ΔT_{max} and ΔL_{max} the maximum temperature rise and maximum expansion, $\Theta(t)=\Delta T(t)/\Delta T_{max}$, $l(t)=\Delta L(t)/\Delta L_{max}$. As $\Theta(t)-l(t)>0$, instantaneous thermal stress is hardly avoidable. The thermal stresses have been calculated as 520 MPa and 65 MPa for the situations in section 3.1 [3] and 3.2 [6], respectively. While the electronic wind force calculated by the parameters in section 3.2 is only 0.1 MPa. So it may be possible to promote the structure change of metals under electropulsing of high current density at the so large thermal pressure stresses.

5. Summary

By the inspiration of biomaterials that possess functional adaptability, distinguish structure, system regeneration and self-healing ability, several smart materials and structures have been designed and some examples of improving for properties and structures of engineering metal materials by electropulsing of high current density have been given in this paper.

The probable mechanism of promoting the structure change of metals under electropulsing by using large thermal pressure stresses caused by the nonsychronous effect has been discussed briefly.

Acknowledgments

This work was supported by Special Funds for the Major State Research Project of China (G1999065009) and National Natural Science Foundation of China (59889101 and 59931020).

References

1. Zhou, B. L.: The biomimetic study of composite materials, *JOM* **February**, (1994),57-62.
2. Zhou, B.L.: Improvement of mechanical properties of materials by biomimetic treatment, *Key Engineering Materials*, **145-149** (1998), pp. 765-774.
3. Zhou, Y. Z., Xiao, S. H., Gan, Y., Gao, M., He, G. H., Zhou, B. L.: The healing of quenched crack in carbon steel under electropulsing, *Acta Met. Sinica*, **36** (2000), 43-45.
4. Conrad, H., White, J., Cao, W. D., Lu, X. P., Sprecher, A. F.: Effect of electric current pulse on fatigue characteristic of polycrystalline copper, *Mat. Sci. & Eng.*, **A145** (1991), 1-12.
5. Cao, W. D., Conrad, H.: On the effect of persistent slip band (PSBs) parameters on fatigue life, *Fatigue Fract. Engng. Mater. Struct.*, **33** (1992), pp. 573-583.
6. Xiao, S. H., Zhou, Y. Z., Wu, S. D., Yao, G., Li, S. X., Zhou, B. L.: The effect of high current density electropulsing on persistent slip bands (PSBs) in fatigured copper single crystals, *Acta Met. Sinica*, **36** (2000), (to be published).
7. Yan, H. C., He, G. H., Zhou, B. L., Qin, R. S., Guo, J. D., Shen, Y. F.: The influence of pulse electric discharging on solidified structure of Sn-10 % Pb alloy, *Acta Met. Sinica*, **33** (1997), 352-358.
8. Conrad, H., Sprecher, A. F. (1989) The electroplastic effect in metals, in F. R. N. Nabarro (ed.), *Dislocation in solids*, Elsevier, Amsderdan, pp. 498-541.
9. Zhou, B. L., He, G. H., Gao, Y. J., Zhao, W. L., Guo, J. D.: The microscopic nonequilibrium process in solids under transient heating, *Intern. J. Thermophysics*, **18** (1997), 481-492.
10. Tang, D .W., Zhou, B. L., Cao, H., He, G.H.: Dynamic thermal expansion under transient laser-pulse heating, *Appl. Phys. Lett.*, **59** (1991), 3113-3114.
11. Tang, D. W., Zhou, B. L., Cao, H., He, G. H.: Thermal stress relaxation behavior in thin films under transient laser-pulse heating, *J. Appl. Phys.*, **173** (1993), 3749-3752.

MAO TECHNOLOGY OF NEW ACTIVE ELEMENTS RECEPTION

S.N. ISAKOV, T.V. ISAKOVA, E.S. KIRILLOV*
State Polytechnic University
Frunze str., 21, Kharkiv, Ukraine, 61002
** Innovation Centre "ESCORT" ltd.*
P.O.BOX 10122, Kharkiv, Ukraine, 61002

1. Introduction

Modern technologies create opportunities for the development and creation of new active materials and adaptive constructive elements. This report is devoted to the use of microarc oxidation (MAO) technology [1] for creation of a new generation of active elements.

2. MAO Technology Description

Microarc oxidation technology is based on the formation of microarc discharges on the surfaces of elements and details submerged in aqueous electrolyte solutions. This leads to the development of multifunctional coatings with a unique complex of physical and mechanical properties. This technology works well with valve metals such as Aluminium, Titanium, Magnesium etc., as well as their alloys.

Microarc discharges are used to produce high-temperature crystal oxides on the product surface, so giving the products qualitatively new properties:

- High hardness up to 24000 MPa;
- Low factor of friction f=0,01-0,005;
- High wear resistance $i_p=10^{-12}$;
- Heat resistance up to 800-1200^0 C;
- Piezoelectric properties;
- Corrosion resistance in various media;
- Dielectric durability 10-20 V/micron;
- Thickness up to 300 microns;
- Heat conductivity factor 5-10 W/(m c);
- Surface finish Rz~1-40 microns (without technological layer removing) and Ra~0,04-0,08 (after polishing):
- Grit 1-10 microns;
- etc.

For aluminium alloys the MAO coating is a high temperature modification of aluminium oxides, and has a double layer structure. The external layer (mullet) is low-

U. Gabbert and H.S. Tzou (eds.), IUTAM Symposium on Smart Structures and Structronic Systems, 81–86.

strength, porous and can be removed by polishing. The internal layer is hard, high-strength and consists of α-Al_2O_3 (HV-24000Mpa) and γ-Al_2O_3 (HV-14000Mpa). The relation between α and γ components is determined by the process parameters. It is possible to form the coatings from 100% α-Al_2O_3 to 100% γ-Al_2O_3, any everything between.

The experimental set-up for microarc oxidation with pulse power supply is shown in Figure 1, and the technical characteristics of installation are presented in Table 1.

TABLE 1. Technical characteristics of the experimental set-up

№	Characteristics	Value
1	Voltage, V	
	- positive pulse	0 – 500
	- negative pulse	0 – 500
2	Current density, A/ dm^2	Up to 20
3	Frequency, Hz	300 – 1200
4	Pulse duration, s	
	- positive pulse	$200 \cdot 10^{-6}$-$2 \cdot 10^{-3}$
	- negative pulse	$200 \cdot 10^{-6}$-$2 \cdot 10^{-3}$
3	Maximal consumption power, kW	75
4	Working volume of a bath, l	150
5	Dimensions of a bath, m	0,5*0,5*0,7
6	Cooling system power, kW	50
7	Amount of the components in an electrolyte (on the technology), kg	8 – 15
8	Electrolyte temperature (°C)	20 - 40
9	Cooperative square of a coating surface, dm^2	Up to 10
10	Coating rate, micron/hour	100-300

The voltage and current pulses forms are shown in Figure 2, which shows the quasistatic character of the voltage pulses (its form is similar to square pulse) and dynamic changes of the current in the pulse.

The ratio between positive and negative pulses of current is maintained within the limits 1 – 1,2.

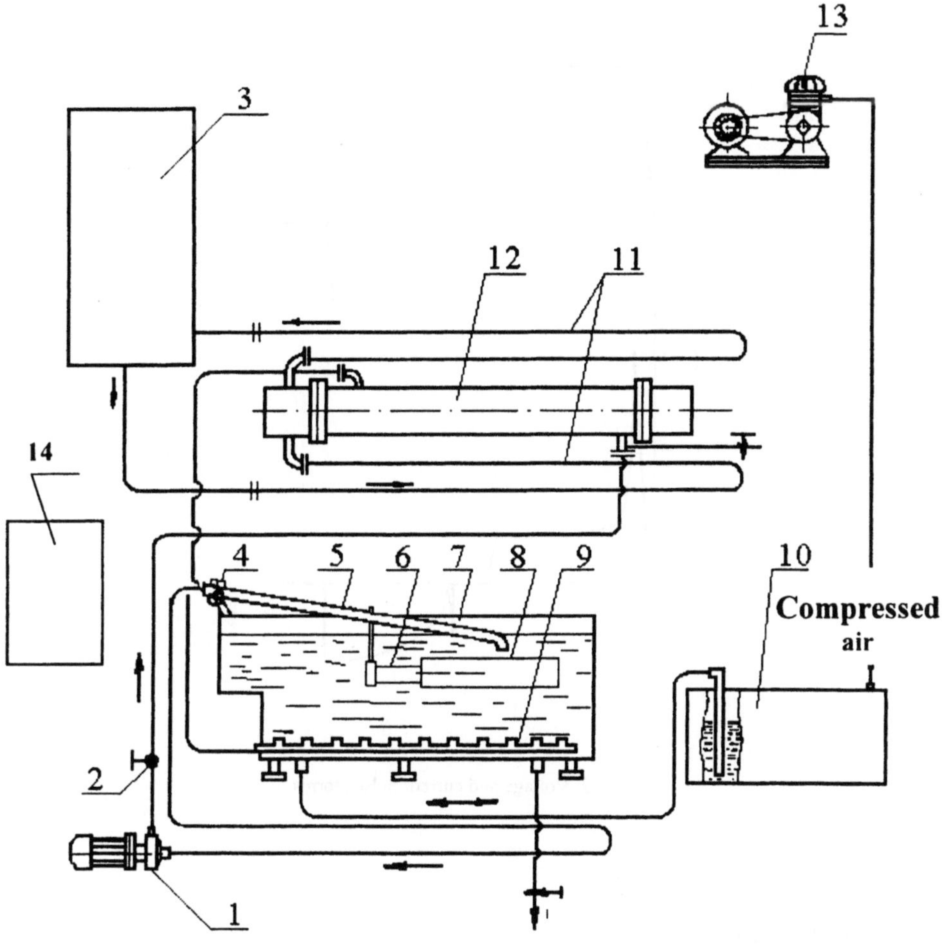

Figure 1. The scheme of installation for microarc oxidation: 1 - electrolyte pump, 2 - cock, 3 - refrigerating machine, 4 - trilling rotary, 5 - intake pipes, 6 - bracket, 7 - electrolytic bath, 8 - detail, 9 - an electrolyte feed tube, 10 - compensating tank, 11 - insulating pipes, 12 - heat exchanger, 13 – compressor; 14 – impact power supply.

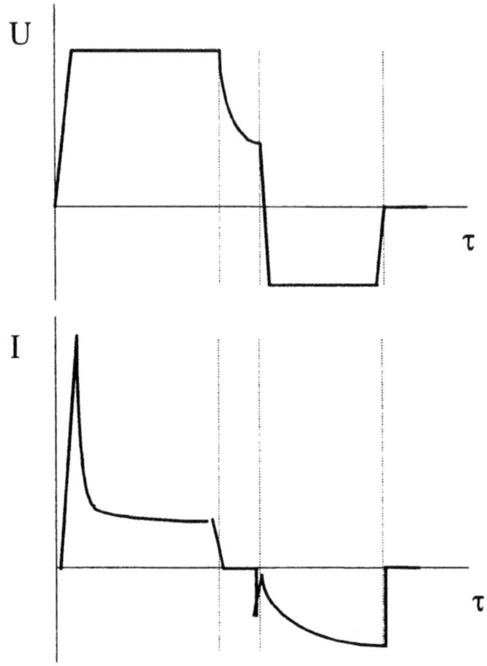

Figure 2. Voltage and current pulses forms.

3. Active Elements

Using various chemical additives in the electrolyte, we can form coatings with additional functionalities. Added into the electrolyte like a fine-dyspersated powder the piezoactive material is included into the coating structure. Using appropriate polarization we can strengthen the piezoelectric properties of a coating at preservation of its durability, its adhesion with a basis, its hardness and other properties.

The technological process of the active element creation consists of the following stages:

- required form element manufacturing from an aluminium (titanium, magnesium, etc.) alloy;
- MAO coating formation (using piezoelectric materials in the electrolyte);
- electrodes evaporation on the external side of a coating;
- coating polarization according to the functional requirements.

It is possible to produce the required form of an active layer on the product surface. Using appropriate masking and form of electrodes we can carry out active layer polarization depending on the functional requirements. That is the important feature of MAO technology. This technology allows creating of various active elements with a wide spectrum of functionalities.

The piezoactive element structure is shown in Figure 3.

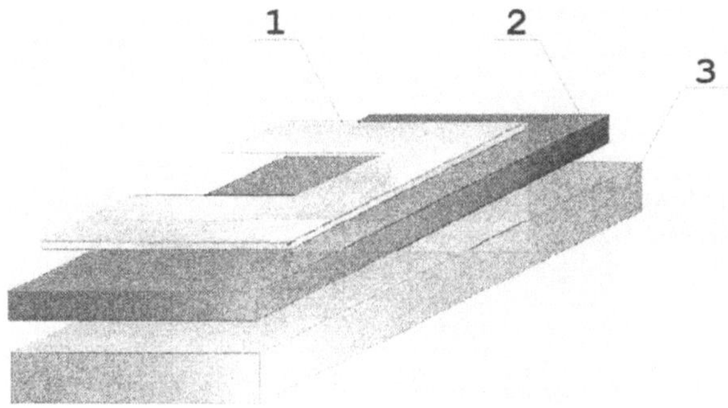

Figure 3. Piezoactive element structure: 1 – electrode; 2 – MAO coating; 3 – base, which is used as second electrode.

It is important to note the following. MAO technology allows us to form different coatings on the segments of processing surface as well as on the external and internal ones. To form special coatings it is necessary to use appropriate masking as well as different electrolyte compositions and electrolysis modes.

For example, it is possible to form an external protective coating which has a high hardness and thickness, and an internal active one which has piezoelectric properties. This three layers structure can be used as an active exterior casing. Using the special form of the second electrode (see Figure 4.) we can obtain information from the casing segments, and apply control pulses to particular places.

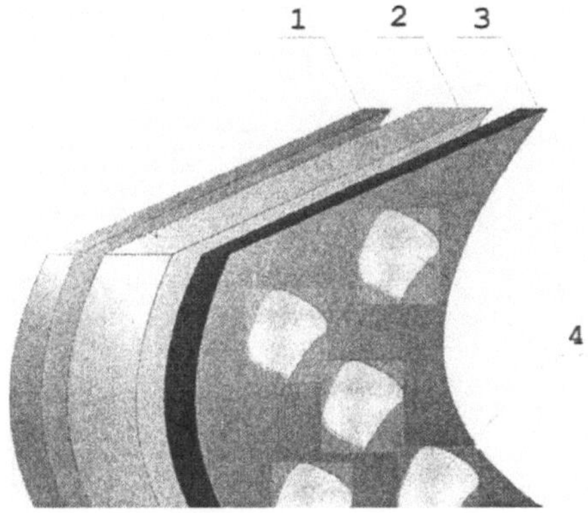

Figure 4. Exterior casing: 1 – protective layer; 2 – casing; 3 – active layer; 4 - second electrodes.

4. Conclusion

These researches are at the beginning stage and some of the hypotheses and ideas stated in this report must still be confirmed and tested. We continue our researches to estimate piezoelectric properties of the different MAO coatings, and to explore their technical applications.

5. References

1. Nicolaev A.V., Markov G.A., Pezchevitchkij B.I. *New phenomena in electrolysis J. Publ. by SO AN SSSR. Chemical Sciences (*1977) T.5, № 12, pp.32-34.

MODELING OF BENDING ACTUATORS BASED ON FUNCTIONALLY GRADIENT MATERIALS

T. HAUKE, A. Z. KOUVATOV, R. STEINHAUSEN, W. SEIFERT, H.T. LANGHAMMER, H. BEIGE
Martin-Luther-Universität Halle-Wittenberg, FB Physik, Friedemann-Bach-Platz 6, D-06108 Halle

1. Introduction

Piezoelectric bending actuators are frequently used when large deflections and low forces are required. Typically, two thin sheets with different piezoelectric coefficients are joined by a glue layer [1]. The glue layer may peel off or crack during high temperature changes. Additionally, during bending large mechanical stresses occur especially at the interface between the sheets. This may lead to a nucleation and propagation of microcracks, or a mechanical depolarization of the material. All these phenomena can reduce the lifetime and reliability of the actuators.

Recently several attempts have been made to overcome this problems by using Functionally Gradient Materials (FGM) with a one dimensional gradient of the piezelectric coefficients. Compared to conventional bimorphs, monolithic ceramics have lower production costs and avoid the problems connected with the glue layer [2-4].

For that purpose, many preparation procedures have been developed to produce ceramics with a gradient of the chemical composition [2-8]. Different mechanisms can be used to transform the gradient of the composition into a gradient of the piezoelectric coefficients during poling. For example, a gradient of the electrical conductivity can be realized by a gradient of dopants or a chemical reduction of one side of the sample [2, 4, 7]. Another idea is to use the dependence of the Curie temperature on the chemical composition as in the $Pb(Ni_{1/3}Nb_{2/3})O_3$-$PbZrO_3$-$PbTiO_3$-system [3, 4, 8].

Much attention has been paid to preparation, but little to modeling the bending behavior of FGM, i.e. the calculation of the main parameters characterizing the actuator performance like deflection and mechanical stress in the actuators. Haertling et al. used a Finite Element Method (FEM) for modeling the resonance modes and frequencies of RAINBOW actuators [9]. Xu and Meng calculated the deflection of a special, experimentally found shape of the gradient of the piezoelectric coefficient d_{31} [5, 10]. Wu et al. argued, that the mechanical stress in the actuators can be reduced by "smoothing" the jumps of d_{31} between the sheets [4]. However, it is still not known whether the use of

U. Gabbert and H.S. Tzou (eds.), IUTAM Symposium on Smart Structures and Structronic Systems, 87–94.

a material with a smooth gradient of the piezoelectric coefficients has special advantages or disadvantages compared to conventional bimorphs or unimorphs.
The aim of this paper is the modeling of the bending behavior of FGM based actuators and comparing the main parameters, deflection and mechanical stress inside the actuator, with conventional bimorph actuators. For that purpose, we develop an analytical solution and compare the results with data obtained from a FEM-model.

2. Basic Assumptions

2.1. STRUCTURE OF THE FGM BASED BENDING ACTUATOR

The model structure for the bending actuators consist of $N \geq 2$ layers of equal thickness $t_L = t/N$, which are numbered from 1 at the upper face to N at the lower face of he actuator. The coordinate system is chosen so, that x corresponds to the length L of the actuator, y corresponds to its width w and z corresponds to its thickness h (Fig. 1).

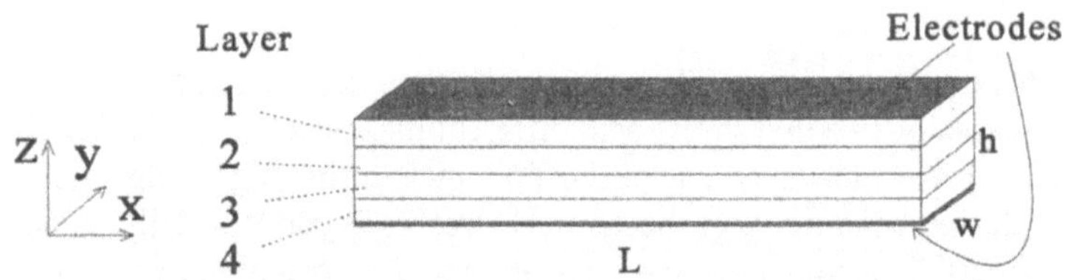

Figure 1 Example of a bending actuator consisting of 4 layers and the coordinate system used for the calculations

It is assumed that the upper and the lower layer are completely poled and therefore have maximum piezoelectric coefficients $+d_{mi}^{max}$ and $-d_{mi}^{max}$, respectively. The jumps of the piezoelectric coefficcients between neighboured layers were kept constant. Thus, the piezoelectric cofficients $d_{mi,k}$ of the layer i are

$$d_{mi,k} = d_{mi}^{max}\left(\frac{2(k-1)}{N-1} - 1\right) \quad . \tag{1}$$

This actuator structure allows us to compare directly the linear continuous gradient of d_{31} ($N \rightarrow \infty$) with a conventional bimorph ($N = 2$). As an example, Figure 2 shows the dependence of the piezoelectric coefficient d_{31} within actuators consisting of two and five layers, and for a continuous gradient of the piezoelectric coefficient, i.e. $N \rightarrow \infty$.
For practically used materials like PZT or $BaTiO_3$, the dielectric and elastic coefficients depend on the poling degree, too. However, this dependence is much less pronounced than that of the piezoelectric coefficients. Therefore, for simplicity, the dielectric and elastic coefficients were kept constant for all layers.

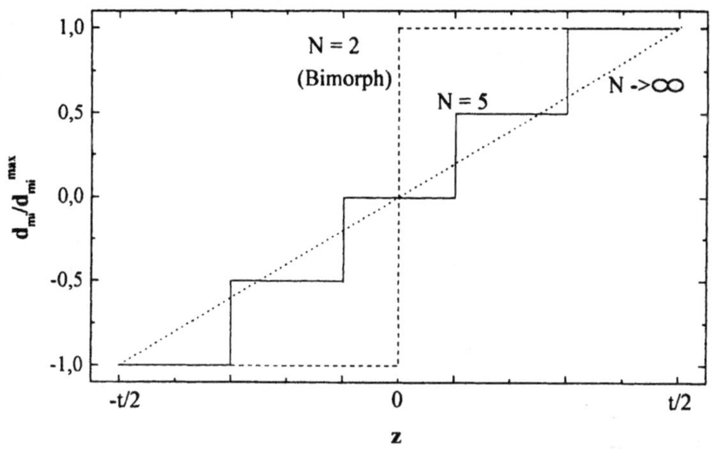

Figure 2 Variation of the piezoelectric coefficient d_{31} within actuators consisting of different numbers of layers

2. 2. ANALYTICAL MODEL

The analytical model is similar to that used by Marcus [11]. It is assumed, that the thickness of the actuator is much smaller than the width, which is much smaller than the length:

$$L \gg w \gg h \ . \tag{2}$$

When a voltage is applied to the electrodes, the actuator bends. Because d_{31} and d_{32} are nonvanishing , bending occurs in the directions of x and y. In the limit of eq. (2), the radii of curvature R_x and R_y are constant, i.e. independent of x, y and z. Using the symmmetry of the tensors of dielectric, elastic and piezoelectric properties, we see that $R_x=R_y=R$. It will be denoted R in the following. From the symmetry it follows also that the neutral plane is situated in the middle of the actuator with respect to the z-axis, and that the components of the mechanical stress T_1 and T_2 are equal. The origin of the coordinate system is now chosen so that the neutral plane corresponds to z = 0. Thus the upper surface has the coordinate z = h/2 and the lower face has the coordinate z = - h/2.
The transversal components of the strain S_1 and S_2 depend on R according to

$$S_1(z) = S_2(z) = \frac{z}{R} \ . \tag{3}$$

It is reasonable to assume that T_1 and T_2 are much higher than the other components of the mechanical stress, which are therefore neglected. Then the stress $T_1(z)$ is related to the components of the strain $S_1(z)$ and S_2 (z) and the applied electric field E_3, which is assumed to be constant inside the actuator:

$$T_1(z)=\frac{1}{Y_1}\left(S_1(z)-d_{31}(z)E_3\right) . \quad (4)$$

Here $Y_1 = 1/(s_{11}{}^E+s_{12}{}^E)$ is taken as an effective Young's modulus. When the external bending moment is zero, for each cross-section perpendicular to the x-axis the bending momentum has to vanish:

$$\int_{-\frac{h}{2}}^{\frac{h}{2}}\int_y zT_1(z)dydz = w\int_{-\frac{h}{2}}^{\frac{h}{2}} zT_1(z)dz = 0 . \quad (5)$$

Finally, the substitution of eqs. (3) and (4) into eq. (5) yields

$$\frac{1}{R}=\frac{\int_{-\frac{h}{2}}^{\frac{h}{2}} Y_1 d_{31}(z)E_3 z dz}{\int_{-\frac{h}{2}}^{\frac{h}{2}} Y_1 z^2 dz} . \quad (6)$$

When the radius of the curvature is large with respect to the beam length, the deflection δ at the end of the cantilever (position x = L) can be expressed by

$$\delta=\frac{L^2}{2R} . \quad (7)$$

With this formalism it is possible to calculate the bending behavior of the structure introduced in section 2.1.

2.3. FEM-MODELING

For the FEM-modeling the commercial package ANSYS 5.4 was used. The geometrical dimensions were chosen to be 14.8 mm × 4 mm × 1.32 mm for all investigated bending actuators. The model structures consist of up to 17,600 elements of the type SOLID5, which is a 3-dimensional, linear coupled-field element with 8 nodes. For the elastic, dielectric and piezoelectric coefficients a data set was used, which was reported for poled $BaTiO_3$ - ceramics [12]. For example d_{31} = -60 pm/V, $\varepsilon_{33}{}^T$ = 1950 ε_0, and $c_{11}{}^E$ =166 GPa were used.

A voltage of 100 V was applied to the actuators. The mechanical stress components were obtained from the element solutions of the FEM modeling.

3. Results

First, the dependence of the deflection δ on the number of layers N will be calculated. Introducing eqs. (1) and (6) in eq. (7) yields

$$\delta = \frac{L^2}{2}\frac{1}{R} = \frac{L^2}{2}\frac{E_3 Y_1 \sum_{i=1}^{N} d_{31,i} z\, dz}{Y_1 \int_{-\frac{h}{2}}^{\frac{h}{2}} z^2\, dz} \quad . \tag{8}$$

After evaluating the integrals and simplifying one obtains the expression

$$\delta(N) = \frac{E_3 L^2 d_{31}^{max}}{h}\left(\frac{N+1}{N}\right) \quad . \tag{9}$$

Figure 3 shows this dependence, and the FEM results.

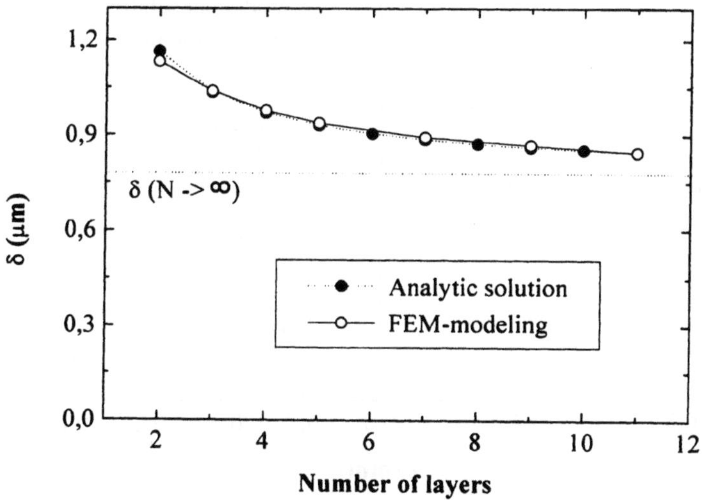

Figure 3 Dependence of the deflection δ on the number of layers for the analytical solution and the FEM modeling

At first glance, the results of the FEM modeling and the analytical solution are in a good agreement. However, the values obtained by the FEM modeling are somewhat smaller than the values obtained by the analytical solution. This may be caused by nonvanishing components of the mechanical stress T_3, T_4, T_5 and T_6, which were assumed to be zero in the analytical solution.

The deflection decreases with increasing number of layers, but also for a high number of layers it is finite. Equation (9) shows that the deflection of an actuator with continuously

changing piezoelectric coefficient ($N \to \infty$) is 2/3 of that for a bimorph ($N = 2$). The limiting value is almost reached even for a quite small number of layers.
Now the mechanical stress T_1 arising within the actuator will be calculated. Introducing eqs. (1) and (5) in eq. (10) yields the expression

$$T_1(N,z) = E_3 d_{31}^{max} Y_1 \left(\frac{2(N+1)}{hN} z - \left(\frac{2}{N-1} INT(N(\frac{z}{h}+\frac{1}{2})) - 1 \right) \right) \quad . \qquad \textbf{(10)}$$

In this expression INT denotes the integer part of the argument. Figure 4 shows the dependence of the mechanical stress T_1 on the coordinate z within actuators consisting with 2 and 4 layers. Again, in the analytical solution eq. (10) for L, h the same values as for the FEM modeling were used. It can be seen that the analytical solution and the FEM modeling are in a good agreement. However, near the interface between neighboured layers the stress calculated by the FEM-modeling is somewhat higher than that obtained by eq. (10). Again, the nonvanishing components T_3 - T_6 may be responsible for these differences.

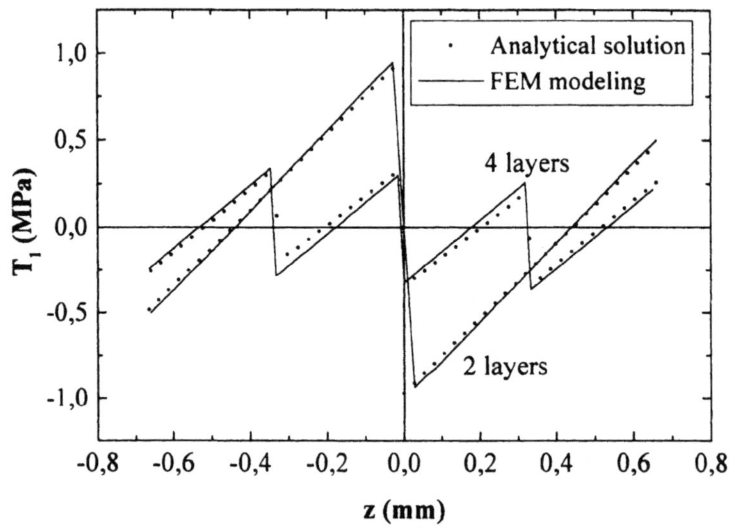

Figure 4 Mechanical stress T_1 within bending actuators consisting of 2 and 4 layers

Figure 4 and eq. (10) show that the maximum stress T_{1max} occurs at the interface between the first and the second layer or at the interface between the last and the last but one layer. Then one obtains from eq. (10) the expression

$$T_1^{max}(N) = E_3 Y_1 d_{31}^{max} \left(\frac{N+2}{N^2} \right) \quad . \qquad \textbf{(11)}$$

Like the deflection δ, the maximum stress decreases with increasing number of layers (Figure 5). However, this decrease is much more pronounced for the maximum stress than for the deflection. Moreover, even if the voltages applied to actuators with different numbers of layers are chosen so, that their deflections are equal, the maximum stress strongly decreases with increasing number of layers [13]. In particular, for a continuously changing piezoelectric coefficient, the maximum mechanical stress T_{1max} tends to zero. This means that such an actuator bends without internal mechanical stress.

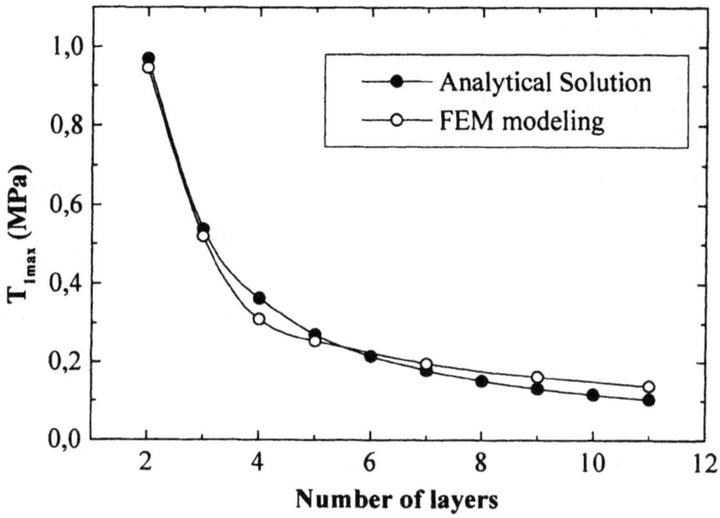

Figure 5 Dependence of the maximum stress within the actuator on the number of layers

4. Summary

The behavior of FGM based bending actuators was calculated by an analytical model and a FEM model. Model structures consisting of N layers with different piezoelectric coefficients.

The analytical calculations show that the deflection at the end of the cantilever and the maximum mechanical stress within the actuator decrease with increasing number of layers. For a continuous gradient of the piezoelectric coefficients, i.e. $N \to \infty$, the deflection remains finite. It will reach still 2/3 of the deflection of the bimorph (N = 2), whereas the maximum stress goes to zero. These results are in a good agreement with the data obtained from the FEM model.

Therefore, FGM seems to be a promising candidate for the development of bending actuators with reduced internal mechanical stresses and hence longer lifetime and greater reliability [4].

5. Acknowledgement

This work was supported by the Ministry of Science and Research of Sachsen-Anhalt.

6. References

1. Uchino, K., *Piezoelectric Actuators and Ultrasonic Motors*, Kluwer Academic Publishers, Boston/Dordrecht/London (1997)

2. Haertling, G.H., "Chemically reduced PLZT ceramics for ultra-high displacement actuators", Ferroelectrics, **154** 1-4 (1994), 101

3. Kawai, T., Miyazaki, S., Araragi, M., "A piezoelectric actuator using functionally gradient material.", Yokogawa Technical Report **14** (1992), 6

4. Wu, C.C.M., Kahn, M., Moy, W., "Piezoelectric ceramics with functional gradients: A new application in material design", Journal of the American Ceramic Society **79** 3 (1996), 809

5. Zhu, X., Wang, Q., Meng, Z., "A functionally gradient piezoelectric actuator prepared by powder metallurgical process in PNN-PZ-PT system", Journal of Material Science Letters **14** (1995), 516

7. Li, G., Furman, E., Haertling, G.H., "Fabrication and properties of PSZT antiferroelectric rainbow actuators", Ferroelectrics **188** (1996), 223

8. Kawai, T., Miyazaki, S., Araragi, M., "A new method for forming a piezoelectric FGM using a dual dispenser system", Proceedings of The First International Symposium of Functionally Gradient Materials, Sendai, Japan, (1990), 191

9. Furman, E., Li, G., Haertling, G.H., "An investigation of the resonance properties of rainbow devices", Ferroelectrics **160** (1994), 357

10. X. Zhu, Z. Meng, "Operational principle, fabrication and displacement characteristics of a functionally gradient piezoelectric ceramic actuator", Sensors and Actuators A, **A48** (1995), 169

11. Marcus, M..A., "Performance characteristics of piezoelectric polymer flexure mode devices", Ferroelectrics **57** (1984) 203

12. Landolt-Börnstein, *Numerical data and functional relationships in science and technology*, Vol. 16, 1981, Springer Verlag Berlin Heidelberg New York

13. Kouvatov, A., Steinhausen, R., Seifert, W., Hauke, T., Langhammer, H.T., Beige, H.., Abicht, H.-P., "Comparison between bimorphic and polymorphic bending devices" Journal of the European Ceramic Society **19** (1999), 1153

FABRICATION OF SMART ACTUATORS BASED ON COMPOSITE MATERIALS

H. Asanuma
Dept. of Electronics & Mechanical Engineering, Chiba University
Yayoicho 1-33, Inage-ku, Chiba-shi, Chiba, 263-8522 Japan

1. Introduction

Advanced material systems such as actuators and/or sensors embedded materials are becoming of worldwide interest because of their new material functions, e.g., actuation, noise reduction, vibration suppression, damage detection, self recovery or repair, and fabrication process monitoring. Fabrication of these new material systems used for smart structures has been realized by embedding sophisticated sensors and/or actuators in host structural materials such as polymer matrix composites.[1,2]

In this paper, a basic concept and elemental developments to realize a new actuator made of active composite material to be used for smart structures without using these sophisticated sensors and/or actuators are introduced as follows: (1) A piece of CFRP prepreg was laminated on a piece of metal plate to form a composite actuator as shown

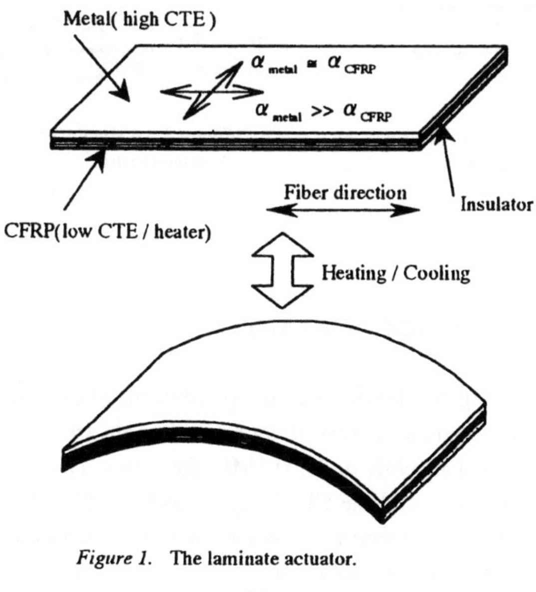

Figure 1. The laminate actuator.

U. Gabbert and H.S. Tzou (eds.), IUTAM Symposium on Smart Structures and Structronic Systems, 95–102.

in Figure 1.[3] (2) This concept was applied to metal matrix composites. As an example, a piece of SiC/Al composite plate was laminated on a piece of unreinforced aluminum plate to try to form a modified type of actuator.[4] (3) In order to develop sensors to be embedded in these actuators, a pre-notched optical fiber filament was embedded in matrix material and was fractured apart in it to try to form a fiber optic strain sensor as shown in Figure 2[5], and also an oxidized nickel wire (NiO/Ni fiber) was embedded in aluminum matrix to try to form a temperature and strain sensor as shown in Figure 3.[6]

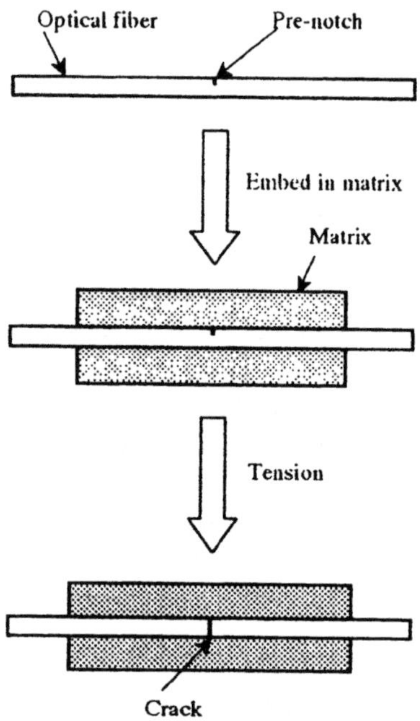

Figure 2. An idea to make an in-situ fiber optic strain sensor in matrix material.

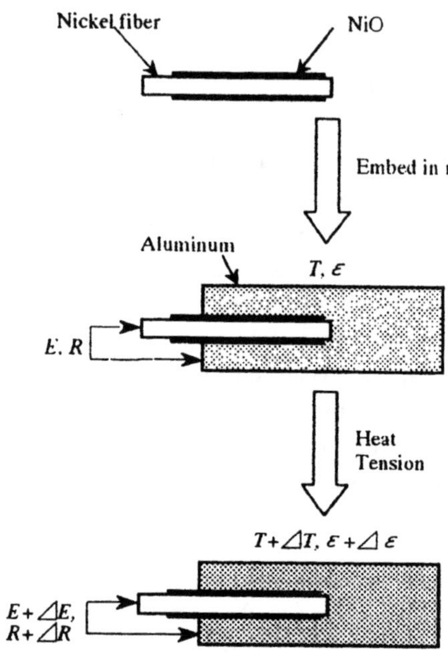

Figure 3. A simple temperature and strain sensor in aluminum matrix.

2. Experimental

2.1 FABRICATION OF LAMINATE ACTUATORS

A piece of CFRP prepreg was laminated on a piece of metal plate with a piece of KFRP prepreg as an insulator and two pieces of copper foil as electrodes as shown in Figure 4. This pile was hot pressed at 393K under 0.5MPa for 3.6ks. The electrodes were connected to a power source and the CFRP layer was heated by electric resistance heating. Displacement of the specimen was measured as a function of temperature and its curvature was calculated from the result.

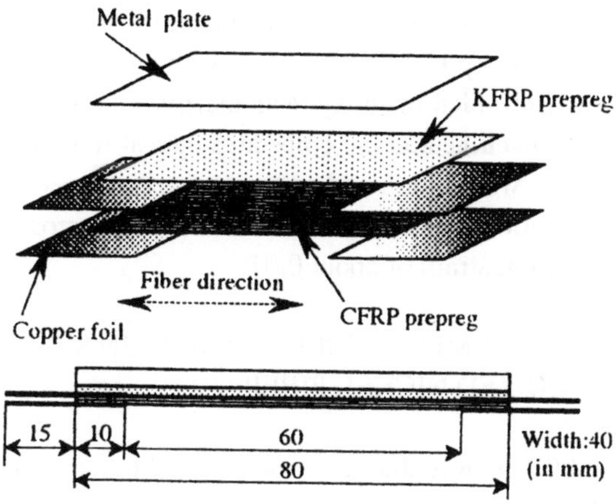

Figure 4. Lamination of materials.

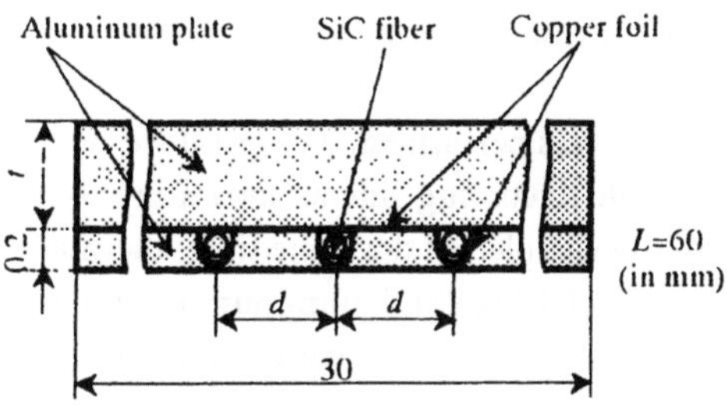

Figure 5. A cross section of the SiC/Al actuator.

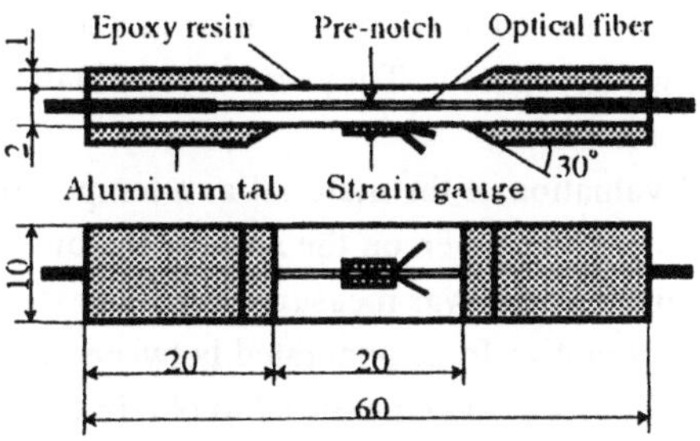

Figure 6. A test piece for measurements of tensile strain and optical transmission.

In order to make a modified type of actuator using a low CTE FRM instead of using polymer matrix composites, a piece of continuous SiC fiber/Al matrix composite plate was laminated on a piece of unreinforced pure aluminum plate as shown in Figure 5 by the interphase forming/bonding method[7-9] using copper insert foil under the conditions of 873K, 2.7MPa and 1.2ks. Effect of thickness of the aluminum plate t and inter-fiber spacing d shown in Figure 5 on curvature of the specimen r^{-1} at room temperature was investigated. The effect of temperature T on the curvature r^{-1} was also investigated.

2.2 FORMATION OF A FIBER OPTIC STRAIN SENSOR IN MATRIX MATERIAL

A commercially available quaurtz type and single-mode optical fiber was used. Epoxy resin was selected as the matrix material because it is widely used as composite matrix and because of its transparency which enables observation of the embedded optical fiber.

Shape and dimensions of the tensile test specimen used in this study are given in Figure 6. A notch was made on an optical fiber filament with an optical fiber cutter and the pre-notched fiber was embedded in epoxy resin matrix. The specimen was attached in an Instron type testing machine and the embedded optical fiber was connected to LD light source of 0.67μm wavelength and power meter. This specimen was tensile tested under monitoring the optical power variation at the constant crosshead speed of 1.7 μm /s up to the maximum strain of about 0.01.

2.3 FORMATION OF A TEMPERATURE AND STRAIN SENSOR IN ALUMINUM MATRIX USING OXIDIZED NICKEL FIBER

A pure nickel wire of 0.15mm in diameter was selected to form a thermocouple with aluminum matrix. Pieces of pure aluminum plates annealed at 623K for 1.8ks were used as matrix, of which sizes were 0.2 to 1.0mm in thickness, 14mm in width and 60mm in length. The nickel wire was oxidized at 1073K for 7.2ks in air to form a uniform NiO layer to electrically insulate it from aluminum matrix. Configuration of the materials is shown in Figure 7. They were consolidated by hot pressing at 798K, under 16.4MPa and for 1.8ks.

Evaluation of the material as a temperature sensor was performed. Temperature gradient was given on the specimen from heating side to cooling side as shown in Figure 7 and it was measured with an external thermocouple. At the same time, thermal electromotive force generated between embedded NiO/Ni fiber and aluminum matrix at x=15mm was also measured to obtain the temperature. These values were compared with each other at this position. Evaluation of the material as a strain sensor was also performed by measuring electrical resistance change of the embedded NiO/Ni fiber during tensile test. Aluminum tabs were put on both ends of the specimen to adjust its gage length as 20mm. Tensile test was carried out by an Instron type tensile test

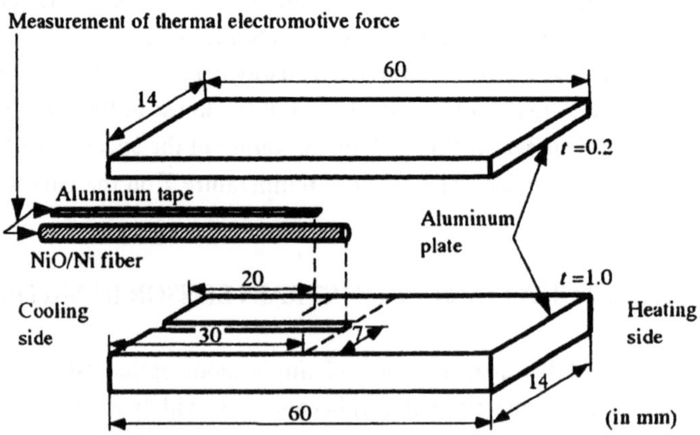

Figure 7. Fabrication of the NiO/Ni fiber embedded aluminum.

machine under the constant crosshead speed of 2μm/s. Strain of the specimen was measured by using a strain gauge. During the tensile test of the composite, electrical resistance change of the embedded NiO/Ni fiber was monitored.

3. Results and Discussion

3.1 ACTUATION OF THE LAMINATES

In the case of the CFRP/KFRP/metal actuators, aluminum and titanium were selected for the metal plate because they are candidates for the skin material of fiber/metal laminates. The laminate actuators are flat when they are kept at the hot pressing temperature and their curvatures increase when they are cooled as shown in Figure 8. Mechanism of the actuation is fundamentally the same as that of bimetal, but its major advantage is that it is not isotropic but strongly anisotropic due to directionality of the reinforcement fibers and their anisotropy of CTE, which enables the directional actuation. Curvature change of the actuator using aluminum is much larger than that using titanium in any temperature range of this experiment.

In the case of the SiC/Al actuator, it was curved by cooling from the hot pressing temperature. Its curvature at room temperature r^{-1} could be increased up to 12.9 m^{-1} by investigating the effect of thickness of the aluminum plate t and the inter-fiber spacing d. When it was heated, its curvature r^{-1} decreased with increasing temperature T and became zero at about 580K as shown in Figure 9. This temperature and the curvature at room temperature after thermal cycles were reproducible and were the same even after ten cycles, so availability of this material as an actuator was ascertained.

3.2 PERFORMANCE OF THE FIBER OPTIC STRAIN SENSOR

Optical transmission loss and trensile strain of the test piece during tensile test as a function of time are given in Figure 10. The optical loss clearly starts to fluctuate at the strain of about 0.66%, where the embedded optical fiber fractured at the position of the notch. This part can work as a strain sensor. So, this simple method is thought to be able to form a strain sensor instead of embedding an expensive and delicate sensor by complicated processes.

3.3 PERFORMANCE OF THE TEMPERATURE AND STRAIN SENSOR

The results of temperature measurements of the specimen are summarized in Figure 11. The curve in this figure indicates the temperature gradient given on the specimen, which shows that the temperatures of specimen surface at the position of the embedded NiO/Ni fiber sensor, that is, at x=30mm is 370K as indicated by the arrow. On the other

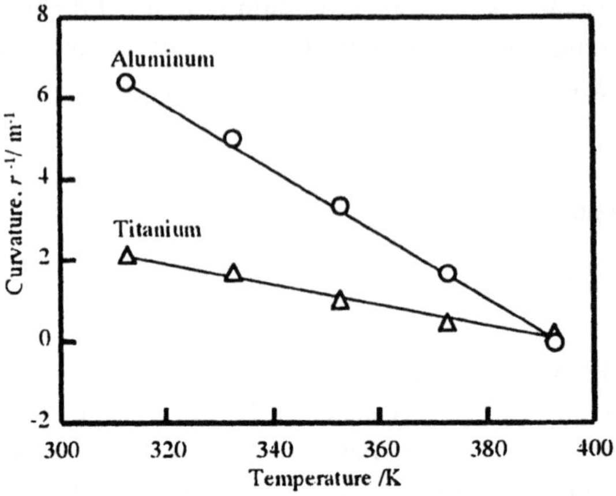

Figure 8. Curvatures of the specimens using aluminum plate and titanium plate as a function of temperature.

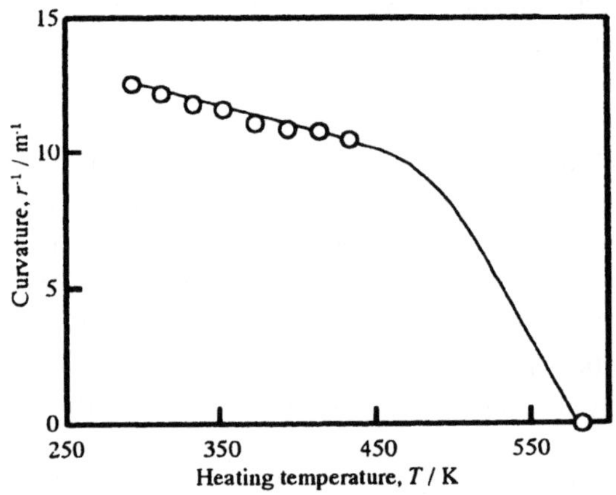

Figure 9. Variation of curvature of the SiC/Al actuator during heating.

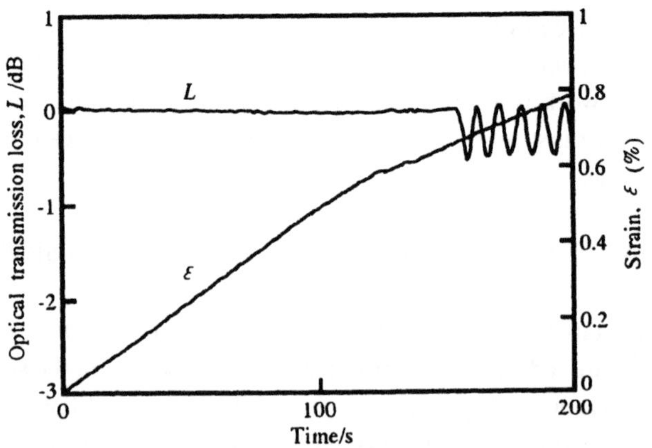

Figure 10. Optical transmission loss and strain of the test piece during tensile test as a function of time.

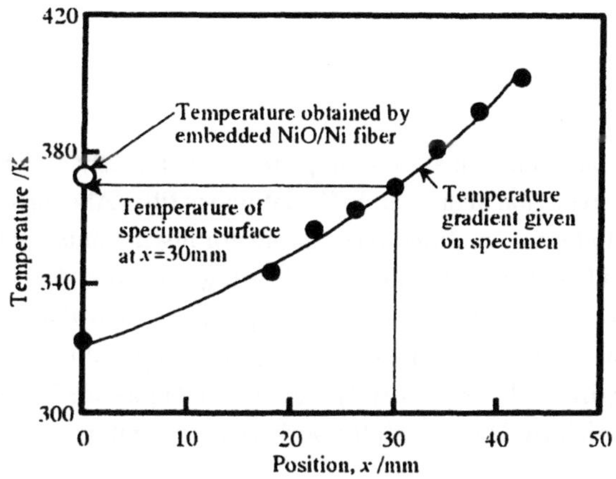

Figure 11. Comparison of the temperatures of the sample measured with the embedded NiO/Ni fiber sensor and an external thermocouple.

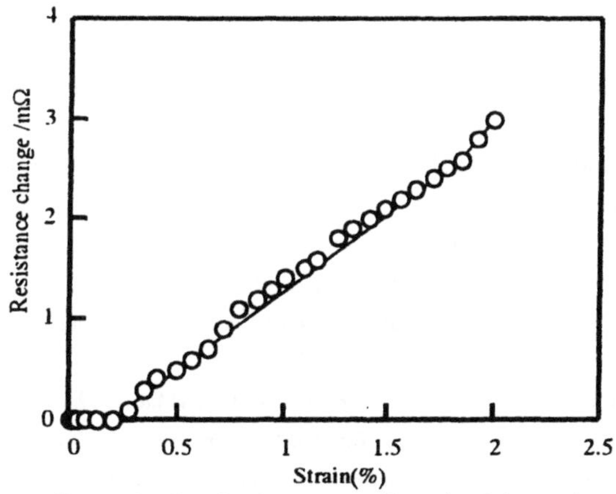

Figure 12. Relation between tensile strain of the specimen and resistance change of the embedded NiO/Ni fiber.

hand, the temperature obtained from measurement of thermal electromotive force generated between the embedded NiO/Ni fiber and the aluminum matrix at x=30mm is 373K which is given by the open circle in the same figure. These values coincide well with each other, which means that the embedded NiO/Ni fiber is working as a temperature sensor.

The relation between tensile strain of the composite and electrical resistance change of the embedded NiO/Ni fiber was obtained by tensile test as shown in Figure 12. According to this figure, the electrical resistance linearly increases with increasing the tensile strain up to around 1.8%. So, it is clear that the NiO/Ni fiber is working as a strain sensor in the aluminum matrix. The steep increase of the electrical resistance from the strain of 1.8% to 2.0% was caused by debonding of the NiO/Ni fiber from the aluminum matrix.

4. Conclusions

CFRP prepreg was laminated on aluminum plate to develop an actuator. This laminate could perform unidirectional actuation. SiC continuous fiber/Al composite thin plate could also be used to form a modified type of actuator instead of using CFRP. As sensors to be embedded in this actuator, the following ones were developed. (1) A pre-notched optical fiber filament was embedded in epoxy matrix without fracture and fractured in it at the notch, which enabled forming of an optical interference type strain sensor. (2) Nickel wire could be uniformly oxidized and embedded in aluminum matrix without fracture, which could successfully work as a temperature sensor and a strain sensor.

References

1. M. V. Gandhi, and B. S. Thompson: *Smart Materials and Structures*. Chapman and Hall, New Yowk. 1992.
2. T. Fukuda, et al.: Smart Composites Ⅰ-Ⅵ, *J. Japan Society for Composite Materials* 22 (1996), pp. 85-90. – 23 (1997), pp. 161-166.
3. H. Asanuma et al.: Development of an Actuator Utilizing Thermal Deformation of Ply Composites, *Proc. Japan Society for Composite Materials* (1996), pp. 19-20.
4. H. Asanuma et al.: Fabrication of Laminated Composite Actuators Utilizing Their Thermal Deformation, *Proc. 1999 JSME Annual Meeting* Ⅰ (1999), pp. 375-376.
5. H. Asanuma and H. Kurihara: Embedment of Pre-Notched Optical Fiber in Matrix Materials for In-Situ Formation of Strain Sensor, *Proc. 6th Japan Intl. SAMPE Symposium* (1999), pp. 1217-1220.
6. H. Asanuma et al., Fabrication of Aluminum Based Composites with a Function of Self-Temperature Monitoring, *94th Conf. of Japan Inst. Light Metals* (1998), pp. 281-282.
7. H. Asanuma, et al.: Lower Temperature Fabrication of Fiber Reinforced Metal Using Insert and Its Secondary Forming, *1st Japan Intl. SAMPE Sympo.* (1989), pp. 979-984.
8. H. Asanuma, et al.: Health Monitoring of a Continuous Fiber Reinforced Aluminum Composite with Embedded Optical Fiber, *J. Intelligent Material Systems and Structures* 7 (1996), pp. 307-311.
9. H. Asanuma and H. Du.: Monitoring of Optical Transmission Loss through Optical Fiber Embedded in SiC Fiber Reinforced Aluminum Composite during Tensile Test, *4th European Conference on Smart Materials and Structures and 2nd MIMR Conference* (1998), pp. 629-634.

ON THE ANALYTICAL AND NUMERICAL MODELLING OF PIEZOELECTRIC FIBRE COMPOSITES

M. SESTER , Ch. POIZAT*
Fraunhofer Institut für Werkstoffmechanik (now at SENSTRONIC, F), Wöhlerstraße 11-13, D-79108 Freiburg, E-mail: se@iwm.fhg.de*

1. Introduction

Piezoelectric composites with relatively high electro-mechanical coupling have been developed in many forms [1], including piezoelectric fibres [2] or particles [3] embedded in a non-piezoelectric polymer matrix. Such ceramic / polymer composites are often a better technological solution in a lot of applications such as ultrasonic imaging, sensors, actuators and damping [4, 5, 6, 7]. The main drawback in comparison to bulk piezoelectric materials consists in the complexity in design and analysis. This is essentially due to the coupled electrostatic behaviour, the highly different phase properties and the relatively complex geometry of the materials with its electrodes. Nevertheless, an increasingly amount of modelling works has been directed towards the study of these new smart materials. It should help in designing optimally the composite for each type of application.

The aim of this article is to provide a critical overview on several methods dedicated to the analysis of piezoelectric composites.

After some definitions and notation conventions about piezoelectricity in general and piezoelectric composites in particular, a brief review of the main homogenisation methods is given. Two types of homogenisation techniques are distinguished: the aperiodic and periodic models. Then, we discuss the advantages and drawbacks of both approaches. We finally focus on the topic of the electrode structure optimisation.

2. Piezoelectricity and Piezoelectric Composites

Piezoelectric composites, shortly piezocomposites, can be modelled by a homogeneous medium provided that the involved vibration wave lengths are greater than the characteristic length of the composite regarded as inhomogeneous. Under this assumption, the set of constitutive equations has the same form as the one of bulk piezoelectric material, local fields and local properties being simply replaced by spatial average fields and effective properties. As in the case of bulk piezoelectric materials, the coupling between mechanical and electrical fields is characterised by piezoelectric coefficients (denoted d, e, g and h). In a compact matrix form, in terms of the stress

U. Gabbert and H.S. Tzou (eds.), IUTAM Symposium on Smart Structures and Structronic Systems, 103–112.

coefficients $e_{ij,}$ the constitutive equations of linear piezoelectricity [8] for piezocomposites are:

$$\begin{Bmatrix} <T> \\ <D> \end{Bmatrix} = \begin{bmatrix} c^{E\,eff} & -e^{eff}{}_t \\ e^{eff} & \varepsilon^{S\,eff} \end{bmatrix} \begin{Bmatrix} <S> \\ <E> \end{Bmatrix} \tag{1}$$

or, in terms of the charge coefficients d_{ij}:

$$\begin{Bmatrix} <S> \\ <D> \end{Bmatrix} = \begin{bmatrix} s^{E\,eff} & d^{eff}{}_t \\ d^{eff} & \varepsilon^{T\,eff} \end{bmatrix} \begin{Bmatrix} <T> \\ <E> \end{Bmatrix} \tag{2}$$

Spatial average values are denoted by brackets <.> and effective material properties are denoted by the superscript ^{eff}. The third direction is associated with the direction of poling. The subscript $(\ )_t$ denotes the conventional matrix transpose. D and E respectively denote the electric displacement and the electric potential gradient. T and S are the stress and strain. $[c^E]$ (respectively $[s^E]$) denotes the stiffness (resp. compliance) matrix under short circuit conditions. $[\varepsilon^S]$ and $[\varepsilon^T]$ are the dielectric matrixes under constant strain or stress.

The effective properties depend on a high number of material parameters and geometrical parameters such as connectivity [1], or distribution function of the inclusions [9, 10, 11, 12].

3. Homogenisation Techniques

As underlined in the introduction, aperiodic as well as periodic methods allow the prediction of the elastic, dielectric and piezoelectric coefficients for a given composite. All approaches have their advantages and drawbacks and especially distinguish each other through their assumptions about the fibre distribution (random or periodic) and the fibre interactions. Independently of the method, the resulting effective properties serve as input material data for the design of adaptive structures [13]. We briefly review both types of approach.

3.1. APERIODIC APPROACHES

Several aperiodic approaches exist and this research field remains opened. The common assumption of these methods consists in the aperiodic distribution of the inclusions. They differ in their way to model the interaction between fibres.

The methods are mostly an extension to piezoelectricity of existing models dedicated to elastic [11, 12, 14] or dielectric materials [15, 16]. They are termed dilute, self-consistent (or effective medium theory), differential, effective field theory. In the case of piezoelectric composites, the reader is referred to several recent articles [17, 18, 19]. In all cases, inclusions and matrix are perfectly bonded and, since all the aforementioned approaches rest on the elementary inclusion problem [20], the inclusions are ellipsoidal.

Dunn and Taya [17, 18] have especially generalised the Mori-Tanaka method, usually applied for thermomechanical analysis of composites. This method is generally believed to be one of the most powerful in the study of the overall behaviour of a composite containing a finite concentration of inclusions. One of the main assumptions deals with the inclusion distribution: the inclusions are aligned and their distribution is the same for the inclusions and their shape. This assumption is known under the following term: ellipsoidal symmetry [21, 22].

The analytical model recently developed by Levin is also suited for random microstructures. In the Levin approach, the fibre interactions are modelled by the effective field approximation (EFA). The well-known Mori-Tanaka approach appears as a special case of this effective field approximation, with aligned inclusions distributed in the same manner as their shape [21, 22]. Details on this effective field approach may be found in [19].

Dunn and Taya [17, 18] and Levin, Rakovskaya and Kreher [19] obtained very interesting results with the developed models. The influence of the fibre volume fraction, the connectivity and the fibre aspect ratio, e.g., are reported in the above references. Among all the results obtained by these authors, we have chosen to reproduce in Fig.1 and comment part of the Fig. 5 from the reference article from Dunn and Taya [17, Fig. 5 p.171]. Fig. 1 reports the results of three of the numerous aperiodic models. The composite is made up with spherical piezoelectric ceramic particles embedded in an epoxy matrix.

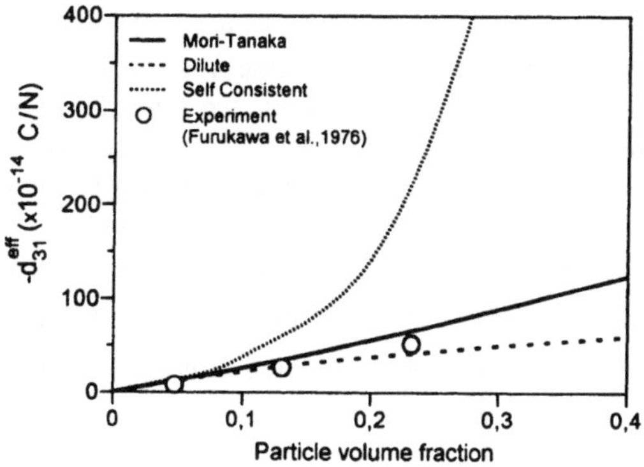

Figure 1. Influence of the fibre volume fraction on the effective transversal strain coefficient (0-3 composite made up with piezoceramic particles embedded in an epoxy matrix, see [17, p.171] for details)

In Fig. 1, the analytical results are quite different above a volume fraction of about 10%. The comparison with one of the rare experiments dealing with 0-3 composites [23] shows that the self consistent approach clearly overestimates the effective properties. The reason for this overestimation is the same as in the case of elastic composite: this approach assumes that an inclusion is surrounded with a medium whose properties are those of the (searched) effective properties. Hence, as said by Hashin, "The self-consistent scheme assumes that a tree sees the forest, but the tree sees only other trees". On the contrary, the dilute approximation is known to underestimate the effective properties when the volume fraction becomes higher and higher. This model does not take fibre interaction into account. Once more, as in the case of elastic composite, the Mori-Tanaka method, here extended to piezoelectricity, leads to a quite good agreement with the experimental results. This method is an effective field method. In this type of approach, the interaction between fibre is not reported onto the surrounded medium, but onto the electromechanical fields [14, 17, 18, 19]. This however does not mean that the effective field method is the best method in all cases. The assumptions given above have to be verified. In the case of other distributions, this method is no longer valuable as it will be shown later.

3.2. PERIODIC APPROACHES

Periodic approaches assume that the inclusion distribution is periodic, for instance with a simple cubic or body centered cubic lattice (Fig. 2). As a consequence, the representative volume element is reduced to unit cells, that are representative of the periodicity. They are mainly solved with the help of the Finite or Boundary Element Methods [9, 24, 25]. The details of the micromodelling method used by the authors, presented elsewhere [9], is not recalled here. Briefly, representative unit cells together with boundary and symmetry conditions are worked out with 3D piezoelectric elements of the FE code ABAQUS. It allows the prediction of the elastic, dielectric and piezoelectric coefficients as a function of several parameters like connectivity, volume fraction of ceramic fibres, fibre and matrix properties or fibre distribution as reported in [10]. We focuse here on the influence of the distribution. The later is also characterised by the distances between fibres. For a fixed volume fraction, the distance between fibres may be highly different. To illustrate this point, we consider a 0-3 composite: short piezoelectric fibres made up with piezoelectric ceramic are embedded in an epoxy matrix (see Appendix A). The corresponding unit cell is schematically represented in Fig. 3 for two lattices (simple cubic, sc, and body centered cubic, bcc). The fibre aspect ratio is fixed at a value of 10. In the FE analysis, the simple cubic (sc) and the body centred cubic (bcc) lattices are considered. In Fig. 4, e_{33}^{eff} is plotted versus the distance between fibres in direction 3 (fibre axes and poling direction, see Fig. 2).

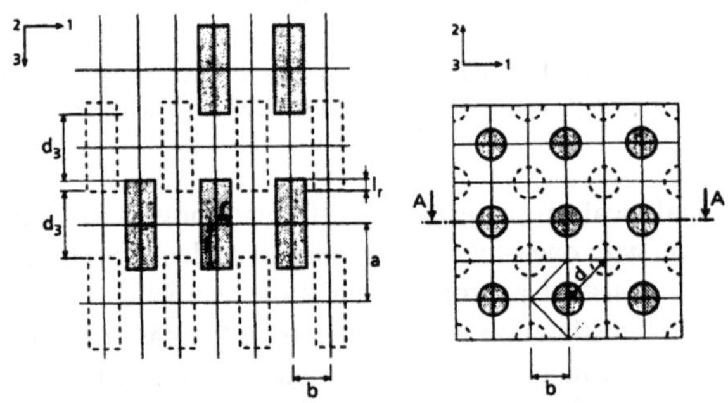

Figure 2. 0-3 composite with a body centred cubic distribution of cylindrical inclusions

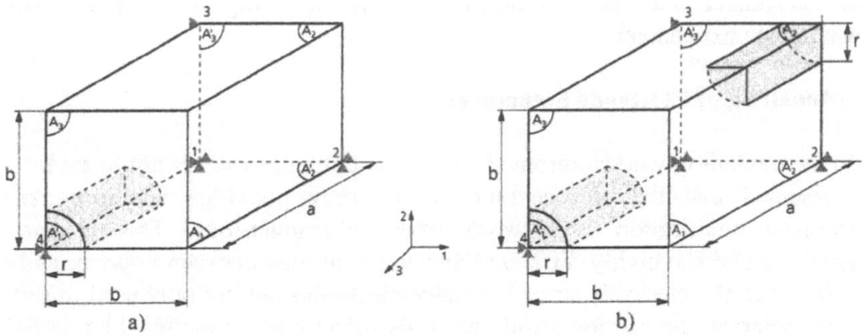

Figure 3. Unit cell of a 0-3 composite with a simple (a) and body centred cubic (b) distribution: geometry

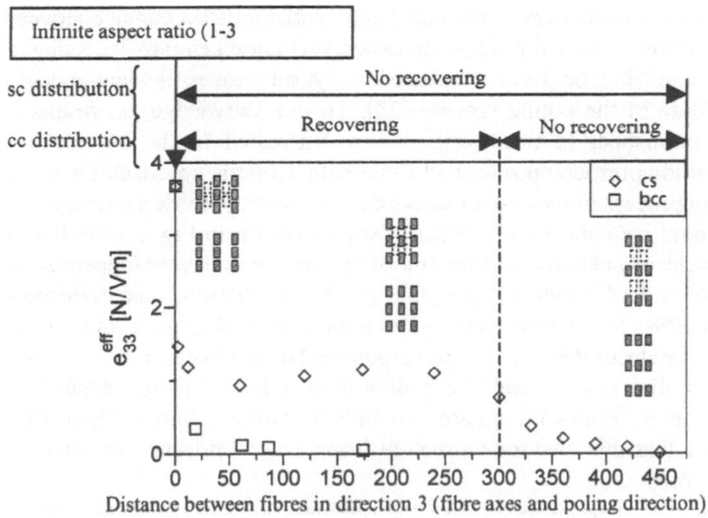

Figure 4. Influence of the fibre distribution on the effective stress coefficient e_{33}^{eff} (0-3 composite made up with PZT-7A cylindrical fibres in an epoxy matrix, see appendix)

With Fig. 4, it is clearly to be seen that not only the type of periodicity (sc or bcc) is a key parameter, but also that, once the type of periodicity is chosen, the distances between fibres is determinant. This result has consequence on the validity of the aperiodic models. As written above, they implicitly assume a ratio between the fibre aspect ratio and the distribution function of the fibre centres (notion of ellipsoidal symmetry, see above). The numerical result (Fig. 4) shows how carefully engineers should use homogenisation models, since real distributions can't be exactly characterised. Homogenisation models, whether periodic or aperiodic, give trends, but quantitative results should be considered carefully. The Finite element method was here helpful to carry out a parametric study. It is however time consuming. In comparison to aperiodic models, the advantages of numerical methods consist in the knowledge of the local mechanical and electrical fields or in the modelling of "material-structure" as shown in the next chapter.

4. Modelling of Electrode Structures

In the modelling works presented above, the poling process is not treated. Instead, it was assumed that the piezoceramic is uniformly poled. Apart from perfect 1-3 composites, polarisation is however often inhomogeneous. This is due to the connectivity and the highly different dielectric constants between polymer and ceramic [13] or due to the electrode structure. Since electrodes are not only used in service, but also to polarise the piezoelectric materials, electrode structures like interdigitated electrodes [26] lead to a non-uniform electric field distribution and consequently to a piezoelectric material with non-uniform polarisation. In [27], Kuna deals with an interdigitated electrode piezoelectric fibre composite, without direct contact between the electrodes and the fibres. The real poling orientation was not taken into account. On the contrary, Ghandi and Hagood have developed a hybrid Finite Element model that enables the modelling of the poling process [28]. To our knowledge, no results using this model were published. In this work, we are interested in the optimisation of interdigitated electrode piezocomposites and show with a simple approach the necessity to model the poling process in the case of active devices with complex geometry.

To reach this aim, we consider a 2-2 composite as shown in Fig. 5. The layers are alternatively piezoelectric ceramic and non-piezoelectric epoxy layers (appendix B). In the following analysis, L=1.2 mm, l=1 mm, $h_{i,i=1,6}$=0.1, h_7=0.01 mm. The corresponding volume fraction is 50%. Two types of electrodes are considered (Fig. 6a and 6b). In the configuration 6a, the electrodes $E_{i,i=1\text{-}6}$, are perpendicular to direction 3. The resulting poling direction is direction 3, and the polarisation state is homogeneous. In the configuration 6b, the electrodes $E'_{i,i=1\text{-}6}$ are perpendicular to direction 2. Finite Element analyses [28] show that this electrode structure leads a non-uniform field distribution during the poling process.

In the absence of a material model like the one described in [28], the polarisation state was approximated as shown in Fig. 7a and b for the electrode configuration 6a and b respectively. In case 6b especially, a local material orientation is attributed to each element under the electrode. Note that, due to the method, the mesh is relatively coarse, but fine enough for our purposes.

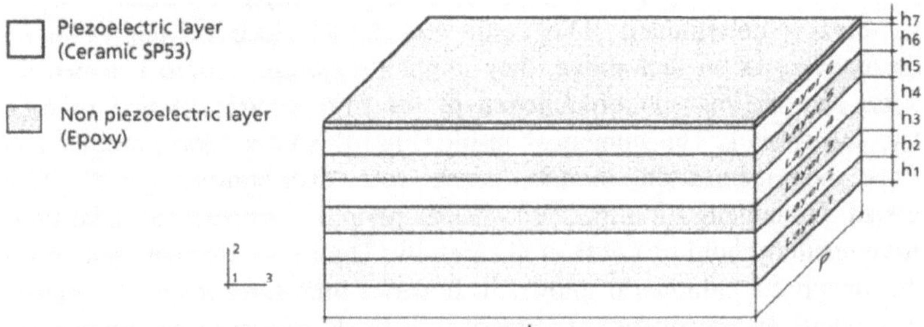

Figure 5. Geometry and structure of the piezoelectric composite

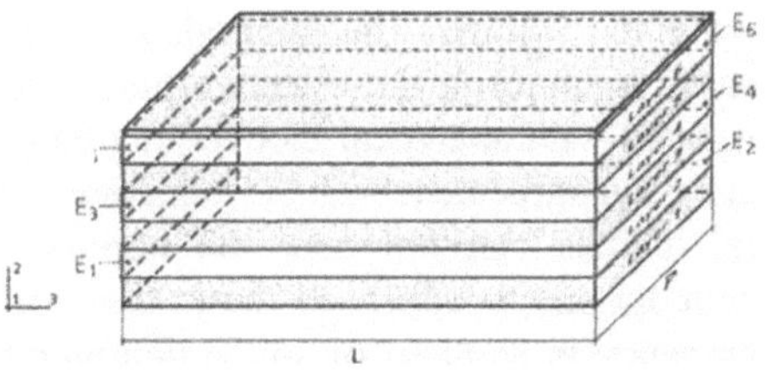

Fig. 6a: electrode perpendicular to direction 3

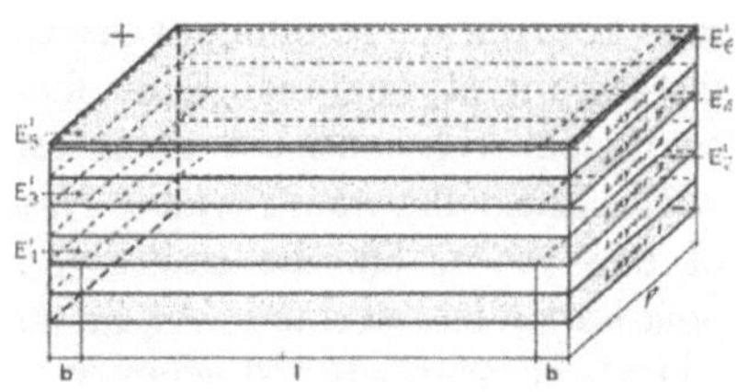

Fig. 6b: electrode perpendicular to direction 2

Figure 6: Electrode configuration 6a and 6b

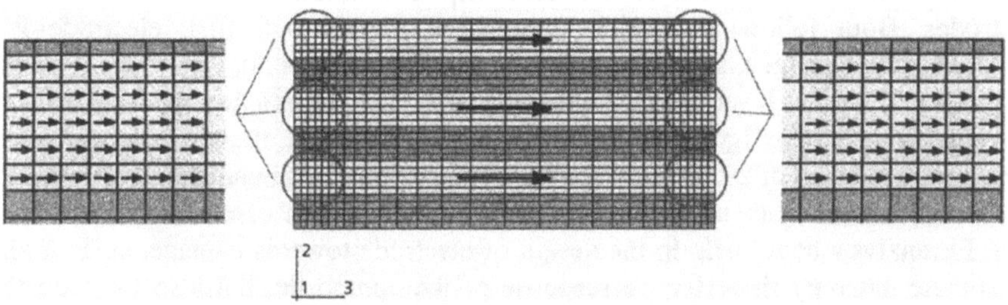

Figure 7a: FE mesh and poling direction for the electrode configuration 6a

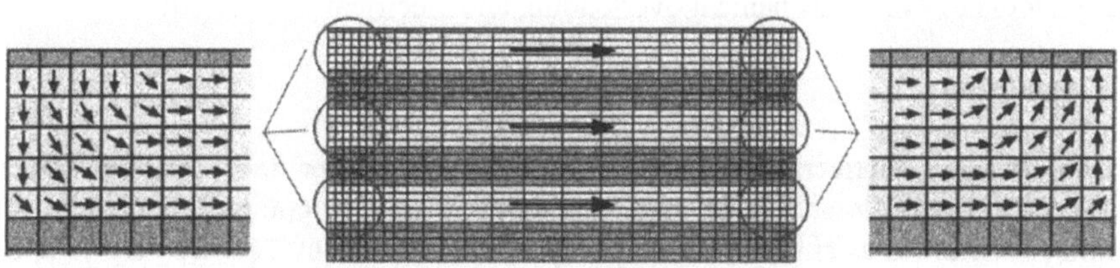

Fig. 7b : FE mesh and poling direction for the electrode configuration 6b

Figure 7. FE mesh and poling direction for the electrode configuration 6a and 6b
(The arrows represent the polarisation direction)

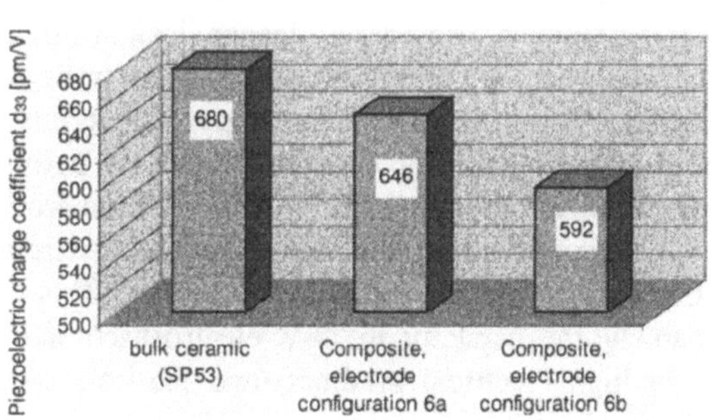

Figure 8. Piezoelectric effect with electrode configuration 6a and 6b.
The piezoelectric coefficient of the bulk ceramic is represented as reference (see Appendix B).

A generalised piezoelectric charge coefficient is plotted in Fig. 8. It is defined as the ratio between the resulting displacement for a given voltage (The electrode potential is the same in all analysis). Since the polarisation is uniform, the effective piezoelectric coefficient with configuration 5a can be estimated with analytical models. It's equal to 95% of the bulk ceramic value. This is in very good agreement with the numerical result of 646 pm/V. On the contrary, analytical models can not take the non-uniform polarisation into account. The FE analysis with our simple approach shows that, for the considered materials and geometry, the discrepancy in comparison to the configuration 6a is of 8.5%. This discrepancy depends furthermore of the ratio α=2b/L between the electrode brightness 2b and the mean distance between two electrodes in direction 3, L. For $0.1<\alpha<0.5$, we found that $2\%<e<10\%$ for the electrode configuration 6b and the polarisation state 7b. FE analysis can also help to estimate the influence of damaged electrodes. Both following analyses have been carried out: first, electrode E'_1 is supposed to be damaged, so that layer 2 is not poled, and second, electrodes E'_1 and E'_4 are damaged, so that layer 1 and 2 are not poled. The polarisation of the poled layers corresponds to the configuration 7b. As physically expected, the piezoelectric effect is respectively reduced of 29.5% and 62.6% in comparison to the undamaged structure: the two values are about equal to the unpoled volume fraction of ceramic. Last, but not least, FE analysis could help in the design of electrode towards damage, since it allows to estimate not only the effective response of the composite, but also to evaluate the stress and electric field level at each material point. Numerical methods are hence efficient tools to avoid mechanical overloading and electrical short circuit.

5. Conclusion

Due to their connectivity, their phase geometry and/or the electrode structure, piezoelectric materials should be regarded as more or less complex structures. If the resulting structure is as simple as a 1-3 composite, aperiodic models are very efficient in predicting the effective properties as a function of, e.g., the phase properties or volume fractions. In the case of 0-3 composites, the method depends highly on the distribution. The main difficulty is that the real distribution is never exactly characterised. Nevertheless, aperiodic methods based on the effective field approximation seem to be more efficient homogenisation techniques than periodic ones. For simple structures, the periodic homogenisation techniques provide a deeper insight in the influence of inclusion distribution. On the contrary, relatively complex electrode structures require the use of numerical analysis. As shown with a simple approach in this contribution, the development of a material model taking into account exactly the poling process should allow an optimisation of the electrode structure. To this end, the work of Ghandi and Hagood [28] could be a good basis for an implementation in commercially available FE code. The use of the FE program ATILA is another possibility. This code dedicated to coupled analysis has namely the capacity to take electrostriction (appearing at the electrode end due to very high electrical gradient) and the polarisation process into account. With such methods, the optimal design of electrodes and materials in order to reduce damages comes into sight.

APPENDIX A : Material properties PTZ-7A (Fibres) and epoxy (matrix). See [17, p. 169] for details.

	c_{11}^E GPa	c_{12}^E GPa	c_{13}^E GPa	c_{33}^E GPa	c_{44}^E GPa	e_{31} Cm^{-2}	e_{33} Cm^{-2}	e_{15} Cm^{-2}	ε_{11}^S $10^{-9} Fm^{-}$	ε_{33}^S $10^{-9} Fm^{-1}$
PZT-7A	148	76.2	74.2	131	25.4	-2.1	9.5	9.2	4.070	2.080
Epoxy	8	4.4	4.4	8	1.8	0	0	0	0.037	0.037

APPENDIX B : Material constants of SONOX®P53 ceramic [29, p. 11]

	s^E_{11} 10^{-12} m^2/N	s^E_{33} 10^{-12} m^2/N	c^D_{33} GPa	c^D_{55} GPa	k_p	k_{31}	k_{33}	k_t	k_{15}	d_{33} pC/N	d_{31} pC/N	d_{15} pC/N	ε^T_3 nF/m	ε^T_1 nF/m	ε^S_3 nF/m	ε^S_1 nF/m
SONOX®53	15.8	22.9	152	61	.65	.38	.74	.51	.73	680	-275	770	33.6	31.6	14.4	14.8

References

[1] Newnham, R.E. et al, Connectivity and piezoelectric-pyroelectric composites, Mat. Res. Bull., 1978, Vol. 13, pp. 525-536.

[2] Sporn, D. et al., Smart Structures by Integration of Piezoelectric Fibres – Present State and Future Applications, 9th CIMTEC (invited lecture), Florence, June 14-19 1998.

[3] Erba, F., Thiebaud, F., Perreux, D., Elaboration de piézo-composites à connectivité 0-3 par centrifugation, Revue des composites et des matériaux avancés, ed. Hermès, to be published.

[4] Challande, P., Optimising ultrasonic transducers based on piezoelectric composite using a finite element method, IEEE Trans. of Ultr., Fer. and Freq. Control, 1990, Vol. 37, 2, pp. 135-140.

[5] Gaudenzi, P., On the electromechanical response of active composite materials with piezoelectric inclusions, Computers and Structures, 1997, Vol. 65, n°2, pp. 157-168.

[6] Bent, A.A. et al., Anisotropic actuation with piezoelectric fiber composites, Journal of Intelligent Material Systems and Structures, 1990, Vol. 6, pp. 338-349.

[7] Lesieutre, G. A. et al., Damped structural composite material using resistively shunted piezoelectric ceramic short fibers, AIAA Journal, 1993, pp. 3238-3243.

[8] Ikeda, T., Fundamentals of piezoelectricity, Oxford University Press, 1996, 263 p.

[9] Poizat, Ch., Sester, M., Effective properties of composites with embedded piezoelectric fibres, Computational Materials Science, 1999, Vol. 16, n°1-4, pp. 89-97.

[10] Poizat, Ch, Modélisation numérique de matériaux et structures composites à fibres piézoélectriques, PhD, Université de Technologie de Troyes (UTT, F) march 2000.

[11] Mura, T. , Micromechanics of Defects in Solids, Martinos Nijhoff Publishers, 1987.

[12] Suquet, P., Continuum Micromechanics. CISM courses Nb. 377, Springer Verlag, 1997, 347 p.

[13] Sester, M., Poizat, Ch., Simulation Techniques for Piezoelectric Composite Materials and Their Applications to Smart Structures, SPIE's 7th International Symposium on Smart Structures and Materials, Session 16-A, Modeling and Control, 5-9 march 2000, Newport Beach, California, USA.

[14] Peyroux, R., Modélisation du comportement élastique de matériaux composites à fibres courtes, Revue des composites et matériaux avancés, 1992, Vol. 2, n°1, pp. 55-78.

[15] Neelakanta P. S., Handbook of electromagnetic materials, Boca Raton, CRC Press, 1995, 231 p.

[16] Sareni, B. et al., Effective dielectric constant of periodic composite materials, J. Appl. Phys., 1996,Vol. 80, n°3, pp. 1688-1696.

[17] Dunn, M. L.; Taya M., Micromechanics predictions of the effective electroelastic moduli of piezoelectric composites, Int. J. of Solids and Structures, 1993, Vol. 30, n°2, pp. 161-175.

[18] Dunn, M. L.; Taya, M., An analysis of piezoelectric composite materials containing ellipsoidal inhomogeneities, Proc. R. Soc. Lond., 1993, 443, pp. 265-287.

[19] Levin, V.M., Rakovskaya, M.I. et Kreher, W.S., The effective thermo-electroelastic properties of microinhomogeneous materials, Int. J. Solids Structures, 36, 1999, pp. 2683-2705.

[20] Eshelby, J.D., The determination of the elastic field of an ellipsoidal inclusion and related problems, 1957, Proc. Roy. Soc., A271, pp. 376-396.

[21] Willis, J.R., Bounds and self consistent estimates for the overall moduli of anisotropic composites, J. Mech. Phys. Solids, 1977, 25, pp. 185-202.

[22] Ponte Castañeda, P., Willis, J.R., The effect of spatial distribution on the effective behaviour of composite materials and cracked media. J. Mech. Phys. Solids, Vol. 43, 12, pp. 1919-1951, 1995.

[23] Furukawa, T. et al., Electromechanical properties in the composites of epoxy resin and PZT ceramics, Jap. J. Appl. Phys., 1976, Vol. 15, pp. 2119-2129.

[24] Pastor, J., Homogenisation of linear piezoelectric media, Mechanics Research Communication, 1997, Vol. 24, n°2, pp. 145-150.

[25] Bowen C.R. et al., Analytical and numerical modelling of 3-3 piezoelectric composites, Proc. of the ISIF'2000 conference, 12-15 march 2000, Aachen, Germany, to be published.

[26] Poizat Ch., Sester M., Thielicke B., Modelling of composites with embedded piezofibers, pp. 137-146. In Gabbert U., Modelling and Control of Adaptive Mechanical Structures, Fortsch.-Ber., VDI Reihe 11 Nr. 268. Düsseldorf: VDI Verlag 1998, Euromech 373 Colloquium, Magdeburg (D), 1998, 468 p.

[27] Kuna, M., et al.., Finite element modelling of adaptive composites with integrated piezoelectric fibres, "FE applications for adaptive structural systems", NAFEMS, Magdeburg (D), 1998.

[28] Ghandi, K., Hagood, N.W., A hybrid finite element model for phase transitions in nonlinear electro-mechanically coupled material, SPIE Paper n°3039-11, SPIE's 4th Annual Symposium on Smart Structures and Materials, San Diego, CA, 1997.

[29] CeramTec AG., « Piezoelektrische Bauteile », CeramTec AG, 1998, pp. 1-32.

ON SUPERELASTIC DEFORMATION OF NiTi SHAPE MEMORY ALLOY MICRO-TUBES AND WIRES — BAND NUCLEATION AND PROPAGATION

Q. P. SUN, Z. Q. LI, K. K. TSE
Department of Mechanical Engineering,
The Hong Kong University of Science and Technology,
Kowloon, Hong Kong, China

1. Introduction

Applications of shape memory alloys (SMAs) in the medical industry are multiplying rapidly. NiTi polycrystalline SMAs, due to their unique superelastic properties and biocompatibity, have been successfully used to manufacture medical devices in recent years. One of the most dramatically demonstrated examples is the utility of NiTi superelastic micro-tubes and wires in minimal access surgery or less invasive operations as guidewires, guidetubes and stents etc.. High flexibility, large recoverable deformation, good fatigue life and outstanding superelastic behavior at or around the body temperature are qualities that make it possible to reduce and to minimize the size of critical medical device and to perform functions impossible with other materials. It offers the best compromise between engineered plastics and traditional metals. High mechanical reliability (such as kink resistance) and controllability of deformation are the two critical requirements for very long and small diameter surgery instruments [1,2]. Though some fundamental studies on the bulk superelastic NiTi wire and strips have been performed in recent years [3], only a few conference reports on tests of NiTi tubes can be found in the open literature [2]. With the decrease in the dimension of medical instruments, size effect on the mechanical behavior of materials becomes another topic of great concern and has received wide attention in last decade. For materials undergoing martensite type phase transformation such as NiTi polycrystalline shape memory alloys, interesting size effect on the macroscopic martensite band nucleation stress of the wire under tension was observed in the authors' recent experiments, i.e., the smaller the diameter of the wire the higher the nucleation stress. The present paper serves as a stage report of the research progress on the above two aspects of SMA by the authors' research group.

The first part of the paper will focus on the recent experimental observation on the nucleation and propagation of martensite band in the NiTi SMA micro-tubes under tension. In the second part, we report the specimen size effect on the nucleation stress (or peak stress) of martensite band in NiTi wire specimen under tension. The results are discussed and the important issues regarding the underlying mechanisms and modelling for the observed phenomena are addressed finally.

U. Gabbert and H.S. Tzou (eds.), IUTAM Symposium on Smart Structures and Structronic Systems, 113–120.

2. Experimental Procedures and Results

2.1 MARTENSITE BAND IN NiTi SUPERELASTIC MICRO-TUBE

The commercial grade (Shape Memory Application, Inc.) polycrystalline NiTi (Ni-49%, Ti-51%) micro-tube with dark oxidized surface layer was used in the present test. The outer and inner diameters of the tube are 1.12mm and 0.9mm respectively. The phase transformation temperatures are M_s=27.94°C, M_f=5.92°C, A_s=2.63°C and A_f=26.25°C. TEM images of the grain size of the polycrystal is shown in Fig. 1 where a strong texture can be seen since the tubes are manufactured by drawing. The experiments were conducted at room temperature (23 °C) in air on a standard testing machine (Sintech 10D) with a 1KN load cell and under displacement control loading condition. The elongation rate is 1.9×10^{-5} mm/s. Figure 2 shows a typical measured nominal stress-strain curve of the micro-tube under tension. To observe the nucleation and the propagation process in micro-tube, a solution of ethanol-rosin is painted on the specimen surface so that a smooth and transparent layer of rosin (thickness about 20 μm) will remain on the surface after volatilization of ethanol. This rather brittle coating tend to form many micro-cracks when subjected to significant transformation strain and the color of the brittle layer will change from transparent to white due to the change of reflectivity by microcracks [4].

Fig. 3 shows schematically the formation of a martensite band in micro-tube at the beginning of A→M transformation. Careful surface observation revealed that it was of the spiral lens-shape. The middle line of the spiral inclined at about 61° to the axis of loading. The tube surface morphologies at different stages of loading (including nucleation and propagation of martensite band) under elongation rate of 1.9×10^{-5} mm/s are shown in Fig. 4. The corresponding positions on the stress-strain curve are marked in Fig. 2. The width of the middle portion of the martensite band was W_a= 0.76 mm with length L_a = 13.1mm at point *a* (immediately after martensite band was nucleated, see Fig. 4 (a)). The corresponding stress drop was 42MPa. After initial nucleation, the stress remained almost constant and the martensite band grew gradually along its length and width direction. At point *b* the maximum width and the length of the band increased to W_b=1.1mm and L_b=13.8mm (Fig. 4 (b)). There is a clear interface between martensite and austenite domains (here referred to as A-M interface). With increase of tube elongation, the middle portion of the band kept widening with the two A–M interfaces remaining parallel to each other. Finally the band started to merge at point d (Fig. 4 (d)). Also the length of the band reached its maximum (about 16mm). After point d, the length of the front (interface) decreased rapidly due to the mergence of the band and as the result the propagation speed of the front increased rapidly. Figure 4 (a)–(d) shows that the A-M interface in the micro-tube is very sharp at different stages of loading. This indicates that there is a very rapid strain change or high strain gradient across the A-M interface. Inside the martensite band, the resin (coating) cracks along the direction perpendicular to the loading axis (principal stress direction). Further study on the band nucleation and propagation in the tube under torsion and tension-torsion is under progress.

2.2 MARTENSITE BAND NUCLEATION IN NiTi WIRE — SIZE EFFECT

The NiTi polycrystal shape memory alloy wires were supplied by Shape Memory Applications, Inc, USA (Ni-54%, Ti-46% (wt %), diameter 1.35mm). The transition temperatures of the material are M_s = -40.95 °C, M_f = -62.05 °C, A_s = -16.02 °C, A_f = 6.34 °C. The testing temperature is 23°C so the initial state of the specimen is austenite (parent phase) and that the stress-induced martensite will be recovered after unloading. Fig. 5 shows the dimensions of the specimens. A 25mm gauge length extensometer was fixed on the edges of the narrow uniform region of the specimen. Chemical etching is used to prepare the samples with uniform middle regions of different diameters. In this experiment, wires with diameters of 1.22, 0.95, 0.86, 0.65, 0.53, 0.43, 0.15, and 0.08mm are obtained by consecutively etching a single piece of 1.35 mm diameter wire. Therefore all these wires have the same microstructure (grain size is shown in Fig. 6).

The typical nominal stress strain curve of the wire is shown in Fig. 7. It is linear at the beginning because of elastic deformation of austenite. When the martensite nucleated as a thin band somewhere in the middle portion of the specimen, the stress dropped from about 500MPa to 415MPa, and then maintained a constant plateau value for the propagation of the band. After the whole length of the specimen was occupied by martensite, the stress increased rapidly due to the elastic deformation of martensite. During the unloading process, there was no stress valley since the reverse transformation was realized by the reverse motion of the A-M interface. This is consistent with the observation in [3] where clear explanations have been provided. Fig. 8 is a schematic description of the surface morphology evolution during loading process corresponding to Fig. 7.

Wires with different dimensions were tested under the same condition. In all the test the plateau stresses in the stress strain curves are nearly the same (about 474MPa) and are independent of wire diameter. But the amount of stress drop (or nucleation stress) is strongly size dependent. Figure 9 shows the enlarged part of the stress strain curve near the peak stress region. It is seen that the stress drop increased with the decrease of the diameter. The relations between the stress drop and the corresponding wire diameter and cross section area are shown in Fig. 10. When the diameter changed from 0.08mm (area = 0.005mm^2) to 1.22mm (area = 1.169mm^2), the corresponding stress drop decreased from 104.6MPa to 59.4MPa. Further test and analysis work to understand and model this size effect is currently under way.

3. Discussion and Concluding Remarks

3.1 PATTERN FORMATION AND EVOLUTION DURING TRANSFORMATION

In this on going research project, it is observed that deformation of superelastic polycrystalline NiTi SMA microtubes and wires during the stress-induced transformation at T >A_f is inhomogeneous and takes the form of localized band nucleation and subsequent propagation (growth). It is again demonstrated that the nucleation of the new phase within a uniform region requires a more "severe" stress (nucleation stress) than the stress required to subsequently propagate the martensite band (propagation stress) [3]. Stress peak during nucleation of band can be observed, which is followed by a steady-state propagation stress plateau. Inside the martensite band, the deformation is uniform and can be characterized by "transformation strain" as measured using high resolution Moiré interference technique [5]. The martensite and austenite regions are separated by a narrow zone which is usually called transition front or macroscopic A-M interface where strain changes very rapidly. According to the nominal stress–strain curve of the circular wire, the amount of stress drop exhibits a strong size effect. The smaller the wire diameter, the higher the stress drops during nucleation of martensite band. Another interesting phenomenon is that before the nucleation of macroscopic martensite band (stress drop), there is a stable martensite transformation stage where macroscopic stress strain curve is monotonic and the strain is homogeneous in the specimen.

The above observation, together with the previous study [3], provides the following brief scenario to the superelastic deformation process of NiTi polycrystal: After initial elastic deformation of the crystal, the nucleation and growth process of martensite happens at a relatively low stress level in discrete grains (grain size 30-50 nm). At the beginning this nucleation and growth process is macroscopically stable, i.e., the macroscopic homogeneous transformation strain increases with increase in the applied stress in a controllable manner (as shown by the monotonic part of the stress strain curve before the peak stress). The volume fraction of martensite is very small and therefore there is very week interaction among the transformed martensite product. With the increase in stress and martensite volume fraction this elastic interaction increases. When the stress reaches the peak stress the autocatalytic or domino effect dominates the process and the former homogeneous transformation develops into a localized transformation in the form of a macroscopic band inside which the polycrystal is almost fully transformed into martensite. The subsequent deformation (transformation) is realized by the growth of the band via the A-M interface propagation.

3.2 AUTOCATALYTIC TRANSFORMATION AND MATERIAL SOFTENING INSTABILITY

From the mechanical point of view, the scientific understanding and modeling of the above phenomena in NiTi microtube and wire can be concluded as the following two continuum mechanics level issues: (1) How and why such strain discontinuity (or transformation localization) happens in an initially uniformly stressed and deformed body? (2) what are the intrinsic constitutive relations that capture the underlying physical mechanism? and (3) What is the overall response of the TiNi specimen as a structure containing such a propagating martensite band and its relation with the material's constitutive law? Since the real stress strain curve is very difficult to measure due to the localized deformation, the second issue sometimes becomes practically more important.

Generally, such kind of inhomogeneous deformation (localization and the Luders-like band propagation) is due to the intrinsic strain softening (during transformation) and rehardening (after exhaust of transformation) of constitutive relations of the material in a so-called up-down-up fashion. For SMAs, the physical origins of such softening are mainly the autocatalytic effects due to the strong elastic interaction of different stress-induced martensite products (variants) in the grain size or even smaller microscopic length scales. Tremendous effort has been made in continuum mechanics modelling of such strain softening, localization and propagation in metals and polymers. For solids capable of undergoing phase transitions such as SMAs, many researchers including the present authors have been actively involved in studying this unstable material behavior. However, quantitative and reliable solutions on the issues such as the peak stress and steady-state propagation stress during pattern formation as well as their control with the key material parameters are quite difficult and have not been acquired. For example, the simple elaboration of the strain softening response during the stress-induced transformation in the sense of the classical rate independent local continuum will lead to spurious mesh-sensitivity in finite element calculations. Therefore microstructure-based non-local gradient-dependent softening constitutive models will be explored and developed so that it not only has the ability to capture the localized deformation pattern formation but also can deal successfully with the mesh-sensitivity in the material softening regime.

3.3 SIZE EFFECT AND STRAIN GRADIENT

The test results of the present paper demonstrate that there is a strong specimen size effect on the nucleation stress of the martensite band. The real physical mechanisms leading to this size effect are not clear so far even though several classical models have potential to simulate this effect. In recent years, strain gradient theory of micro-plasticity has been developed to incorporating the size effect of plasticity at micron scale [6]. Like various instability phenomena in plasticity, pattern formation during phase transformation indeed involves high strain gradient at different length scales (for example, from the lattice variant at the nanometer and micrometer scales to the macroscopic band at millimeter scale). However, whether the non-local gradient theory developed so far can describe the "size effect" phenomena reported here remains to be an open issue which is, without doubt, a research topic of great interest for both mechanics of materials and intelligent microsystems such as MEMS communities.

Another important aspect of the material behavior is the effect of texture on the pattern-forming instability since the microtubes and wires are obtained by drawing during which strong texture develops in the sample especially for tubes. The resulting anisotropy in constitutive behavior and their effect on the deformation of the tubes (including band nucleation and propagation) must be taken into account in the future design.

In general, applications of shape memory alloys in the medical industry and smart systems have provided many research topics and there are many possible avenues to explore in mastering the materials behavior during instability. In such a development careful designed model test with compelling experimental evidence will play a critical role in understanding the mechanics and physics of the deformation process. The present paper is only part of a series experimental studies on the material instability in typical phase transforming shape memory alloys. Results from the uniaxial tension are very encouraging but more work is needed to studying the situation under complex stress state such as torsion and tension-torsion test, an investigation that is currently under way.

Acknowledgment
The work described in this paper was fully supported by a grant from the Research Grants Council of the Hong Kong Special Administrative Region of China (Project No. HKUST6074/00E) and the National Natural Science Foundation of China (Project no. 19825107).

References

1. Duerig, T. W.: Recent and future applications of shape memory and superelastic materials, *Mat. Res. Soc. Symp. Proc.*, Vol. 360, (1995), pp. 497-506. Material Research Society.
2. Pelton, A. et al., (eds.), *Proc. of Second Int. Conf. on Shape Memory and Superelastic Technologies* (SMST-97), California, 2-6 March, 1997.
3. Shaw, J.A. and Kyriakides (1995), J. Mech. Phys. Solids, 43, 1243-1281.
4. Li, Z. Q. and Q. P. Sun: Some deformation features of polycrystalline superelastic NiTi shape memory alloy thin strip and wire under tension, *Key Engineering Materials*, Vol. 177-180, (2000), pp. 455-460.
5. Tse, K. K. and Q. P. Sun: Superelastic NiTi memory alloy micro-tube under tension — Nucleation and propagation of martensite band, *Key Engineering Materials*, Vol. 177-180, (2000), pp. 561-166.
6. Hutchinson, J. W.: Plasticity at the micron scale, *Int. J. Solids and Structures*, Vol 37 (2000), pp. 225-238.

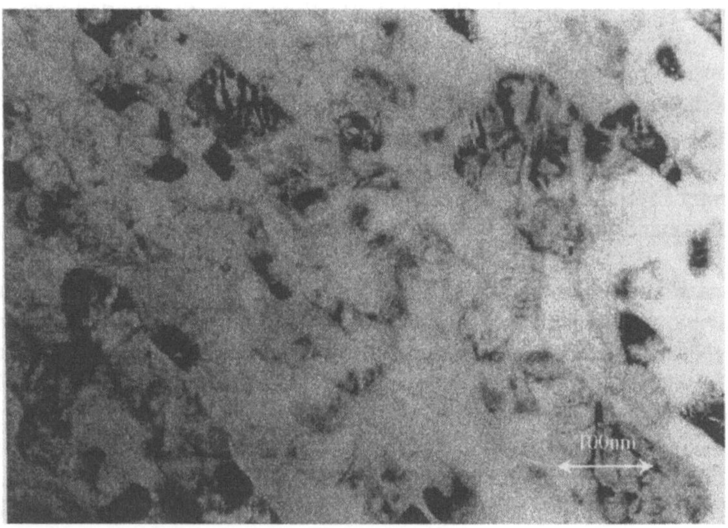

Fig. 1 TEM image of the grain morphology in cross sections along the rolling direction of microtube where a strong texture can be seen.

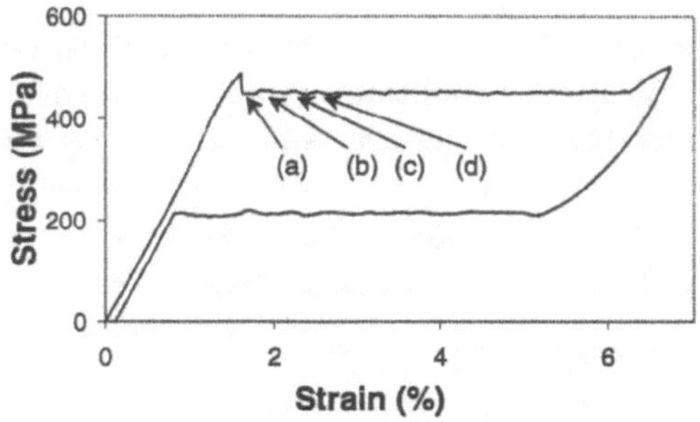

Fig. 2 Typical nominal stress strain curve of microtube under tension

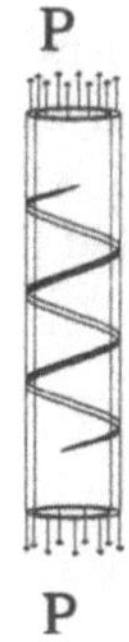

Fig. 3 Nucleation of martensite band in microtube (schematic)

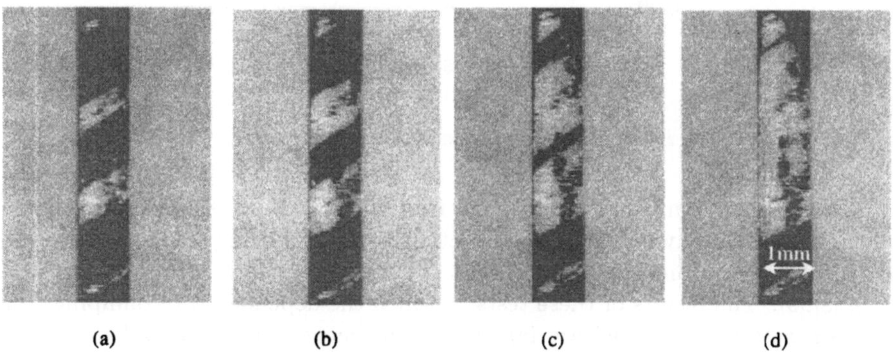

Fig. 4 Nucleation and growth of martensite band in micro-tube (photo): (a) - (d) corresponding to points (a) – (d) in Fig.2

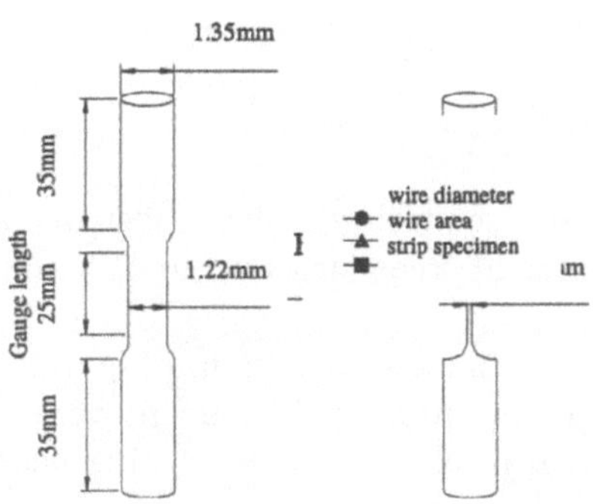

Fig. 5 Wire specimen of different diameter obtained by consecutive etching

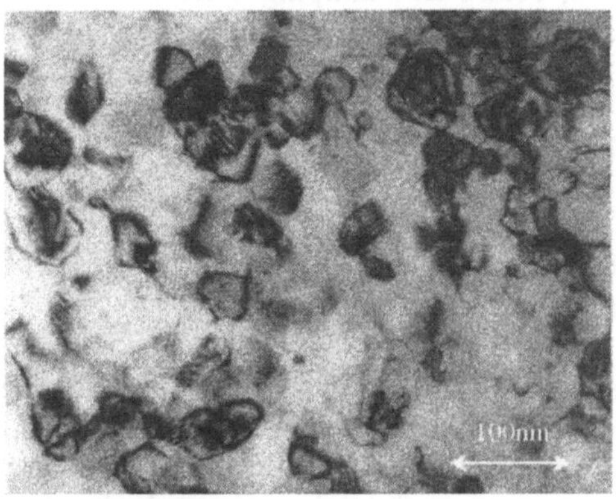

Fig. 6 TEM image of the grain morphology in polycrystalline wire

stress strain diagram of a NiTi wire. In every cycle the wire is loaded to a strain one percent less than in the cycle before. Up to a strain of 0.8% the material behaves linear elastic. Without any pre-stress, amplitudes of about 1% have to be applied in order to generate considerable damping effects in dynamic systems with integrated SMAs.

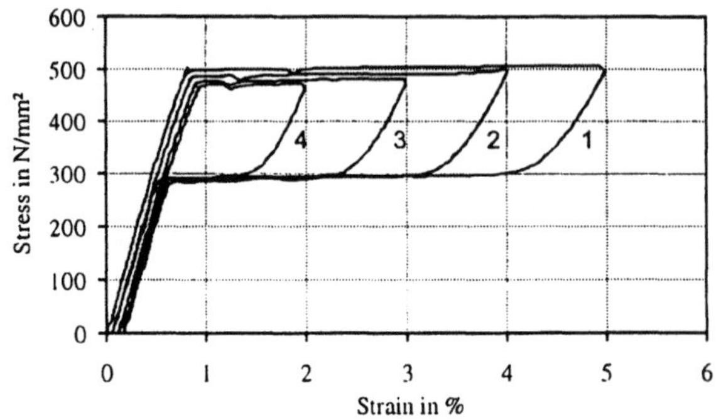

Figure 1. Internal loops of cycling NiTi wires

In the literature, most of the internal friction research of SMAs has been concentrated on the study of the internal friction peaks appearing during heating and cooling [3],[4],[5]. Moreover a few work is done in tension tests at low strain rates [6],[7], [8],[9]. The results of these tests in which the dependency of damping from the strain rate is considered are far away from being consistent. Whereas some work has been done in tension tests, only a few papers are concerned with the damping behavior of SMA components and their damping behavior [5],[10],[11].

In front of this background, the present paper is mainly concerned with basic components made of pseudoelastic SMAs and their damping behavior. Nevertheless, before the damping behavior of a NiTi spring is presented, dynamic tensile tests of binary NiTi specimens are covered within the next Section.

2. Cyclic Tensile Tests

The investigations on the dynamic behavior of pseudoelastic shape memory alloys started with tensile tests on NiTi specimens at frequencies up to 4Hz and a maximum strain of 3% [12].

For these tests, superelastic NiTi wires with a diameter of 0.9mm were used in strain-controlled cyclic tensile tests at room temperature. The specimens were put into heavy clamps of steel without any thermal insulation for generating nearly isothermal conditions. The distance between the clamps was about 26mm. The tests were conducted on a servohydraulic machine by subjecting the specimens to an initial static strain of 1.5% and then harmonically oscillating them around this offset level with a strain amplitude of 1.5%. During cycling the grips were moved

THE DAMPING CAPACITY OF SHAPE MEMORY ALLOYS AND ITS USE IN THE DEVELOPMENT OF SMART STRUCTURES

R. LAMMERING, I. SCHMIDT
Institute of Mechanics, University of the Federal Armed Forces Hamburg, Holstenhofweg 85, 22043 Hamburg, Germany

1. Introduction

The mechanical behavior of shape memory alloys shows various properties that can be of valuable use in the development of smart structures, e.g. the shape memory effect, the pseudoelastic behavior, and the high damping capacity. Whereas a lot of scientific work is concerned with the shape memory effect and with pseudoelasticity, the damping behavior of SMAs has drawn less attention.

In the present study the damping capacity of SMAs in the pseudoelastic range is under consideration. When austenitic material is loaded beyond a critical value, dependent on temperature, stress induced martensite is built. In this transformation process to detwinned martensite, the stress raises only slightly and a large, apparently plastic strain is achieved. This large strain is recovered by unloading in a considerable hysteresis loop since the martensitic material is unstable without stress. The area of the hysteresis loop represents the amount of energy which is transformed into heat due to internal friction and is responsible for the damping capacity of the material.

The properties of SMAs depend strongly on the microstructure with an interaction of phase boundaries, dislocations, precipitates, impurities and point defects [1]. Hence, the damping capacity depends on a great variety of internal and external parameters. The main influence results from the temperature, but also frequency, strain amplitude, time, alloy composition, grain size and density of phases play an important role [2]. Because of this complexity, experimental results often seem to be contradictory and not all relationships have been completely understood until now. Of course, also different materials showing the shape memory effect have different properties in detail and among them some may have a stronger effect with respect to damping than others. Since the goal of the investigations presented in the following is to study the phenomenon of pseudoelastic SMA damping, the experiments are performed solely with a single material, which is binary NiTi in the superelastic range. The aim of this work, however, is not to optimize the damping effect with respect to the material composition.

Beside others, own experiments show that the area of hysteresis depends on the strain amplitude of the specimen. Figure 1 depicts internal hystereses in the

U. Gabbert and H.S. Tzou (eds.), IUTAM Symposium on Smart Structures and Structronic Systems, 121–128.

stress strain diagram of a NiTi wire. In every cycle the wire is loaded to a strain one percent less than in the cycle before. Up to a strain of 0.8% the material behaves linear elastic. Without any pre-stress, amplitudes of about 1% have to be applied in order to generate considerable damping effects in dynamic systems with integrated SMAs.

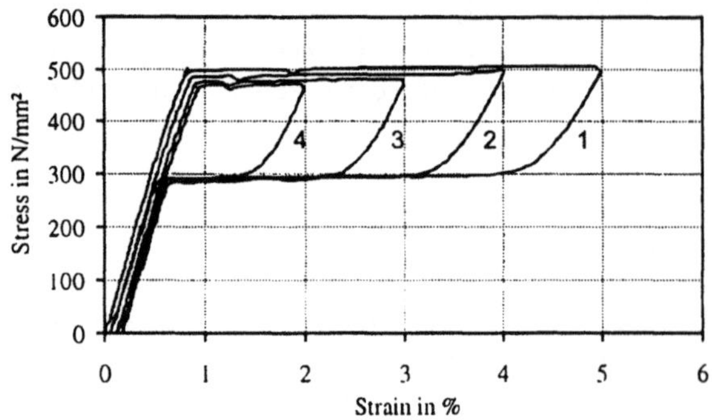

Figure 1. Internal loops of cycling NiTi wires

In the literature, most of the internal friction research of SMAs has been concentrated on the study of the internal friction peaks appearing during heating and cooling [3],[4],[5]. Moreover a few work is done in tension tests at low strain rates [6],[7], [8],[9]. The results of these tests in which the dependency of damping from the strain rate is considered are far away from being consistent. Whereas some work has been done in tension tests, only a few papers are concerned with the damping behavior of SMA components and their damping behavior [5],[10],[11].

In front of this background, the present paper is mainly concerned with basic components made of pseudoelastic SMAs and their damping behavior. Nevertheless, before the damping behavior of a NiTi spring is presented, dynamic tensile tests of binary NiTi specimens are covered within the next Section.

2. Cyclic Tensile Tests

The investigations on the dynamic behavior of pseudoelastic shape memory alloys started with tensile tests on NiTi specimens at frequencies up to 4Hz and a maximum strain of 3% [12].

For these tests, superelastic NiTi wires with a diameter of 0.9mm were used in strain-controlled cyclic tensile tests at room temperature. The specimens were put into heavy clamps of steel without any thermal insulation for generating nearly isothermal conditions. The distance between the clamps was about 26mm. The tests were conducted on a servohydraulic machine by subjecting the specimens to an initial static strain of 1.5% and then harmonically oscillating them around this offset level with a strain amplitude of 1.5%. During cycling the grips were moved slightly apart to adapt the offset to the residual strain. As depicted in Figure 2, quasistatic measurements were performed for comparison.

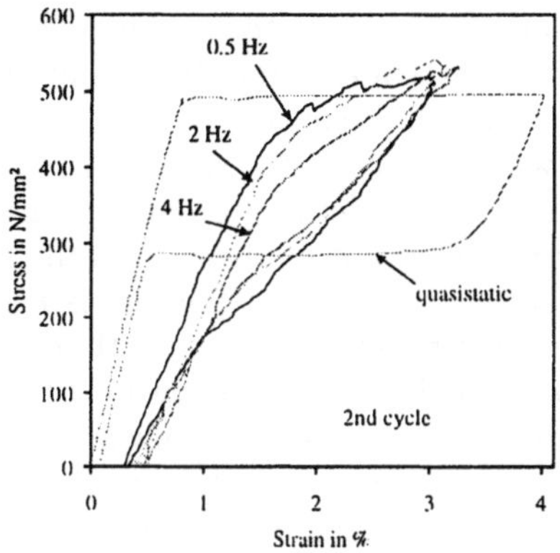

Figure 2. Cyclic tests of NiTi wires at different frequencies

Figure 2 shows, that the beginning of the transformation process is difficult to observe from the stress-strain-curves in dynamic experiments. But from the curves it becomes clear, that the critical stress value for the transition to martensite becomes obviously smaller at higher strain rates. Therefore, dynamic loads can give rise to considerable damping at higher stain rates even if the load is below the level which causes the transformation process in the quasistatic case. Furthermore, in contrast to the quasistatic cycling process, a continuous increase of stress is necessary to maintain the transformation from austenite to martensite after its beginning. As a consequence, the stress plateau characterizing the transformation from martensite to austenite disappears. The higher the strain rate rises the higher is the slope of the formerly horizontal lines. Finally it is observed, that the area of the hysteresis loop which indicates the dissipated energy per cycle shows a strong decrease with increasing loading frequency.

Within a small range of the strain rate, cf. 0.5Hz curve in Figure 2, the stress-strain curve becomes jagged during transformation. This distinctive feature was already described by Wolons et al. [9]. At higher strain rates, this phenomenon is not observed any longer.

In order to investigate the damping behavior at higher strain rates, further experiments are performed on a spring-mass system.

3. Vibration of a Spring-Mass System

A stiff beam on which additional masses are mounted is supported by two coil springs, so that the whole assembly can be considered as a spring-mass system, cf. Figure 3. Two types of coil springs have been used in the experiments: hollow NiTi springs and solid steel springs. The NiTi springs were made from tubes with an outer diameter of 1mm and a wall thickness of 0.1mm. From these tubes the coil springs were built with a diameter of 8mm and 6 windings. For comparison, coil springs (diameter 8mm, 7 windings) made from solid spring steel wires (diameter 0.6mm) were used. The load carried by the springs consists of the beam and the additional masses as well as a force transducer and an accelerometer. The system is excited by an electrodynamic shaker. In order to generate high excitation amplitudes a lever is used which amplifies the shaker displacement amplitudes ten times. A leaf is fixed to the beam in order to prevent unwanted eigenforms (rocking modes) of the system at frequencies not far away from the frequency range under consideration. In the following, the mass of the leaf is neglected. A pestle fills exactly the distance between lever and leaf. The length of the pestle as well as the distance between the beam and the masses are two other parameters which can be used for tuning the system in order to avoid unwanted oscillations.

The excitation frequency is chosen around the resonance frequency of the system with a step-width of 0.01Hz for small damping values and 0.025Hz at higher damping. These experiments are driven in a closed loop process, i.e., during the sine sweep the amplitude of acceleration is kept constant.

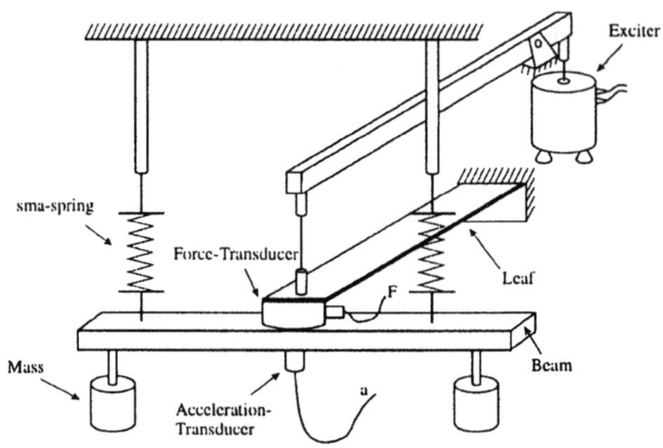

Figure 3. Experimental setup of the spring-mass-system

The frequency response function (FRF) is calculated as the inertance from the acceleration of the oscillating beam and the force acting on the beam. The tests are carried out at room temperature, ranging from 20°C to 21.5°C. As expected, the steel springs show a linear load-deflection curve whereas the highly non-linear load deflection curves of the NiTi springs before and after cycling are depicted in Figure 4. The experiments discussed in the following have been performed with different masses mounted to the beam and, as a result, at different working points on the load-deflection curve of the springs.

In the experiments with steel springs, the maxima of the FRFs varies slightly with the amplitudes of acceleration, whereas the eigenfrequency is nearly constant, cf. Figure 5. Furthermore, the narrow peaks indicate a very small damping ratio.

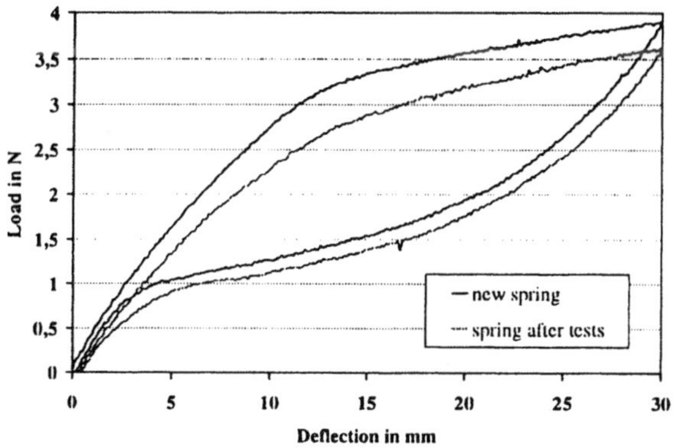

Figure 4. Load - deflection curve of springs made from NiTi tubes

In contrast to steel springs, NiTi springs show a strong rise in damping at higher levels of excitation as it can clearly be seen from the FRFs in Figure 6 that drop sharply with increasing acceleration amplitude. Moreover, there is a remarkable decrease in the characteristic frequency.

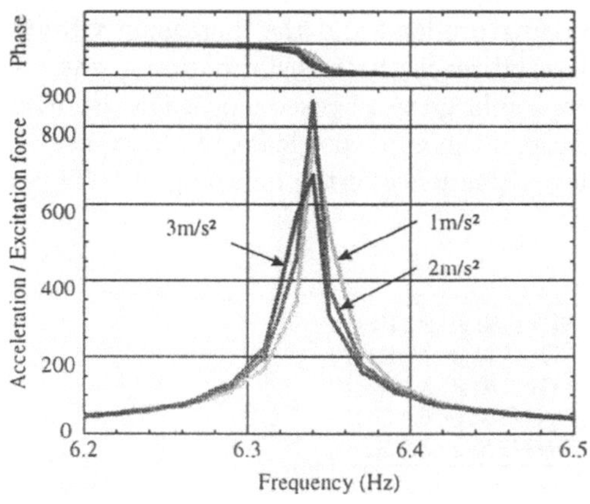

Figure 5. Inertance of steel springs (Load 560 gr)

It is emphasized that the experiments on steel springs have only be performed for the purpose of comparison with NiTi springs and not for an exact determination of the material damping of steel. In the experiments with both materials, the influences from outside that cause damping are the same: cables of the force transducer and the accelerometer as well as the surrounding air. Since the damping ratio of steel is within the range known from the literature, the damping parameters that have been determined for NiTi are considered to be reliable.

Figure 7 summarizes a lot of the experimental investigations on the spring experiments and comprises results obtained from steel and NiTi springs. In Figure

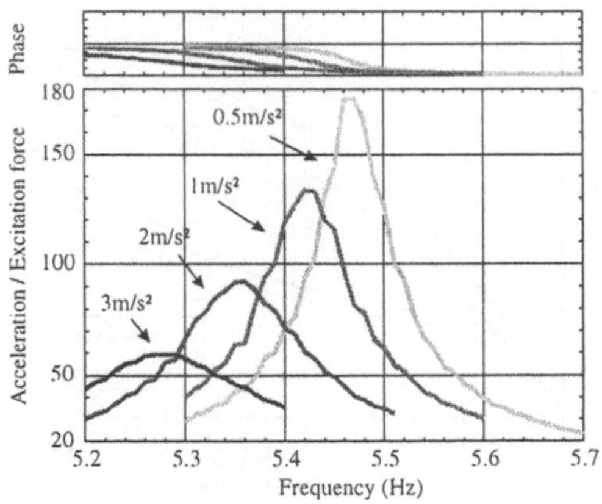

Figure 6. Inertance of NiTi springs (Load 560 gr)

7 the total mass carried by the springs is given as curve parameter. Steel springs show nearly no dependency of the damping ratio on the acceleration and the pre-stress, whereas an increase of both the mass as well as the acceleration results in a stronger damping in the experiments with NiTi springs. These results are strong indications that the rising pre-load and the increasing vibration amplitude make more parts of the material reach the transformation stress during oscillation and that the hysteresis loops enlarge as a consequence. On the other hand, it should be noticed, that an increase of the mass also leads to a decrease of the eigenfrequency. Therefore, a frequency dependence of the damping of NiTi cannot be excluded.

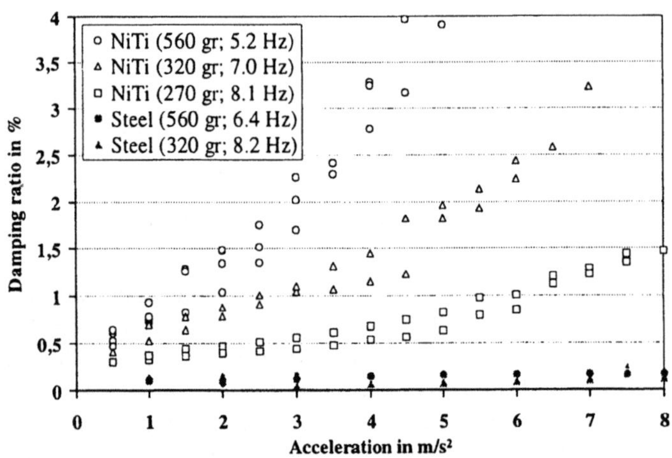

Figure 7. Damping coefficients from measurements with springs

4. Conclusions and Outlook

Experiments on NiTi components have been performed in order to investigate the damping capacity of NiTi in the pseudoelastic range. These experiments show clearly that shape memory alloys are well suited for structural integration in order to improve the damping properties when they are loaded dynamically within the transformation range from austenite to martensite. Since the critical stress at which the transformation from austenite to martensite takes place depends on the temperature, this parameter can be used to adjust the system under consideration to the individual requirements, e.g. in the case of a non-constant mass.

The design of structural systems which make use of pseudoelastic shape memory alloy damping requires a suitable constitutive law. In recent years the literature on shape memory alloy constitutive modelling has grown enormously and cover a lot of aspects of the various effects, cf. the review papers of Birman [14] and Brinson and Huang [15]. On the other hand, the influence of the loading rate on the shape memory alloy mechanical response has drawn little attention especially as the physical reason of damping is not well understood yet. One assumption is, cf. Saadat et al. [2] that not the strain rate but the self-heating of the material due to phase transformations changes the pseudoelastic behavior. In contrast, Auricchio and Sacco [16] describe the strain rate dependency by introducing two internal variables, ξ and ξ_{ST} representing the martensite fraction for dynamic and for static loading conditions, respectively. In a similar way, other constitutive equations could be expanded in order to model properly the damping behavior of shape memory alloys.

References

[1] van Humbeeck J 1996 Damping Properties of Shape Memory Alloys During Phase Transformation *Journ. de Physique IV* **6** C8 371-80

[2] Saadat S, Salichs J, Duval L, Noori M N, Hou Z, Bar-on I and Davoodi H 1999 Utilization of Shape Memory Alloys (SMAs) for Structural Vibration Control (SVC) http://me.wpi.edu/~clpsi/...

[3] Liu Y, van Humbeeck J, Stalmans R and Delaey L 1997 Some aspects of the properties of NiTi shape memory alloys *Journ. of Alloys and Compounds* **247** 115-21

[4] Morin M and Bigeon M J 1994 Austenitic-martensitic interface damping measured in shape memory alloys by a cyclic tensile machine. *Journal of Alloys and Compounds* **211/212** 632-5

[5] Lin H C, Wu S K and Yeh M T 1993 Damping Characteristics of TiNi Shape Memory Alloys *Metallurgical Transactions A* **24** 2189-94

[6] Wu K, Yang F, Pu Z and Shi J 1996 The Effect of Strain Rate on Detwinning and Superelastic Behavior of NiTi Shape Memory Alloys *Journ. of Intell. Mater. Syst. and Struct.* **7** 138-44

[7] Lin P-H, Tobushi H, Tanaka K, Hattori T and Ikai A 1996 Influence of Strain Rate on Deformation Properties of TiNi Shape Memory Alloy *JSME Int. Journ. Ser. A* **39** 117-23

[8] Piedboeuf M C, Gauvin R and Thomas M 1998 Damping behavior of Shape Memory Alloys: Strain Amplitude, Frequency and Temperature Effects *Journ. of Sound and Vibration* **214** 885-901

[9] Wolons D, Gandhi F and Malovrh B 1998 An Experimental Investigation of the Pseudoelastic Hysteresis Damping Characteristics of Nickel Titanium Shape Memory Alloy Wires *AIAA Structures, Structural Dynamics & Materials Conf.* **4** pp 2821-33

[10] Li D Z and Feng Z C 1997 Dynamic Properties of Pseudoelastic Shape Memory Alloys *Proceedings of SPIE* **3041** pp 715-25

[11] Graesser E J 1995 Effect of Intrinsic Damping on Vibration Transmissibility of Nickel-Titanium Shape Memory Alloy Springs *Metallurg. and Mat. Transactions A* **26** 2791-6

[12] Schmidt I and Lammering R 2000 Pseudoelastic NiTi - Alloys under Cyclic Loading *Zeitschrift für Angewandte Mathematik und Mechanik Proc. of GAMM 1999* pp 453-4

[13] Schmidt I and Lammering R 2000 Dynamisches Verhalten einer superelastischen NiTi - Legierung *Technische Mechanik* **20** 51-60

[14] Birman V 1997 Review of mechanics of shape memory alloy structures *Appl. Mech. Rev.***50** 629-45

[15] Brinson L C and Huang M S 1996 Simplification and Comparisons of Shape Memory Alloy Constitutive Models *Journ. of Intell. Mater. Syst. and Struct.* **7** 108-14

[16] Auricchio F and Sacco E 1999 Modelling of the Rate-Dependent Superelastic Behavior of Shape Memory Alloys *Proc. of ECCM '99*

PREDICTION OF EFFECTIVE STRESS-STRAIN BEHAVIOR OF SM COMPOSITES WITH ALIGNED SMA SHORT-FIBERS

J. WANG, Y. P. SHEN
Department of Engineering Mechanics, Xi'an Jiaotong University
Xi'an, Shaanxi Province, 710049, People's Republic of China

1. Introduction

Shape memory alloys (SMAs) have found increasing applications as candidate smart materials and components of adaptive structures[1]. One potential application of shape memory alloys is their use as fibrous sensors or actuators within a composite material system. Schetky and Wu[2], Schetky[3], and White et al[4] had discussed some technical details of such applications. The use of shape memory alloys as sensors may be related to the dependence of their properties on the martensite fraction. Conversely, the use of SMA fibers as distributed micro-actuators is also very attractive because of the relatively high strains that can be generated.

The analysis of composites seems a rather open-ended research area involving solution of several problems including admissible micromechanics and the nonlinear constitutive relations of SMAs. Because of the fiber-matrix interaction in composites, the developed in situ stresses will depend on phase transformation strains, thermal strains, and the constituent properties. This interdependence results in a nonlinear thermomechanical response, which makes an iterative analysis and incremental approach seemingly necessary, as shown in present paper. Numerous micromechanical theories may be used to formulate the present problem, yet limited work has been reported in the area of SM composites [5-7]. For the composites reinforced in SMA short fibers, hitherto, the research was still very limited. Wang and Shen[8-9] had discussed the transformation strain coefficient and thermal expansion coefficient of SMA short-fiber composite on the basis of he Eshelby's inclusion method by use of the incremental approach.

In this paper, the response of SM composite is presented in isothermal longitudinal loading and unloading under various temperature conditions. As well, the phase transformation critical stresses of SM composites are also discussed.

2. Thermomechanical Properties of Materials

2.1. SMA FIBER

Considering that the austenite and marstensite phase in an SM alloy are assumed to

U. Gabbert and H.S. Tzou (eds.), IUTAM Symposium on Smart Structures and Structronic Systems, 129–136.

behave thermoelastically, the incremental constitutive relation is derived as the following form when neglecting the difference of material properties within a short time period δt and the last time is always considered as the referential state.

$$\Delta\sigma_{ij}^{f} = C_{ikjl}^{f}(\xi)\cdot\left(\Delta\varepsilon_{kl} - \alpha_{kl}(\xi)\Delta T - \omega_{kl}\Delta\xi\right) \tag{1}$$

in which the symbol Δ stands for the increment related to the last time. where $\varepsilon_{kl}, \Delta T, \varepsilon_{kl}^{tr}$ stand for the total strains, the temperature change and transformation strains as a result of phase transformation□respectively. $\sigma_{ij}^{f}, \alpha_{kl}, C_{ikjl}^{f}$ are stress vector, thermal expansion coefficient vector and elasticity tensor, respectively. During thermoelastic martensitic transformation, the martensite fraction depends on the free chemical energy in the material, which, in turn, includes the thermal and strain energy components. This relation may be represented by phenomenological equations proposed by Liang and Rogers[10] that use the effective stress to represent the three-dimensional stress in SMA materials. The transformation strain coefficient ω_{kl} have been discussed by Wang and Shen [9] and given by

$$\omega_{22} = \omega_{33} = -\frac{1}{2}\omega \qquad \omega_{11} = \omega \tag{2}$$

2.2. SM COMPOSITE

The incremental equivalent constitutive equations for a composite medium with SMA fibers undergoing a martensitic or reverse transformation can be postulated to have the following form:

$$\Delta\sigma_{ij}^{c} = C_{ikjl}^{c}(\xi)\cdot\left(\Delta\varepsilon_{kl}^{c} - \alpha_{kl}^{c}(\xi)\Delta T - \omega_{kl}^{c}\Delta\xi\right) \tag{3}$$

in which the subscript c indicates the homogenized SM composite material. The main difference with Eq. (1) is that the resultant composite will exhibit orthotropy in the LT plane(see figure 1), including elastic stiffness, thermal expansion coefficient and transformation strain. The transformation laws remain valid during the martensitic or reverse phase transformation for SMA fiber in composite system, so the transformation behavior of composite is determined by the thermomechanical behavior of SMA fiber.

3. Micromechanics Approaches

Three different micromechanical approaches are presented in this section to predicate the effective elastic stiffness of short-fiber composite, the effective transformation strain coefficients, and the response of SM composite and the corresponding micro-stresses in

SMA fiber.

Good approximations of the equivalent properties of SM composite may be obtained using the self-consistent approach [11] According to the micromechanics model, the equivalent elastic stiffness matrix is defined by the following equations.

$$C^c = C^m + \sum_{r=1}^{n} m_r (C^r - C^m) A^r \tag{4}$$

in which, based on the self-consistent theory, A^r is obtained.

$$A^r = S[(C^c - C^r)(S - I) - C^r]^{-1}(C^c - C^r) + I \tag{5}$$

where S, I are the Eshelby tensor[12] in transversely isotropic medium and unit tensor, A^r is related to the properties of matrix and the fiber. Noting that the equivalent elastic stiffness C^c of composite should be solved by an iterative method.

Under the temperature change, Wang and Shen[9]extend the equivalent inclusion method to solve the thermal expansion coefficient and transformation strain coefficient of SM composite. The equivalent thermal expansion coefficient is

$$\alpha^c = \alpha^m + \gamma_D^{\alpha} / \Delta T \tag{6}$$

and the transformation strain coefficient ω^c is

$$\omega^c = \gamma_D^{tr} / \Delta \xi \tag{7}$$

where γ_D^{α} and γ_D^{tr} denote the volume average strain induced due to the mismatch thermal strain and the mismatch transformation strain under the temperature change condition in domain D (see figure 1 and [8,9]). The solutions address that the thermal expansion coefficients and the transformation strain coefficients are dependent on the fiber volume fraction and fiber geometric ratio L/D; as well as, the transformation strain coefficients of SM composites can be justified by changing fiber volume fraction and fiber geometric ratio.

Our analytical model consists of aligned SMA short fibers that are embedded in an infinite isotropic matrix as shown in figure 1. In order to consider the interaction between fibers and matrix, the analytical model assumes that the fiber is surrounded by the homogenized composite medium with its equivalent properties. So, let the domain of the composite body and fiber, be denoted by D and Ω, respectively. The surrounding homogenized composite medium is $D - \Omega$.

When the uniform stress increment σ^0 is applied to the composite system, according to Eshelby's inclusion model, the disturbed stress $\tilde{\sigma}$ and the disturbed strain $\tilde{\varepsilon}$ will be induced. So the local stress in the homogenized composite medium, $D - \Omega$, is the sum of the applied stress and the disturbed stress,

$$\sigma^c = \sigma^0 + \tilde{\sigma} = C^c(\varepsilon^0 + \tilde{\varepsilon}) \tag{8}$$

and the stress increment in fiber, Ω, can be written as

$$\sigma^f = \sigma^0 + \tilde{\sigma} = C^f(\varepsilon^0 + \tilde{\varepsilon} + \varepsilon - \varepsilon^p) \tag{9}$$

or ,

$$\sigma^f = C^c(\varepsilon^0 + \tilde{\varepsilon} + \varepsilon - \varepsilon^p - \varepsilon^*) \tag{10}$$

The strain ε is the induced elastic strain in domain Ω due to the sum of the mismatch 'eigenstrain' ε^p and the equivalent 'eigenstrain' ε^*.

$$\varepsilon = S\left(\varepsilon^p + \varepsilon^*\right) \tag{11}$$

S is the Eshelby tensor in transversely isotropic medium. The mismatch 'eigenstrain' ε^p arise from the mismatch transformation strain,

$$\varepsilon^p = \varepsilon^{ftr} - \varepsilon^{ctr} = (\omega^f - \omega^c)\Delta\xi \tag{12}$$

Using Mori-Tanaka method[13], the average stress increment within domain D should be equal to the applied stress increment. After simply derivation, there is

$$\tilde{\varepsilon} + f(S - I)\varepsilon + f(S - I)\varepsilon^p = 0 \tag{13}$$

As seen in the previous analytical formulations [9], when the uniform stress increment σ^0 is applied to the composite system, only three unknown vectors $\sigma^f, \tilde{\varepsilon}, \varepsilon^*$ need to be solved from the above equations.

$$\begin{bmatrix} A1 & A2 & A3 \\ B1 & B2 & B3 \\ C1 & C2 & C3 \end{bmatrix} \begin{Bmatrix} \tilde{\varepsilon} \\ \varepsilon^* \\ \sigma^f \end{Bmatrix} = \begin{Bmatrix} A4 \\ B4 \\ C4 \end{Bmatrix} \tag{14}$$

the coefficient matrices are related to the properties of fiber and matrix, and the fiber geometry and volume fraction. After the unknown variables are solved, the strain increment ε^c in domain D is leaded due to the applied stress increment σ^0 on composite system,

$$\varepsilon^c = \varepsilon^0 + \tilde{\varepsilon} + f(S - I)(\varepsilon^p + \varepsilon^*) + (1 - f)\varepsilon^{ctr} + f\varepsilon^{ftr} \tag{15}$$

Noting that the stress increment σ^f plays an important role in forming the governing equation and determining whether the phase transformation occurs in fiber but it is unknown. So the iterative technique is necessary in solution procedure.

4. Numerical results and discussions

Because mechanical responses of SMA fiber are various with the change of temperature condition, numerical cases are carried out under three temperature zones, that is, $M_s < T < A_s$ □ $A_s < T < A_f$ and $T > A_f$. The predicated loading and unloading response of SM composites are shown in figure 2 under different temperature conditions. When the temperature is satisfied with $M_s < T < A_s$ and $A_f < T$, the stress-strain curves of SM composites are very similar to one of SMA fiber. If temperature belongs to the region $A_s < T < A_f$, there is an obvious difference between SMA fiber and SM composites during unloading, i.e., SM composite firstly undergoes elastic unloading, then reverse transformation, finally completely elastic unloading. The above phenomena occur because of the change of the local stress in SMA fiber. If temperature is larger than the transformation point A_f, the SMA fiber will undergo a completely martensitic transformation and reverse transformation, so the local stresses in fiber are zero after completely removing load, and the residual strains of composite vanish.

However, the temperature belongs to $M_s < T < A_s$ and $A_s < T < A_f$, SMA fiber undergoes a completely martensitic transformation during loading, but no reverse transformation when $M_s < T < A_s$ and partial reverse transformation when $A_s < T < A_f$ during unloading. Therefore, the residual stress in SMA fiber is larger when $M_s < T < A_s$ and smaller when $A_s < T < A_f$.

The effects of fiber volume fraction or fiber geometric on response of composite are addressed here. It is found that the hysteresis zone of SM composite is larger with an increase in fiber volume fraction. The effective stress of fiber during loading and unloading and the critical stress of martensitic and reverse phase transformation of SM composite, $\sigma_S^A, \sigma_F^A, \sigma_S^M, \sigma_F^M$, are presented in figure 3. The results show that the start and finish critical stress of martensitic transformation and the finish critical stress of austenite transformation increase with the fiber volume fraction, but the start critical stress of austenite transformation declines. The phase transformation stress zone of composites declines with an increase in the fiber volume fraction. The equivalent properties of composite depend on the fiber volume fraction in composite, so the local stresses in SMA fiber are various even under the same outer loading. Therefore, the critical stress of martensitic and reverse phase transformation is various for composite with the change of fiber volume fraction, although the phase transformation critical stress of SMA fiber is changeless for the given temperature.

As be well known, the critical stress of martensitic and reverse phase transformation of SMA materials linearly depends on temperature[1,14]. The above analysis addresses the dependences of the critical stresses upon the fiber volume fraction and fiber geometric ratio, the figure 4 presents the typical curves of the critical stresses and temperature.

5. Summary

Micromechanics for the evaluation of the properties and mechanical response in composite systems with aligned short SMA fibers in an elastic matrix are presented. The stress-strain curves, the internal stress in the fiber and the critical stress of martensite and reverse phase transformations are discussed by use of the equivalent inclusion model and Mori-Tanaka method.

The numerical results demonstrate the longitudinal response of the composite under isothermal loading and unloading conditions. The ranges of stress, strain, and temperature where martensite and reverse transformations occur are also shown. Especially, the effects of fiber volume fraction and fiber geometric ratio on the above mentioned factors are discussed in detail. In closing, this work only represents an effort towards understanding and modeling the complex, yet promising, thermomechanical behavior of SM composites. Additional analytical and experimental work should be conducted in the future.

Acknowledgments: This work was supported by FML (the Failure Mechanics Laboratory of Tsinghua University) and the State Key Laboratory of Industry Equipment and Structural Analysis in DaLian University of Polytechnology. The authors would like to express thanks to them.

6. References

1. Duerig, T. W., Melton, K. N., Stokel, D., and Wayman, C. M., Engineering Aspects of Shape Memory Alloys, Butterworth-Heinemann, London, 1990.
2. Schetky, L. M., and Wu, M.H., The properties and processing of shape memory alloys for use as actuators in intelligent composites materials, Smart Structures and Materials, Edited by G.K. Haritos and A.V. Srinivasan, American Society of Mechanical Engineering, New York, 1991, pp65-71
3. Schetky, L. Mcd., The role of shape memory alloys in smart/adaptive structures, Shape Memory Materials and Phenomena—Fundamental Aspects and Applications, edited by C. T. Liu, H. Kunsmann, K. Otsuka, and M.Wutting, Vol.246, Materials Research Society Symposium Proceedings, Materials Research Society, Pittsburgh, PA, 1992, pp. 299-307
4. White, S.R., Whitlock, M.E., Ditman, J.B., and Hebda, D.A.., 1993, Manufacturing of adaptive Graphite/Epoxy structures with embedded Nitinol fibers, Adaptive Structures and Material Systems, edited by G. P. Carman and E. Garcia, American Society of Mechanical Engineers, New York, 1993, pp.71-79
5. Body, J.G., and Lagoudas, D.C., 1993, Thermomechanical response of shape memory composites, Smart Structures and Intelligent Systems, edited by N. W. Hagood and G. J. Knowles, Vol. 1917, Pt.2, Society of Photo-Optical Instrumentation Engineers, 1993, Bellingham, WA, pp. 774-790
6. Sullivan, B. J., and Buesking, K.W., 1994, Structural integrity of intelligent materials and structures, Final Rept., U.S. Air Force Office of Scientific Research, AFOSR-TR 94 0388, Bolling AFB, Washington, DC.

7. Birman, V., Saravanos, D.A., and Hopkins, D.A., 1996, Micromechanics of composites with shape memory alloy Fibers in Uniform thermal fields, AIAA, Vol. 34, No.9, pp.1905-1911
8. Wang Jian, Shen Yapeng, 1999, Transformation Behavior of Composites Reinforced in SMA Short Fibers under Thermal Load Conditions, Acta Mechanics Scinia, (in printing)
9. Wang Jian, Shen Yapeng, 2000, Micromechanics of Composites Reinforced in the Aligned SMA Short Fibers in Uniform Thermal Fields, Smart Materials and Structures, Vol. 9(1), pp.69-77
10. Liang, C., and Rogers, C. A., 1991, The multi-dimensional constitutive relations of shape memory alloys, AIAA, Paper91-1165
11. Chou, T.W., Nomura, S., Taya, M., 1980, A self-consistent approach to the elastic stiffness of short fiber composites, J. Composite Materials, Vol.14, pp.178-188
12. Eshelby, J. D., 1957, The determination of the elastic field of an ellipsoidal inclusion, and related problems, Proceedings of the Royal Society of London, Series A, Vol.241, pp.376-396
13. Mori, T., and Tanaka, K., 1973, Average stress in matrix and average elastic energy of materials with misfitting inclusions, Acta Metallurgica, Vol. 21, pp. 571-574
14. Shaw, J. A. and Stelios, K., 1995, Thermomechanical aspects of NiTi, J. Mech. Phys. Solids, Vol.43, No.8, pp.1243-1281

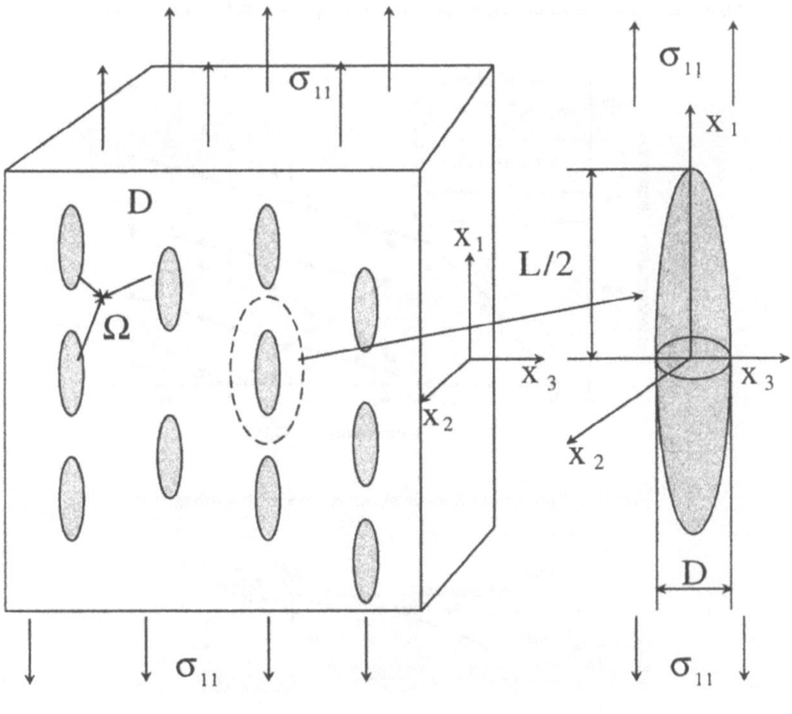

Figure 1. Micromechanical model and coordinate

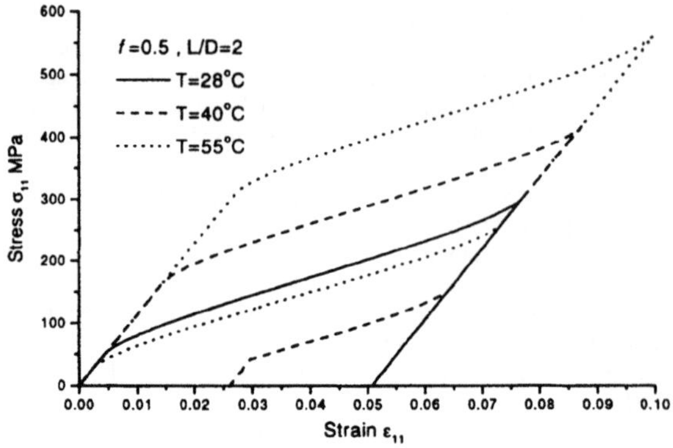

Figure 2. The stress-strain responses of SMA composites under various temperatures

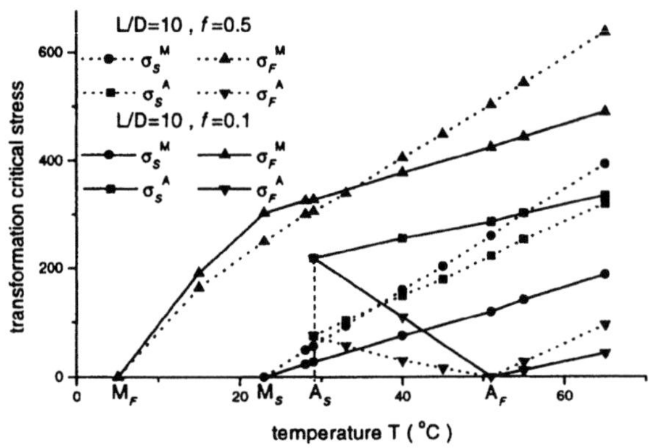

Figure 3. Various of effective stress in fiber with loading stress

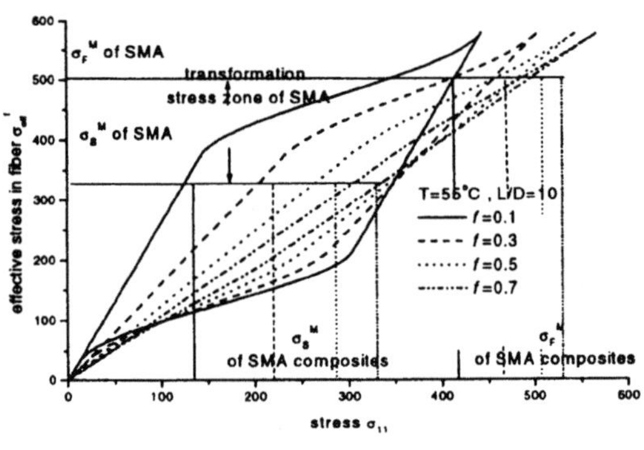

Figure 4. The phase transformation critical stress of SMA composites under different temperatures

MODELING AND NUMERICAL SIMULATION OF SHAPE MEMORY ALLOY DEVICES USING A REAL MULTI-DIMENSIONAL MODEL

X. GAO, W. HUANG AND J. ZHU
Centre for Advanced Numerical Engineering Simulations
School of MPE, Nanyang Technological University
Nanyang Avenue, Singapore 639798

1. Introduction

It is well known for some years that Shape Memory Alloys (SMAs) have great potential in a wide range of applications. The commercial market of SMA based devise is expanding very quickly in the past few years.

Despite decades of investigation on the thermo-mechanical behavior of SMAs, a review of the available literature shows the luck of reliable computational tool for design of SMA devices, in particular, under multi-dimensional stress state. This is mainly due to the many difficulties in applying traditional elastic/inelastic models to SMAs.

In one way or the other, most of the proposed multi-dimensional models are based on the theory of classic plasticity (Liang and Rogers 1992, Brinson and Lammering 1993), which fails in providing an adequate framework on description of many unique behaviors of SMAs. For instance, the apparent asymmetry in tension and compression in many SMAs (Gall et al. 1999, Berg 1997, Lim and McDowell 1999) is one among others. Some models do take asymmetry into consideration. However, it is most likely done by introducing new parameters that have to be empirically determined without much clear physical foundation (e.g., Auricchio and Lubliner 1997) or otherwise using the measured strain-stress curve directly (e.g., Pelton et al. 1994).

Patoor et al. (1994) derived a complex micro-mechanical model to predicate the nonsymmetrical phenomenon with some success. Recently, Huang (1999a) proposed a simple method to link the asymmetry with the lattice structure of each particular SMA.

Based on the previous work (Huang 1999a, 2000), a multi-dimensional constitutive model, which is compatible with finite element analysis, is developed. The study on two SMA devices demonstrates the importance of using real multi-dimensional model.

2. One-Dimensional Model

It is not the intention of this paper to develop a totally new SMA model. Current approach is more on the side to develop a framework for multi-dimensional analysis based on the platform that the behavior of SMA under a given stress state, for instance, uni-axial tension, is known either by experiment or by theoretical analysis.

U. Gabbert and H.S. Tzou (eds.), IUTAM Symposium on Smart Structures and Structronic Systems, 137–144.

Many one-dimensional models are available in the literature. Among others, the models proposed by Tanaka (1986), Liang and Rogers (1990) are theoretical ones, while the model proposed by Huang (1999b) is derived directly from experiment.

According to Tanaka (1986), stress σ, strain ε, temperature T, and martensite fraction ξ are internal variables. The mechanical constitutive equation of SMA may be expressed as

$$\sigma = D\varepsilon + \Theta T + \Omega\xi \tag{1}$$

where D is the Young's modulus, Θ the thermoelastic constant, and Ω the transformation constant, a metallurgical quantity which represents the change of strain during phase transformation. For simplicity, we assume that the Young's modulus of austenite is the same as that of martensite.

3. Construction of multi-dimensional model

Similar to classic plasticity, the construction of multi-dimensional model needs to define yield criterion and flow rule first. We assume that there is no non-recoverable deformation. Therefore, *yield* is induced by phase transformation only.

3.1 YIELD CRITERION

Martensitic transformation and its reverse transformation correspond to different yield surfaces. Hear we may follow the method proposed by (Huang 1999a, 2000) to get the real yield surface of various SMAs. We may define the equivalent yield stress as

$$\sigma_{eq} = f(\{\sigma\}) = \frac{1}{\alpha(\{\sigma\})}\sqrt{\frac{3}{2}\{s\}^T\{s\}} \tag{2}$$

where $\{s\}$ is the deviatoric stress vector, and $\alpha(\sigma)$ is the ratio of the real transformation start stress over that given by von Mises criterion which is determined from the real yield surface. In the case of the uni-axial tensile, $\alpha(\sigma)=1$. If this condition is satisfied for all the stress states, the equivalent stress becomes into von Mises equivalent stress.

It is assumed here that the subsequent yield surface is similar to the initial yield surface. For phase transformational, the yield criterion may be expressed by

$$F(\sigma,T,\xi,\xi_t,\xi_r) = f(\sigma) - K(T,\xi,\xi_t,\xi_r) = 0 \tag{3}$$

where ξ_t and ξ_r denote the maximum and minimum martensitic fractions in the previous martensitic transformation and the reverse transformation, respectively. K measures the size of yield surface.

For convenience, we may split Eqn. (3) into two yield criteria, i.e.

$$F_t(\sigma,T,\xi,\xi_r) = f(\sigma) - K_t(T,\xi,\xi_r) = 0 \tag{4}$$

$$F_r(\sigma,T,\xi,\xi_t) = f(\sigma) - K_r(T,\xi,\xi_t) = 0 \tag{5}$$

which correspond to the martensitic transformation and its reverse transformation, respectively. They are associative with each other. The corresponding yield surfaces are schematically plotted in Figure 1.

Inside of the most inner surface, it is pure austenite (ξ=0), while outside of the most outer yield surface corresponds to pure martensite (ξ=1).

In Figure 1, deformations in zones I, III and V are elastic. K_r and K_t may be determined experimentally.

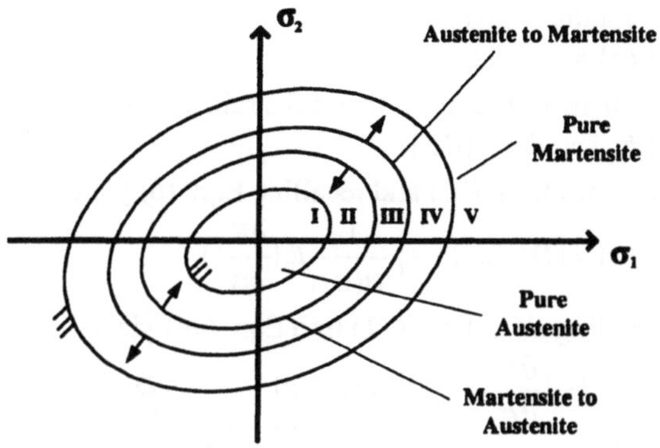

Figure 1 Schematic relationship of yield surfaces of martensitic and reverse transformation.

3.2 FLOW RULE

Similar to Prandtl-Reuss incremental flow theory, it can be proved that the transformation strain is along the normal direction of the yield surface (Huang 2000). Accordingly, we take the negative normal direction of the yield surface as transformation strain direction in reverse transformation, i.e.,

$$\{d\varepsilon_t^M\} = \frac{1}{\alpha(\sigma)} d\xi\varepsilon_L\{\frac{\partial F}{\partial \sigma}\}, \qquad \{d\varepsilon_t^A\} = \frac{-1}{\alpha(\sigma)} d\xi\varepsilon_L\{-\frac{\partial F}{\partial \sigma}\} \tag{6}$$

where ε_L is the maximum transformation strain in uni-axial tension.

If the applied equivalent stress exceeds the elastic limitation, phase transformation starts. We may derive the corresponding transformation strain increment accordingly. Rewrite the yield criterion, Eqn. (4), as

$$F(\sigma, K) = f(\sigma) - K = 0 \tag{7}$$

where K is the hardening parameter.

$$K = K_0(\xi_t, \xi_r, T) + \int \{\sigma\}^T \{d\varepsilon_t\} \tag{8}$$

Substituting Eqns. (6) into Eqn. (8) yields

$$K = K_0(\xi_t, \xi_r, T) + \int \{\sigma\}^T \{\frac{\partial F}{\partial \sigma}\} \varepsilon_L \frac{1}{\alpha(\sigma)} d\xi \tag{9}$$

Equation (7) can be differentiated, so that

$$dF = \frac{\partial F}{\partial \sigma} d\sigma + \frac{\partial F}{\partial K} dK = 0 \tag{10}$$

which is similar to the plastic consistency condition.

Differentiation Eqn. (9) and substitution dK into Eq. (10) yield the martensitic fraction increment as

$$d\xi = \frac{\{\frac{\partial f}{\partial \sigma}\}^T\{d\sigma\} - \frac{\partial K}{\partial T}dT}{\{\sigma\}^T\{\frac{\partial F}{\partial \sigma}\}\varepsilon_L \frac{1}{\alpha(\sigma)}} \tag{11}$$

Elastic stress-stain relation is given by

$$\{d\sigma\} = [D]\{d\varepsilon_e\} \tag{12}$$

where [D] is the elastic stiffness matrix. If thermal strain can be ignored,

$$\{d\varepsilon_e\} = \{d\varepsilon\} - \{d\varepsilon_t\} \tag{13}$$

Substituting Eqn.(13) into Eqn. (12) and recalling Eqn. (6) yield

$$\{d\sigma\} = [D](\{d\varepsilon\} - \varepsilon_L \frac{1}{\alpha(\sigma)} d\xi\{\frac{\partial F}{\partial \sigma}\}) \tag{14}$$

Subsequently, substituting $d\xi$ in Eqn. (11) into Eqn. (14) yields

$$\{d\sigma\} = \frac{[D]([I] - \{\frac{\partial F}{\partial \sigma}\}\{\frac{\partial F}{\partial \sigma}\}^T[D])\{d\varepsilon\}}{\{\frac{\partial F}{\partial \sigma}\}^T[D]\{\frac{\partial F}{\partial \sigma}\} + \{\sigma\}^T\{\frac{\partial F}{\partial \sigma}\}} + \frac{[D](\frac{\partial K}{\partial T}dT)\{\frac{\partial F}{\partial \sigma}\}}{\{\frac{\partial F}{\partial \sigma}\}^T[D]\{\frac{\partial F}{\partial \sigma}\} + \{\sigma\}^T\{\frac{\partial F}{\partial \sigma}\}} \tag{15}$$

Rewrite Eqn. (15) into a tidy form,

$$\{d\sigma\} = [D_{et}]\{d\varepsilon\} + [D_T]dT\{\frac{\partial F}{\partial \sigma}\} \tag{16}$$

We may call $[D_{et}]$ as effective stiffness matrix, which does not dependent upon the temperature, while $[D_T]$ as temperature deviation matrix.

Equation (16) is incremental form of stress-strain-temperature relationship. In isothermal case, Eqn.(16) may be simplified as

$$\{d\sigma\} = [D_{et}]\{d\varepsilon\} \tag{17}$$

Eqn. (17) is similar to elasto-plastic stiffness matrix of traditional elasto-plasticity.

4. Finite Element Approach

The expression for finite element approach can be derived from the principle of virtual work. The details can be found in any book of finite element method, for example, Owen and Hinton (1980). Following the principle of virtual work the incremental residual force vector is given as,

$$\Delta\psi = [K_{et}]\{\Delta d\} - (\{\Delta f\} + \int_\Omega [N]^T\{\Delta b\}d\Omega) + \{K_T\}dT \tag{18}$$

where

$$K_{et} = \int_\Omega [B]^T[D_{et}][B]d\Omega \tag{19}$$

$$\{K_T\} = \int_\Omega [B]^T[D_T]\{\frac{\partial F}{\partial \sigma}\}d\Omega \tag{20}$$

Here, $[N]$ and $[B]$ are the matrixes of shape function and elastic strain, respectively. $\{\Delta b\}$, $\{\Delta f\}$, and $\{\Delta d\}$ are the incremental of distributed loads, external applied forces and nodal displacements, respectively.

5. Simulation

The constitutive model developed in Sections 3 and 4 can be integrated with any commercial FEM software. One-dimensional model proposed by Liang and Rogers (1990), in which ξ is taken as cosine function, is chosen in current study. Readers may refer to Liang and Rogers (1990) for details of this model. Following the exact same approach described in Sections 3 and 4, one can easily implement other one-dimensional model in finite element analysis.

In the course of this study, we have successfully reproduced all shape memory related phenomena using commercial finite element package: *ANSYS*.

Following two simple cases demonstrate the capability of the proposed approach and the importance of using real multi-dimensional model in simulation.

Assume the SMA studied in both cases is non-textured polycrystalline NiTi. The corresponding yield surface is calculated following the average scheme proposed by Huang (1999a). The only difference is that the phase transformation strain is derived from the Green strain tensor of phase transformation (Bhattacharya and James 1999) which is defined by

$$E_g = \tfrac{1}{2}(F^T F - I) \tag{21}$$

where F is the lattice shape change due to phase transformation.

Parameters of NiTi SMA are list in Table 1.

TABLE 1 Material parameters.

E	M_f	M_s	A_f	A_s	C_A	C_M	ε_L	ν
46.65GPa	9°C	18.4°C	49°C	34.5°C	10.3 MPa/°C	10.3 MPa/°C	6.7%	0.3

5.1 NITI PLATE UNDER UNI-AXIAL TENSION/COMPRESSION

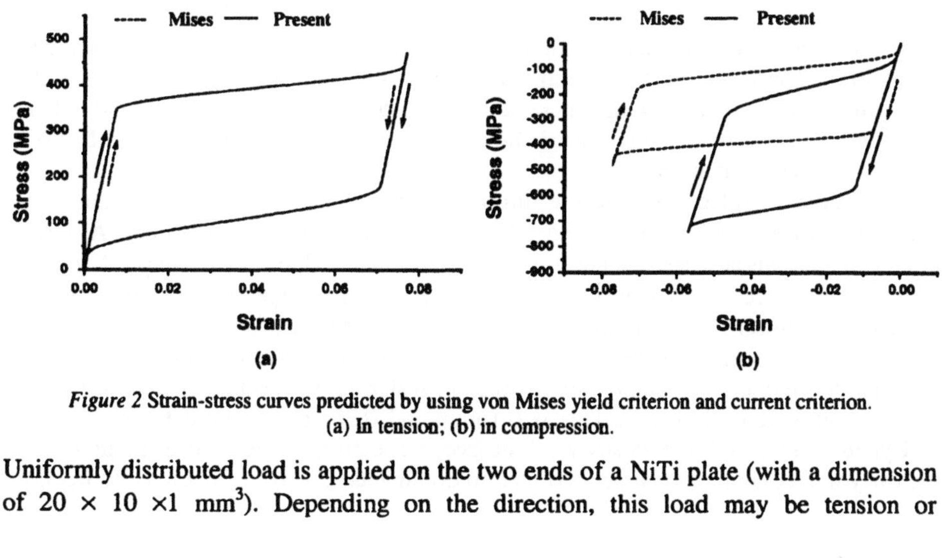

Figure 2 Strain-stress curves predicted by using von Mises yield criterion and current criterion. (a) In tension; (b) in compression.

Uniformly distributed load is applied on the two ends of a NiTi plate (with a dimension of 20 × 10 ×1 mm^3). Depending on the direction, this load may be tension or compression. For simplicity, we assume the ambient temperature T=52°C, which is slightly higher than A_f. Thus, the material is initially pure austenite, and it can fully recover the original shape after releasing the applied load (superelasticity).

Figure 2 shows the strain-stress curves in tension and compression using von Mises yield criterion and current criterion. As expected, the strain-stress curves in tension are the same, while significant difference is observed upon compression. On comparison with the result of von Mises criterion, the recovery strain predicated by current model is much smaller, but the transformation stress is much higher. Apparent nonsymmetrical behavior is reproduced.

5.2 NITI TUBE UNDER CYCLIC INNER PRESSURE

SMA tube has been used in a various fields as novel connector or coupler to replace many traditional assembly methods. Compared with the situation in the previous plate case, its stress status is far more complex.

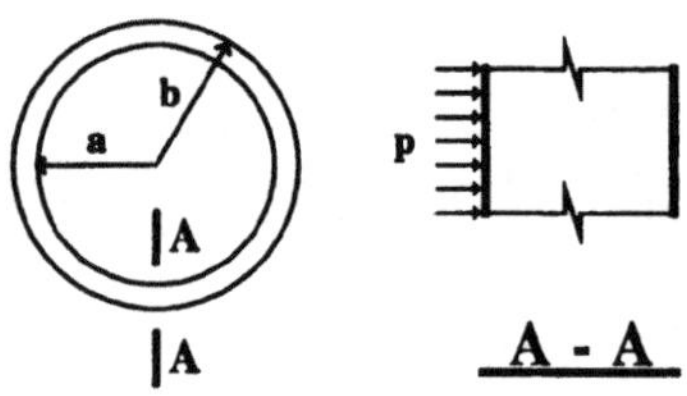

Figure 3 Schematic Dimension of the tube.

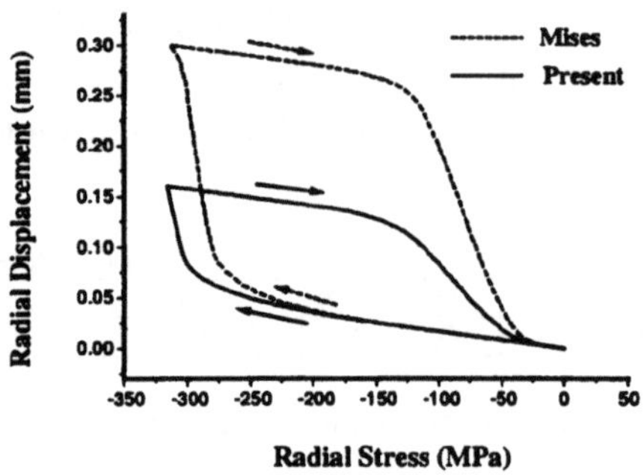

Figure 4 Radial displacement at inner surface of NiTi tube in a complete cycle.

We simulate the response of a long NiTi tube under cyclic inner pressure p as shown in Figure 3. The inner radius a=4mm, and outer radius b=8mm. Again, for simplicity, we assumed the ambient temperature T=52°C. Therefore, the tube is initially pure austenite. 2-dimensional axi-symmetric 8-node structural solid element (*ANSYS* 1999) is used in finite element analysis.

The radial displacement at the inner surface of the tube under a complete loading/unloading cycle produces close hysteresis loop as shown in Figure 4. At lower pressure, the whole cross section is pure austenite. Upon increasing of p, the displacement increases slowly in a linear manner at early stage and then dramatically due to the occurrence of martensitic transformation. The transformation start pressure predicted by the real yield surface is about 10% higher than that of von Mises surface. On the other hand, given same maximum p, in which all the section is martensite, the radial displacement at inner surface is 0.16mm and 0.30mm for simulations using the real yield surface and von Mises yield surface, respectively. The ratio is about 1:2. In the early unloading path, elastic deformation dominants. Change of martensitic fraction is small. Upon further unloading, the displacement decreases dramatically due to reverse transformation.

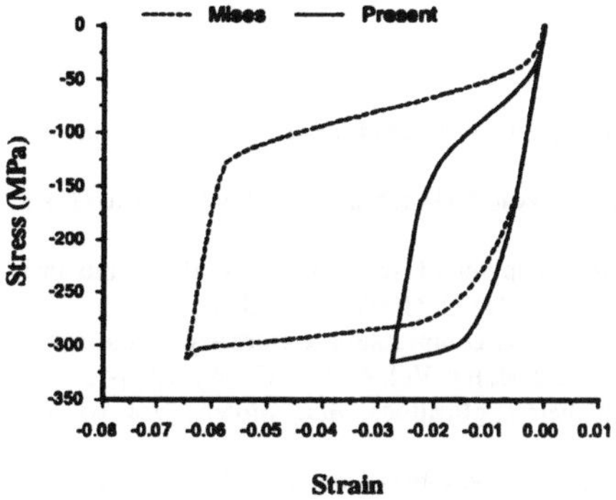

(a)

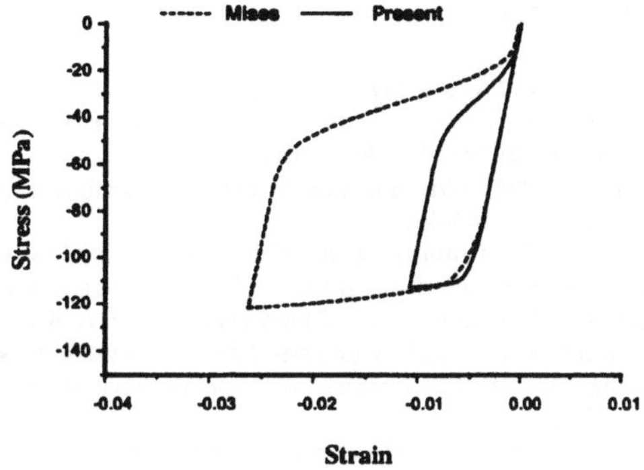

(b)

Figure 5 Comparison of radial stress vs. radial strain relationships using different yield surface. (a) At inner surface; (b) at middle point.

Figure 5 illustrates the radial stress vs. radial strain curves of NiTi tube. The maximum radial strain at the inner surface is 0.0276 given by the real yield surface, which is significantly lower than 0.0647 obtained from von Mises surface. At the middle point, both the maximum radial stress and the maximum radial strain are different.

After one complete cycle, full recovery is observed in all curves. Besides the differences in displacement, strain and stress, hysteresis loop produced by using the real yield surface is much narrower. It appears that the simulation using von Mises yield surface may not be adequate, in particular, the displacement may be overestimated.

6. Conclusions

A multi-dimensional constitutive model for SMAs is constructed based on the real yield surface, which is derived from the particular microstructure of a given SMA. Yield criterion and flow rule are derived. The start and finish of martensitic transformation and its reverse transformation correspond to two pairs of yield surfaces. All other surfaces are assumed to be similar to the initial yield surface in shape but different in size. Based on this assumption the incremental σ - ε - T relationship is obtained. On comparison with the finite element method of classical plasticity, current model has a temperature deviation matrix, which represents the effect of temperature. The corresponding finite element approach is set up. All shape memory phenomena are realized by ANSYS.

Two simple SMA devices, namely, a NiTi plate and a NiTi long tube, are investigated by using von Mises method and the new model. For simplicity, we assume SMA to be superelasticity. Asymmetry in tension and compression is reproduced. It is also found that the traditional von Mises method may not be suitable for the simulation of SMA device. A real multi-dimensional model, such as the one we proposed here, is required to provide a far more reliable prediction.

References

Ansys, *Ansys Element Reference*, Ansys Inc. 1999.

Auricchio, F., Lubliner L.: A uni-axial model for shape-memory alloys, *International Journal of Solids and Structures*, 34:27(1997), 3601-3618.

Berg, B. (1997) Twist and stretch: combined loading of pseudoelastic NiTi tubing, *SMST-97*, Pacific Grove, California, pp. 443-448.

Brinson, L.C., Lammering, R.: Finite element analysis of the behavior of shape memory alloys and their applications. *International Journal of Solids and Structures*, Vol. 30, No. 23(1993), 3261-3280.

Gall, K., H. Sehitoglu, Y.I. Chumlyakov and I.V. Kireeva: Tension-compression asymmetry of the stress-strain response in aged single crystal and polycrystalline NiTi, *Acta Mater*, Vol. 47 No.4(1999), 1203-1217.

Huang, W.: "Yield" surfaces of shape memory alloys and their applications, *Acta Mater.*, Vol. 47, No. 9(1999a), 2769-2776.

Huang, W.: Modified shape memory alloy (SMA) model for SMA wire-based actuator design, *Journal of Intelligent material systems and structures*, Vol. 10, No. 3(1999b), 221-231

Huang, W.: To predict the behavior of shape memory alloys under proportional load, (2000) *submitted*.

Bhattacharya, K., James, R.D.: A theory of thin films of martensitic materials with applications to microactuators, *Journal of the Mechanics and Physics of Solids*, 47(1999), 531-576.

Liang, C. and Rogers, C.A.: One dimensional thermomechanical constitutive relations for shape memory materials, *Journal of Intelligent material systems and structures* 1(1990), 207-234.

Liang, C. and Rogers, C.A.: A multi-dimensional constitutive model for shape memory alloys, *Journal of Engineering Mathematics* 26(1992), 429-443.

Lim T.J. and D.L. McDowell: Mechanical behavior of an Ni-Ti shape memory alloy under axial-torsional proportional and nonproportional loading, *Journal of Engineering Materials and Technology*, 121:1(1999), 9-18.

Owen D. R. J. and E. Hinton: *Finite Elements in Plasticity: Theory and Practice*, Pineridge Press Ltd, Swansea, U.K., 1980, 235-242.

Patoor, E., A. Eberhardt and M. Berveiller: Micromechanical modeling of the shape memory behavior, Mechanics of phase transformations and shape memory alloys, *ASME*, AMD Vol. 189(1994), pp. 23-38.

Pelton, A.R., Rebelo, N., Duerig, T.W. and Wich A. (1994) Experimental and FEM analysis of the bending behavior of superelastic tubing, *SMST94*, Pacific Grove, California, USA, pp.353-364.

Tanaka, K.: A thermomechanical sketch of shape memory effect: One-dimensional tensile behavior, *Res. Mechanica* 18(1986), 251-263.

THE ROLE OF THERMOMECHANICAL COUPLING IN THE DYNAMIC BEHAVIOR OF SHAPE MEMORY ALLOYS

OLAF HEINTZE, OLIVER KASTNER, HARSIMAR-SINGH SAHOTA
Institute of Thermodynamics, TU Berlin, Sekr. HF2
Straße des 17. Juni 135, D-10623 Berlin, Germany

STEFAN SEELECKE
Department of Mechanical & Aerospace Engineering,
North Carolina State University, Box 7910
Raleigh, NC 27695-7910
email: seelecke@thermodynamik.tu-berlin.de

1. Introduction

In recent years, shape memory alloys (SMAs) have started to attract increasing attention due to some of their dynamic properties. The hysteretic phase transformation between austenite and martensite at high temperature and between different twins of the martensite phase at low temperature constitutes an intrinsic dissipation mechanism, which results in a considerable damping capacity. Graesser and Cozarelli [1] suggested the use of SMAs as novel damping materials and Clark *et al* [2] demonstrated the feasibility of the concept for a Nitinol wire device. Potential applications are seen for example in civil structures like buildings and bridges needing an efficient seismic base isolation, see Wilde *et al* [3].

In [4] and [5], the last author has studied the dynamic behavior of a single degree of freedom (SDOF) system with a shape memory element. The system consists of a rigid mass suspended by a thin-walled SMA tube. Due to its material properties the SMA tube acts as a complex temperature-dependent spring and a damping mechanism at the same time. Simulations of the free and forced vibration behavior under torsional loading have been performed on the basis of an improved version of the Müller-Achenbach model. This model is based on the description of the phase transformations as thermally activated processes with evolution laws for the martensite phase fractions. The specific form of these evolution laws is derived using ideas from statistical thermodynamics and relies on a suitable formulation of the free energies of the phases. The mathematical structure of the model is given by a set of ordinary differential equations in time and thus blends naturally into the equations of motion for the rigid mass. A standard numerical integration algorithm can be applied for the solution of the entire system.

Although the calculations in [5] are thermodynamic in the sense that they treat several different temperatures, they are still isothermal. However, in a number of experimental investigations, a strain rate dependency of the material has been observed at already very low rate levels, see Shaw and Kyriakides [6] or Tobushi *et al* [7]. Careful

U. Gabbert and H.S. Tzou (eds.), IUTAM Symposium on Smart Structures and Structronic Systems, 145–152.

analysis of the temperature in the SMA has revealed that this effect is due to self-heating and self-cooling of the material which is the consequence of the rate-dependent release and absorption of latent heats during the deformation. This couples into the mechanics of the material through the temperature-dependence of the load-deformation behavior and strongly influences the shape of the hysteresis loops. It is clear that, particularly in a dynamic application utilizing SMAs, this is an important effect.

In the present work, the isothermal setting described above is extended to the full thermomechanical case by inclusion of the balance of energy for the SMA. The simulations of the free vibration cases are re-performed with a focus on the role of the latent heats and the heat exchange with the environment. Special attention is paid to their influence on the resulting forces and the damping behavior.

2. Basic Equations

In this section, a brief review of the basic model equations is given together with their implementation into the equations of motion. The presentation closely follows the one from the two above mentioned papers [4] and [5]. For details about the model, the interested reader is referred to [8] and [9] and the web page [10], which features a Java-based online implementation of the SMA model.

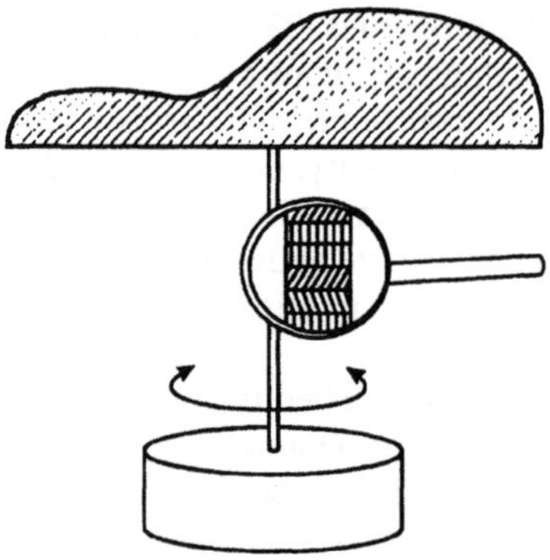

Figure 1. Oscillating rigid mass suspended by a thin-walled SMA tube consisting of austenite/martensite ring sections.

In the following, attention is confined to a simple SDOF system consisting of a rigid mass suspended by a thin-walled SMA tube. The angular motion of the rigid mass is determined by the balance equation of angular momentum, which reads

$$\Theta\ddot{\Phi} = -M\left(\Phi, T, x_{A,} x_{\pm}\right) . \tag{1}$$

The moment of inertia of the mass is denoted by Θ, $\ddot{\Phi}$ is its angular acceleration, and M is the torque exerted onto the mass by the shape memory tube. M is a complex function of the angle Φ, temperature T and the phase fractions of austenite and martensite[1], x_A and $x_{\pm}$, and will be derived in the following.

The shear of the layers causes the cross sections to rotate, thus twisting the wire. Obviously, the total angle of rotation at the lower end of the wire equals the deflection of the mass Φ, and it is given by the sum of rotation angles φ_α over all sections α,

$$\Phi = \sum_{\alpha=1}^{N} \varphi_\alpha = \sum_{A} \varphi N_\varphi^A + \sum_{+} \varphi N_\varphi^+ + \sum_{-} \varphi N_\varphi^- \ . \tag{2}$$

The number of layers with shearing angle φ are denoted by N_φ^A and $N_\varphi^\pm$. After division by the length of the tube L, the rotation angle of the mass can be calculated from

$$\frac{\Phi}{L} = x_A \varphi_A + x_+ \varphi_+ + x_- \varphi_- \ . \tag{3}$$

The quantities φ_A and $\varphi_\pm$ are the expectation values of the twist in the individual phases. They follow from partial differentiation of the Gibbs free energies of the phases with respect to the torque. The Gibbs free energy is the central quantity determining the kinetics of the phase transformations and the load-deformation behavior. It is assumed to be a nonconvex function of twist, and again, details can be found in the above mentioned references. The expectation values depend on the applied torque, and thus the inverted form of Equation (3) can be considered as the load-deformation relation needed for the RHS of Equation (1). However, in order to evaluate it, the phase fractions have to be known. They follow from a set of evolution equations which are derived from the theory of activated processes,

$$\begin{aligned} \dot{x}_+ &= -x_+ \overset{+A}{p} + x_A \overset{A+}{p} \ , \\ \dot{x}_- &= -x_- \overset{-A}{p} + x_A \overset{A-}{p} \ . \end{aligned} \tag{4}$$

In the above equation, $\overset{\alpha\beta}{p}$ are the transition probabilities for the phase transition from phase α to phase β. They depend on torque M and temperature T through the Gibbs free energies and can be calculated using methods from statistical thermodynamics. The evolution of the temperature in the SMA tube is calculated from the balance of internal energy,

$$mc\dot{T} = -k\left(T - T_0(t)\right) - \dot{x}_+ h^+(M) - \dot{x}_- h^-(M) \ . \tag{5}$$

---

[1] Due to the constraint $x_A + x_+ + x_- = 1$, x_A can be expressed by the other two phase fractions and is dropped from the list of variables. From here on, it is merely used as an abbreviation.

Two mechanisms thus influence the temperature change. At first, there is the heat exchange with the environment at temperature $T_0(t)$, characterized by the coefficient k. Secondly, there are the latent heats of the phase transformations $h^{\pm}(M)$, which can also be derived from the Gibbs free energy.

The computation of the rotational motion of the rigid mass proceeds as follows. Introducing the angular velocity Ω as a new variable, we re-write the balance of angular momentum in the form

$$\dot{\Phi} = \Omega \ ,$$
$$\dot{\Omega} = -M(\Phi, T, x_A, x_\pm) \ , \qquad (6)$$

and, with (4) and (5) it combines to give a system of five first order ordinary differential equations. The inverted form of Equation (3) provides the RHS of (6) and, together with appropriate initial conditions, the system serves to calculate the motion of the mass $\Phi(t)$.

3. Comparison of Isothermal and Nonisothermal Vibrations

The above system is solved numerically for the free vibration case by use of an implicit Runge-Kutta scheme. The mass is initially at rest and starts from an angle of $\Phi(0) = 324°$, at which the SMA tube is pretwisted to be completely in the martensite phase M_+. Two different environmental temperatures are considered; the first one is $T_0 = 293K$, a rather low temperature, at which the material is martensitic and exhibits the so-called *quasiplastic* behavior. The second case investigates the behavior at an elevated temperature, $T_0 = 353K$, at which the material is austenitic in the unloaded state; the corresponding behavior is termed *pseudoelastic*.

The objective of the present investigation is to point out the role of the thermomechanical coupling due to the latent heats released and absorbed during the phase transformations. To this purpose, all calculations are performed twice. At first, a purely mechanical model is used, assuming the SMA temperature to remain constant and equal to the environmental temperature. This is achieved by simply ignoring the balance of internal energy. For the second case, this balance equation is included, and both cases are then plotted in the following diagrams for comparison. The material data for the simulations are selected so as to match those of a typical *NiTi* alloy, the moment of inertia of the mass is $0.0113\ kg\,m^2$ and the length of the tube is $L = 0.05m$.

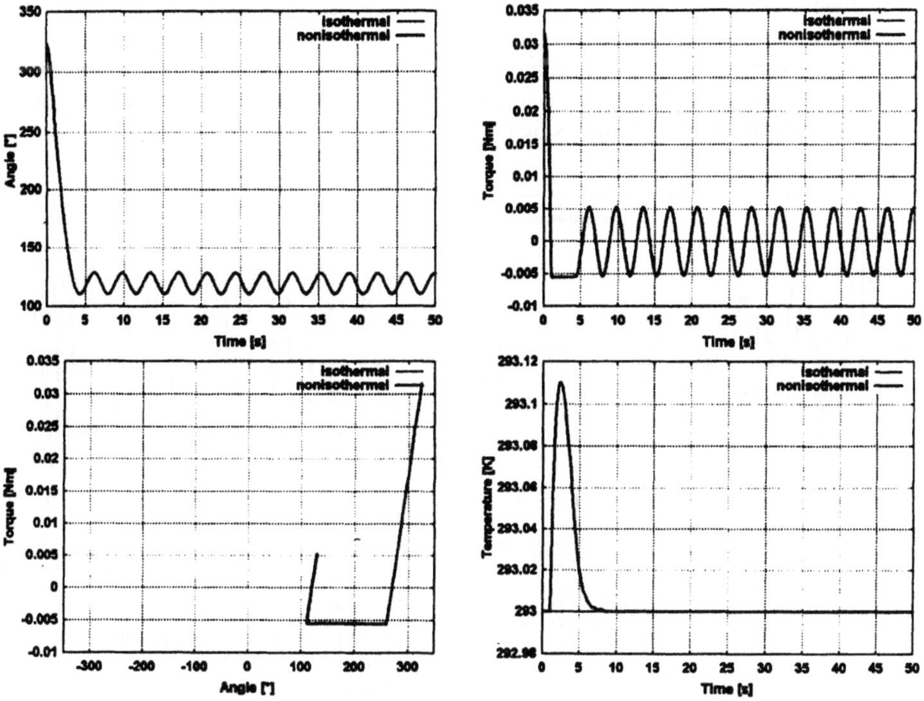

Figure 2. Free vibration at 293K (quasiplastic behavior). Isothermal and nonisothermal case are almost identical.

Figure 2 shows the behavior at room temperature. The motion is strongly damped in the beginning and then exhibits an undamped, small-amplitude oscillatory behavior about an angle of $\Phi = 120°$, see the upper left diagram. This limit cycle behavior is somewhat artificial and would not be observed in an experiment. It is due to the fact that the model does not include the internal damping mechanism due to lattice imperfections, impurities etc., which is present in every metal. This is a contribution which is in good approximation proportional to the velocity, and, of course, it could easily be included in the simulation. However, it is orders of magnitude smaller than the dissipation due to the hysteretic phase transformations. Moreover, it would make it more difficult to reveal the difference between the isothermal and nonisothermal case, thus it is omitted here.

In the upper right hand diagram, the torque in the SMA tube is plotted as function of time, and the lower left hand diagram shows what could be called a dynamic hysteresis plot of torque vs. angle. This diagram is important for the understanding of the strong damping, because it indicates that the SMA tube behaves like a simple linear spring in the beginning of the motion. However, it quickly enters the regime where a phase transition occurs. Due to the withdrawal of kinetic energy, an angle is reached where the driving force is not sufficient any more for the phase transformation to proceed. A harmonic oscillation at fixed phase fraction about this angle is the result, see the lower left hand diagram for illustration. In an experiment, the motion would come to a rest at this point due to the always present velocity-dependent damping.

In the lower right hand diagram, it can be seen that the temperatures are almost identical in both cases. This is a consequence of the latent heats being only very small for the transitions between martensitic twins.

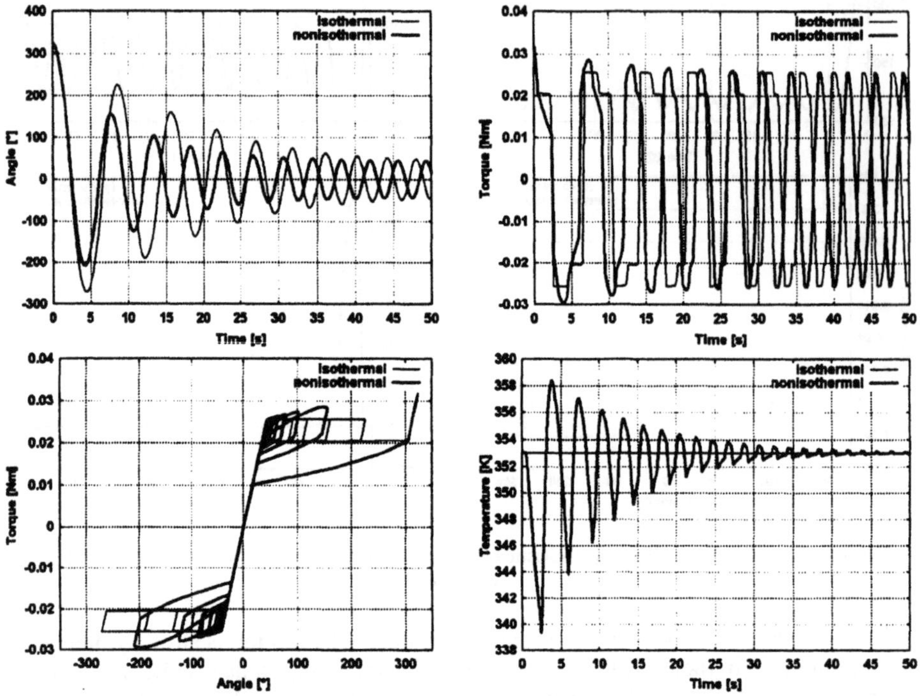

Figure 3. Free vibration at 353K (pseudoelastic behavior). Isothermal and nonisothermal case are drastically different due to the latent heats.

While for the quasiplastic case, it does not appear to make a big difference whether the energy balance is included or not, the situation is drastically different at elevated temperatures, where the behavior is pseudoelastic. The motion is noticeably more strongly damped in the nonisothermal case, and furthermore, the torque in the shape memory alloy reaches a higher maximum value during the first few cycles compared to the isothermal case. In the lower right hand diagram of Figure 3, the temperature in the SMA tube is plotted, and a pronounced deviation from isothermal behavior can be observed. The temperature is equal to the environmental temperature of 353K initially, but cools down to less than 340 K during the first quarter of the first cycle. The reason is that the material transforms from its initial martensitic state (100% M_+) to austenite during this part of the motion, and this transition wants to absorb heat from the environment. However, for the large amount required, the heat exchange coefficient is too small. The energy is taken from the material itself instead, thus lowering its temperature. During the second quarter of the cycle, a transformation from austenite to the other martensite variant M_- takes place, which releases heat and leads to an increase in temperature again. Due to the damping induced by the phase transition, the amplitude of the motion decreases continuously, causing the heating not to be as strong as the cooling during the first quarter. Further, one observes an oscillation in the temperature dictated by the two transitions from austenite to martensite and vice versa. This oscillation is slowly decaying until it reaches a steady state at environmental temperature. Finally, in the lower left hand corner of the figure, the path of the process is plotted in load-deformation space, illustrating the differences in the hysteresis curves for both cases.

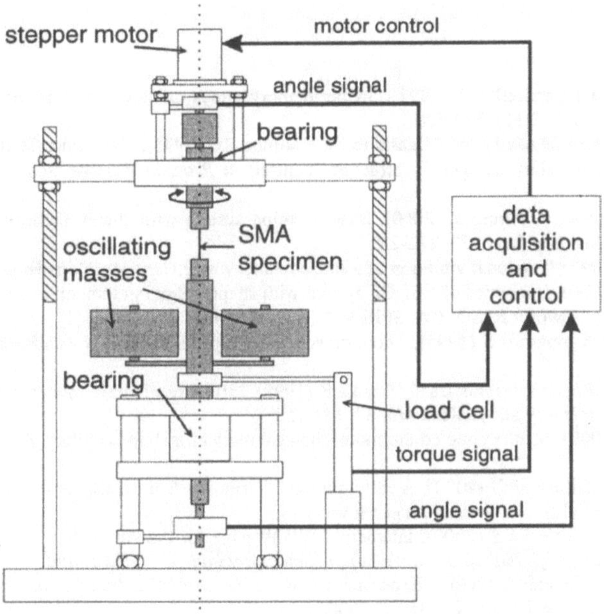

Figure 4. Torsion pendulum for the investigation of free and forced vibration of SMA tubes.

4. Conclusions and Perspectives

The paper has studied the dynamic behavior of a simple SDOF system with a shape memory element by means of a numerical simulation. Two different models have been used, one which is purely mechanical, i.e. one which considers the temperature in the SMA to remain constant during the vibration. A second – thermomechanical -model takes the heat exchange with the environment and the latent heats into account through the inclusion of the energy balance for the SMA.

Models of the first kind are predominantly being used in the literature; however, the calculations clearly show that, while this may be appropriate for quasiplastic behavior, for the case of pseudoelasticity, the obtained results may lead to a considerable underestimation of damping and resulting forces in the system.

To further investigate these effects, a torsion pendulum has been designed and built as the result of a final year project at TU Berlin, see [11]. It permits the experimental investigation of the free vibration case and the effects that have been pointed out above. Moreover, it is possible to study the case of forced vibration and deformation-controlled experiments at different strain rates by means of an included stepper motor, see Figure 4 above.

The support of the Deutsche Forschungsgemeinschaft (DFG) sponsoring this work through project Mu 313/16-2 is gratefully acknowledged.

5. References

1. Graesser EJ and Cozarelli FA (1991), Shape memory alloys as new materials for seismic isolation, *J Eng Mech ASCE*, 117(11), 590-608
2. Clark PW, Aiken ID, Kelly JM, Higashino M, Krumme RC (1995), Experimental and analytical studies of shape memory alloy damper for structural control, in *Proc. of Passive Damping*, San Diego, CA, 1995
3. Wilde K, Gardoni P, Fujino Y (2000), Base isolation system with shape memory device for elevated highway bridges, *Eng Struct* 22, 222-229
4. Seelecke S (1997), Torsional vibration of a shape memory wire,*Cont Mech Thermodyn* 9, 165-173
5. Seelecke S (2000), Dynamics of a SDOF system with shape memory element, in *Proc. Smart Structures and Materials*, Newport Beach, CA, SPIE Vol. 3992, 474-481
6. Shaw JA and Kyriakides S (1995), Thermomechanical aspects of NiTi, *J Mech Phys Solids* 43, 1243-1281
7. Tobushi H, Shimeno I, Hachisuka T, Tanaka K (1998), Influence of strain rate on superelastic properties of NiTi shape memory alloy, *Mech Mat* 30, 141-150
8. Seelecke S (2000), A fully coupled thermomechanical model for shape memory alloys, Part I: Theory, in preparation
9. Seelecke S and Kastner O (2000), A fully coupled thermomechanical model for shape memory alloys, Part II: Numerical simulations, in preparation
10. Heintze O and Seelecke S (2000), Interactive WWW page for the simulation of shape memory alloys, *http://www.thermodynamik.tu-berlin.de/haupt/simulation/Sma_Sim_Home.html*
11. Sahota H and Heintze O (2000), Experimental investigation of the dynamic behavior of SMAs under torsional loading, Tech. Rep. TU Berlin, 2000

DYNAMIC INSTABILITY OF LAMINATED PIEZOELECTRIC SHELLS

X. M. YANG, Y. P. SHEN, X. G. TIAN
Department of Engineering Mechanics, Xi'an Jiaotong University
Xi'an, Shaanxi Province, 710049, People's Republic of China

1. Introduction

In recent years, smart structures with piezoelectric sensors and actuators have attracted serious attention for they can sense and alter the mechanical response during in-service operation. On the other hand, light-weight shell type structures may be one of the most popularly used structures in space vehicles. For this reason, shell type smart structures have become the focus of study for many researchers. Tzou and his coworkers[1-3] studied piezoelectric shell type continua using finite element method and analytical analysis method. Chen and Shen[4,5] performed the study of exact studies of piezoelectric circular cylindrical shells and piezothermoelastic shells, also, they studied the stability of piezoelectric circular cylindrical shells[6].
Dynamic instability of circular cylinders has been studied by many researchers. The parametric resonance of cylindrical shells under axial loads was first treated by Bolotin[7], Yao[8,9] and Tamura[10], etc. For thin cylindrical shells under periodic axial loads, the method of solution is usually first to reduce the equation of motion to a system of Methieu's equations. Bert and Birman[11] studied the parametric instability of thick, orthotrpic, circular cylindrical shells using first-order shear deformable shell theory. The perturbation method is employed by Argento and Scott[12,13] in the study of dynamic instability of layered anisotropic circular cylindrical shells. In recent study, Ng, Lam and Reddy[14] investigated the effects of different lamination schemes of antisymmetric cross-ply laminates on the instability regions of the laminated cylindrical shells. Lam and Ng [15] studied the dynamic stability of thin isotropic cylindrical shells using four common thin shell theories; namely, the Donnell, Love, Sanders and Flugge shell theories. Up to now, to the authors' knowledge, however, there is no paper published about the dynamic instability of piezoelectric circular cylindrical shells.
In this paper, dynamic instability of a simple supported, finite-length, laminated circular cylindrical shells covered by piezoelectric materials, which are distributed symmetrically, is studied. The shell is subjected to uniformly distributed cyclic axial loading and including the effects of transverse shear deformation. The effect of the piezoelectric effect on the dynamic instability is considered. The result can be useful in an active control.

U. Gabbert and H.S. Tzou (eds.), IUTAM Symposium on Smart Structures and Structronic Systems, 153–160.

2. Assumptions

The analysis is based on the following assumptions, most of which can be found in[11]:

1. The analysis is linear, including both linear constitutive relations for the material and linear strain-displacement relations.
2. The shell is circular cylindrical without initial imperfections.
3. The in-surface and rotary inertia are neglected.
4. The loading considered is axial, assumed to be uniformly distribution and to consist of a constant portion and a simple harmonic excitation over the end sections.
5. The perfect bonding is considered between different layers.
6. The directions of polarization of piezoelectric layers are in the same direction.
7. All damping effects are neglected.
8. The unimodal approximation of solution is adopted and several isolated dynamic instability mode shapes are analyzed in numerical examples in order to take into account the effect of piezoelectricity.

3. Analysis

Consider an orthotropic circular cylindrical cross-ply laminated shell covered by piezoelectric material symmetrically. The shell is subjected to uniformly distributed, parametric, time-dependent loads of intensity $N_1(t)$ along axial direction. According to Love's and Loo's theory, the equations of motion of such a shell are (the effect of in-surface and rotary inertias has been neglected),

$$N_{1,x}+N_{6,y}=0 \tag{1}$$

$$N_{6,x}+N_{2,y}+\frac{1}{R}Q_4=0 \tag{2}$$

$$Q_{5,x}+Q_{4,y}-\frac{N_2}{R}+N_1^{\,0}(t)w_{,xx}=\rho h w_{,tt} \tag{3}$$

$$M_{1,x}+M_{6,y}-Q_5=0 \tag{4}$$

$$M_{6,x}+M_{2,y}-Q_4=0 \tag{5}$$

The charge equilibrium equation is:

$$D_{z,z}+D_{y,y}+D_{x,x}+\frac{1}{R}D_z=0 \tag{6}$$

where M_i and N_i denote the stress couples and in-surface stress resultants respectively; Q_{ij} are the shearing stress resultants; u, v, w are displacements along the axis x (axial), axis y (circumferential), axis z (radial); respectively, ψ_1 and ψ_2 are the bending slopes in the x-z and y-z planes, t is the time, $(\ldots)_{,i}\equiv\partial(\ldots)/\partial i$, ρ is the material density, and D_i is the electric displacement along the i-axis.

Supposed the voltage distribution is $\phi=z\phi_0(x,y)$. The constitutive equations of a specially material are:

$$\begin{Bmatrix} N_1 \\ N_2 \\ N_6 \\ M_1 \\ M_2 \\ M_6 \end{Bmatrix} = \begin{bmatrix} A_{11} & A_{12} & 0 & 0 & 0 & 0 \\ A_{12} & A_{22} & 0 & 0 & 0 & 0 \\ 0 & 0 & A_{66} & 0 & 0 & 0 \\ 0 & 0 & 0 & D_{11} & D_{12} & 0 \\ 0 & 0 & 0 & D_{12} & D_{22} & 0 \\ 0 & 0 & 0 & 0 & 0 & D_{66} \end{bmatrix} \begin{Bmatrix} \varepsilon_1{}^0 \\ \varepsilon_2{}^0 \\ \varepsilon_6{}^0 \\ \chi_1 \\ \chi_2 \\ \chi_6 \end{Bmatrix} - \begin{bmatrix} c_{31} \\ c_{32} \\ 0 \\ f_{31} \\ f_{32} \\ 0 \end{bmatrix} \{-\phi_0\} \tag{7}$$

The extensional, coupling, and bending stiffness are defined as

$$(A_{ij}, D_{ij}) = \int_{-h/2}^{h/2} (1, z^2) Q_{ij} dz + \int_{h/2}^{h/2+h_p} (1, z^2) E_{ij} dz + \int_{-h/2-h_p}^{-h/2} (1, z^2) E_{ij} dz \tag{8}$$

$$\begin{aligned} (c_{ij}, f_{ij}) &= \int_{h/2}^{h/2+h_p} (1, z) e_{ij} dz + \int_{-h/2-h_p}^{-h/2} (1, z) e_{ij} dz \\ \bar{\varepsilon}_{ij} &= \int_{h/2}^{h/2+h_p} z\varepsilon_{ij} dz + \int_{-h/2-h_p}^{-h/2} z\varepsilon_{ij} dz \end{aligned} \tag{9}$$

where, E_{ij} are the modulus of the piezoelectric film PVDF, and h_p is the thickness of the PVDF film, Q_{ij} are the plane-stress transformed reduced elastic stiffness, and ε_{ij}, e_{ij} are the dielectric constant and piezoelectric stress coefficient respectively. The shear stress resultants are:

$$\begin{Bmatrix} Q_4 \\ Q_5 \end{Bmatrix} = \begin{bmatrix} S_{44} & 0 \\ 0 & S_{55} \end{bmatrix} \begin{Bmatrix} \varepsilon_4 \\ \varepsilon_5 \end{Bmatrix} + \begin{bmatrix} 0 & f_{24} \\ f_{15} & 0 \end{bmatrix} \begin{Bmatrix} \phi_{0,x} \\ \phi_{0,y} \end{Bmatrix} \tag{10}$$

where the thickness shear stiffness are

$$S_{ii} = k_i^2 \int_{-h/2}^{h/2} Q_{ii} dz \qquad (i = 4,5) \tag{11}$$

and k_i^2 is the shear correction coefficient. The electric displacements are:

$$\begin{Bmatrix} D_x \\ D_y \\ D_z \end{Bmatrix} = \begin{bmatrix} 0 & 0 & 0 & e_{15} & 0 \\ 0 & 0 & e_{24} & 0 & 0 \\ e_{31} & e_{32} & 0 & 0 & 0 \end{bmatrix} \begin{Bmatrix} \varepsilon_1^0 \\ \varepsilon_2^0 \\ \varepsilon_4 \\ \varepsilon_5 \\ \varepsilon_6^0 \end{Bmatrix} + \begin{bmatrix} \varepsilon_{11} & & \\ & \varepsilon_{22} & \\ & & \varepsilon_{33} \end{bmatrix} \begin{Bmatrix} -z\phi_{0,x} \\ -z\phi_{0,y} \\ -\phi_0 \end{Bmatrix} \tag{12}$$

In equations above, $\varepsilon_i{}^0$ are the middle-surface engineering strain components, ε_4 and ε_5, the transverse shear strains, and χ_i, the curvature and twist changes given by

$$\begin{aligned} &\varepsilon_1^0 = u_{,x}, \ \varepsilon_2^0 = v_{,y} + w/R \\ &\varepsilon_6^0 = u_{,y} + v_{,x} \\ &\varepsilon_4 = \psi_2 + w_{,y} - \frac{v}{R}, \ \varepsilon_5 = \psi_1 + w_{,x} \\ &\chi_1 = \psi_{1,x}, \ \chi_2 = \psi_{2,y} \\ &\chi_6 = \psi_{1,y} + \psi_{2,x} + (v_{,x} - u_{,y})/2R \end{aligned} \tag{13}$$

Because of the symmetric arrangement of PVDF film, the relationship of voltage of the two piece of PVDF film is:

$$\phi^{up} = -\phi^{down} \tag{14}$$

So, the even term of z which related with ϕ will be zero when the integral is along z-axis.

The shell is freely supported. The boundary conditions considered in [10] are used:

$$\begin{aligned} &N_1(0,y) = N_1(L,y) = 0 \qquad v(0,y) = v(L,y) = 0 \\ &M_1(0,y) = M_1(L,y) = 0 \qquad \psi_2(0,y) = \psi_2(L,y) = 0 \\ &w(0,y) = w(L,y) = 0 \end{aligned} \tag{15}$$

and the electric boundary conditions:

$$\phi(0,y) = \phi(L,y) = 0 \tag{16}$$

If the shell is a cylindrical curved panel (open shell) freely supported, which requires that the following additional boundary conditions are satisfied[10]:

$$\begin{aligned} &N_2(x,0) = N_2(x,b) \quad M_2(x,0) = M_2(x,b) \\ &w(x,0) = w(x,b) \quad u(x,0) = u(x,b) \\ &\psi_1(x,0) = \psi_1(x,b) \quad \phi(x,0) = \phi(x,b) \end{aligned} \tag{17}$$

The boundary conditions (15) and (16) are satisfied if

$$\begin{aligned} &u = U(t)h\sin\alpha x\cos\beta y \quad \psi_1 = X(t)\cos\alpha x\sin\beta y \\ &v = V(t)h\cos\alpha x\sin\beta y \quad \psi_2 = Y(t)\sin\alpha x\cos\beta y \\ &w = W(t)h\sin\alpha x\sin\beta y \quad \phi_0 = \Phi(t)h\sin\alpha x\sin\beta y \end{aligned} \tag{18}$$

where, $\alpha \equiv m\pi / L$ and $\beta \equiv n / R$ for a complete cylinder and $n\pi / b$ for a panel.

The effect of in-surface and rotary inertias on vibration of the shell at frequencies near the fundamental frequency are negligible. Since dynamic stability is most important in case in which the excitation frequencies are of the same order as the fundamental frequency, these inertias are neglected here. Substituting Eqs(7), (10), (12) and (18) into Eqs. (1)-(6) and integrating the Eq.(6) along z-axes, then one can obtain the following set of equations:

$$[C_{ij}]\{U,V,W,X,Y,\Phi\}^{\mathrm{T}} = \{0,0,\frac{\rho h^2}{E_{\mathrm{T}}}W_{,tt},0,0,0\}^{\mathrm{T}} \tag{19}$$

Representing the nondimensional x-axes force by

$$\overline{N}^0{}_1 = \overline{N}_0 + \overline{N}_1\cos 2\omega t \tag{20}$$

where $\overline{N}_0$ is a constant portion and always set to zero in next sections. $\overline{N}_1$ is the swing of the simple harmonic excitation. Then the set (19) can be reduced to a single linear second-order differential equation as following:

$$W_{,\tau\tau} + (a_0 - 2q\cos 2\tau)W = 0 \tag{21}$$

This is the well-known Mathieu's equation. Where, τ is a nondimensional time parameter ($= \omega t$), and the Mathieu parameters are

$$\begin{aligned} &a_0 = (1/\varpi^2)(L/h)^4 K_w, q = (1/2\varpi^2)(L/h)^4\overline{N}_1\alpha^2 \\ &\varpi^2 = (\rho L^4 / E_{\mathrm{T}}h^2)\omega^2 \end{aligned} \tag{22}$$

The boundary of the instability regions of the solutions of the Mathieu's equation are tabulated. The boundary of the first instability region are given by Mclachchlan[16]

$$a_0 = 1 \mp q - \frac{1}{8}q^2 \pm \frac{1}{64}q^3 - \cdots \tag{23}$$

If q is small enough so that the series converges. The higher instability regions are not always realized if the shell vibrates with limited amplitudes due to damping.

It is convenient to show the boundary of the instability regions on the frequency-load plane where the horizontal axis corresponds to the squared nondimensional frequency $\overline{\omega}^2$ and the vertical axis represents the nondimensional amplitude of the load $\overline{N}_1$.

If the term q is so small that nonlinear terms in K_w can be neglected, the boundaries of the first instability region are represented by the following relations:

$$\overline{\omega}^2 = (L/h)^4 (K_w \mp \frac{1}{2}\overline{N}_1 \alpha^2) \tag{24}$$

4. Results and Discussion

A two-layers cross-ply shell is considered. The material properties taken were the following:

$$E_L / E_T = 40, G_{LT} / E_T = G_{LZ} / E_T = 0.6, G_{TZ} / E_T = 0.5, \nu_{LT} = 0.25, k_4^2 = k_5^2 = \frac{5}{6}$$

$$E_T = 6.85 \times 10^9 \, Pa$$

The dimension of the shell

$$h = 0.02m, \qquad b/R = \theta = \frac{\pi}{6}$$

For the shell without piezoelectric material, the dimensionless stretching, bending, and transverse shear stiffness are as reference[11]. The material properties of the PVDF are taken as $e_{31} = 0.044c/m^2$, $\varepsilon_{11} = 0.1 \times 10^{-9} F/m$, $\varepsilon_{22} = 0.1 \times 10^{-9} F/m$, $\rho = 1800 kg/m^3$, $E_{11} = E_{22} = 2 \times 10^9 \; Pa$, $G_{12} = G_{13} = 0.775 \times 10^9 Pa$, $G_{23} = 0.775 \times 10^9 Pa$. The thickness of PVDF film, $h_p = 50 \mu m$. Let $\beta = n\pi / b$, the method in Bert and Birman[11] can be used for the calculation of the dynamic instability of a cylindrical panel. For comparison with results in [11], the axisymmetric dynamic instability regions(n=0) of cylindrical shells are also calculated and shown in Tables and Figures. Fig. 1 and Fig. 2 present the boundaries of instability regions corresponding to different shells. In these figures the boundaries of instability regions which consider the effect of PVDF film and which don't consider the effect of PVDF film are almost overlapped, this claims that the effect of PVDF film is very slight.

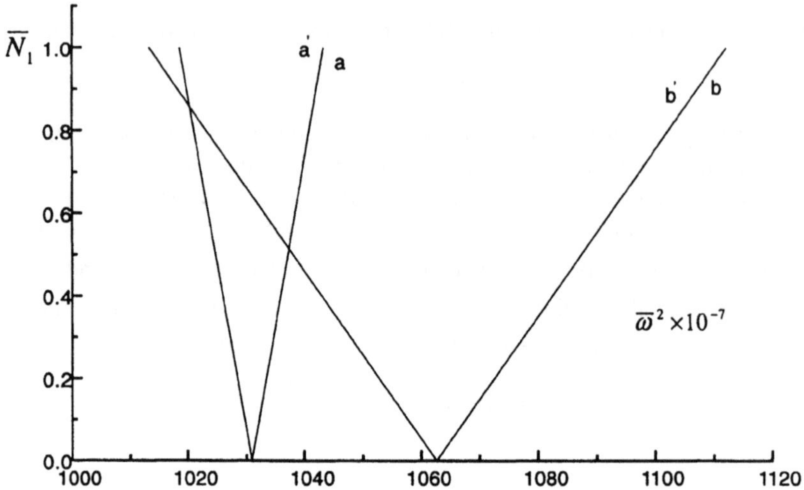

Fig. 1 L/R=1, R/H=5, a, b(solid) for (m,n)=(1,5), (2,5) without PVDF film respectively and a' b'(dash) for the case with PVDF film

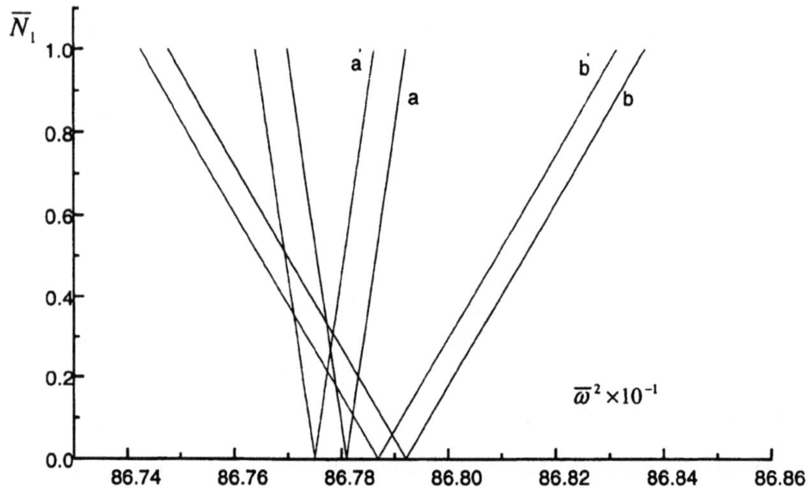

Fig. 2 L/R=10, R/H=15, a,b(solid) for (m,n)=(1,5), (2,5) without PVDF film respectively and a' ,b' (dash) for the case with PVDF film

Table 1 presents a comparable result of a short thick cylindrical panel (L/R=1, R/H=5), and Table 2 presents the result of a long thin cylindrical panel (L/R=10, R/H=15). Four mode shapes are considered here, it is m(the number of half waves along the shell axis)=1 and 2, n(the number of circumferential half waves)=5. Where L stands for including piezoelectric layers but ignoring piezoelectric effect, and P stands for considering piezoelectric effect.

From the result of calculation, one can draw a conclusion: To short thick shell, the PVDF film hardly affects the dynamic instability. If the precision is improved enough, then, to long thin shell, the dynamic instability frequency is slightly waned due to the piezoelectric effect.

When using PVDF film as actuators, it is equivalent to apply the load. Theoretically, the size of instability zone and instability frequency can be altered freely. There is, however, almost no variation of dimensionless dynamic instability frequency until the voltage reaches 10^{11}V, the voltage will cause the electric field which will damage the PVDF film.

If by using of piezoelectric ceramic PZT, the result is different from using PVDF film because its modulus and piezoelectric constant are greater than PVDF's.

Table 1 Effect of PVDF film on a short, thick shell (L/R=1, R/H=5)

M	n	$\bar{\omega}_0^2 \times 10^{-3}$ ($\bar{N}_1 = 0$)		$\bar{\omega}_1^2 \times 10^{-3}$ ($\bar{N}_1 = 1$)		$\bar{\omega}_2^2 \times 10^{-3}$ ($\bar{N}_1 = 1$)	
		L	P	L	P	L	P
1	5	10.3094	10.3095	10.1860	10.1861	10.4328	10.4328
1	0	0.57975	0.57975	0.45638	0.45638	0.70312	0.70312
2	5	10.6268	10.6269	10.1333	10.1335	11.1203	11.1204
2	0	0.89906	0.89906	0.40558	0.40558	1.39254	1.39254

Table 2 Effect of PVDF film on a long thin shell (L/R=10, R/H=15)

M	n	$\bar{\omega}_0^2 \times 10^{-7}$ ($\bar{N}_1 = 0$)		$\bar{\omega}_1^2 \times 10^{-7}$ ($\bar{N}_1 = 1$)		$\bar{\omega}_2^2 \times 10^{-7}$ ($\bar{N}_1 = 1$)	
		L	P	L	P	L	P
1	5	86.7810	86.7751	86.7699	86.7640	86.7921	86.7862
1	0	4.61183	4.61184	4.60073	4.60073	4.62293	4.62294
2	5	86.7921	86.7869	86.7477	86.7425	86.8365	86.8313
2	0	4.61208	4.61208	4.56767	4.56767	4.65649	4.65649

Table3 Effect of PZT on a long thin shell (L/R=10, R/H=15)

m	n	$\bar{\omega}_0^2 \times 10^{-7}$ ($\bar{N}_1 = 0$)		$\bar{\omega}_1^2 \times 10^{-7}$ ($\bar{N}_1 = 1$)		$\bar{\omega}_2^2 \times 10^{-7}$ ($\bar{N}_1 = 1$)	
		L	P	L	P	L	P
1	5	143.379	143.281	143.368	143.270	143.390	143.292
1	0	5.04993	5.03446	5.03882	5.02336	5.06103	5.04556
2	5	143.442	143.316	143.397	143.271	143.486	143.360
2	0	5.05026	5.03049	5.00584	4.98607	5.09467	5.07490

Table 3 presents a comparable result of different modes. The comparison clearly shows that the effect of PZT is slight. Material properties of the PZT used in Table 3 are as follows:

$$[c] = \begin{bmatrix} 13.90 & 7.78 & 7.43 & & & \\ 7.78 & 13.90 & 7.43 & & & \\ 7.43 & 7.43 & 11.50 & & & \\ & & & 2.56 & & \\ & & & & 2.56 & \\ & & & & & 3.06 \end{bmatrix} \times 10^{10}\, Pa$$

$$e_{31} = e_{32} = -5.20 C/m^2$$
$$\rho = 7.5 \times 10^3 Kg/m^3$$
$$h_p = 0.001 m$$
$$\varepsilon_{11} = \varepsilon_{22} = 730\varepsilon_0$$
$$\varepsilon_{33} = 635\varepsilon_0$$

where $\varepsilon_0 = 8.854 \times 10^{-12} F/m$ and other terms not shown here are assumed to be zero.

When using PZT as actuators, the effect of piezoelectric is notable compared with PVDF due to the larger piezoelectric constant. But the voltage still reaches $10^8 V$. It's obviously too high to work for PZT.

5. Conclusion

Comparison of the effect of PZT and PVDF film shows that PZT has a stronger effect than PVDF film. However, Neither PZT nor PVDF film has strong effect on the unstable region of the structure. Dynamic instability regions mainly depend on the mechanic load.

Acknowledgment: This work was supported by National Natural Science Foundation of China. The authors would like to express thanks to them.

Reference

1. Tzou, H.S. and Tseng, C.L.: Distributed modal identification and vibration control of continua: piezoelectric element formation and analysis, *J. of Dynamic Systems Measurement and control* 113(1991), 500-505.
2. Tzou, H.S. ,Zhong, J.P. and Holkamp, J.J.: Spatially distributed orthogonal piezoelectric shell actuators: theory and application, *J. Sound Vibration* 117(1994), 363-378.
3. Tzou, H.S. and Bao, Y.: Nonlinear Piezothermoelasticity and Multi-Field Actuations, part 1: Nonlinear Anisotropic Piezothermoelastic shell Laminates, ASME *J. of Applied Mechanics* 119(1997), 374-381.
4. Chen, C.Q. and Shen,Y.P.: Piezothermoelasticity analysis for circular cylindrical shell under the state of axisymmetric deformation, *Int. J. Engineering Science* 34(1996), 1585-1600.
5. Chen, C.Q., Shen, Y.P. and Wang, X.M.: Exact solution of orthotropic cylindrical shell with piezoelectric layers under cylindrical bending, *Int. J. Solid and Structures* 33(1996), 4481-4494.
6. Chen, C.Q. and Shen, Y.P.: Stability analysis of piezoelectric circular cylindrical shells, ASME, *J. of Applied Mechanics* 64(1997), 847-862.
7. Bolotin, V. V.: *The Dynamic Stability of Elastic Systems*, Holden-Day, San Francisco, 1964
8. Yao, J. C.: Dynamic stability of cylindrical shells under static and periodic axial and radial loads, *AIAA J.* 1(1963), 1391-1396
9. Yao, J. C.: Nonlinear elastic bucking and parametric excitation of a cylinder under axial loads, ASME *J. of Applied Mechanics*,29(1965), 109-115.
10. Tamura. Y.S. and Babcock, C.D.: Dynamic stability of cylindrical shells under step loading, ASME *J. of applied mechanics*, 42(1975), 190-194.
11. Bert, C.W. and Birman,V.: Parametric instability of thick, orthotropic, circular cylindrical shells, *Acta Mechanics* 71(1988), 61-76.
12. Argento, A and Scott, R. A.: Dynamic instability of layered anisotropic circular cylindrical shells, Part I: theoretical development, *J. of Sound and Vibration* 162(1993), 311-322.
13. Argento, A and Scott, R. A.: Dynamic instability of layered anisotropic circular cylindrical shells, Part II: numerical results, *J. of Sound and Vibration* 162(1993), 323-332.
14. Ng, T.Y., Lam, K.Y. and Reddy, J.N.: Dynamic stability of cross-ply laminated composite cylindrical shells, *Int. J. Mechanics Science*, 40(1998), 805-823.
15. Lam, K.Y. and Ng, T.Y.: Dynamic stability of cylindrical shells subjected to conservative periodic axial loads using different shell theories, *J. of Sound and Vibration* 207(1997), 497-520.
16. McLachlan, N.W.: *Theory and Application of Mathieu Functions*, Dover Publications, New York, 1964.

FLEXURAL ANALYSIS OF PIEZOELECTRIC COUPLED STRUCTURES

Q. WANG and S.T. QUEK
Department of Civil Engineering, National University of Singapore,
1 Engineering Drive 2, Singapore 117576.

1. Introduction

The analysis of a coupled piezoelectric structure has recently been keenly researched because piezoelectric materials are more extensively used either as actuators or sensors. The challenge of developing a basic mechanics model for the piezoelectric coupled structure has been met by many researchers. Crawley and deLius (1987) developed a uniform strain model for a beam with surface bonded and embedded piezoelectric actuator patches accounting for the shear lag effects of the adhesive layer between the piezoelectric actuator and the beam. A model to account for the coupling effect was later proposed based on the Euler beam assumption (Crawley and Anderson, 1989). Leibowitz and Vinson (1993) derived a model based on Hamilton's principle in which the elastic layers, soft-core layers or piezoelectric layers are included. Recently, Zhang and Sun (1999) presented their research on the analysis of a sandwich beam and plate structure containing a piezoelectric core. In most published literature on the mechanics model for the analysis of the coupled structure, the distribution of the electric potential is assumed to be uniform in the longitudinal direction of the piezoelectric actuator and linear in its thickness direction, which may violate the Maxwell static electricity equation. To satisfy Maxwell equation, a constant electrical potential distribution in the longitudinal direction and linear distribution in flexural direction cannot be assumed but rather obtained by solving the coupled governing equations by assuming a certain distribution of electric potential in the thickness direction (Wang and Quek, 2000). Recently, the distribution of electric potential for the case where the electrodes are shortly connected has been investigated by researchers. Krommer and Irschik (1999) suggested a parabolic distribution for the electric potential in wave propagation problems of piezoelectric coupled plate structures. Gopinathan et al (2000) adopted the assumption that the electric displacement is null in the longitudinal direction and showed that for long laminae, the electromechanical fields predicted by their first-order shear deformation model agree well with the elasticity solution. However, their model cannot predict the potential distribution inside the sensor layers accurately.

The objective of this paper is to present a basic mechanics model for the dynamic analysis of beam embedded with piezoelectric layers, formulated based on Euler beam

U. Gabbert and H.S. Tzou (eds.), IUTAM Symposium on Smart Structures and Structronic Systems, 161–168.

theory. Two different ways of connecting the electrodes of piezoelectric layer are considered, namely, the electrodes are shortly connected (Case I) and the electrodes are left open (Case II). For Case I, the half-cosine distribution is shown theoretically to be the solution of a simple piezoelectric beam subjected to uniform moment and numerically verified by the FEM considering two boundary conditions, namely simply-supported and propped cantilever. For Case II, the semi-analytical method is used to obtain the coupled governing equation for electrical and mechanical variations by assuming an expression for the electric potential in the piezoelectric layers.
In the second part of this paper, the free vibration analysis, static response, and the dynamic response of a piezoelectric coupled circular plate are presented based on the above assumption on the electric potential distribution in the thickness direction of the plate.

2. Model for Dynamic Analysis of Sandwich Piezoelectric Beam

Consider the beam embedded with two piezoelectric layers. To formulate the dynamical equations, the Euler beam assumption is adopted and is valid when the ratio of the length to height of the beam is more than 20. In this case, the shear deformation and rotary inertia effect in the structure are omitted.
The two piezoelectric layers are placed symmetrically with poling direction oriented in the transverse direction of the beam. When voltages of equal magnitude but opposite sign are applied to the upper and lower piezoelectric layers of the beam, a differential strain is induced resulting in flexural action. The amplitude of the strain ε and stress σ in the beam and piezoelectric layer can be written correspondingly.
The piezoelectric sensors and actuators are lined with a thin layer of silver electrode on both surfaces to facilitate providing external voltage to actuate the structure or measure the electrical output to sense the internal deformation of the host structure. There are two possible electrical boundary conditions depending on how the electrodes are connected. The first is when the two electrodes are shorted (or shortly connected), herein known as Case I, and the second is when they are left open (Case II). The free vibration analysis is formulated for both cases.

2.1 WITH ELECTRODES SHORTLY CONNECTED

When the electrodes at the two surfaces of the piezoelectric layer are shortly connected, the electric potential is zero throughout the whole of both surfaces. Within the piezoelectric layer, different assumptions for the distribution of the electric potential in the thickness direction will yield different mechanics models for the piezoelectric coupled structures and result in diverse possible solutions. It was recently pointed out by Gopinathan et al. (2000) and Wang and Quek (2000) that the most used assumption of a uniform distribution of the electric potential in the longitudinal direction and a linear distribution in the flexural direction may violate Maxwell static electricity equation. Krommer and Irschik (1999) suggested a parabolic distribution model in the longitudinal direction whereas Gopinathan et al. (2000) assumed the nullity of electric displacement in the longitudinal direction. However, in the limit that the entire structure is of piezoelectric material, it can be shown that the distribution of the electric potential in the flexural direction is a half-cosine distribution when a uniform moment is applied. This assumption is verified numerically by FEM for the case of uniform moment applied to the piezoelectric beam structure. The electric potential distribution in the thickness direction of the piezoelectric layer obtained is shown in Figure 1. This model is thus adopted in this paper for Case I and used as the basic mode in the dynamic analysis of the piezoelectric coupled plate structure.

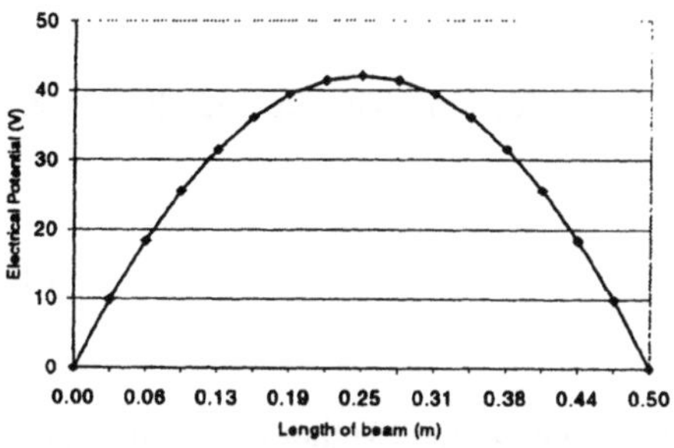

Fig. 1 Electrical Potential Distribution at Section 1-1

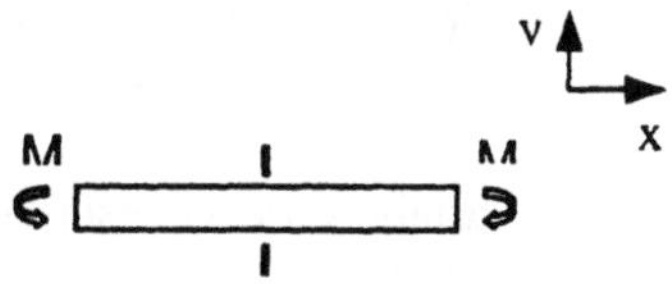

Material: PZT4
Length: 3m
Height: 0.5m

For the general case, the electric potential is assumed as a combination of a half-cosine and linear variation,

$$\phi = \phi(x, y, z, t) = -\cos\frac{\pi z_l}{h_1} \cdot \bar{\phi}(x, y)e^{i\omega t} + \frac{2z_l}{h_1}\phi_a e^{i\omega t} \tag{1a}$$

where z_l is measured from the centre of the piezoelectric layer in the global z-direction, h_1 the thickness of the layer, $\bar{\phi}(x, y)$ the spatial variation of the electric potential in the global x-direction, ϕ_a the value of external electric voltage applied to the electrodes and ω the natural frequency of the plate. As with the geometric displacement $u(x,y)$, ϕ must correspondingly satisfy the electric boundary conditions. Since only wave propagation analysis is addressed in this paper, equation (1a) can be simplified as

$$\phi = -\cos\frac{\pi z_l}{h_1} \cdot \bar{\phi}(x, y)e^{i\omega t} \tag{1b}$$

By studying the state of force balance in an infinitesimal element of the beam structure, we can get beam equilibrium equation coupled with mechanical and electrical variables. By satisfying Maxwell equation yields another coupling equation. These two equations are the base for analysis of the piezoelectric coupled structures.

2.2 WITH ELECTRODES LEFT OPEN

For the case when the surfaces of the electrode layer are not connected, the electric charge at the surfaces of the piezoelectric layer vanishes. A semi-analytical approach is adopted to obtain a solution to this problem. The potential function within the piezoelectric layer is assumed as follows:

$$\phi(x,z,t)=\left[-\frac{h_1}{\pi L}\sin\frac{\pi z_l}{h_1}\bar{\phi}(x)+e_{31}\frac{z}{2\Xi_{33}}\int_0^L\bar{\varepsilon}_x dx/L\right]e^{i\omega t} \tag{2}$$

where the variables are as described in the previous section. The electric charge can then be obtained as

$$Q=\int_0^L\bar{D}_z dx=\Xi_{33}\cos\frac{\pi z_1}{h_1 L}\int_0^L\bar{\phi}dx \tag{3}$$

where Q is the electric charge. It can be seen from equation (3) that by assuming the potential function given by equation (2), the condition that the charge on the surfaces of the piezoelectric layer vanishes is satisfied. The coupling equation can also be obtained for this open circuit case.

2.3 VALIDATION STUDY

To illustrate the implications resulting from the proposed mechanics model, a beam with two embedded piezoelectric actuator layers is considered. Direct finite difference methods are employed to solve the coupled equations.

The fundamental frequency of a coupled steel beam with two different mechanical boundary conditions for Case I, Case II and the case when piezoelectric effects are neglected are listed in Table 1 as well as those from the ABAQUS FE software. The length of the beam is $1m$ and the thickness is $0.04m$. The two mechanical boundary conditions studied correspond to a simply supported beam and a propped cantilever. The two piezoelectric layers are embedded symmetrically h/5 from the surfaces of the beam. The piezoelectric layers used are Piezoceramics C-82 with properties obtained from a commercial catalogue supplied by Fuji. The adhesive layer is very thin and hence neglected in the model. It is seen from Table 1 that good agreement between the results of the proposed models and those by ABAQUS are obtained. Similar conclusions were made by Krommer and Irschik (1999) for Case I where the influence of electromechanical coupling is very small. The results for Case II show a stronger influence in this study.

TABLE 1. First non-dimensional resonant frequency for S-S and C-S beam ($\omega_i / \pi^2 \sqrt{\frac{EI}{\rho A}}$)

Cases	S-S	C-S
Proposed Model (no piezo effects)	0.87	1.351
ABAQUS (no piezo effects)	0.87	1.357
Proposed Model (Case I)	0.87	1.351
ABAQUS (Case I)	0.87	1.357
Proposed Model (Case II)	0.89	1.365
ABAQUS (Case II)	0.89	1.363

3. Model for Dynamic Analysis of Circular Piezoelectric Plate

In most practical applications, the ratio of its radius to the thickness of the plate is more than 10, and the Kirchhoff assumption for thin plate is applicable, whereby the shear deformation and rotary inertia can be omitted.

For free vibration analysis, a sinusoidal variation of the electric potential in the transverse direction is assumed, and the potential function can be written as

$$\phi = -\cos\frac{\pi z_l}{h_1} \cdot \hat{\phi}(r) e^{i(p\theta - \omega t)} \tag{4}$$

where p is the wave number.

For static analysis, the time component can be removed from Eq. (4). In addition, it is assumed that there is no variation of the electric potential in the circumferential direction. An additional term to represent the influence of the external applied voltage $\tilde{\phi}$ must be included. Hence, the assumed potential function is given as

$$\phi = -\cos\frac{\pi z_l}{h_1} \cdot \hat{\phi}(r) + \frac{2z_1}{h_1}\tilde{\phi} \tag{5}$$

Potential distribution in the longitudinal direction, $\hat{\phi}(r)$, can be obtained by satisfying Maxwell static electricity equation. In most published papers, only the second term is considered, which implies that the Maxwell electricity equation is not satisfied.

When a time-varying voltage is applied to the piezoelectric layer, Eq. (4) can be modified as

$$\phi = -\cos\frac{\pi z_l}{h_1} \cdot \hat{\phi}(r) e^{i(p\theta - \omega t)} + \frac{2z_1}{h_1}\tilde{\phi}\, e^{i(p\theta - \omega t)} \tag{6}$$

where $\tilde{\phi}$ is the magnitude of the time-varying voltage applied to the piezoelectric layer.
The coupling equation for the mechanical and electrical variables can again be obtained similar with that in the beam structure logically, but more complicated in formulation.

3.1 STATIC ANALYSIS

For static analysis, the external potential applied to the electrodes is equivalent to a uniform moment applied to this plate. Finally, the flexural deflection is obtained as:

$$\hat{w} = -\frac{2B\tilde{\phi}}{D(1+\upsilon)}\left(r^2 - a^2\right) \tag{7}$$

where $B = 2e^*{}_{31}h$, D and υ are stiffness and Poisson of the plate structure, $e^*{}_{31}$ is the piezoelectric coefficient in plane stress problem.

3.2 DYNAMIC ANALYSIS

When a time-varying external potential is applied across the electrodes, the dynamic response based on Eq. (6) can be estimated by the modal superposition method.
The equation for modal coordinate variable $q_i(t)$ can be obtained herein below,

$$q_i(t) + \omega_i^2 q_i(t) = \left(\frac{M_{rr}(t) + M_{\theta\theta}(t)}{1+\nu}\right)\phi_i'(a) \tag{8}$$

where $\phi_i(r)$ is the mode shape of the coupled plate structure.
Since $M_{rr}(t) = M_{\theta\theta}(t) = B\tilde{\phi}e^{i\omega t}$, we have the vibration response due to the external force as,

$$q_i(t) = \frac{2B\tilde{\phi}}{\omega_i^2 - \omega^2}\frac{\phi_i'(a)}{1+\nu}e^{i\omega t} \tag{9}$$

So the dynamic response of the plate will be expressed in terms of Eq. (9) together with the term due to the initial condition of the plate.
By observing Eq. (9), another conclusion is obtained as a by-product:

- Based on Kirchhoff plate theory, external voltage will only have effect for a simply supported or free edge circular plate system. This is because the external electric force acts as a uniform normal moment in plate structure; hence, a clamped circular plate cannot be actuated by external voltage. "Locking" phenomenon will be experienced for a clamped plate, and there will be no sensor output when the piezoelectric layer is used as sensor.

3.3 FREE VIBRATION ANALYSIS

For clarity of the presentation, we only list the main results in this session. The mode shapes of the mechanical and electrical potential are obtained for different types of beams by direct difference method.

3.3.1 Clamped circular plate

Table 2 lists the first four frequencies of a pure host plate structure and the corresponding piezoelectric coupled structure with the same host material by this theoretical model and finite element analysis for the purpose of comparison. In this example the thickness ratio of the piezoelectric layer and the host plate is 1/10. It can be seen that the results by our model agree well with those from finite element analysis, and the error is within 1.1% for all the first four modes. Another finding is that the piezo-effect plays a role in the vibration of this piezoelectric plate, and the percentile of the increase is around 3.6% for all the four modes. To verify the proposed transverse variation of electric potential, the electric potential distribution obtained from finite element analysis along line at $r=r_0/2$ and $\theta=0$ for the first mode is plotted in Figure 2 after being normalized. The normalized proposed quadratic distribution is also plotted in the same coordinate system for comparison and a close match is obtained.

TABLE 2 - Comparison of the first four resonance frequencies between pure structure and piezoelectric coupled structure (clamped boundary condition)

Mode No.	FEA Results			Analytical Results			Error (%) (Piezoelectric coupled)
	Pure Structure	Piezoelectric coupled	Increments (%)	Pure Structure	Piezoelectric coupled	Increments (%)	
1	868.06	900.15	3.70	869.691	902.479	3.77	0.26
2	1801.3	1866.9	3.64	1809.87	1878.17	3.77	0.60
3	2945.1	3050.9	3.59	2969.34	3081.08	3.76	0.99
4	3355.7	3475.2	3.56	3385.71	3513.43	3.77	1.1

3.3.2 Simply supported circular plate

Table 3 lists the first four frequencies of the piezoelectric coupled structures for different thickness of piezoelectric layers. The piezo-effect is obvious when thicker piezoelectric layer is covered on the host material. Figures 3 plots the first four mode shapes of the electric potential in piezoelectric layer $\hat{\phi}(r)$. Such mode shape of the electric potential is obtained by solving the Maxwell equation based on the assumed distribution of the electric potential in the thickness direction.

TABLE 3 - The first four resonance frequencies under simply supported boundary condition with different piezoelectric layers

Thickness ratio ($h_1/2h$)	First mode	Second mode	Third mode	Fourth mode
1/12	432.192	1217.69	2244.33	2604.22
1/10	435.635	1227.49	2262.44	2625.23
1/8	441.420	1243.94	2292.80	2660.48
1/5	462.337	1303.26	2402.29	2787.56

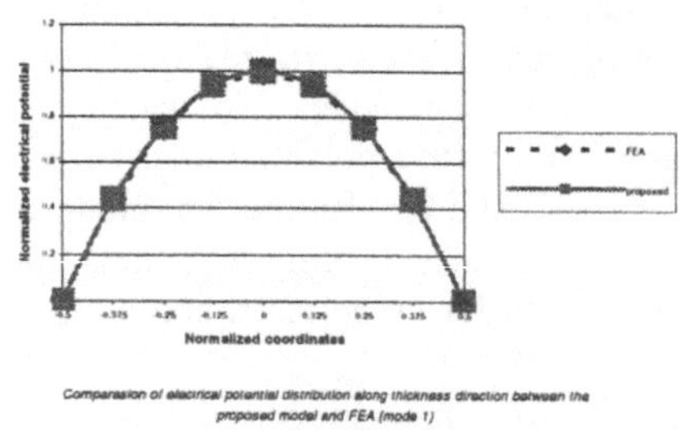

Fig 2. Comparison of electrical potential distribution along thickness direction between the proposed model and FEA (mode 1)

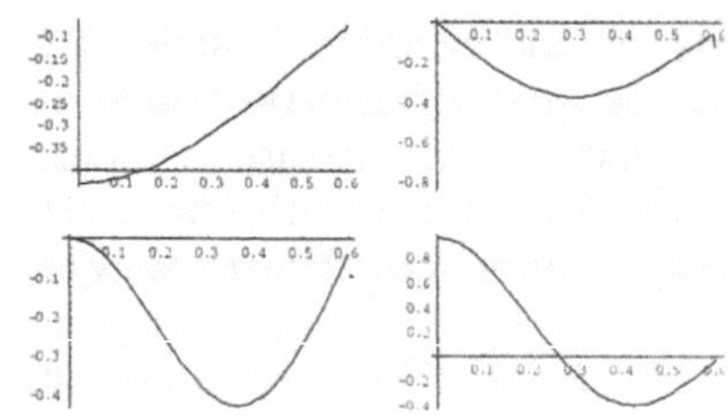

Fig 3. The first four mode shapes of electrical potential in radial direction

4. Conclusion

The coupled governing equations based on Euler model for a long and thin beam and a thin plate structure with surface bonded piezoelectric actuators are investigated. The assumption of electric potential distribution in the thickness direction for the piezoelectric layer is presented and supported by FEM results. Based on this assumption, the piezoelectric effects are concisely investigated, and some numerical simulations for plate structures are conducted.

5. References

1. Crawley, E.F. and Javier de Luis: Use of piezoelectric actuators as elements of intelligent structures, *AIAA Journal* 25(1987), 1373-1385.
2. Crawley, E.F. and Anderson, E.: Detailed model of piezoelectric actuation of beams, *Proceedings of 30th AIAA/ASME/SAE Structures, Structural Dynamics, and Material Conference* (1989). 2000-2010.
3. Leibowitz and Vinson.: The use of Hamilton's principle in laminated piezoelectric and Composite Structures, *Adaptive Structures and Material Systems*, AD-35.(1993).
4. Zhang, X.D. and Sun, C.T.: Analysis of sandwich plate containing a piezoelectric core, *Smart Materials and Structures* 1(1999), 31-40.
5. Wang, Q. and Quek, S.T.:. Flexural vibration analysis of sandwich beam coupled with piezoelectric actuator, *Smart materials and Structures* 9(2000), 103-109
6. Krommer, M and Irschik, H.: On the influence of the electric field on free transverse vibration of smart beams, *Smart Materials and Structures* 8(1999)., 401-410.
7. Gopinathan, S.V., Varadan, V.V., and Varadan, V.K.: A review and critique of theories for piezoelectric laminates, *Smart Materials and Structures* 9(2000), 24-48.

ACTIVE NOISE CONTROL STUDIES USING THE RAYLEIGH-RITZ METHOD

S. V GOPINATHAN, V. V. VARADAN, V. K. VARADAN
Center for the Engineering of Electronic and Acoustic Materials,
The Pennsylvania State University, University Park, PA 16802, USA

1. Introduction

The use of piezoelectric materials in controlling the vibration of continuous structures has grown significantly in recent years. A number of studies using Finite Element (FE) method [1-4] have been made on noise transmission studies for rectangular enclosures through flexible smart panels. In these models, the host plate, actuators and sensors are modeled using 2D and 3D elements, which are later coupled, to the cavity in which the pressure is expressed in terms of rigid cavity modes. Although these earlier FE models predict the behavior of the structural panel and the fluid-structure interaction accurately at low frequencies, at high frequencies the size of the model increases resulting in very long computational time. Further, optimal sensor/actuator placement studies, involve repeated FE remeshing during the iterations, hence a simple model like the RR approach is preferred. The potential and kinetic energies of the panel with surface bonded discrete piezoelectric patches are estimated and the equations of motion for the smart panel are derived using Hamilton's principle. The electric potential inside the piezoelectric patches are assumed to be a quadratic function of thickness coordinate. Classical laminated plate theory is used for modeling the host plate and the electroelastic theory is used to model the surface bonded patches. In electroelastic theory, reduced charge equation is satisfied inside both sensor and actuator patches. For the acoustic enclosure, the cavity pressure is expressed in terms of rigid cavity modes [5]. For the numerical study and to validate the RR approach, the frequencies obtained using RR approach are compared with the FE results for a smart aluminum plate backed cubic cavity.

2. Piezoelectric Patches: Modeling And Formulation

Figure 1 shows a rectangular laminate on which the piezoelectric patches are bonded. The length and width of the patch are 'a' and 'b'. L_x and L_y are the dimensions of the laminate. The piezoelectric patches are polarized in the thickness direction and the major surfaces are covered with electrodes of negligible thickness. The laminate is assumed to lie on the $z = 0$ plane. The classical laminated plate theory, which is an extension of the Kirchoff's (classical) plate theory to laminated composite plates assume the following displacement fields:

$$u(x,y,z,t)=u_0(x,y,t)-z\frac{\partial w_0}{\partial x};\ v(x,y,z,t)=v_0(x,y,t)-z\frac{\partial w_0}{\partial y};\ w(x,y,z,t)=w_0(x,y,t) \quad (1)$$

U. Gabbert and H.S. Tzou (eds.), IUTAM Symposium on Smart Structures and Structronic Systems, 169–177.

where u, v, w are the displacements along the x, y, z coordinate directions, respectively, of a point on the midplane (i.e., z = 0). The Kirchoff's assumption amounts to neglecting both transverse shear and transverse normal effects, i.e., deformation is due to entirely bending and in-plane stretching. This leads to zero transverse strain components (ε_{zz}, γ_{xz}, γ_{yz}). Thus for small strains and moderate rotations the strain-displacement relations take the form

$$\varepsilon_{xx}=\frac{\partial u_0}{\partial x}-z\frac{\partial^2 w_0}{\partial x^2};\ \varepsilon_{yy}=\frac{\partial v_0}{\partial y}-z\frac{\partial^2 w_0}{\partial y^2};\ \gamma_{xy}=\frac{\partial u_0}{\partial y}+\frac{\partial v_0}{\partial x}-2z\frac{\partial^2 w_0}{\partial x\partial y} \tag{2}$$

or shortly $\varepsilon=\varepsilon^{(0)}+z\varepsilon^{(1)}$, where $\varepsilon^{(0)}=\begin{Bmatrix}\varepsilon_{xx}^{(0)}\\ \varepsilon_{xx}^{(0)}\\ \gamma_{xy}^{(0)}\end{Bmatrix}=\begin{Bmatrix}\frac{\partial u_0}{\partial x}\\ \frac{\partial v_0}{\partial y}\\ \frac{\partial u_0}{\partial y}+\frac{\partial v_0}{\partial x}\end{Bmatrix}$ and $\varepsilon^{(1)}=\begin{Bmatrix}\varepsilon_{xx}^{(1)}\\ \varepsilon_{xx}^{(1)}\\ \gamma_{xy}^{(1)}\end{Bmatrix}=\begin{Bmatrix}-\frac{\partial^2 w_0}{\partial x^2}\\ -\frac{\partial^2 w_0}{\partial y^2}\\ -2\frac{\partial^2 w_0}{\partial x\partial y}\end{Bmatrix}$

The linear constitutive relations for the host laminate in the principal material directions are

$$\begin{Bmatrix}\sigma_{xx}\\ \sigma_{yy}\\ \tau_{xy}\end{Bmatrix}^{(k)}=\begin{bmatrix}\overline{Q}_{11} & \overline{Q}_{12} & \overline{Q}_{16}\\ \overline{Q}_{21} & \overline{Q}_{22} & \overline{Q}_{26}\\ \overline{Q}_{16} & \overline{Q}_{26} & \overline{Q}_{66}\end{bmatrix}^{(k)}\begin{Bmatrix}\varepsilon_{xx}\\ \varepsilon_{yy}\\ \gamma_{xy}\end{Bmatrix}^{(k)} \tag{3}$$

where $\overline{Q}_{ij}^{(k)}$ are the transformed stiffness coefficients. The linear constitutive relations written in terms of nonzero stress and strain components for the attached piezoelectric patches are

$$\begin{Bmatrix}\sigma_{xx}\\ \sigma_{yy}\\ \tau_{xy}\end{Bmatrix}^{(a),(s)}=\begin{bmatrix}C_{11}^{(a),(s)} & C_{12}^{(a),(s)} & 0\\ C_{12}^{(a),(s)} & C_{22}^{(a),(s)} & 0\\ 0 & 0 & C_{66}^{(a),(s)}\end{bmatrix}\begin{Bmatrix}\varepsilon_{xx}\\ \varepsilon_{yy}\\ \gamma_{xy}\end{Bmatrix}^{(a),(s)}-\begin{bmatrix}0 & 0 & e_{13}^{(a),(s)}\\ 0 & 0 & e_{23}^{(a),(s)}\\ 0 & 0 & 0\end{bmatrix}\begin{Bmatrix}E_x\\ E_y\\ E_z\end{Bmatrix}^{(a),(s)}$$

$$\begin{Bmatrix}D_x\\ D_y\\ D_z\end{Bmatrix}^{(a),(s)}=\begin{bmatrix}0 & 0 & 0\\ 0 & 0 & 0\\ e_{13}^{(a),(s)} & e_{23}^{(a),(s)} & 0\end{bmatrix}\begin{Bmatrix}\varepsilon_{xx}\\ \varepsilon_{yy}\\ \gamma_{xy}\end{Bmatrix}^{(a),(s)}+\begin{bmatrix}\epsilon_{11}^{(a),(s)} & 0 & 0\\ 0 & \epsilon_{22}^{(a),(s)} & 0\\ 0 & 0 & \epsilon_{33}^{(a),(s)}\end{bmatrix}\begin{Bmatrix}E_x\\ E_y\\ E_z\end{Bmatrix}^{(a),(s)} \tag{4}$$

where $C_{ij}^{(a),(s)}$, $e_{ij}^{(a),(s)}$ and $\epsilon_{ii}^{(a),(s)}$ are the stiffness coefficients at constant electric field, piezoelectric constants and dielectric permittivities at constant state of stress of the piezoelectric material. σ_i and D_i are the stress and electric displacement components. Apart from satisfying the equilibrium equations the piezoelectric patch should satisfy the charge equation $\nabla.\mathbf{D}=0$ everywhere inside. If the patches are thin, we can assume that x and y component electric displacement fields are constant along x and y directions within the patch then the charge equation reduces to $\frac{dD_z}{dz}=0$.

2.1 ACTUATOR PATCH MODEL

The electric potential inside the actuator patch (ϕ^a) is assumed to vary quadratically along the thickness direction as:

$$\phi^a=\phi_0^a+z\phi_1^a+z^2\phi_2^a \tag{5}$$

It is assumed, that for the actuator patches the electrodes located at the laminate patch interface (i.e., @ $z=z_1$) is always grounded and a potential of V^a is applied at the other electrode surface (i.e., @ $z=z_0$). Using these conditions in eqn. (5) we obtain

$$\phi_1^a = \frac{V^a}{h^a} - 2h_a^m \phi_2^a \tag{6}$$

where $h^a = (z_0 - z_1)$ is the thickness of the actuator patch and $h_m^a = \frac{(z_0 + z_1)}{2}$ is the distance of the actuator middle surface from the laminate middle surface. Using the constitutive equation and the reduced charge equation we obtain

$$\phi_2^a = -\frac{e_{31}}{2\epsilon_{33}}\left(\varepsilon_{xx}^{(1)} + \varepsilon_{yy}^{(1)}\right)^{(a)} \tag{7}$$

So from eqns. (5), (6) and (7) the electric potential and electric field inside the actuator patch is:

$$E_z^a = -\frac{d\phi^a}{dz} = -\frac{V^a}{h^a} + e_{31}\left(\frac{z - h_m^a}{\epsilon_{33}}\right)\left(\varepsilon_{xx}^{(1)} + \varepsilon_{yy}^{(1)}\right)^{(a)} \tag{8}$$

Substituting eqn. (8) into eqn. (4) we obtain the stress-strain relation for the actuator patches:

$$\begin{aligned} \sigma_{xx}^a &= C_{11}^a \varepsilon_{xx}^a + C_{12}^a \varepsilon_{yy}^a + \frac{e_{31}V^a}{h^a} + \frac{e_{31}^2 (z - h_m^a)}{\epsilon_{33}}\left(\varepsilon_{xx}^{(1)} + \varepsilon_{yy}^{(1)}\right)^{(a)} \\ \sigma_{yy}^a &= C_{12}^a \varepsilon_{xx}^a + C_{22}^a \varepsilon_{yy}^a + \frac{e_{31}V^a}{h^a} + \frac{e_{31}^2 (z - h_m^a)}{\epsilon_{33}}\left(\varepsilon_{xx}^{(1)} + \varepsilon_{yy}^{(1)}\right)^{(a)} ;\ \tau_{xy} = C_{66}^a \gamma_{xy}^{(a)} \end{aligned} \tag{9}$$

2.2 SENSOR PATCH MODEL

From the reduced charge equation, we observe that the z component electric displacement should be constant along the thickness (z) direction of the sensor patch. Also for the sensor there is no external supply of charge into the patch, so the total charge that appears on the sensor patch electrode surfaces should be zero. This condition for the present case can be written as $D_z\big|_{z=z_n} = -D_z\big|_{z=z_{n-1}}$. This condition along with the reduced charge equation forces D_z to be zero everywhere inside the patch. So from eqn. (4) we can write

$$E_z^s = -\frac{e_{31}}{\epsilon_{33}}\left(\varepsilon_{xx} + \varepsilon_{yy}\right)^{(s)} = -\frac{e_{31}}{\epsilon_{33}}\left\{\left(\varepsilon_{xx}^{(0)} + \varepsilon_{yy}^{(0)}\right)^{(s)} + z\left(\varepsilon_{xx}^{(1)} + \varepsilon_{yy}^{(1)}\right)^{(s)}\right\} \tag{10}$$

If the electric potential inside the sensor patch is also assumed to vary quadratically with z then we write:

$$\phi^s = \phi_0^s + z\phi_1^s + z^2\phi_2^s \tag{11}$$

The electric field inside the patch is then obtained as

$$E_z^s = -\frac{d\phi^s}{dz} = -\phi_1^s - 2z\phi_2^s \tag{12}$$

Comparing eqns. (10) and (12) we obtain $\phi_1^s = -\frac{e_{31}}{\epsilon_{33}}\left(\varepsilon_{xx}^{(0)} + \varepsilon_{yy}^{(0)}\right);\ \phi_2^s = -\frac{e_{31}}{2\epsilon_{33}}\left(\varepsilon_{xx}^{(1)} + \varepsilon_{yy}^{(1)}\right)$

The average sensor potential (i.e., the voltage that appears between the sensor electrodes is

$$V^s = \frac{1}{A_s}\int_{As}\left(\phi^s\big|_{z=z_1} - \phi^s\big|_{z=z_2}\right)dA = \frac{h^s e_{31}}{A_s \in_{33}}\int_{As}\left\{\left(\varepsilon_{xx}^{(0)} + \varepsilon_{yy}^{(0)}\right) - h_m^s\left(\varepsilon_{xx}^{(1)} + \varepsilon_{yy}^{(1)}\right)\right\}dA \quad (13)$$

where $h^s = (z_{n+1} - z_n)$ is the thickness of the sensor patch and $h_m^s = \frac{(z_{n+1} + z_n)}{2}$ is the distance of the sensor patch middle surface from the laminate middle surface. Substituting (13) into (4) we obtain the stress-strain relation for the sensor patch as:

$$\sigma_{xx}^s = \left(C_{11} - \frac{e_{31}^2}{\in_{33}}\right)\varepsilon_{xx}^s + \left(C_{12} - \frac{e_{31}^2}{\in_{33}}\right)\varepsilon_{yy}^s;\ \sigma_{yy}^s = \left(C_{21} - \frac{e_{31}^2}{\in_{33}}\right)\varepsilon_{xx}^s + \left(C_{22} - \frac{e_{31}^2}{\in_{33}}\right)\varepsilon_{yy}^s;\ \tau_{xy}^s = C_{66}\gamma_{xy}^s \quad (14)$$

3. Potential and Kinetic Energies of the System

3.1 COMPOSITE PLATE

Using classical thin plate theory, the bending strain energy and the kinetic energy for a symmetrically layered composite panel is given by

$$U_p = \frac{1}{2}\int_0^{lx}\int_0^{ly}\int_{-h/2}^{h/2} \varepsilon'\sigma \, dx\, dy\, dz\ ;\ T_p = \frac{1}{2}\int_0^{lx}\int_0^{ly} \rho h\left(\frac{\partial w}{\partial t}\right)^2 dx\, dy \quad (15)$$

3.2 PIEZOELECTRIC PATCHES

The strain energy and kinetic energy stored in a piezoelectric patch is estimated in a similar way and is given by

$$U_{pz}^w = \frac{1}{2}\sum_{i=1}^{Na+Ns}\int_{x1i}^{x2i}\int_{y1i}^{y2i}\int_{z2}^{z3} \varepsilon'\sigma \, dx\, dy\, dz\ ;\ T_p = \frac{1}{2}\sum_{i=1}^{Na+Ns}\int_{x1i}^{x2i}\int_{y1i}^{y2i} \rho^i h^i \left(\frac{\partial w}{\partial t}\right)^2 dx\, dy \quad (16)$$

Apart from strain energy, the potential energy stored in the piezoelectric patches has one more component called electrical energy due to the presence of electric and electric displacement fields in the piezoelectric material. The electric energy stored in a piezoelectric material is expressed as

$$U_{pz}^e = \frac{1}{2}\sum_{i=1}^{Na}\int_{x1i}^{x2i}\int_{y1i}^{y2i}\int_{z2}^{z3} \mathbf{E}'\mathbf{D}\, dx\, dy\, dz = \frac{1}{2}\sum_{i=1}^{Na}\int_{x1i}^{x2i}\int_{y1i}^{y2i}\int_{z2}^{z3} D_3 E_3 \, dx\, dy\, dz \quad (17)$$

Due to the assumption of zero D field inside the sensors, there is no electric energy stored in the sensors. Substituting the expressions for the electric displacement and electric field from eqns. (4) and (8) we obtain the electric energy stored in the actuators as

$$U_{piezo}^e = \frac{1}{2}\sum_{i=1}^{Na}\int_{x1i}^{x2i}\int_{y1i}^{y2i}\left(-e_{31}^a V^a(\varepsilon_{xx}^a + \varepsilon_{yy}^a) + \in_{33}^a \frac{(V^a)^2}{h^a}\right) dx\ dy \quad (18)$$

3.3 ACOUSTIC CAVITY

The potential energy associated with the acoustic cavity is given by

$$U_{aco} = \frac{1}{2}\int_0^{lx}\int_0^{ly}\int_0^{lz}\left(\nabla p.\nabla p - \frac{1}{c^2}\dot{p}^2\right) dx\, dy\, dz \quad (19)$$

and the work done by the flexible walls due is given by $Q_{pr} = -\int_0^{lx}\int_0^{ly} p.\frac{\partial p}{\partial n}\, dx\, dy = -\int_0^{lx}\int_0^{ly} p.\rho_0\, \ddot{w}\, dx\, dy$. where ρ_0 is the density of air in the cavity and c is the sound velocity.

4. Dynamic Equation of Motion: Smart Panel-Cavity System

After estimating all the energies associated with the host composite plate and attached piezoelectric patches, the dynamic equations of motion are obtained using Hamilton's principle. The Hamilton's principle for the smart plate can be stated as

$$\delta\int_{t_1}^{t_2}(T-U+W)\,dt=0 \tag{20}$$

where T and U are the total kinetic and strain energies for the plate and patches and W is the work done by the external forces on the smart plate. The total kinetic and strain energies for the present smart plate is given by $T=T_p+T_{piezo}$ and $U=U_p^w+U_{piezo}^w+U_{piezo}^e$. Before applying the Hamilton's principle we assume a series solution for the transverse displacement 'w' containing approximation functions $f_{ij}(x,y)$ multiplied by the modal coordinates $W_{ij}(t)$. In view of rectangular geometry and clamped boundary conditions of the panel, the eigenfunctions of beam with clamped ends are the good choice for $f_{ij}(x,y)$. The transverse displacement 'w' can be written as

$$w=\sum_{i=1}^{m}\sum_{j=1}^{n}f_{ij}(x,y)W_{ij}(t)=\sum_{i=1}^{m}\sum_{j=1}^{n}X_i(x)Y_j(y)W_{ij}(t)=\mathbf{F}^T\,\mathbf{W} \tag{21}$$

where
$$X_i(x)=\sin\lambda_i x-\sinh\lambda_i x+\alpha_i\,(\cosh\lambda_i x-\cos\lambda_i x)\quad for\ i=1,2,\ldots,m$$
$$Y_j(y)=\sin\beta_j y-\sinh\beta_j y+\alpha_j\,(\cosh\beta_j y-\cos\beta_j y)\quad for\ j=1,2,\ldots,n$$

in which λ_i ,β_j and are the roots of the characteristic equation

$$\cos\lambda_i l_x\cosh\lambda_i l_x+1=0\,;\ \cos\beta_i l_y\cosh\beta_i l_y+1=0\ \text{ and }\ \alpha_i=\frac{\sinh\lambda_i l_x+\sin\lambda_i l_x}{\cosh\lambda_i l_x+\cos\lambda_i l_x}$$

Substituting the expressions derived for the kinetic and strain energies in the Hamilton's equation and taking the variation with respect to the generalized coordinates W_{mn}, we obtain the equation of motion for the smart plate as

$$(\mathbf{M}_p+\mathbf{M}_{pz})\ddot{\mathbf{W}}+(\mathbf{K}_p+\mathbf{K}_{pz}^w)\mathbf{W}+\mathbf{K}_{pz}^{wv\,T}\mathbf{V}^a=\mathbf{Q}_p+\mathbf{Q}_m \tag{22}$$

where

$$\mathbf{M}_p=\int_0^{l_x}\int_0^{l_y}\rho h\mathbf{F}\mathbf{F}^T\,dx\,dy\,;\mathbf{M}_{pz}=\sum_{i=1}^{Na}\int_{x1_i}^{x2_i}\int_{y1_i}^{y2_i}\rho_i h_i\,\mathbf{F}\mathbf{F}^T\,dx\,dy+\sum_{j=1}^{Ns}\int_{x1_j}^{x2_j}\int_{y1_j}^{y2_j}\rho_j h_j\,\mathbf{F}\mathbf{F}^T\,dx\,dy$$

$$\mathbf{K}_p=\int_0^{lx}\int_0^{ly}D_{11}\left(\frac{\partial^2\mathbf{F}}{\partial x^2}\frac{\partial^2\mathbf{F}^T}{\partial x^2}\right)+2D_{12}\left(\frac{\partial^2\mathbf{F}}{\partial x^2}\frac{\partial^2\mathbf{F}^T}{\partial y^2}\right)+D_{22}\left(\frac{\partial^2\mathbf{F}}{\partial y^2}\frac{\partial^2\mathbf{F}^T}{\partial y^2}\right)+4D_{66}\left(\frac{\partial^2\mathbf{F}}{\partial x\partial y}\frac{\partial^2\mathbf{F}^T}{\partial x\partial y}\right)dx\,dy$$

$$\mathbf{K}_{pz}^w=\sum_{i=1}^{Na+Ns}\int_{x1_i}^{x2_i}\int_{y1_i}^{y2_i}D_{11}^i\left(\frac{\partial^2\mathbf{F}}{\partial x^2}\frac{\partial^2\mathbf{F}^T}{\partial x^2}\right)+2D_{12}^i\left(\frac{\partial^2\mathbf{F}}{\partial x^2}\frac{\partial^2\mathbf{F}^T}{\partial y^2}\right)+D_{22}^i\left(\frac{\partial^2\mathbf{F}}{\partial y^2}\frac{\partial^2\mathbf{F}^T}{\partial y^2}\right)+4D_{66}^i\left(\frac{\partial^2\mathbf{F}}{\partial x\partial y}\frac{\partial^2\mathbf{F}^T}{\partial x\partial y}\right)dx\,dy$$

$$\mathbf{K}_{pz}^{wv}=\frac{1}{2}\sum_{i=1}^{Na}\int_{x1_i}^{x2_i}\int_{y1_i}^{y2_i}e_{31}^i\left(\frac{\partial^2\mathbf{F}^T}{\partial x^2}+\frac{\partial^2\mathbf{F}^T}{\partial y^2}\right)dx\,dy\,;$$

$$\mathbf{Q}_{pr}=\int_0^{lx}\int_0^{ly}p\,\mathbf{F}\,dx\,dy;\quad \mathbf{Q}_{f'}=\int_0^{lx}\int_0^{ly}f'\mathbf{F}\,dx\,dy$$

where $D_{ij}^{(a)}=\frac{1}{3}C_{ij}^{a}(z_3^3-z_2^3)$; $D_{ij}^{(s)}=\frac{1}{3}(C_{ij}^{s}-\frac{e_{31}^2}{\in_{33}})(z_1^3-z_0^3)$ $D_{ij}=\frac{1}{3}\sum_{k=1}^{n}(\overline{Q}_{ij})_k(z_k^3-z_{k-1}^3)$

$\mathbf{Q}_{pr}$ is the work done by the cavity acoustic pressure on the vibrating panel and p is the cavity pressure and f' is the external forces (i.e., point forces or distributed forces like incident noise etc.,) acting on the top surface of the panel. The sensor voltage can be obtained from eqn (13) and the matrix form of this equation is $\mathbf{V}^s=\mathbf{K}_s^{wv}\mathbf{W}$ where

$$\mathbf{K}_s^{wv}=-\sum_{i=1}^{Ns}\frac{h^s h_m^s e_{31}}{A_i\in_{33}}\int_{x1i}^{x2i}\int_{y1i}^{y2i}\left(\frac{\partial\mathbf{F}^T}{\partial x^2}+\frac{\partial\mathbf{F}^T}{\partial y^2}\right)dx\,dy.$$

For the acoustic cavity, the cavity pressure is expressed as a sum of a finite number of rigid cavity modes

$$p^c=\sum_{l=1}^{Nx}\sum_{m=1}^{Ny}\sum_{n=1}^{Nz}\psi_{lmn}(x,y,z)P_{lmn}(t)=\mathbf{\Psi}^T\mathbf{P},\ \text{where}\ \psi_{lmn}(x,y,z)=\cos\left(\frac{l\pi x}{l_x}\right)\cos\left(\frac{m\pi y}{l_y}\right)\cos\left(\frac{n\pi z}{l_z}\right)$$

are the rigid body cavity modes. The plate cavity coupling occurs via the force vector $\mathbf{Q}_{pr}$. Substituting the function expression assumed for the cavity pressure in the force vector term we obtain $\mathbf{Q}_{pr}=\mathbf{K}_{ac}^{w}\mathbf{P}$, where $\mathbf{K}_{ac}^{w}=\int_0^{lx}\int_0^{ly}\mathbf{\Psi}(x,y,0)\mathbf{F}^T dx\,dy$. Substituting the assumed function for the pressure in the energy expression for the cavity and applying the Hamilton's equation and taking the variation with respect to the generalized pressure $P_{lmn}(t)$, we obtain the governing equation for the panel backed cavity as

$$\mathbf{M}_{ac}\ddot{\mathbf{P}}+\mathbf{K}_{ac}\mathbf{P}+\mathbf{K}_{ac}^{w}\ddot{\mathbf{W}}=0 \tag{23}$$

where $\mathbf{M}_{ac}=\frac{1}{c^2}\int_0^{lx}\int_0^{ly}\int_0^{lz}\mathbf{\Psi}\mathbf{\Psi}^T dx\,dy\,dz$

$$\mathbf{K}_{ac}=\int_0^{lx}\int_0^{ly}\int_0^{lz}\frac{\partial\mathbf{\Psi}}{\partial x}\frac{\partial\mathbf{\Psi}^T}{\partial x}+\frac{\partial\mathbf{\Psi}}{\partial y}\frac{\partial\mathbf{\Psi}^T}{\partial y}+\frac{\partial\mathbf{\Psi}}{\partial z}\frac{\partial\mathbf{\Psi}^T}{\partial z}dx\,dy\,dz$$

Combining eqns. (22) and (23), we obtain the coupled equations for the smart panel backed cavity as

$$\begin{bmatrix}\mathbf{M}_p+\mathbf{M}_{pz} & \mathbf{0}\\ \mathbf{K}_{ac}^{w} & \mathbf{M}_{ac}\end{bmatrix}\begin{Bmatrix}\ddot{\mathbf{W}}\\ \ddot{\mathbf{P}}\end{Bmatrix}+\begin{bmatrix}\mathbf{K}_p+\mathbf{K}_{pz}^{w} & \mathbf{K}_{ac}^{w}\\ \mathbf{0} & \mathbf{K}_{ac}\end{bmatrix}\begin{Bmatrix}\mathbf{W}\\ \mathbf{P}\end{Bmatrix}=\begin{Bmatrix}\mathbf{K}_{pz}^{wv}\mathbf{V}^a\\ \mathbf{0}\end{Bmatrix}+\begin{Bmatrix}Q^{ext}\\ 0\end{Bmatrix} \tag{24}$$

with the sensor equation $\mathbf{V}^s=\mathbf{K}_s^{wv}\mathbf{W}$

5. Results

To validate the RR method, the resonant frequencies of an aluminum panel structure with five sensor/actuator patches bonded to its surface are estimated using NASTRAN and CEEAM[†] developed FE analysis code for smart structures. The details of the CEEAM developed FE model are given in ref. 4. The geometric and material properties of this smart panel are given in table 1 (also refer fig. 1). The resonant frequencies of this smart plate and cubic cavity estimated using FE method and RR method are shown in table 2. Once the structural model for the panel is developed then it is coupled with the cavity and the coupled frequencies of the system are determined using eqn. (24). In the NASTRAN FE analysis, 36

† *Center for the Engineering of Electronic and Acoustic Materials*

shell elements and 512 acoustic elements are used to model the panel-cavity system. The actuators and sensors are modeled with the 10 non-piezoelectric brick elements. The coupled panel-cavity system is also modeled using NASTRAN FE software where the cavity is modeled using acoustic brick (CHEXA) elements. The frequencies of the panel-backed cavity obtained from NASTRAN FE analysis and RR method are given in table 3. From the table it is observed that the simple RR approach is capable of predicting the dynamics of the smart panel-cavity system accurately. For RR method, first 7 mode shapes in each direction of the plate are included for the displacement and first 6 rigid cavity modes in each direction of the cavity for the cavity pressure are used. For noise control studies, a uniform harmonic pressure load of 2 Pa, which excites the panel at its first two resonant frequencies, is applied on the panel surface. The pressure at the center of the cavity ($l_x/2, l_y/2, l_z/2$) is computed with and without velocity feedback control and is given in table. 4. A simple velocity feedback control law $\mathbf{V}_a = -gain \times \mathbf{I} \times \dot{\mathbf{V}}_s$ is used. Here *gain* is the feedback gain value and **I** (5×5) is an identity matrix. From table. 4, we observe that the velocity feedback control using discrete sensor/actuator patches reduce the transmitted pressure to 14 dB for the first mode excitation and 19 dB for the second mode excitation.

6. Conclusion

The Rayleigh-Ritz method is used to model an active noise control system using discrete sensor/actuator patches. A smart aluminum plate backed cavity is studied using this method and the results are compared with FE analysis results. From the results we conclude that the RR approach is quite accurate in predicting the dynamic behavior of the vibroacoustic system with active noise control system. Results of studies showed that the RR model, which can predict the first few frequencies at the same level of accuracy, can execute four times faster than the corresponding FE model. At high frequencies, the FE method involves larger matrices, which are computationally expensive. Controller design and optimal placement of sensor / actuator pairs using the RR approach is preferred over the FE method due its simplicity and size. For problems involving irregular boundaries and complicated boundary conditions, it is not possible to define a global shape function. In such cases the RR approach loses some of its advantages over the FE method.

7. References

1. Craggs A., "The transient response of a coupled plate–acoustic system using plate and acoustic finite elements", *J. Sound Vib.*, 15, 1971, 509–28.
2. W Shields, J. Ro, and A. Baz, "Control of sound radiation from a plate into an acoustic cavity using active piezoelectric-damping composites", *Smart Mater. Struct.*, 7, 1998, 1–11.
3. Jaehwan Kim, Bumjin Ko, Joong-Keun Lee and Chae-Cheon Cheong, "Finite element modeling of a piezoelectric smart structure for the cabin noise problem", *Smart Mater. Struct.*, 8, 1999, 380–389.
4. Y-H Lim, S. Gopinathan, V. V Vardan, and V. K. Varadan, "Finite element simulation of smart structures using an optimal output feedback controller for vibration and noise control", *Smart Mater. Struct.*, 8, 1999, 324–337.
5. Dowell, E. H., Gorman, G. F. and Smith, D. A., "Acoustoelasticity: General theory, acoustic natural modes and forced response to sinusoidal excitation, including comparisons with experiment," *J. of Sound and Vib.*, 52(4), , 1977, 519-542.
6. Rogacheva, N. N., 1994, Theory of Piezoelectric Shells and plates, Boca Raton, FL: CRC Press.

TABLE 1: Material Properties

Properties	Host Plate (Al)	PZT	AIR
Youngs Modulus (E) N/m^2	68.0e+09	12.6e10	
Density (ρ) Kg/m^3	2800	7500	1.293
Thickness (h) m	0.8e-3	1.0e-3	
Poission's Ratio (ν)	0.32	0.2	
Piezoelectric Constant (e_{31}) N/Vm	-	-6.5	
Size m x m	0.305 x 0.305	Refer fig.1	
Velocity of sound 'C' m/sec			330.0

TABLE 2: Uncoupled resonant frequencies

Cavity			Smart Panel			
Type	R-R Method	NASTRAN FE	Type	R-R Method	CEEAM	NASTRAN
(0,0,0)	0.0	0.0	(1,1)	71.8	70.4	69.9
(0,0,1)	540.9	544.0	(1,2)	151.7	150.5	148.4
(0,1,1)	765.0	770.1	(2,2)	222.5	220.2	216.1
(1,1,1)	937.0	951.0	(1,3)	258.9	254.4	249.8
(0,1,2)	1081.1	1178.0	(2,3)	288.1	290.1	280.0

TABLE 3: Coupled frequencies of the smart panel-cavity

Excitation Frequency (Hz)	Pressure at the center (dB)	
	Without Feedback	With velocity Feedback
71.0	108	94
259.0	92	73

TABLE 4: Transmitted noise inside the cavity

R-R Method	NASTRAN/FE
0.0	0.0
87.1	85.3
156.7	151.9
217.4	213.42
315.4	304.3

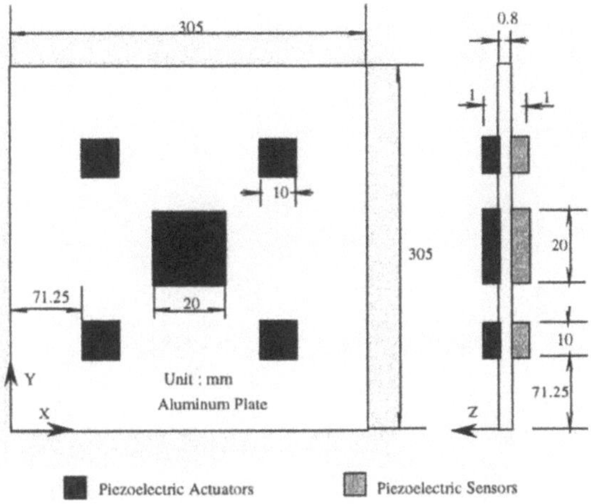

Figure 1. Clamped smart panel with piezoelectric sensors/actuators

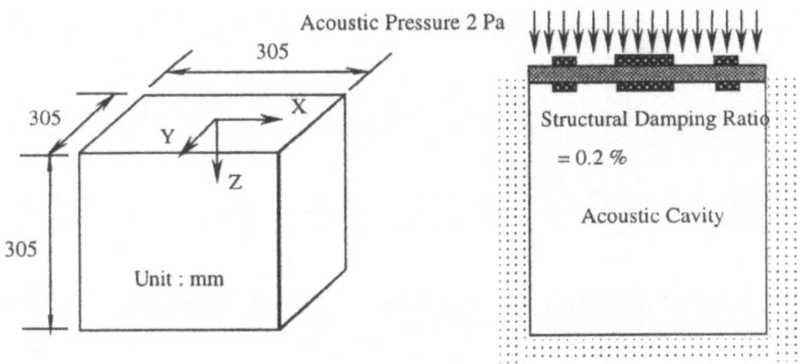

Figure 2. Cavity with smart panel and acoustic excitation

A WAVELET-BASED APPROACH FOR DYNAMIC CONTROL OF INTELLIGENT PIEZOELECTRIC PLATE STRUCTURES WITH LINEAR AND NON-LINEAR DEFORMATION

You-He ZHOU , Jizeng WANG, Xiao Jing ZHENG
Department of Mechanics, Lanzhou University
Lanzhou, Gansu 730000, P.R. China

1. Introduction

Piezoelectric devices present an important new group of sensors and actuators for active vibration control systems [1-4]. Indeed, this technology allows to developing spatially distributed devices, which requires special control techniques to improve the dynamical behavior of this kind of smart structure. Especially to geometrically nonlinear plates with piezoelectric sensing and actuating, there is a little of numerical results in the literature to quantitatively analyze the behavior of the vibration control of the structures [5-7]. This paper is concerned with the mathematical model of this kind of vibration control for a geometrically nonlinear plate with piezoelectric sensors and actuators by means of the scaling function transform of the wavelet theory [8,9]. Based on the generalized Gaussian integral to the scaling function transform, an explicit formula or algorithm of identification for the deflection of plates from the measured electric signals, i.e., electric charges and currents, on piezoelectric sensors is established. When a control law of negative feedback of the identified signals is employed, the applied voltages on piezoelectric actuators are determined by the wavelet Galerkin method. Finally, some typical examples, e.g., beam-plates with either small deflection or geometrically nonlinear deformation, of simulation are taken to show the feasibility of this control approach. It is found that this control model may auto-avoid those undesired phenomena of control instability generated from the interaction between measurement and controller with spilling over of high-order signals since the scaling function transform is low-pass.

2. Basic Equations

Consider the beam-plates with the geometrical nonlinearity of the von Karman type and attached by piezoelectric layers, e.g., PVDF, on the top and the bottom surfaces of plate used as sensors and actuators. Take the coordinate plane oxy to be coincident with the mid-plane of the plates, and the x- and y- axes along the longitudinal and width directions, respectively. Denote the control voltage by $V(x, t)$ applied on the piezoelectric actuators. From the theory of laminated plates, we can write the governing equations of the plates:

U. Gabbert and H.S. Tzou (eds.), IUTAM Symposium on Smart Structures and Structronic Systems, 179–187.

DYNAMIC EQUATION OF PLATE

$$\rho h w_{,tt} + D w_{,xxxx} - N_x w_{,xx} = -r_a e_{31} V_{,xx} \qquad 0 < x < L \tag{1}$$

in which

$$\rho h = \sum_{i=1}^{3} \rho_i (z_i - z_{i-1}) \tag{2}$$

$$D = \frac{1}{3} \sum_{i=1}^{3} \frac{Y_i}{1-\mu_i^2} (z_i^3 - z_{i-1}^3) \tag{3}$$

For the cantilever beam-plate, we have the boundary conditions:

$$N_x = 0 \tag{4}$$

$$x = 0 : w = \frac{\partial w}{\partial x} = 0 \tag{5}$$

$$x = L : -D \frac{\partial^2 w}{\partial x^2} - r_a e_{31} V = \frac{\partial^3 w}{\partial x^3} = 0 \tag{6}$$

For the beam-plates with un-movable simply supports at the ends, we can write

$$N_x = \frac{1}{L} \int_0^L (\frac{A}{2} w_{,x}^2 - e_{31} V) dx \tag{7}$$

$$x = 0 : w = -D \frac{\partial^2 w}{\partial x^2} - r_a e_{31} V = 0 \tag{8}$$

$$x = L : w = -D \frac{\partial^2 w}{\partial x^2} - r_a e_{31} V = 0 \tag{9}$$

where

$$A = \sum_{i=1}^{3} \frac{Y_i}{1-\mu_i^2} (z_i - z_{i-1}) \tag{10}$$

SENSING EQUATIONS

Let $\Omega_k = [x_{k-1}, x_k]$ (*k=1,2,...,M*) be the regions occupied by piezoelectric elements. When the plate is deformed with deflection *w=w(x,t)* at instant *t*, the electric charge $q_k(t)$ and electric current $I_k(t)$ measured from the *k*th piezoelectric sensor can be expressed by Lee [3]

$$q_k(t) = e_{31}^s \int_{\Omega_k} (\frac{1}{2} w_{,x}^2 - r_s w_{,xx}) dx \tag{11}$$

$$I_k(t) = \dot{q}_k(t) = e_{31}^s \int_{\Omega_k} (w_{,x} \dot{w}_{,x} - r_s \dot{w}_{,xx}) dx \tag{12}$$

in which $r_s(x)$ denotes the distance from the mid-plane to the piezoelectric sensors at point x. In order to simplify the discussions later, we introduce the following dismensionless quantities:

$$\overline{w} = w/r_s\,,\ \overline{x} = x/L,\ \overline{t} = t/\sqrt{\rho h L^4/D}\,,\ \overline{V} = V/(\frac{Dr_s}{e_{31}L^2 r_a})$$

$$\overline{q} = q/(\frac{e_{31}r_s^2}{L}),\ \overline{N}_x = N_x/(\frac{D}{L^2}),\ \overline{\Omega}_k = [\overline{x}_{k-1}, \overline{x}_k] \tag{13}$$

Then equations (1), (7), (11) and (12) are non-dimensionlized by the form

$$\overline{w}_{,\overline{t}\overline{t}} + \overline{w}_{,\overline{x}\overline{x}\overline{x}\overline{x}} - \overline{N}_{\overline{x}}\overline{w}_{,\overline{x}\overline{x}} = -\overline{V}_{,\overline{x}\overline{x}} \tag{14}$$

$$\overline{N}_{\overline{x}} = \int_0^1 (r_s^2 \frac{A}{2D} \overline{w}_{\overline{x}}^2 + \overline{V}) d\overline{x} \tag{15}$$

$$\overline{q}_k(\overline{t}) = \int_{\overline{\Omega}_k} (\frac{1}{2}\overline{w}_{,\overline{x}}^2 - \overline{w}_{,\overline{x}\overline{x}}) d\overline{x} \tag{16}$$

$$\overline{I}_k(\overline{t}) = \int_{\overline{\Omega}_k} (\overline{w}_{,\overline{x}}\dot{\overline{w}}_{,\overline{x}} - \dot{\overline{w}}_{,\overline{x}\overline{x}}) d\overline{x} \tag{17}$$

When the plate is disturbed by deflection $w(x,t)$ and velocity $\dot{w}(x,t)$, the electric charges $q_k(t)$ and electric current $I_k(t)$ will be generated on the piezoelectric sensors. After an appreciate control law has to be chosen to feed the measurable signals to the piezoelectric actuators through applying voltage across the actuators, the deflection of plate will be changed.

From here on, we use only the dimensionless variables and parameters in the theoretical analysis. For simplicity, we will drop the bar over each dimensionless quantity.

3. Identification of Deflection

Here, we use the scaling function transform of the Daubechies wavelet theory to develop an identification and control approach of the control system. From Zhou and Wang [8,9], we can write the approximation of a function $f(x)$ in terms of the basic scaling function $\phi_{n,k}(x) = 2^{n/2}\phi(2^n x - k)$ in the form of generalized Gaussian integral with algebraic accuracy of order 3 :

$$f(x) \approx 2^{-n/2} \sum_{k=k_1}^{k_2} f(\frac{x_1^* + k}{2^n})\phi_{n,k}(x) \tag{18}$$

3.1. FOR LINEAR CASE:

For the linear case of the plate with small deflection, the nonlinear terms in the equations (16) and (17) may be neglected, and $N(x) \equiv 0$ .In this case, Zhou and Wang [9] gave an explicit formulas of identification of deflection of the form

$$w(x,t) = w(0,t) + x\frac{\partial w(0,t)}{\partial x} - \sum_{j=1-2N}^{2^n} H(\frac{[x_1^*]+j}{2^n},t)\int_0^x \phi(2^n x + [x_1^*] - k)dx \tag{19}$$

in which

$$H(x_k, t) = \sum_{j=1}^{k} q_j(t) \tag{20}$$

Taking the differentiation of Eqs. (19) and (20) with respect to time variable t, there is no difficulty for one to write the identification formula of velocity $\dot{w}(x,t)$.

3.2. FOR NONLINEAR CASE:

Denote

$$H(x,t) = \int_0^x \left(\frac{1}{2} w_{,x}^2 - w_{,xx}\right) dx \tag{21}$$

$$\dot{H}(x,t) = \int_0^x (w_{,x}\dot{w}_{,x} - \dot{w}_{,xx}) dx \tag{22}$$

$$H(x_k, t) = \sum_{j=1}^{k} q_j(t) \tag{23a}$$

$$\dot{H}(x_k, t) = \sum_{j=1}^{k} I_j(t) \tag{23b}$$

Then equation (21) may be reduced into

$$\frac{1}{2} w_{,x}^2 - w_{,xx} = H_{,x}(x,t) \tag{24}$$

Applying equation (18) to the functions $H(x, t)$ and $w(x, t)$, we get

$$H(x,t) \approx \sum_{j=1-2N}^{2^n} H\left(\frac{[x_1^*]+j}{2^n}, t\right) \phi(2^n x + x_1^* - [x_1^*] - j) \tag{25}$$

and

$$w(x,t) \approx \sum_{k=1-2N}^{2^n} w_k \phi(2^n x - k) \tag{26}$$

where $w_k = w\left(\frac{x_1^* + k}{2^n}, t\right)$. Further,

$$w_{,x} \approx \sum_{k=1-2N}^{2^n} w_k 2^n \phi'(2^n x - k) \tag{27}$$

$$w_{,x}^2 \approx \sum_{k=1-2N}^{2^n} \sum_{i=1-2N}^{2^n} \sum_{j=1-2N}^{2^n} w_i w_j 4^n \phi'(x_1^* + k - i)\phi'(x_1^* + k - j)\phi(2^n x - k) \tag{28}$$

Substituting of equations (26)-(28) into equation (24), and taking the base scaling function of the Haar's wavelet as the weighted functions of the weighted residual method to the resulting equation, we obtain

$$\sum_{k=1-2N}^{2^n} \left[2^{n-1} \sum_{i=1-2N}^{2^n} \sum_{j=1-2N}^{2^n} w_i w_j \phi'(x_1^* + k - i)\phi'(x_1^* + k - j) \right] \cdot$$

$$\left[\phi^{\int}(\alpha+1-k)-\phi^{\int}(\alpha-k)\right]-\sum_{k=1-2N}^{2^n} w_i 2^n\left[\phi'(\alpha+1-k)-\phi'(\alpha-k)\right]$$
$$=H\left(\frac{\alpha+1}{2^n},t\right)-H\left(\frac{\alpha}{2^n},t\right) \tag{29}$$

where $\phi^{\int}(x)\equiv\int_{-\infty}^{x}\phi(x)dx$, $\alpha=2-2N,3-2N,\ldots,2^n-1$. Consider the boundary conditions of the plate deflection

$$w(0,t)=w(1,t)=0 \tag{30}$$

for the cantilever beam-plate, and

$$w(0,t)=w_{,x}(0,t)=0 \tag{31}$$

for the simply supported plate, we can obtained the solution of unknowns w_k so as to that the deflection of the plates is measured by equation (26). In order to get the measurable velocity, we take a differentiation of equation (24) with respect to x, i.e.

$$w_{,x}\dot{w}_{,x}-\dot{w}_{,xx}=\dot{H}_{,x}(x,t) \tag{32}$$

Denote $u(x,t)=w_{,x}(x,t)$. Then, equation (32) may be re-written by

$$u\dot{u}-\dot{u}_{,x}=\dot{H}_{,x}(x,t) \tag{33}$$

According to the theory of ordinary differential equation, we can get

$$\dot{u}=e^{\int_0^x udx}\left[C_0-\int_0^x\dot{H}_{,x'}(x',t)\,e^{-\int_0^{x'}u(x'',t)dx''}dx'\right]$$
$$=e^{w(x,t)}\left[C_0-\int_0^x\dot{H}_{,x'}(x',t)\,e^{-w(x',t)}dx'\right] \tag{34}$$

in which $C_0=\dot{u}(0,t)$. For the cantilever beam-plate, we have $C_0=\dot{u}(0,t)=0$, thus,

$$\dot{u}=-e^{w(x,t)}\int_0^x\dot{H}_{,x'}(x',t)\,e^{-w(x',t)}dx' \tag{35}$$

Integration of equation (35) with respect to x and consideration of $\dot{w}(0,t)=0$ lead to

$$\dot{w}(x,t)=-\int_0^x\left[e^{w(x',t)}\int_0^{x'}\dot{H}_{,x''}(x'',t)\,e^{-w(x'',t)}dx''\right]dx' \tag{36}$$

in which

$$\dot{H}_{,x}\approx\sum_{j=1-2N}^{2^n}2^n\dot{H}\left(\frac{[x_1^*]+j}{2^n}\right)\phi'(2^n x+x_1^*-[x_1^*]-j) \tag{37}$$

By the similar steps of above derivation to the simply supported beam-plate, we have

$$\dot{w}(x,t)=\int_0^x e^{w(x',t)}\left[C_0-\int_0^{x'}\dot{H}_{,x''}(x'',t)\,e^{-w(x'',t)}dx''\right]dx' \tag{38}$$

where

$$C_0=\int_0^1 e^{w(x,t)}\int_0^x\dot{H}_{,x'}(x',t)\,e^{-w(x',t)}dx'\Big/\int_0^1 e^{w(x,t)}dx \tag{39}$$

Thus, there is no difficulty for one to get identification of velocity $\dot{w}(x,t)$ from equations (36) and (38).

4. Applied Voltage on Piezoelectric Actuators

To control system with piezoelectric sensing and actuating, we choose a control law of negative feedback of the measured deflection and velocity of the form

$$-\frac{\partial^2 V(x,t)}{\partial x^2} = -G_1 w^*(x,t) - G_2 \dot{w}^*(x,t) \tag{40}$$

where $w^*(x,t)$ and $\dot{w}^*(x,t)$ are the measured deflection and velocity given in the previous section, and G_1 and G_2 are the gains. In order to determine the control voltage applied on actuators from Eq. (5), we add the boundary conditions:

$$V(0,t) = V(1,t) = 0 \tag{41}$$

Taking the approximation of equation (18) to $V(x,t)$, $w^*(x,t)$ and $\dot{w}^*(x,t)$, i.e.,

$$V(x,t) \approx 2^{-n/2} \sum_{k=1-2N}^{2^n} V_k(t)\phi_{n,k}(x) \tag{42}$$

$$w^*(x,t) \approx 2^{-n/2} \sum_{k=1-2N}^{2^n} w_k^*(t)\phi_{n,k}(x) \tag{43}$$

$$\dot{w}^*(x,t) \approx 2^{-m/2} \sum_{k=1-2n}^{2^n} \dot{w}_k^*(t)\phi_{m,k}(x) \tag{44}$$

where $V_k(t) = V(\frac{x_1^* + k}{2^n}, t), w_k^*(t) = w^*(\frac{x_1^* + k}{2^n}, t)$ and $\dot{w}_k^*(t) = \dot{w}^*(\frac{x_1^* + k}{2^n}, t)$, and applying the weighted residual method to the boundary-value problem of differential equation (40) with the additional boundary conditions (41), we get a system of algebraic equations on unknowns $V_k(t)$ of the form

$$\sum_{k=1-2N}^{2^n} 2^n \left[\phi'(j-k) - \phi'(j-k-1)\right] V_k(t) =$$
$$\sum_{k=1-2N}^{2^n} \frac{1}{2^n} \left[G_1 w_k^*(t) + G_2 \dot{w}_k^*(t)\right] \left[\phi^J(j-k) - \phi^J(j-k-1)\right] \tag{45}$$

$$\sum_{k=1-2N}^{2^n} V_k(t)\phi\ (-k) = 0, \quad \sum_{k=1-2N}^{2^n} V_k(t)\phi\ (2^n - k) = 0 \tag{46}$$

in which $j = 1-2N+1, 1-2N+2, \ldots 2^n - 1$. By solving equations (51) and (52) for $V_k(t)$, we can get $V(x,t)$ through equation (42), a continuous distribution of control voltage. Since the applied voltage across a piezoelectric actuator, is a constant without varying with x in practice, we take the control voltage applied on the kth piezoelectric actuator, $V_k^*(t)$, to be the average of the continuous distribution in its sub-region, i.e.,

$$V_k^*(t) = 2^n \int_{(k-1)/2^n}^{k/2^n} V(x,t)dt = \sum_{j=1-2N}^{2^n} V_j(t)\left[\phi^J(k-j) - \phi^J(k-j-1)\right] \quad (47)$$

Thus, the design of controller is completed.

5. Numerical Simulations and Discussions

According to the identification and the control voltage introduced in the precious sections, a program is established to simulate the evolution of deflection of the controlled plates to two cases, i.e., one is the plate with a clamped and a free ends, another is that with two un-movably simple supports. The parameters of geometric and materials employed here are listed in table 1.

TABLE 1. Parameters of Materials and Geometry in simulation

Materials	Y GPa	ρ (Kg/m^3)	μ	thickness Mm	Length Mm	width Mm	e_{31} $(N.(V.m)^{-1})$
Stainless steel	210	8000	0.3	1	300	20	
PVDF	2	1780	0.3	0.12	300	20	0.06

Fig.1 shows a comparison of the identifications between the linear and nonlinear deformation of the case plates. Their relative error between the linear identification and the exact deflection is plotted in Fig.2. From figure1, it is found that the nonlinear identification is almost coincided with the original or exact deflection. With increasing of the deflection, the relative error of the linear identification increases. In Fig.3 displays the responses of the maximum deflection of the plates when the control voltages are applied by means of the manner introduced in the previous sections. Here, an initial disturbance of deflection of free vibration modal of order 1 is chosen. When the magnitude of the initial disturbance increases, the effect of nonlinearity on duration of responses becomes notable. Fig.4 indicates the responses of the control voltage applied on the piezoelectric actuators. When we choose the disturbance by high order modal of vibration in the study, it is found that same phenomena are obtained, and the control ability of the system to the number of order is dependent upon the number of piezoelectric layers. As the number of piezoelectric sensors/actuators increases to 2^n, the numerical results show that the disturbance of initial deflection with vibration modal of the first 2^n orders can be suppressed. At the same time, this control approach may auto-avoid the phenomenon of control instability from the spilling over of measurement and controller since the scaling function transform has the low-pass behavior employed in the identification and control voltage calculations.

Acknowledgement: The authors sincerely appreciate the supports to this research from the National Natural Science Foundation of China, and the National Science Foundation of China for Outstanding Young Researchers.

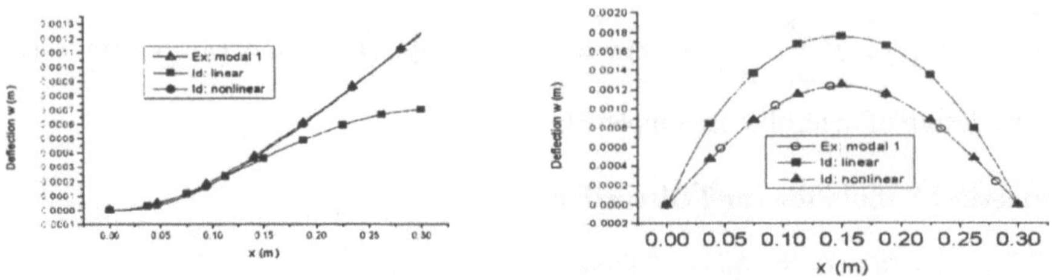

(a) (b)

Fig.1. A comparison between the identifications of deflection by means of the nonlinear and linear approaches to the deflection of modal 1. (*n=3; N=5;* w_{max}/h=1.0). (a): cantilevered beam-plate ; (b) beam-plates with un-movably simple supports at ends.

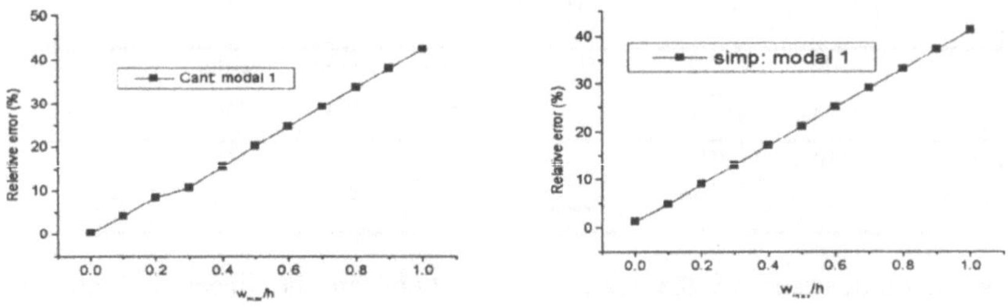

(a) (b)

Fig.2. Relative error of the maximum deflection from the identification of linear approach to the original/exact deflection of modal 1. (*n=3; N=5*). (a) for the tip deflection of the cantilevered beam-plate; (b) for the deflection at the middle point of the beam-plate with un-movably simple supports at ends. (Cant. = Cantilevered beam-plate; simp. = simply supported beam-plate).

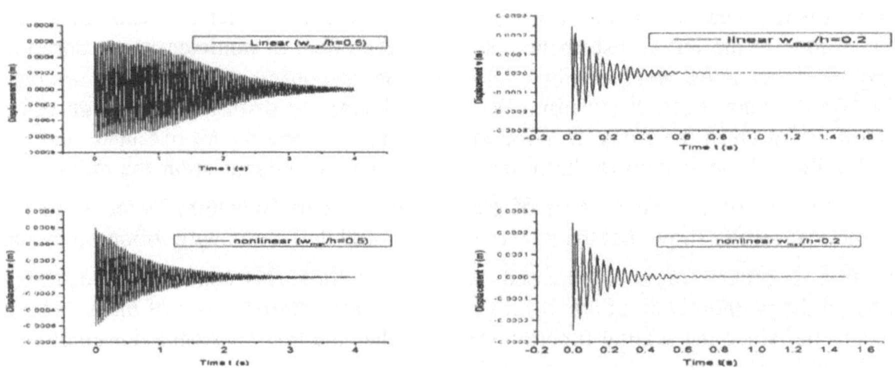

(a) (b)

Fig.3. Response of deflection at the middle point of the beam plate with un-movably simple supports at ends. (a) w_{max}/h=0.5 ; *n=3; N=5;* $G_1=0$; $G_2=0.1$. (b) $w_{max}/h=0.2$, *n=3; N=5;* $G_1=0$; $G_2=0.5$. (linear = linear control approach; nonlinear = nonlinear control approach).

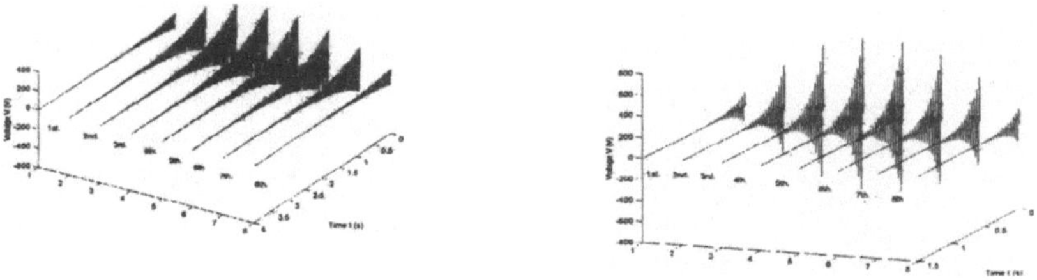

(a) (b)

Fig.4. Responses of the control voltage applied on piezoelectric actuators to the nonlinear beam-plates with un-movably simply supported ends (*n=3; N=5;* $G_1 = 0$). (a) $w_{max} / h = 0.5$; $G_2 = 0.1$; (b) $w_{max} / h = 0.2$; $G_2 = 0.5$.

References

1. Tzou, H.S., and Anderson, G.L. (eds.), *Intelligent Structural Systems*, Kluwer Academic Publications, Boston, 1992.
2. Zhou, Y.H., Wang, J.Z., Zheng, X.J., and Jiang, Q., Vibration control of variable thickness plates with Piezoelectric sensors and actuators based on wavelet theory, *J. of Sound and Vib.*, (2000), (in press)
3. Lee C. K.(1992), Piezoelectric laminates: theory and experiments for distributed sensors and actuators, Intelligent Structural Systems, in H.S. Tzou and G.L. Anderson (eds.), *Intelligent Structural Systems*, Kluwer Academic Publishers, Boston, pp. 75~167.
4. Yu, Y.Y., Some recent advances in linear and nonlinear dynamical modeling of elastic and piezoelectric Plates, *Adaptive Structures and Material Systems*, 35 (1992), 185-195.
5. Tzou, H.S., and Zhou, Y.H., Nonlinear piezothermoelasticity and multi-field actuation, Part 2: control and nonlinear deformation, buckling and dynamics, *ASME J. of Vibrations and Acoustics*, 119(1997), 382-389.
6. Zhou, Y.H., and Tzou, H.S., Active control of nonlinear piezoelectric Spherical shallow shells, *Int. J. of Solids and Struct.*, 37(2000), 1663-1677.
7. Tzou, H.S., and Zhou, Y.H., Dynamics and control of piezoelectric circular plates with geometrical non-nonlinearity, *J. Sound and Vib.*, 188(1995), 189-207.
8. Zhou, Y.H., and Wang, J.Z., Generalized Gaussian integral method for calculations of scaling function Transform of wavelets and its applications, *Acta Mathematical Scientia* , 19 (1999), 293-300, (in Chinese)
9. Zhou, Y.H., and Wang, J.Z., A dynamic control model of piezoelectric cantilevered beam-plate based on wavelet theory, *Acta Mechanica Sinica* (Chinese edition), 30 (1998), 719-727.

ON FINITE ELEMENT ANALYSIS OF PIEZOELECTRIC CONTROLLED SMART STRUCTURES

H. BERGER, H. KÖPPE, U. GABBERT, F. SEEGER
Institut für Mechanik
Otto-von-Guericke-Universität Magdeburg
Universitätsplatz 2
D-39106 Magdeburg
Germany

1. Introduction

The increasing engineering activities in the development and industrial application of piezoelectric smart structures require effective and reliable simulation and design tools [9]. Even if significant progress has been observed over the past years most of such developments are restricted to special requirements and applications [2]. In our opinion the finite element method (FEM) is an excellent basis to develop overall software tools which meet the engineering requirements. Consequently, such a general purpose software tool has been designed by the authors and realized step by step over the last few years [1], [3], [4]. Recently, this tool was completed by new thin shell type elements as well as a data interface to connect controller design tools with our finite element analysis tool. In the paper the focus is on these new developments. First the theoretical basis of our finite element software tool is presented briefly. Then the new electromechanical coupled layered thin shell elements are prescribed which are very efficient to simulate the global structural behavior of thin-walled structures controlled by piezoelectric wafers and fibers. After that our concept to connect finite element analysis and controller design is given which result in an computer based overall design and simulation strategy for smart structure. Finally, the active vibration suppression of an excited plate structure is presented as a test example to demonstrate the applicability of our finite element based overall design and simulation software.

2. Finite Element Analysis of Smart Structures

The constitutive relations between stresses σ_{ij}, dielectric displacements D_i, strains ε_{ij}, electric fields E_i and temperature θ can be written as

$$\begin{aligned} \sigma_{ik} &= c^E_{iklm}\varepsilon_{lm} - e_{lik}E_l - \lambda^E_{ik}\theta, \\ D_i &= e_{ilm}\varepsilon_{lm} + \kappa^\varepsilon_{il}E_l + p^\varepsilon_i\theta. \end{aligned} \qquad (1)$$

U. Gabbert and H.S. Tzou (eds.), IUTAM Symposium on Smart Structures and Structronic Systems, 189–196.

These linear constitutive equations are quite accurate in lower electric field applications, and give sufficient results in most design processes of engineering smart structures. The material tensors of inhomogeneous materials such as smart composites with embedded thin piezoelectric fibers or wafers must be determined experimentally or calculated by homogenization methods.

The mechanical balance equations and the electric balance equation (4th Maxwell equation) together with stress boundary conditions and the charge boundary condition can be written in a weak form as

$$\delta\chi = \int_V \left\{ \left(\sigma_{ij,j} + \rho\overline{p}_i - \rho\ddot{u}_i\right)\delta u_i + \left(D_{i,i}\right)\delta\phi \right\} dV \\ + \int_O \left\{ (\overline{q}_i - \sigma_{ij}n_j)\,\delta u_i - (\overline{Q} + D_i n_i)\,\delta\phi \right\} dO = 0. \tag{2}$$

The Equations (1) and (2) together with the strain displacement relation $\varepsilon_{ij}=\frac{1}{2}(u_{i,j}+u_{j,i})$ and the relation between the electric field and the electric potential $E_i=-\phi_{,i}$ are the general basis for the development of any piezoelectric finite element. Using an elementwise approximation of the displacements u_i and the electric potential ϕ and following the standard finite element procedure, the semidiscrete form of the equations of motion can be derived as

$$\mathbf{M}_e\ddot{\mathbf{x}}_e + \mathbf{D}_{de}\dot{\mathbf{x}}_e + \mathbf{K}_e\mathbf{x}_e = \mathbf{F}_e, \tag{3}$$

where $\mathbf{x}_e$ contains the degrees of freedom (*dof's*) of a finite element (displacements and electric potentials). On this basis a comprehensive library of multi-field finite elements (linear and quadratic 1D, 2D, 3D elements, layered composite shell elements, etc.) as well as numerical methods to simulate static and dynamic structural behavior of smart structures have been developed and by the authors (see e.g. Berger et al., 2000). Because smart structures contain in general only a few piezoelectric actuators or sensors with coupled electromechanical *dof's*, it is advantageous to separate mechanical and piezoelectric components in different substructures. This results in a considerable reduction of computer time in complex engineering applications. The solution of the fully coupled thermoelectromechanical three field problem including heat generation caused by electric resistance is based on an staggered strategy. In each time step the two sets of equations – the heat conduction problem and electromechanical problem – are solved separately taking into account the coupling terms as force vectors on the right hand side of the corresponding equations [5].

3. The New *SemiLoof* Type Thin Shell Elements

For the analysis of smart shell structures we developed at first quadrilateral and triangular multilayer thick shell elements on the basis of a layerwise linear approximation in thickness direction. These 3D type elements are an extension of an triangular element first published by Tzou and Ye [7]. From a computational point of view these elements are too expensive in calculating the global structural behavior of thin shell applications. Here finite shell elements on the basis of the classical *Kirchhoff-Love* hypothesis are much more effective. Among the huge amount of different types of thin shell elements we preferred the *SemiLoof* element family, originally proposed by Iron [6], as basis for the development of thin active shell elements. This preference results from a long time of practical experience in several fields of application, where the *SemiLoof* elements have shown a good overall accuracy and robustness in comparison with other shell elements. The quadrilateral and triangular elements with 8 and 6 nodes, respectively, contain the displacements *u*, *v*, and *w* at all nodes and additionally the rotations in tangential direction at the two *Gaussian* integration points at the edges. Two families of shape functions are used: i) *Lagrangian* polynomials for the displacements and ii) *Legendre* polynomials for the rotations. Due to the definition of the rotations and the displacements the quadrilateral and the triangular elements include 32 *dof's* and 24 *dof's*, respectively. The classical laminate theory is applied to model thin composite structures. Consequently, the stresses are assumed to vary linear over the thickness direction. Similar to these restrictions in the mechanical field assumptions have to be introduced for the electric field. We used two different models for the reduction of the electric field which are based on the following assumptions:

1. The electric field parallel to the shell mid surface is neglected and only the field in normal direction is taken into account.
2. The electric field normal to the shell mid surface is neglected and the electric field parallel to the shell mid surface is taken into account.

The implementation of model 2 results in shell elements which can be used for the simulation of piezoelectric fibers or other electromechanical structures with an in-plane poling (e.g. interdigital electrodes, see [4]). In the following the shell elements based on model assumptions 1 are presented only. In this case it is assumed that any piezoelectric layer of a smart composite is polarized normal to the mid-surface with electrodes on top and bottom. The constitutive equations of such a thin active layer can be derived from Equation (1) by taking into account thin shell assumptions which result in

$$\begin{aligned} \begin{Bmatrix} \sigma_{11} \\ \sigma_{22} \\ \tau_{12} \end{Bmatrix} &= \begin{bmatrix} Q_{11} & Q_{12} & 0 \\ Q_{12} & Q_{22} & 0 \\ 0 & 0 & Q_{33} \end{bmatrix} \begin{Bmatrix} \varepsilon_{11} \\ \varepsilon_{22} \\ \gamma_{12} \end{Bmatrix} + \begin{bmatrix} \bar{e}_{11} \\ \bar{e}_{12} \\ 0 \end{bmatrix} \{E_3\}, \\ \{D_3\} &= \begin{bmatrix} \bar{e}_{11} & \bar{e}_{12} & 0 \end{bmatrix} \begin{Bmatrix} \varepsilon_{11} \\ \varepsilon_{22} \\ \gamma_{12} \end{Bmatrix} + [\bar{\kappa}_{33}]\{E_3\}, \end{aligned} \tag{4}$$

where $\mathbf{Q}$, $\bar{\mathbf{e}}$ and $\bar{\boldsymbol{\kappa}}$ are the stress reduced stiffness matrix, the reduced piezoelectric constants and the reduced dielectric constants, respectively.

Due to the small thickness h_i of an active patch in the ith layer and the high difference of the electric potential $\Delta\phi_i$ between the electrodes the electric field E_{3i} in normal direction can be assumed as constant over the thickness direction which results in

$$E_{3i} = \Delta\phi_i / h_i \,. \tag{5}$$

Consequently, the electric potential varies linear over the thickness as the mechanical stress. Thus the electric field of an active layer depends on the difference of the electric potential between top and bottom of the layer only.

In the simplest case it is possible to model the influence of the electromechanical coupling by forces and moments only. This forces and moments are calculated such that they result in the same strain field that is caused by the electric field. But to calculate the sensor signals accurately additional *dof*'s have to be introduced into the element. This electric *dof's* are the difference of the electric potential $\Delta\phi_i$ between bottom and top of each of the active layers (i=1,...,n), where n is the number of active layers in the composite. Consequently, in this case no shape functions are required to approximate the electric field in the finite element. The element is enhanced by additional electric potential differences $\Delta\phi_i$ per each active layer. These additional *dof*'s can be added to the midpoint of the element as additional dof's (see Figure 1). The stiffness matrix of the extended *SemiLoof* element consists of the original pure mechanical stiffness matrix $\mathbf{K}_e^{(mm)}$ of the original passive *SemiLoof* element, the new pure electric stiffness matrix $\mathbf{K}_e^{(ee)}$ and the coupling matrix $\mathbf{K}_e^{(me)}$. These matrices can be derived on the basis of the classical *Lagrangian* finite element formulation as

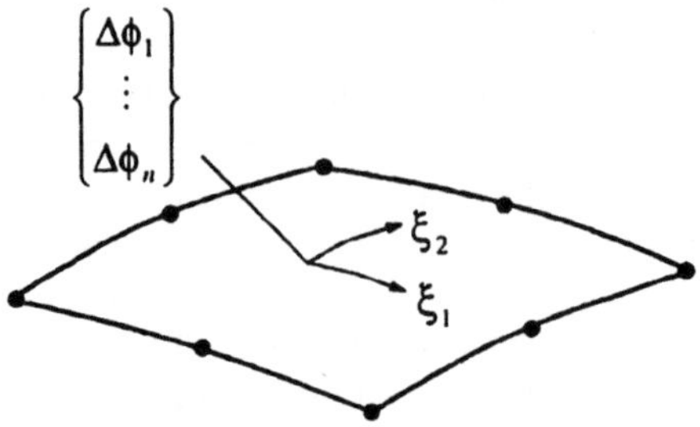

Figure 1: Configuration of the active SemiLoof shell element based on model assumption 1

$$\begin{bmatrix} \mathbf{K}_e^{(mm)} & \mathbf{K}_e^{(me)} \\ \mathbf{K}_e^{(me)T} & \mathbf{K}_e^{(ee)} \end{bmatrix} \begin{Bmatrix} \mathbf{u}_e \\ \Delta\boldsymbol{\varphi}_e \end{Bmatrix} = \begin{Bmatrix} \mathbf{f}_e^{(m)} \\ \mathbf{f}_e^{(e)} \end{Bmatrix}, \tag{6}$$

$$\mathbf{K}_e^{(mm)} = \int_{A_e} \mathbf{B}^{(m)^T} \mathbf{c} \mathbf{B}^{(m)} dA, \; \mathbf{K}_e^{(me)} = \int_{A_e} \mathbf{B}^{(m)^T} \mathbf{e} \mathbf{B}^{(e)} dA, \; \mathbf{K}_e^{(ee)} = \int_{A_e} \mathbf{B}^{(e)^T} \boldsymbol{\kappa} \mathbf{B}^{(e)} dA. \tag{7}$$

In Equation (6) $\Delta\boldsymbol{\varphi}_e$ contains the voltage differences of each active element layer. Consequently, the quadrilateral *SemiLoof* element consists of (32+n) *dof's* finally. But there is no coupling in the electric *dof's* between adjacent elements. Consequently, in all cases where an active layer of a shell structures is meshed with more than one finite element, identical potential differences of this active layer have to be taken into account by constraint conditions which is a standard technique in our finite element package.

4. Control

The numerical simulation of smart structures within the finite element frame requires a overall model which includes the passive structure, active elements for sensors and actors as well as appropriate models for controllers. In general the development and testing of control algorithms for complex smart structures is an sophisticated and complex process which requires special knowledge and experiences. Today comprehensive design tools such as Matlab/Simulink are available to support this design process. Consequently, in our opinion there is no point in implementing this control design process directly into the finite element code. But, on the one hand data from the finite element model such as the mass matrix, the stiffness matrix, the damping matrix as well as sensor and actuator positions are useful to design the controller and on the other hand the controller matrix (or subroutines calculating the controller parameters) are needed in the finite element package to simulate the controlled structural behavior. For this exchange of data and information between a finite element package and a controller design tool a general data exchange interface is required. Recently, such a data interface has been developed to couple our finite element software with MatLab/Simulink. This software package was preferred due to its excellent development environment as well as its extensive library of different controller design tools.

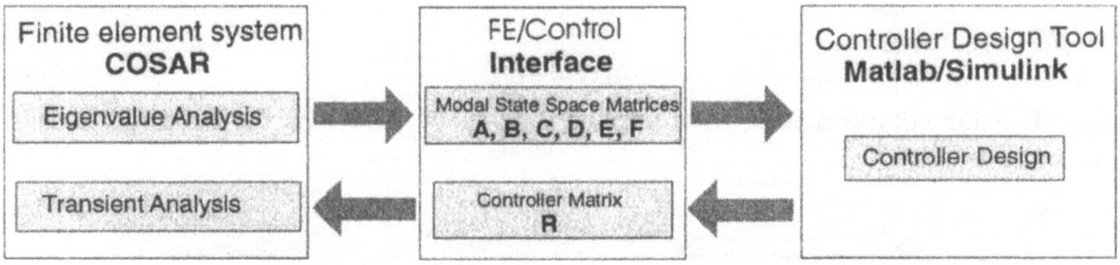

Figure 2. Data exchange between FE system and control design package

The communication concept between our finite element software COSAR and Matlab/Simulink is shown in Figure 2.

In general the control of flexible mechanical structures is a *multiple input - multiple output* (MIMO) problem, where the application of large scale finite element models is infeasible for the controller design. Therefore, an appropriate model reduction technique is required to reduce the number of the finite element equations. One of the best-known model reduction techniques is the modal truncation. This technique has been combined with investigations of the dominance behavior of different modes [8]. The modal truncation seems to be best suited for the controller design of structures based on a finite element discretization, since flexible structures possess a low-pass characteristics, which allows to neglect high-frequency dynamics. Usually, in spite of the system reduction the classical controller design methods in the frequency domain can not be applied. Therefore, the controller description is given in a time discrete standard form. The semi-discrete form of the equation of motion for a controlled structure can be written then as

$$\mathbf{M}\ddot{\mathbf{x}}+\mathbf{D}_d\dot{\mathbf{x}}+\mathbf{K}\mathbf{x}=\overline{\mathbf{E}}\mathbf{f}(t)+\overline{\mathbf{B}}\mathbf{u}(t)\,, \tag{8}$$

where $\mathbf{M}$, $\mathbf{D}_d$ and $\mathbf{K}$ are the mass, damping and stiffness matrix respectively and $\mathbf{f}$(t) and $\mathbf{u}$(t) are the external disturbances and the controller influence on the structure. The matrices $\overline{\mathbf{E}}$ and $\overline{\mathbf{B}}$ describe the positions of the forces and the control parameters in the finite element structure, respectively. The solution of the linear eigenvalue problem

$$(\mathbf{K}-\lambda_i\mathbf{M})\mathbf{\Phi}_i=\mathbf{0} \tag{9}$$

results in the (n×r) modal matrix $\mathbf{\Phi}=[\mathbf{\Phi}_1 \mid \mathbf{\Phi}_2 \mid \cdots \mid \mathbf{\Phi}_r]$ and the (r×r) spectral matrix $\mathbf{\Lambda}=\mathrm{diag}(\lambda_i)$, where $\mathbf{\Phi}$ is ortho-normalized with $\mathbf{\Phi}^T\mathbf{M}\mathbf{\Phi}=\mathbf{I}=\mathrm{diag}(1)$ and $\mathbf{\Phi}^T\mathbf{K}\mathbf{\Phi}=\mathbf{\Lambda}$.

Consequently, inserting the modal co-ordinates $\mathbf{x} = \mathbf{\Phi q}$ into the Equation (8) results in the modal truncated system of (r×r) differential equations which can be written as

$$\ddot{\mathbf{q}} + \Delta\dot{\mathbf{q}} + \Lambda\mathbf{q} = \mathbf{\Phi}^T\overline{\mathbf{E}}\mathbf{f}(t) + \mathbf{\Phi}^T\overline{\mathbf{B}}\mathbf{u}(t)\,, \tag{10}$$

where Δ is the modal damping matrix. With the state space vector

$$\mathbf{z}^T = [\mathbf{q} \quad \dot{\mathbf{q}}] \tag{11}$$

Equation (10) can be rewritten as

$$\dot{\mathbf{z}} = \begin{bmatrix} \mathbf{0} & \mathbf{I} \\ -\Lambda & -\Delta \end{bmatrix}\mathbf{z} + \begin{bmatrix} \mathbf{0} \\ \mathbf{\Phi}^T\overline{\mathbf{B}} \end{bmatrix}\mathbf{u}(t) + \begin{bmatrix} \mathbf{0} \\ \mathbf{\Phi}^T\overline{\mathbf{E}} \end{bmatrix}\mathbf{f}(t) = \mathbf{Az} + \mathbf{Bu}(t) + \mathbf{Ef}(t) \tag{12}$$

Together with the measurement equation

$$\mathbf{y} = \mathbf{Cz} + \mathbf{Du}(t) + \mathbf{Ff}(t) \tag{13}$$

and the control law

$$\mathbf{u}(t) = -\mathbf{Rz}\,, \tag{14}$$

where **R** is the control matrix, the complete state space model for the controller design is given.

5. Overall Design and Simulation of an Active Plate Structure

The following example is used to demonstrate the finite element analysis, the controller design as well as the simulation of the controlled structural behavior. The smart plate attached with eight piezoelectric patches (four on bottom and four on top of the plate) is shown in Figure 3. For an active damping of the plate in a frequency range up to 50 Hz a LQR controller should be designed. At first a finite element model was performed with a mesh of 892 passive and 8 active *Semiloof*

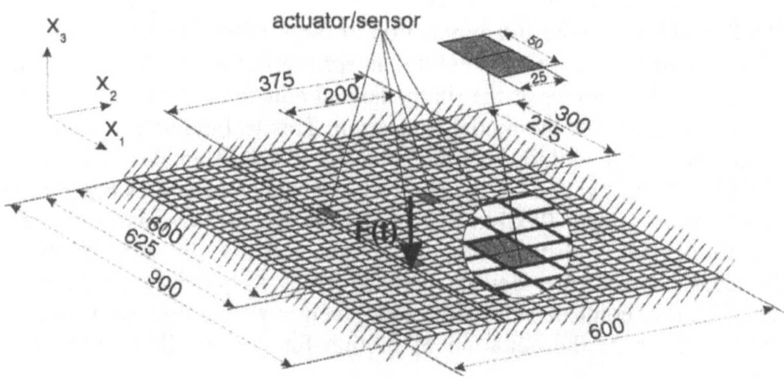

Figure 3. Smart Plate Structure, Material properties: *plate:* E=2.06 10^5 N/mm, ν=0.3, ρ=7.86 10^{-9} Ns^2/mm, t=0.9 mm (thickness), *actuator/sensor:* $E_{11}=E_{22}$=3.77 10^4 N/mm^2, G_{12}=1.3 10^4 N/mm^2, ν=0.38, ρ=7.85 10^{-9} Ns^2/mm^4, d_{31}=2.1 10^{-7} mm/V, κ_{33}=3.36 10^{-9} F/m, t=0.4 mm (thickness)

TABLE 1. Frequency

Number	Frequency [Hz]
1	16.68
2	25.89
3	41.02
4	41.04
5	49.62

shell elements. Based on this mesh the eigenfrequencies and eigenmodes were calculated. Table 1 shows that 5 eigenfrequencies occur in the interesting frequency domain which belong to bending modes of the plate. Then in our finite element software the plate model was modally reduced and transformed into the state space model. Via our data exchange interface this model was then exported into Matlab/Simulink, where a LQR controller was designed and tested. Then the controller matrix **R** was transferred

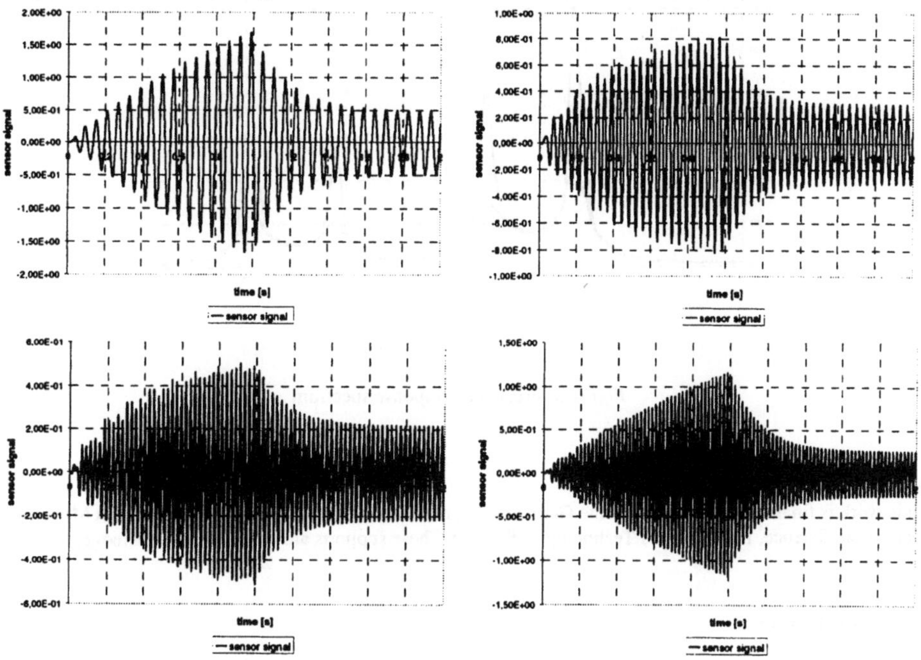

Figure 4. Plate deflection (sensor signal) for different loads

back into our finite element software COSAR via the data exchange interface (see Figure 2). Finally, in the finite element software a dynamic transient simulation was carried out, where different load functions $F(t) = A\sin(\omega_i t)$ are used to excite the plate. After t=1,0s, i.e. before the steady state vibration has been reached, the controller was switched on. The controlled behavior can be seen in Figure 4 and Figure 5, where the first eigenfrequencies of the plate were used as excitation frequencies. The time response of the plate deflection (Figure 4) is measured close to the excitation force (see Figure 3). In Figure 5 the frequency response spectrum of is given.

6. Conclusion

The paper presents a general finite element based overall simulation and design tool of piezoelectric controlled smart structures. This tool can be used i) to simulate both static and dynamic structural problems, ii) to design controllers by Matlab/Simulink based on a general bilateral data exchange interface between finite element analysis software and controller design tools, and iii) to simulate the controlled structural behavior. To demonstrate the capability of the new software tool the overall design and simulation of a smart plate structure have been presented.

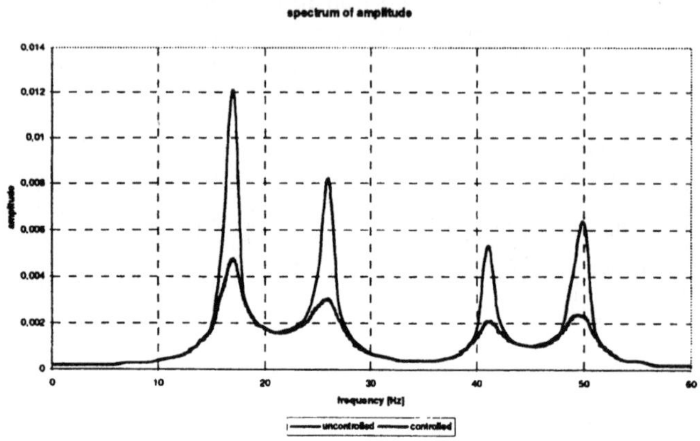

Figure 5: Frequency response spectrum

Acknowledgement

This work is financially supported by the German Research Foundation (DFG) and by the German Ministry of Education, Science, Research and Technology (BMBF). These supports are gratefully acknowledged.

7. References

1. Berger, H., Gabbert, U., Köppe, H., Seeger, F. (2000): Finite Element Analysis and Design of Piezoelectric Controlled Smart Structures. *Journal of Theoretical and Applied Mechanics*, 3, **38**, pp. 475-498
2. Chee, C. Y., Tong, L., Steven, G. P. (1998): A Review on the Modelling of Piezoelectric Sensors and Actuators Incorporated in Intelligent Structures, *J. of Intelligent Material Systems and Structures*, Vol. 9, pp. 3-19.
3. Gabbert, U., Berger, H., Köppe, H., Cao, X. (2000): On Modelling and Analysis of Piezoelectric Smart Structures by the Finite Element Method. *Journal of Applied Mechanics and Engineering*, Vol. 5, No 1, pp.127-142.
4. Gabbert, U., Köppe, H., Fuchs, K., Seeger, F. (2000): Modeling of Smart Composites Controlled by Thin Piezoelectric Fibers, in Varadan, V.V. (Ed.): *Mathematics and Control in Smart* Structures, SPIE Proceedings Series, Vol. 3984, pp. 2-11.
5. Gömandt, A., Gabbert,U. (2000): Finite Element Analysis of Thermopiezoelectric Smart Structures. *Acta Mechanica* (submitted); Uni Magdeburg, Fakultät für Maschinenbau,. Preprint Nr. 1, 2000
6. Irons, B. M. (1976): The Semiloof shell element, in Ashwell D. G. and Gallagher R. H. (Eds.): *Finite Elements for Thin Shells and Curved Members*, J. Wiley, London.
7. Köppe, H., Gabbert, U., Tzou, H. S. (1998): On Three-Dimensional Layered Shell Elements for the Simulation of Adaptive Structures. in Gabbert, U. (Ed.): *Modelling and Control of Adaptive Mechanical Systems*. VDI-Fortschrittberichte Reihe 11, Schwingungstechnik 268, VDI Verlag Düsseldorf, pp. 385-395.
8. Litz, L. (1979): *Ordnungsreduktion lineare Zustandsraummodelle mittels modaler Verfahren*. Hochschulverlag Stuttgart.
9. Tzou, H.-S., Guran, A. (Eds.), (1998): *Structronic Systems: Smart Structures, Devices and Systems*, World Scientific, 1998.

A STUDY ON SEGMENTATION OF DISTRIBUTED PIEZOELECTRIC SECTORIAL ACTUATORS IN ANNULAR PLATES

A. TYLIKOWSKI
Institute of Machine Design Fundamentals
Warsaw University of Technology
Narbutta 84 02-524 Warszawa Poland

1. Introduction

Distributed piezoelectric layers can be used as distributed sensors and actuators for structural monitoring and control of elastic structures. In this paper distributed vibration control of structures using single-piece and multi-piece segmented actuators is examined. The study is motivated by finding an active control strategy to reduce vibrations in circular saw blades and noise transmission in a circular acoustic ducts. Piezoelectric actuators have been applied successfully in the closed-loop control of distributed two-dimensional systems (cf. Dimitriades, Fuller and Rogers [1] for rectangular plates and Van Niekerk et al. [2] for circular plates). Tzou and Fu [6] analysed models of a plate with segmented distributed piezoelectric sensors and actuators, and showed that segmenting improves the observability and the controllability of the system. The dynamic model for an axially symmetrical plate with a piezoceramic actuator was presented in [3]. The capacitively shunted annular plate with piezoelectric elements as a distributed vibration absorber was analysed by the present author [4]. The method proposed in [6] is incorporated to study segmentation of annular and sectorial piezoelements glued to circular and annular plates. The main problem is a role of the actuator segmentation in vibration excitation of annular plates driven by harmonic voltage. In the model developed so far the thickness of the actuator is ignored and the piezoceramics do not add any mass to the structure. The Kirchhoff annular plate is clamped at the inner radius and free at its outer edge. Between the radial coordinates a and b piezoelectric layers are attached. The analysis is performed to axially symmetric modes for an annular actuator and to axially nonsymmetrical solutions for sectorial actuators. The plate loading is defined by distributed moments proportional to free piezoelectric strains and the actuator characteristic function. The proportionality constant depends on geometry and mechanical properties of the plate and the actuator. Annular piezoelectric actuators enable, obviously, obtaining only the axisymmetrical vibration modes [3]. They are very useful in generating required

U. Gabbert and H.S. Tzou (eds.), IUTAM Symposium on Smart Structures and Structronic Systems, 197–204.

eigenforms selectively. It is yet to be emphasized that the use of annular piezoactuators can occur insufficient if such selective approach toward vibration control is not desired, i.e. when all vibration modes need to be affected. This is why sector annular actuators should be analysed in detail. In their case analytical solutions are no longer easy to be obtained as in the case of full annular actuators. The model is able to predict the response of the plate driven by the piezoelectric actuators glued to lower and upper plate surfaces. The actuators are driven by a pair of electrical fields with the same amplitude and in opposite phase. The actuators were used to excite steady-state harmonic vibrations in the plate. Mathematical models of an annular plate with a single-piece distributed annular actuator and multi-piece segmented sectorial actuators are formulated and performances are simulated. Simulations carried out prove that the sector actuators

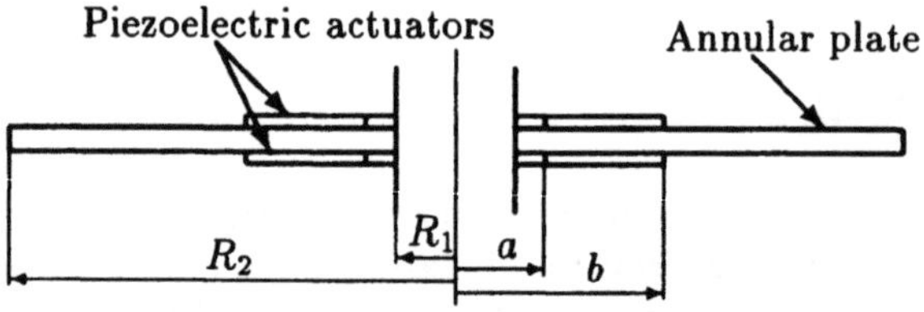

Figure 1. Geometry of plate with actuators

are effective in exciting general vibration modes. The results demonstrate that modes can be selectively excited and the geometry and the electrical wiring of the actuator and sensor segments significantly affect the distribution of the response among modes. It is possible to design the segment shape to either excite or suppress particular modes leading to improved control behavior.

2. Axially Nonsymmetrical Excited Vibrations of Annular Plates

Consider annular-sectorial actuators shown in Fig. 2. Piezoelectric properties have the axially symmetry with respect to the plate axis. The identically polarized actuators are placed to the both opposite sides of the plate and are driven by a pair of electrical fields with the same amplitude in opposite phase. It causes an antisymmetrical state of strains in the plate, the upper plate surface is tensioned while the lower one is compressed. The actuator in the shape of annulus is segmented into four equal sectors. It is assumed that the seperation is very small so that it is continuous elastically and open-circuit electrically [6]. Depending on the wiring the actuators excite different vibration modes. Case 1 corresponds to the in-phase wiring of all four segments (the annulular actuator). In the case 2 the phase of upper and left actuators is opposite to the phase of botoom and right ones. In the case 3 the phase of each neighboring segments is opposite. The first nine natural frequencies are calculated and taken into account in a simulation procedure. Distributed moments m_r and m_φ generated by annular-sectorial actuators are written in the form.

Case 1

$$m_r = m_\varphi = C_o\Lambda(t)[H(r-a) - H(r-b)] \tag{1}$$

Case 2

$$m_r = m_\varphi = C_o\Lambda(t)[H(r-a)-H(r-b)]\left[H(\varphi+\frac{\pi}{2})-2H(\varphi-\frac{\pi}{2})+H(\varphi-3\frac{\pi}{2})\right] \tag{2}$$

Case3

$$m_r = m_\varphi = C_o\Lambda(t)[H(r-a) - H(r-b)]\times$$
$$\times\left[H(\varphi+\frac{\pi}{4}) - 2H(\varphi-\frac{\pi}{4}) + 2H(\varphi-3\frac{\pi}{4}) - 2H(\varphi-5\frac{\pi}{4}) + H(\varphi-7\frac{\pi}{4})\right] \tag{3}$$

where C_o is a constant depending on geometry and mechanical properties of piezoelement and plate [1]. The strain induced by applying a voltage to an unconstrained piezoelectric element is expressed by

$$\Lambda(t) = \frac{d_{31}}{t_{pe}}V \tag{4}$$

where V is the applied voltage, d_{31} is the constant of transverse piezolectric effect, t_{pe} is the thickness of the piezoelement. Due to the electric symmetry $d_{31} = d_{32}$ and the strain Λ used in calculation of the radial and the transversal moment is the same. The twisting moment $m_{r\varphi}$ is equal to zero.

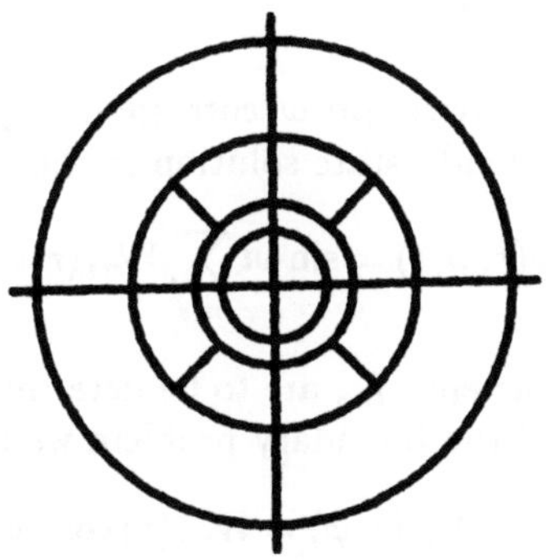

Figure 2. Segmentation of annular actuator

Excited vibrations of annular plate are described by the following dynamic equation

$$D\nabla^4 w + \rho_p t_p \frac{\partial^2 t}{\partial t^2} = \frac{\partial^2 m_r}{\partial r^2} + \frac{2}{r}\frac{\partial m_r}{\partial r} + \frac{1}{r^2}\frac{\partial^2 m_\varphi}{\partial \varphi^2} - \frac{1}{r}\frac{\partial m_\varphi}{\partial r} \tag{5}$$

where ∇^4 is calculated in the polar coordinates r and φ, D is the bending stiffness, ρ_p and t_p denote the density and the thickness of the plate, respectively. Using Eq. (1), (2) and (3) we obtain the nonuniform partial differential equations describing the transverse plate motion, e.g the plate motion excited the two half-annular segments out-of-phase (case 2) is given as follows

$$D\nabla^4 w + \rho_p t_p \frac{\partial^2 t}{\partial t^2} = C_o \Lambda \Big\{ \Big[\delta'(r-a) - \delta'(r-b) + \frac{1}{r}(\delta(r-a) - \delta(r-b)) \Big] \times$$

$$\times [H(\varphi + \frac{\pi}{2}) - 2H(\varphi - \frac{\pi}{2}) + H(\varphi - 3\frac{\pi}{2})] +$$

$$+ \frac{1}{r^2} \Big[\delta'(\varphi + \frac{\pi}{2}) - 2\delta'(\varphi - \frac{\pi}{2}) + \delta'(\varphi - 3\frac{\pi}{2}) \Big] [H(r-a) - H(r-b)] \Big\} \quad (6)$$

Boundary conditions corresponding to the clamped inner edge and the free outer edge have the form

$$w(R_1) = 0 \qquad \frac{\partial w}{\partial r}(R_1) = 0 \quad (7)$$

$$V_r = T_r(R_2) + \frac{1}{R_2}\frac{\partial M_{r\phi}}{\partial \phi}(R_2) = 0 \qquad M_r(R_2) = 0 \quad (8)$$

3. Steady-State Solution

For the single frequency excitation ω corresponding to harmonic voltage $V = V_o \sin \omega t$ we look for the steady state solution in the form

$$w(r, \varphi, t) = \sin \omega t \sum_{n,s} \mathcal{W}_{ns}(r, \varphi) f_{ns} \quad (9)$$

where the expansion coefficients f_{ns} are to be determined. Separating variables of the eigenfunctions $\mathcal{W}_{ns}$ of the boundary problem we have

$$\mathcal{W}_{ns}(r, \varphi) = W_{ns}(r) \cos n\varphi \quad (10)$$

Substituting, Eq. (6) becomes the fourth order ordinary differential equation with respect to space variable r. The right-hand-side excitation denoted by q_{ns} is calculated in a standard way

$$q_{ns} = \frac{1}{\gamma_{ns}^2} \int_{R_1}^{R_2} \int_0^{2\pi} q(r, \varphi, t) \mathcal{W}_{ns}(r, \varphi) r dr d\varphi \quad (11)$$

where γ_{ns} is the normalization factor

$$\gamma_{ns}^2 = \int_{R_1}^{R_2} \int_0^{2\pi} \mathcal{W}_{ns}^2(r, \varphi) r dr d\varphi \quad (12)$$

The Fourier expansion coefficients q_{ns} of external loading in the case 1, the case 2 and the case 3 are equal respectively

$$q_{ns}^{(1)} = 2\pi C_o \Lambda(t)\left[bW'_{0s}(b) - aW'_{0s}(a)\right]\delta_{0n} \tag{13}$$

$$q_{ns}^{(2)} = C_o\Lambda(t)\left\{\frac{bW'_{ns}(b) - aW'_{ns}(a)}{n} - n\int_a^b \frac{W_{ns}}{r}dr\right\}[3\sin n\frac{\pi}{2} - \sin 3n\frac{\pi}{2}] \tag{14}$$

$$q_{ns}^{(3)} = C_o\Lambda(t)\left\{\frac{bW'_{ns}(b) - aW'_{ns}(a)}{n} - n\int_a^b \frac{W_{ns}}{r}dr\right\}\times$$
$$\times\,[\sin n\frac{\pi}{4} - 2\sin 3n\frac{\pi}{4} + 2\sin 5n\frac{\pi}{4} - \sin 7n\frac{\pi}{4}] \tag{15}$$

where δ_{ij} is the Kronecker symbol. Using a solution of the appropriate Bessel equation for the clamped inner edge and the free outer edge

$$W_{ns}(r) = C_1 J_n(\kappa_{ns}r) + C_2 Y_n(\kappa_{ns}r) + C_3 I_n(\kappa_{ns}r) + C_4 K_n(\kappa_{ns}r) \tag{16}$$

The eigenvalues are defined as follows

$$\kappa_{ns}^2 = \omega_{ns}\sqrt{\frac{\rho_p t_p}{D}} \tag{17}$$

The constants C_i for 1=1,2,3,4 are obtained from boundary conditions (7) and (8). for a given excitation frequency ω. As the system of linear equations is homogeneous the equations corresponding to the transverse force is omitted and C_4 is taken to be one. The rest of constants is calculated from the nonhomohenous linear system of equations. It should be stressed that the form of each system of equations n=1,2,.. is different. We can write the steady-state solution of dynamic equation in the following infinite series

$$f_{ns} = \frac{1}{\rho_p t_p}\frac{q_{ns}}{\omega_{ns}^2 - \omega^2} \tag{18}$$

In the case 1 (axisymmetrical vibration) the eigenform and the normalization factor are given as follows

$$W_{0s}(r) = C_1 J_0(\kappa_{0s}r) + C_2 Y_0(\kappa_{0s}r) + C_3 I_0(\kappa_{0s}r) + C_4 K_0(\kappa_{0s}r) \tag{19}$$

$$\gamma_{0s}^2 = 2\pi \int_{R_1}^{R_2} W_{0s}^2(r) r dr \tag{20}$$

4. Simulation

Behaviour simulations based on the formulas presented in Section 3 are performed for angular frequency corresponding to the resonances $n = 0, 1, 2$ and $s = 0, 1$. The dimensions of the steel plate are $R_1 = 20\ mm$, $R_2 = 150\ mm$, thickness $t_p = 2.5\ mm$. The dimensions of PZT actuators $a = 30\ mm$, $b = 70\ mm$, thickness $t_a = 0.2\ mm$. Piezoelectric constants are $d_{31} = d_{32} = 1.9 \times 10^{-10}$ V/m. Mechanical properties of the plate are as follows $E_p = 21.6 \times 10^{10}$ N/m^2, $\rho_p = 7800$ kg/m^3. The plate bending stiffness is $D = 300.48$ m^2/s^2. To include the internal damping the Voigt-Kelvin model is assumed, and the complex Young's modulus with the retardation time $\lambda = 2 \times 10^{-5}\ s$ is taken in numerical calculations

$$E_p(\omega) = E_p(1 + i\lambda\omega) \tag{21}$$

The three lowest frequencies corresponding to $n = 0, 1, 2$, $s = 0$ are equal to $\omega_{00} = 785.61/\text{s}$, $\omega_{10} = 680.31/\text{s}$, $\omega_{20} = 301.61/\text{s}$.

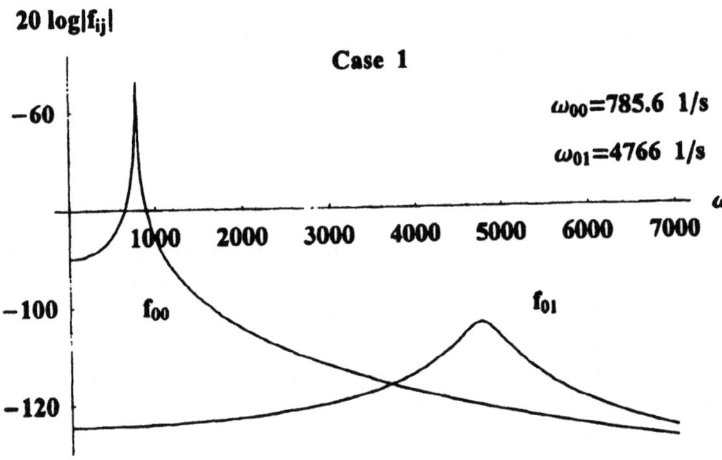

Figure 3. Frequency characteristics of plate with annular actuator

Exemplary frequency characteristics of the transverse plate displacement at the free edge are shown in Fig. 3, Fig. 4 and Fig. 5. If vibrations are excited by the full annular actuators or the in-phase segmented annular actuator we observe axially symmetrical motion (Fig. 3) and the frequency characteristics of nonsymmetrical forms are equal to zero. Due to internal damping the second form (corresponding to $s = 1$) is significantly damped.

If actuator consists of two out-of-phase annular segments (case 2) we observe antysymmetrical vibrations corresponding to $n = 1, s = 0, 1$, Fig. 4). The actuator does not excite the symmetrical forms ($n = 0$) and the forms with two nodal lines ($n = 2$). The similar properties are observed in the case 3, when the actuator excites the form with two nodal lines without the nodal circles (Fig. 5). The excitation of eigenforms $f_{02}, f_{12}, f_{22}, \ldots$ is negligible mainly due to damping effect.

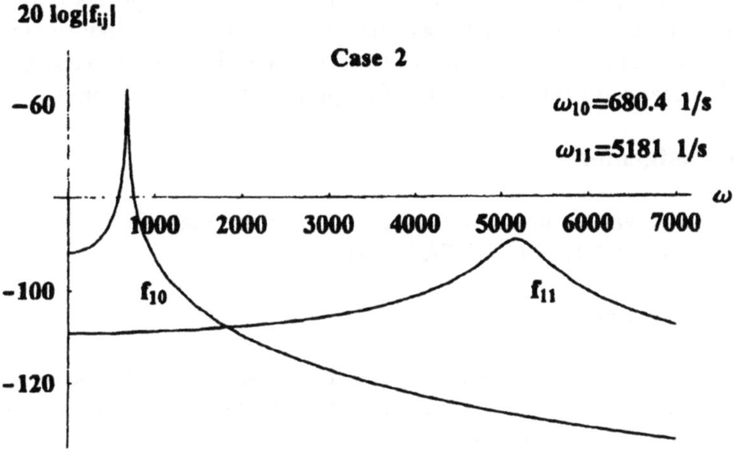

Figure 4. Frequency characteristics of plate with two half-annular actuator

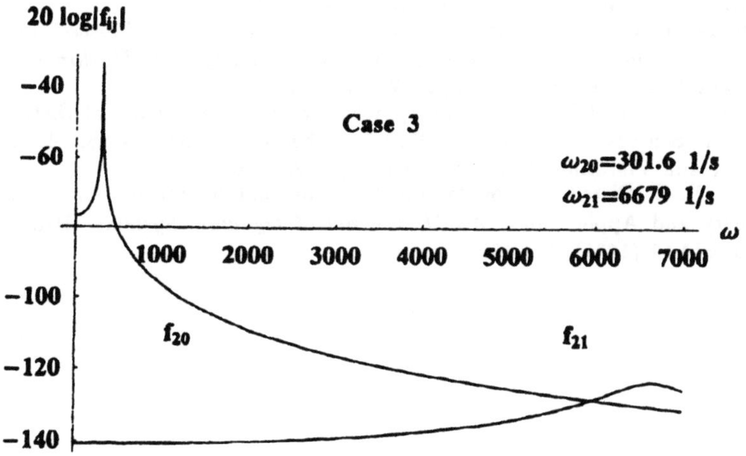

Figure 5. Frequency characteristics of plate with four quarter-annular actuator

5. Conclusions

Mathematical models of an annular plate with a single-piece distributed annular actuator and multi-piece segmented sectorial actuators are formulated and performances are simulated. The results demonstrate that modes can be selectively excited and the that the geometry and the electrical wiring of the actuator segments significantly affect the distribution of the response among modes. It is possible to design the segment shape to either excite or suppress particular modes leading to improved control behaviour. The author would like to stress the conveniency and efficiency of using the mathematical software packages, especially valuable because of the possibility of employing symbolic calculus.

Acknowledgement

This research was supported by the grant from the State Committee for Scientific Research (Grant KBN Nr 7T07A04414).

6. References

1. Dimitriadis, E. K., Fuller, C. R., and Rogers, C. A.: Piezoelectric Actuators for Distributed Vibration Excitation of Thin Plates, *ASME Journal of Vibration and Acoustics* **113** (1991), 100-107.
2. Van Niekerk, J. L., Tongue, H. H., and Packard, A. K.: Active Control of a Circular Plate to Reduce Transient Noise Transmission, *Journal of Sound and Vibration* **183** (1995), 643-662.
3. Tylikowski, A. (1999) Simulation Examination of Annular Plates Excited by Piezoelectric Actuators, in: J. Holnicki-Szulc and J. Rodellar (eds.), *Smart Structures*, Kluwer Academic Publishers, Dordrecht, pp.365-372.
4. Tylikowski, A. (1999) Piezoelectric Vibration Absorbers, in: *Proceedings of the 10-th Polish - German Seminar Development Trends in Design of Machines and Vehicles*, Warsaw University of Technology, Warsaw, pp. 135-142.
5. Tzou, H. S., and Fu, H. Q. (1992) A Study on Segmentation of Distributed Piezoelectric Sensors and Actuators; Part 1 - Theoretical analysis, in: *Active Control of Noise and Vibration, DSC - 38 ASME*, pp. 239-246.
6. Tzou, H. S. Distributed Modal Identification and Vibration Control of Continua: Theory and Applications, *ASME Journal of Dynamic Systems, Measurement and Control* **113** (1991), 494-499.

THIN-WALLED SMART LAMINATED STRUCTURES: THEORY AND SOME APPLICATIONS

N.N. ROGACHEVA
Institute for Problems in Mechanics, Russian Academy of Sciences, Prospekt Vernadskogo 101-1, Moscow 117526, Russia

1. Introduction

Though much research is devoted to smart laminated structures e.g. [1-3] there is large discrepancy between great utility and theoretical treatment of corresponding problems since this field of mechanics is new, and the needed theory is developed more slowly than practical applications.

Arbitrary laminated thin-walled smart structures as beams, plates, and shells composed piezoceramic, PVDF film , elastic, and linear viscoelastic layers are studied.

In this paper the theories of smart laminated beams, plates, and shells are constructed from 3D equations by asymptotic method without using any assumptions by investigating the limiting transition from three- dimensional problems to one and two-dimensional ones.

Some dynamic problems are considered.

2. The Theory of Arbitrary Laminated Shells

We choose curvilinear coordinates α_1 and α_2 so that they coincide with the curvatures of the special neutral surface parallel to the shell faces and positioned at the distance z_0 to the internal shell face, and a linear coordinate γ along to the normal to the neutral surface. The distance z_0 we will select later.

Constitutive relations of electroelastic layers (piezoceramics and PVDF film) can be written in the form

$$e_{ij} = s^E_{ijkl}\sigma_{kl} + d_{kij}E_k \ , \quad D_i = d_{ikl}\sigma_{kl} + \varepsilon_{ik}E_k \tag{1}$$

The behavior of the electric field is described by electrostatic equations in the vector form

$$divD = 0\,, \qquad E = -grad\phi \tag{2}$$

U. Gabbert and H.S. Tzou (eds.), IUTAM Symposium on Smart Structures and Structronic Systems, 205–212.

Here e_{ij} and σ_{ij} are the strains and stresses, D is the electrical induction vector with components D_1, D_2 and D_3, E is the electrical strength vector with omponents E_1, E_2 and E_3. The equations and notations utilized in the paper agree with those used in shell theory [3].

For elastic layers we use Hooke's law as the constitutive relations. For viscoelastic layers' modulus of elasticity in Hooke's law is replaced by complex value with maginary part proportional to the first derivative with respect to the time.

The development of 2D theory of laminated shells is made precisely the same as the development of the piezoelectric shell theory in [3]: we

1) perform a scale extension for the 3D equations with respect to the coordinates and the time in a usual way for the asymptotic analysis,
2) define asymptotic representation of the desired quantities for each layer,
3) integrate each of 3D equations with respect to the thickness coordinate γ, as a result all needed quantities are presented as polynomials in variable γ,
4) neglect in the polynomials small quantities of order ε where $\varepsilon = O(\eta^1 + \eta^{2-2s})$, ($\eta$ is the shell relative half-thickness, s is index of variability of the electroelastic state). As a result we receive the linear law with respect to the thickness for principal stresses and square law for the electrical potential.
5) write the derived equations in the terms of the shell theory on integrating the polynomials for general stresses with respect to γ for each layer.

As a result of asymptotic analysis we get the following equations of the theory for electroelastic laminated shells:

Equilibrium equations

$$\frac{1}{A_1A_2}\frac{\partial}{\partial\alpha_i}(A_iT_i) + k_iS_{ji} + \frac{1}{A_1A_2}\frac{\partial(A_iS_{ij})}{\partial\alpha_j} - k_jT_j - \frac{N_i}{R_i} - \rho_0\frac{\partial^2 u_i}{\partial t^2} + X_i = 0$$

$$\frac{T_1}{R_1} + \frac{T_2}{R_2} + \frac{1}{A_1A_2}\left[\frac{\partial}{\partial\alpha_1}(A_2N_1) + \frac{\partial}{\partial\alpha_2}(A_1N_2)\right] - \rho_0\frac{\partial^2 w}{\partial t^2} + Z = 0 \qquad (3)$$

$$\frac{1}{A_1A_2}\frac{\partial}{\partial\alpha_i}(A_iH_{ji}) + k_iG_{ji} - \frac{1}{A_1A_2}\frac{\partial(A_iG_j)}{\partial\alpha_j} + k_jH_{ij} + N_j + \rho_a\frac{\partial^2 u_i}{\partial t^2} = 0$$

$$k_i = \frac{1}{A_1A_2}\frac{\partial A_i}{\partial\alpha_j}, \quad \rho_0 = \sum_{k=1}^{N}\rho_k h_k, \quad \rho_a = \frac{1}{2}\sum_{k=1}^{N}\rho_k(z_k^2 - z_{k-1}^2)$$

Here G_i, H_{ji} are the bending and twisting moments, T_i, S_{ji}, N_i are the forces, u_i, w are the tangential displacement and the deflection of the neutral surface points, ρ_k is the material density of the kth layer, h_k is the thickness of the kth layer, z_k is the distance of the kth layer upper face from the neutral surface. The shell is composed of N layers.

Each of the electroelasticity relations for arbitrary laminated shell contains the tangential strains ε_i, ω and the flexural strains κ_i, τ simultaneously

$$T_i = A_{11}\varepsilon_i + A_{12}\varepsilon_j + C_{11}\kappa_{ii} + C_{12}\kappa_{ij} + P_i, \qquad S_{12} = S_{21} = A\omega + 2R\tau \tag{4}$$

$$G_i = M_{11}\kappa_i + M_{12}\kappa_j + K_{11}\varepsilon_{ii} + K_{12}\varepsilon_{ij} + Q_i, \quad H_{12} = H_{21} = M\tau + R\omega \tag{5}$$

where

$$A_{11} = \sum_{k=1}^{N} a_{11}^{(k)} h_k \;, \qquad A_{12} = \sum_{k=1}^{N} a_{12}^{(k)} h_k$$

$$C_{11} = \frac{1}{2}\sum_{k=1}^{N} c_{11}^{(k)}(z_k^2 - z_{k-1}^2), \quad C_{12} = \frac{1}{2}\sum_{k=1}^{N} c_{12}^{(k)}(z_k^2 - z_{k-1}^2) \tag{6}$$

$$M = \frac{2}{3}\sum_{k=1}^{N} a^{(k)}(z_k^3 - z_{k-1}^3), \; P_1 = 2\sum_{k=1}^{N} p_1^{(k)} V^{(k)}, \quad P_2 = 2\sum_{k=1}^{N} p_2^{(k)} V^{(k)}$$

$$M_{11} = -\frac{1}{3}\sum_{k=1}^{N} m_{11}^{(k)}(z_k^3 - z_{k-1}^3), \; M_{12} = -\frac{1}{3}\sum_{k=1}^{N} m_{12}^{(k)}(z_k^3 - z_{k-1}^3)$$

$$K_{11} = -\frac{1}{2}\sum_{k=1}^{N} m_{11}^{(k)}(z_k^2 - z_{k-1}^2), \; K_{12} = -\frac{1}{2}\sum_{k=1}^{N} m_{12}^{(k)}(z_k^2 - z_{k-1}^2) \tag{7}$$

$$R = \frac{1}{2}\sum_{k=1}^{N} a^{(k)}(z_k^2 - z_{k-1}^2)$$

$$Q_1 = \sum_{k=1}^{N} p_1^{(k)} V^{(k)}(z_k + z_{k-1}), \qquad Q_2 = \sum_{k=1}^{N} p_2^{(k)} V^{(k)}(z_k + z_{k-1})$$

The coefficients $a_{11}^{(k)}$, ..., $p_2^{(k)}$ are written in [4]. They depends on materials properties and electrical conditions on piezoelectric layers faces. For example for the kth elastic layer they are

$$a_{11}^{(k)} = m_{11}^{(k)} = c_{11}^{(k)} = \frac{E^{(k)}}{1-\nu_k{}^2}, \quad a^{(k)} = \frac{E^{(k)}}{2(1+\nu_k)}$$

$$a_{12}^{(k)} = m_{12}^{(k)} = c_{12}^{(k)} = \nu_k a_{11}^{(k)}$$

and for the kth piezoceramic layer with electrodes

$$a_{11}^{(k)} = \frac{1}{s_{11}^{E(k)}(1-\nu_k^2)}, \quad a_{12}^{(k)} = \frac{\nu_k}{s_{11}^{E(k)}(1-\nu_k^2)}, \quad a^{(k)} = \frac{1}{s_{66}^{E(k)}}$$

$$c_{11}^{(k)} = B^{(k)}(1-k_\sigma^{(k)})^2, \; c_{11}^{(k)} = B^{(k)}(1-k_\sigma^{(k)})^2, \; p_1^{(k)} = p_2^{(k)} = \frac{d_{31}^{(k)}}{s_{11}^{E(k)}(1-\nu_k)}$$

$$m_{11}^{(k)} = B^{(k)}\left(1 - \frac{3h_k(z_k+z_{k-1})^2(1+\sigma^{(k)})}{8(z_k^3-z_{k-1}^3)}(k_p^{(k)})^2\right)$$

$$m_{12}^{(k)} = B^{(k)}\left(\sigma^{(k)} - \frac{3h_k(z_k+z_{k-1})^2(1+\sigma^{(k)})}{8(z_k^3-z_{k-1}^3)}(k_p^{(k)})^2\right)$$

$$k_p^2 = \frac{2k_{31}^2}{1-\nu_k}, \quad 2k_\sigma^{(k)} = (1+\sigma^{(k)})(k_p^{(k)})^2, \quad \sigma^{(k)} = \frac{2\nu_k + (1-\nu_k)(k_p^{(k)})^2}{2-(1-\nu_k)(k_p^{(k)})^2}$$

$$B^{(k)} = \frac{2-(1-\nu_k)(k_p^{(k)})^2}{2s_{11}^{E(k)}(1-\nu_k^2)(1-(k_p^{(k)})^2)}$$

Here $2V^{(k)}$ is the electrical potential difference applied to electrodes of the kth layer's faces.

The position of the neutral surface is defined by the following equation:

$$K_{11} = 0 \tag{8}$$

When K_{11} equals zero the neutral surface agrees with the middle surface for the symmetrically laminated structures. In addition, only in this case the complete 2D problem for arbitrary laminated plate is separated into plane and bending problems [4].

The equations for arbitrary laminated plates and beams can be written as special cases of the arbitrary laminated shell theory.

3. Some Dynamic Problems

3.1 FORCED VIBRATIONS OF A TWO-LAYERED CYLINDRICAL SHELL

A circular cylindrical shell composed of an external metal layer and an internal piezoceramic layer vibrates under the action of electrical loading obey the law $e^{-i\omega t}$, where ω is the angular frequency of the vibrations and t is the time.

Consider a shell whose piezoceramic layer faces are covered with electrodes with given electrical potential. The shell radius is R, the length is L, the thickness of the piezoceramic layer is h_1, the thickness of the metal layer is h_2.

To solve the problem we use an approximate method of partitioning of the electroelastic state into a sum of the following simpler stressed-strained states: the principal electroelastic state with small variability and the electroelastic state with great variability.

The principal electroelastic state changes very slowly along the shell coordinates. In this case the tangential displacements are the same order as the deflections. After asymptotic analysis of the shell theory equations much as it was done in [3] we get the following equations for the principal state:

$$\frac{1}{R}\frac{dT_1}{d\xi} + \rho_0\omega^2 u = 0, \quad \frac{T_2}{R} + \rho_0\omega^2 w = 0 \tag{9}$$

$$T_i = A_{11}\varepsilon_i + A_{12}\varepsilon_j + P \quad (i = 1,.2), \quad G_i = Q, \quad \varepsilon_1 = \frac{1}{R}\frac{du}{d\xi}, \quad \varepsilon_2 = -\frac{w}{R}$$

where the ξ - line coincides with the generatrix of the cylinder.

$$z_0 = -\frac{s_{11}^E E h_2^2 (1-\nu_1^2) + h_1(h_1 + 2h_2)(1-\nu_2^2)}{2s_{11}^E E h_2 (1-\nu_1^2) + h_1(1-\nu_2^2)}, \quad \rho_0 = \rho_1 h_1 + \rho_2 h_2$$

$$A_{11} = \frac{h_1}{s_{11}^E(1-\nu_1^2)} + \frac{Eh_2}{1-\nu_2^2}, \qquad A_{12} = \frac{\nu_1 h_1}{s_{11}^E(1-\nu_1^2)} + \frac{\nu_2 E h_2}{1-\nu_2^2}$$

$$P = \frac{d_{31}V}{s_{11}^E(1-\nu_1)}, \quad Q = -\frac{d_{31}(z_1 + z_2)V}{2s_{11}^E(1-\nu_1)}$$

System of equations (9) with constant coefficients is second -order. The solution of the system is simplest. It contains two arbitrary integration constants , which we find from conditions at the edge of the cylinder. For example the force T_1 equals zero at free edges.

The second problem describes the state with great variability. For statics and quasistatics it is a simple edge effect. For dynamics the solution is not located near edges. It extends over the shell.

Asymptotics for the quantities of the second problem is

$$\left(\frac{N_1}{A_{11}},\frac{u}{R}\right)\approx O(\eta^{1/2}),\quad \left(\frac{w}{R},\frac{T_2}{A_{11}},R\kappa_2,\varepsilon_1,\varepsilon_2\right)\approx O(\eta^0),\quad (R\kappa_1)\approx O(\eta^{-1})$$

$$\left(\frac{RG_1}{A_{11}},\frac{RG_2}{A_{11}}\right)\approx O(\eta^1),\quad \left(\frac{T_1}{A_{11}}\right)\approx O(\eta^2),\quad \eta=\frac{h}{R},\quad h=h_1+h_2$$

For simplicity we assume that the index of variability is $1/2$.

Within accuracy $O(\eta^{-1/2})$ the system of equations for the second problem can be written as

$$\frac{T_2}{R}+\frac{1}{R}\frac{dN_1}{d\xi}+\rho_0\omega^2 w=0,\quad N_1=\frac{1}{R}\frac{dG_1}{d\xi}$$

$$T_2=A_{11}\varepsilon_2+A_{12}\varepsilon_1+C_{12}\kappa_1,\quad G_1=M_{11}\kappa_1+K_{12}\varepsilon_2 \tag{10}$$

$$\varepsilon_2=-\frac{w}{R},\quad \kappa_1=\frac{1}{R^2}\frac{d^2w}{d\xi^2},\qquad \varepsilon_1=-\frac{1}{A_{11}}(A_{12}\varepsilon_2+C_{11}\kappa_1)$$

The resultant equation for the system (10) and its solution has the form

$$M_{11}\frac{d^4w}{d\xi^4}+R(C_{12}-\nu C_{11}-K_{12})\frac{d^2w}{d\xi^2}+(\rho_0\omega^2R^4-(1-\nu^2)A_{11}R^2)w=0$$

$$\nu=\frac{A_{12}}{A_{11}},\qquad w=\sum_{i=1}^{4}c_ie^{-k_i}$$

where k_i are the roots of the characteristic equation.

The constants c_i can be found from the conditions at the edges. For example we meet at free edges the conditions

$$G_1+Q=0,\quad N_1=0$$

As in the classical shell theory after solving the problem we can find all 3D quantities. Figure 1 shows distribution of dimensionless stresses σ_{i*} along the shell thickness for the piezoelectric layer (PZT-5, $h_1=0.0007m$) and the steel layer ($h_1=0.0003m$), the shell radius is $0.01m$, the length is $0.02m$, frequency of vibrations is 80 kHz, $\sigma_{i*}=\sigma_ih/A_{11}$, $P/A_{11}=1$.

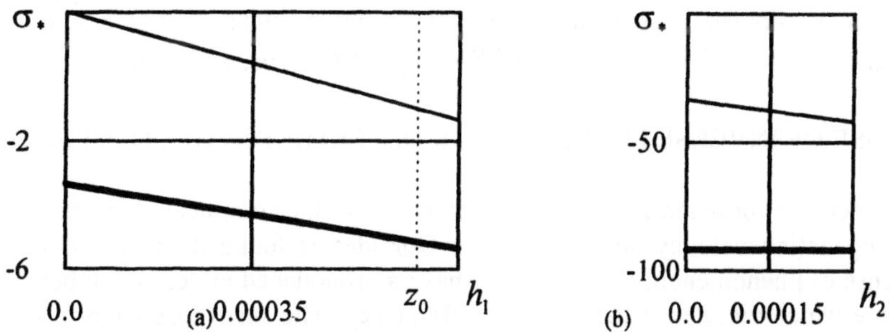

Figure 1. Distribution of dimensionless stresses σ_1 (thin line) and σ_2 (thick line) along thickness of piezoceramic (a) and metal (b) layers

3.2 PIEZOELECTRIC DIAPHRAGM FOR TELEPHONE TRANSDUCER

The structure is composed of two layers, one of them is a piezoceramic circular plate with thickness polarization and the another one is a metal circular plate. The piezoceramic layer's thickness and radius is h_2, R_2. The metal layer's thickness and radius is h_1, R_1, $R_1 > R_2$. The edge of the plate $r = R_1$ is rigidly clamped (Fig.1).

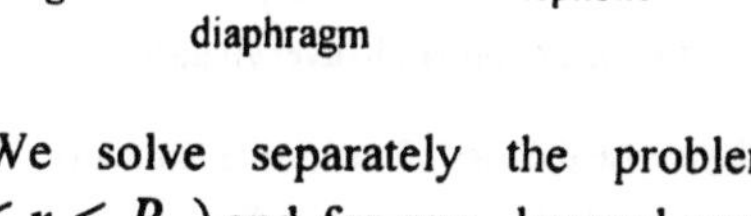

Figure 2. Cross-section of a telephone diaphragm

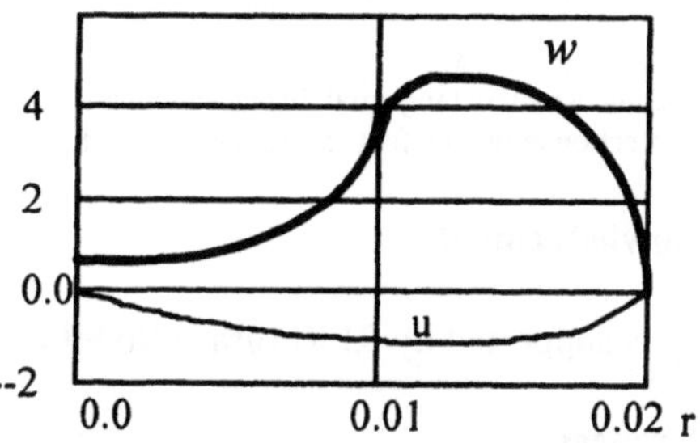

Figure 3. The distribution dimensionless displacements along a plate's radius

We solve separately the problems for two-layered part of the structure $(o \le r \le R_2)$ and for one -layered metal part $(R_2 \le r \le R_1)$. The solution for the two-layered plate was obtained in [4]. The solution for circular metal plate is known.

The arbitrary integration constants are found from the boundary conditions at the interface $r = R_2$ ($u^{(m)} = u^{(c)} - z\gamma_1^{(c)}$, $w^{(m)} = w^{(c)}$, $\gamma_1^{(m)} = \gamma_1^{(c)}$, $T_1^{(m)} = T_1^{(c)}$, $N_1^{(m)} = N_1^{(c)}$, $G_1^{(m)} = G_1^{(c)} + zT_1^{(c)}$) and the conditions at the rigidly clamped edge $r = R_1$ ($u^{(m)} = 0$, $w^{(m)} = 0$, $\gamma_1^{(m)} = 0$), where z is the distance between the neutral plane of the two-layered plate and the middle plane of the metal plate. Figure 3 depicts the dimensionless displacement for diaphragm is made of PZT-5 and Al , $R_1 = 0.02m, R_2 = 0.012m, h_1 = 0.0008m, h_2 = 0.0012m, \Omega = 10kHz$.

3.3 THE USE OF INVERSE PROBLEMS FOR DETERMINING OF LOSS FACTOR

Traditional methods for determining of loss factor involve the complicated measurement of wave propagation velocity and attenuation. Our idea is founded on the following physical sense of phenomenon: the loss factor have a pronounced effect on the behavior of a structure which vibrates at the resonance frequency. The small loss factor away at resonance changes the elastic or electroelastic state of structure moderately. As an example we use into solutions of problems for longitudinal vibrations of PZT-5 bar (Figure 4a) and flexural vibrations PZT-5 of bimorphic plate (Figure 4b) [3,6] complex modulus of elasticity and calculate amplitude values of desired quantity at resonance frequency as a function of loss factor. The result may be used to find determining the loss factor after measuring of resonance amplitude value of desired quantities.

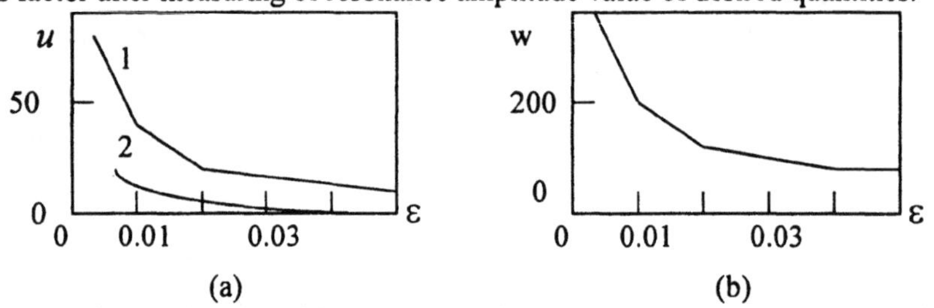

Figure 4. Dimensionless tangential displacement at the first (1) and the second (2) resonance of the bar (a) and deflection at the the first resonance of the bimorphic plate (b) as a function of the loss factor

4. Acknowledgements

The study is supported by RFRF(grant No.99-01-01123) and INTAS(grant No.96-2113)

5. References

1. Tzou, H.S.: Piezoelectric shells (Distributed Sensing and Control of Continua), Kluver Academic Publishers, Dordrecht, 1993
2. Reddy, J.N.: An evaluation of equivalent single-layer and layerwise theories of composite laminates, Computers & Structures 25 (1993), 21-35.
3. Rogacheva, N.N.: The Theory of Piezoelectric Shells and Plates ,CRC Press, Boca Raton, 1994.
4. Rogacheva, N.N. (1998) Modelling and Control of Adaptive Mechanical Structures, Ulrich Gabbert (ed.) Piezoelectric adaptive mirrors, Fortschr.-Ber.VDL Reihe 11 Nr. 268, Dusseldorf, pp.165-174.
5. Yutaka Ichinose: Optimum design of a piezoelectric diaphragm for telephone transducers, JASA, 90,3 (1991), 1246-1252.
6. Rogacheva, N.N., Chang, S.H. and Chou, C.C. : Electromechanical analysis of a symmetric piezoelectric elastic laminated structures: theory and experiment, IEEE Trans. Ultrasonics, Ferroelectrics and Frequency Control, 45, 2 (1998), 285-294.

PRECISION ACTUATION OF MICRO-SPACE STRUCTURES

SHYH-SHIUH LIH[1], GREGORY HICKEY[1],
D. W. WANG[2], H. S. TZOU[2]
[1]*Jet Propulsion Laboratory*
California Institute of Technology, Pasadena, CA 91109, USA
[2] *Department of Mechanical Engineering, StrucTronics Lab*
University of Kentucky, Lexington, KY 40506-0108, USA

1. Introduction

Space exploration and communication satellites and space structures need deployable precision mirrors, reflectors, and antennas. Conventional deployable space structures require motors and kinematic mechanisms to assist the deployment process. These mechanisms have the potential to jam or tangle and thus jeopardize the entire space mission. Recent development of smart structures and structronic systems opens many new design options in precision structures and systems. This paper reports a study of precision actuation and control for micro-shell laminated space structures made of smart materials. New design concepts of deployable micro-shell laminated structures are discussed and conceptual models fabricated. Analysis of precision parabolic struts is carried out and actuation authorities of proposed configurations are evaluated.

Modern space exploration and communication require large reflectors, solar panels, antennas, and mirrors that are lightweight and deployable. Conventional lightweight deployable space structures often utilize kinematic mechanisms driven by motors in the deployment process. Figure 1 shows two large deployable antennas: Galileo Space craft antenna and Next Generation Space Telescope (NGST) structures. Note that deployable struts that provide structural stability and strength are used in all these two designs.

Recent research and development of smart structures and structronic systems reveal many new design opportunities in sensors, actuators, precision systems, mechatronic systems, and adaptive structures (see [1] and [2]). Accordingly, incorporating smart materials to lightweight deployable flexible micro-space structures is a natural trend and this technology needs to be developed quickly. The new design concepts utilize smart materials: piezoelectric materials and shape memory alloys in the design of new "smart" deployable micro-space structures. In these designs shape memory materials are used in the deployment process and piezoelectric materials (bimorphs or laminated structures) are used for precision control and final shape tuning. This paper evaluates the precision control of struts of the micro-space shell structure. Two design concepts showing

U. Gabbert and H.S. Tzou (eds.), IUTAM Symposium on Smart Structures and Structronic Systems, 213–222.

parabolic shells supported by deployable precision struts are illustrated in Figures 2 and 3. Since piezoelectric laminates are associated with the precision adjustment, accurate modeling and analysis of precision laminates is essential to the successful design of deployable precision lightweight shell structures. The organization of this paper is the modeling and analysis of precision piezoelectric struts will be presented first, followed

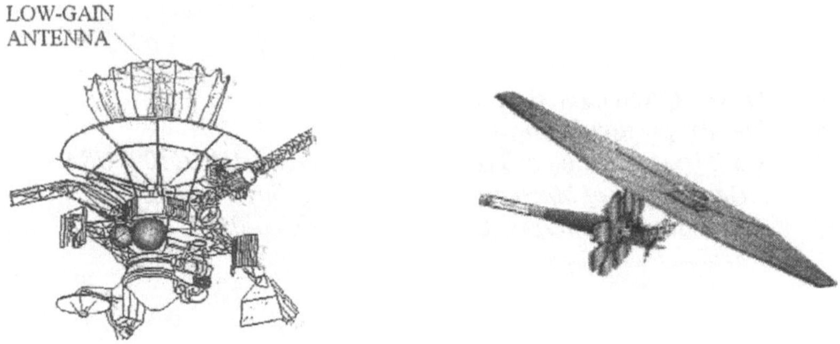

by detailed analysis of precision piezoelectric bimorph and laminated struts.

Figure. 1. NASA Galileo Space Probe (left) and NGST (right).

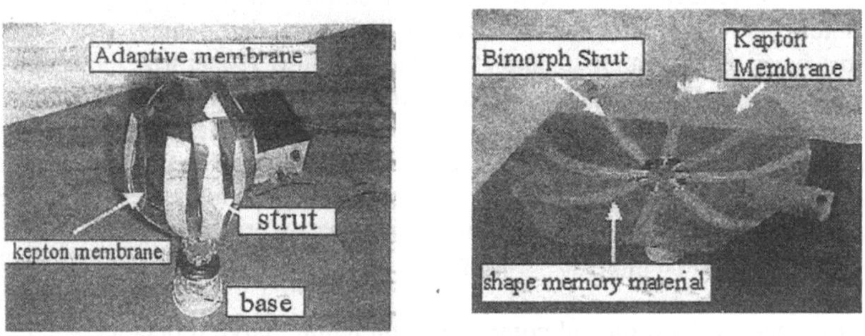

Figure. 2. Proposed stowed and deployed flexible structure.

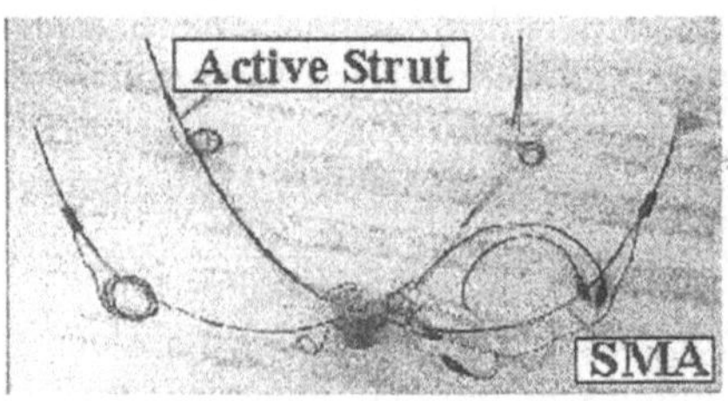

Figure. 3. Conceptual model with SMA and piezoelectric bimorph or laminated struts.

2. Modeling of Parabolic Precision Struts

This study considers two precision piezoelectric parabolic beams: the bimorph parabolic beam and the parabolic laminated beam, which are the constituent elements for the baseline precision micro-space shell structures.

2.1 PARABOLIC BIMORPH STRUTS

Piezoelectric bimorph beams have been shown by Tzou for precision microactuation, positioning, indication, and manipulation in high-precision operations [3]. A piezoelectric bimorph strut is made of two piezoelectric layers laminated together with opposite polarity. Parabolic struts have found uses in many practical applications. Figure 4 illustrates the model used for cantilever piezoelectric parabolic bimorph strut (Note that the mesh is used in the finite element modeling and analysis presented later.) In the development of piezoelectric parabolic bimorph beam, two pieces of piezoelectric materials are bonded together forming a basic bimorph beam structure [3]. It should be noted that the polarized directions of the two layers are opposite to each other in order to introduce a bending effect in the bimorph beam when an external control voltage is applied across the thickness of the parabolic bimorph. Figure 5 illustrates the bending effect due to the voltage induced converse effect.

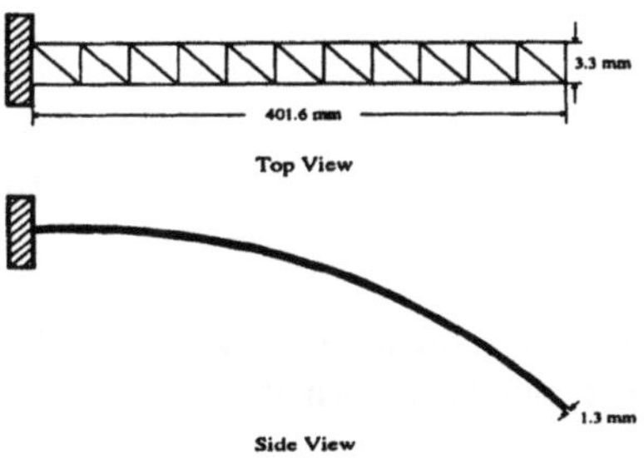

Figure 4. A curved piezoelectric bimorph strut

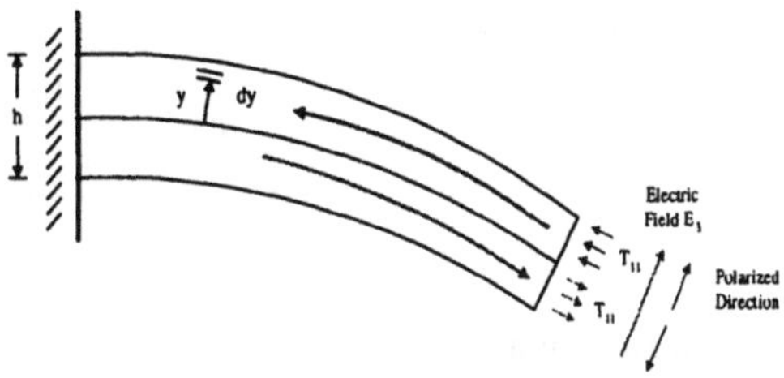

Figure 5. Bending effect of a piezoelectric parabolic bimorph subjected to external voltage.

The normal stress T11 in the upper layer is in compression while the lower layer in tension when a positive voltage E_3 is applied. Stress multiplied by a cross-section area (width b x infinitesimal height dy) gives an equivalent force dF:

$$dF = be_{31}E_3dy \tag{1}$$

And this force multiplied a moment arm y yields a bending moment. Thus, the bending moment M with respect to the neutral axis can be calculated as

$$dM = be_{31}E_3ydy \tag{2-a}$$

$$M = \int_{-h/2}^{h/2} be_{31}E_3ydy = e_{31}E_3\left(\frac{bh^2}{4}\right) \tag{2-b}$$

where h is the thickness; b is the width; y is the moment arm; e_{31} is the piezoelectric stress coefficient; and E_3 is strength of electric field and assumed that it is the uniformly distributed over the surface. According to Castigliano's theory on deflection, the angular displacement β of the point of application of the moment M is determined by the equation (e.g. see [4]).

$$\beta = \frac{\partial C}{\partial M} \tag{3}$$

where the complementary energy C can be expressed as

$$C = \int \frac{M^2 R}{2YI} d\psi \tag{4}$$

Thus, the angular displacement can by calculated by

$$\beta = \int \frac{MR}{YI} d\psi \tag{5}$$

where I is the area moment of inertia ($I = bh^3/12$); Y is the Young's modulus; R is the curvature radius and ψ is the central angle; and $E_3 = V/h$ for a uniformly distributed field. If V is only a function of time and independent of space, the equation above can be further written as

$$\beta = \int \frac{e_{31}\left(\frac{V}{h}\right)\left(\frac{bh^2}{4}\right)}{\frac{bh^3}{12}} R d\psi = \frac{3e_{31}VR}{Yh^2}\psi \tag{6}$$

The above equation provides a relationship between the angular deflection and the applied voltage V of the cantilever bimorph beam. The angular deformation of the free end of the beam, β^*, can be calculated by

$$\beta^* = \frac{3e_{31}VR}{Yh^2}\psi_0 = \frac{3e_{31}L}{Yh^2}V \tag{7}$$

where ψ_0 is the original central angle and L is the length of the beam. From the geometric relationship shown in Figure 6, one can derive the central angle after deformation, ψ_1,

$$\psi_1 = \psi_0 - \beta^* \tag{8}$$

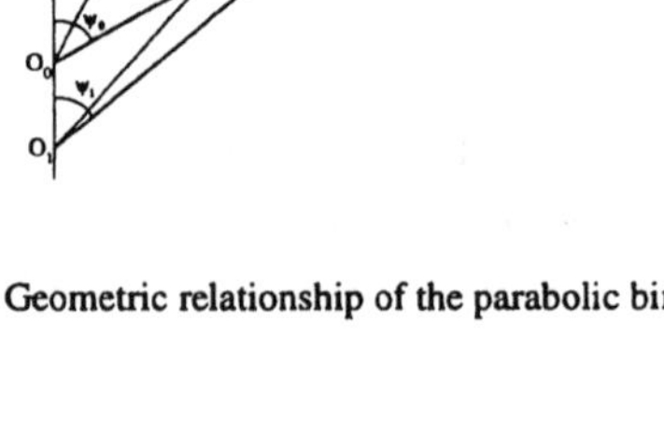

Figure. 6. Geometric relationship of the parabolic bimorph beam.

Then, the transverse, horizontal and the absolute displacements at the free end of the beam are:

$$W = (R_0 - R_1) + (R_1 \cos\psi_1 - R_0 \cos\psi_0) \tag{9-a}$$

$$U = R_0 \sin\psi_0 - R_1 \sin\psi_1 \tag{9-b}$$

where $S = \sqrt{W^2 + U^2}$, and R_0 and R_1 are the original curvature radius and the curvature radius after deformation, respectively; O_0, O_1 are the original center of curvature and the center of curvature after deformation, respectively.

2.2. PARABOLIC LAMINATED STRUTS

Parabolic beams have a major radius and a minor radius. However, if these radii are equal, parabolic beams become semicircular beams. For verification purpose, analytical solutions of semicircular beams are calculated in this section. The laminated beam is an elastic beam sandwiched between two layers of piezoelectric materials acting as actuators/sensors respectively. For this analysis polyvinylidene diflouride (PVDF) is used as the piezoelectric material. Since the thickness of the two pieces of the piezoelectric layer is very small (here we ignore the effect of the PVDF's to the static deformation), it is convenient to use polar coordinates. For a cross section of the semicircular beam located at angle θ from the section on which a concentrated force P is applied at the tip end as shown in Figure 7, we have

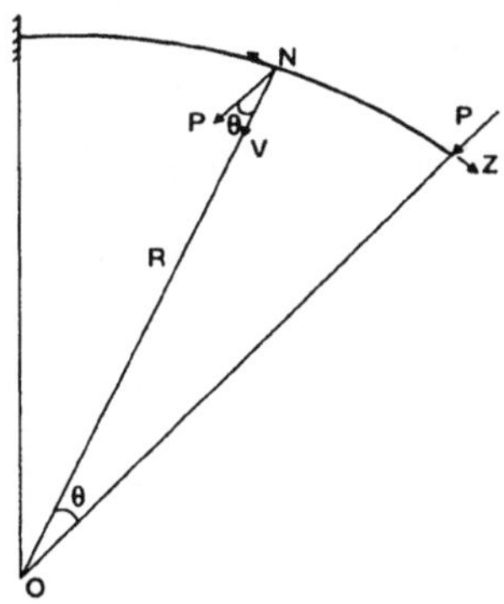

Figure. 7. A semicircular beam subjected to the concentrated force.

$$N = P\sin\theta, \quad \frac{\partial N}{\partial P} = \sin\theta; \quad V = P\cos\theta, \quad \frac{\partial V}{\partial P} = \cos\theta \tag{10-a,b}$$

$$M = PR\sin\theta, \quad \frac{\partial M}{\partial P} = R\sin\theta \tag{10-c}$$

where N, V and M are the normal force, shear force and moment which is generated by P at angle θ. According to Castigliano's theory on deflection, the displacement w in the direction of which force P is applied is determined by the equation

$$w = \frac{\partial U}{\partial P} = \int \frac{N}{EA}\frac{\partial N}{\partial P}dz + \int \frac{kV}{GA}\frac{\partial V}{\partial P}dz + \int \frac{M}{EI}\frac{\partial M}{\partial P}dz \tag{11}$$

The equation above can be further written as:

$$w = \int_0^{\theta} \frac{P\sin\theta}{EA}\sin\theta R d\theta + \int_0^{\theta} \frac{1.5P\cos\theta}{GA}\cos\theta d\theta + \int_0^{\theta} \frac{PR\sin}{EI} R\sin\theta R d\theta \tag{12}$$

Vibration control of the laminated struts is discussed next. For a parabolic beam strut, the curvilinear coordinates α_1 and α_2 are specified by ϕ and x respectively. The Lamé parameters A_i and radii of curvatures R_i are respectively defined by

$A_1 = \dfrac{b}{\cos^3\phi}, A_2 = 1,\ R_1 = \dfrac{b}{\cos^3\phi}, R_2 = \infty$ where b is constant. The closed-form control equations are

$$-\frac{\partial \tilde{N}_{11}}{\partial\phi} - \frac{\partial(\tilde{N}_{21}A_1)}{\partial x} - \frac{1}{R_1}\left[\frac{\partial \tilde{M}_{11}}{\partial\phi} + \frac{\partial(\tilde{M}_{21}A_1)}{\partial x}\right] + A_1\rho h\ddot{u}_1 = A_1F_1 \tag{13}$$

$$-\frac{\partial}{\partial\phi}\left\{\frac{1}{A_1}\left[\frac{\partial \tilde{M}_{11}}{\partial\phi} + \frac{\partial(\tilde{M}_{21}A_1)}{\partial x}\right]\right\}$$

$$-\frac{\partial}{\partial x}\left\{\frac{\partial \tilde{M}_{12}}{\partial\phi} + \frac{\partial(\tilde{M}_{22}A_1)}{\partial x}\right\} + A_1\frac{\tilde{N}_{11}}{R_1} + A_1\rho h\ddot{u}_3 = A_1F_3 \tag{14}$$

where the ~ terms include the elastic component and the feedback control component induced by the converse piezoelectric effect, e.g., $\tilde{N}_{11} = N_{11} + \tilde{N}_{11}^a$, $\tilde{M}_{11} = M_{11} + \tilde{M}_{11}^a$ where N_{11} and M_{11} are the elastic components . The in-plane effective forces $\tilde{N}_{ii}^a$ and moment $\tilde{M}_{ii}^a$ induced by the imposed actuator voltage ϕ^a can be expressed as (see [5])

$$\tilde{N}_{11}^a = d_{31}Y_p\phi^a\ ;\ \tilde{N}_{22}^a = d_{32}Y_p\phi^a \tag{15}$$

$$\tilde{M}_{11}^a = r_1^a d_{31}Y_p\phi^a\ ;\ \tilde{M}_{22}^a = r_2^a d_{31}Y_p\phi^a \tag{16}$$

where Y_p is Young's modulus of the piezoelectric actuator; r_i^a is the effective moment arm; A_i is Lamé parameter; R_i is curvature radius. Note that the control algorithms determine the imposed actuator voltage ϕ^a. For the velocity feedback, the sensor signal ϕ^s used in control components is replaced by $\dot{\phi}^s$, i.e.,

$$\dot{\phi}^s = \frac{\partial}{\partial t}\left[\phi^s(u_1, u_2, u_3, t)\right] = \phi^s(\dot{u}_1, \dot{u}_2, \dot{u}_3, t)$$

aand amplified by the gain factor G. Thus,

$$\tilde{N}_{11}^{a} = -G \cdot d_{31} Y_p \dot{\phi}^s ; \quad \tilde{N}_{22}^{a} = -G \cdot d_{32} Y_p \dot{\phi}^s \tag{17}$$

$$\tilde{M}_{11}^{a} = -G \cdot r_1^a d_{31} Y_p \dot{\phi}^s ; \quad \tilde{M}_{22}^{a} = -G \cdot r_2^a d_{32} Y_p \dot{\phi}^s \tag{18}$$

For independent modal control, the parabolic beam equations are transferred to the modal domain $u_i(\theta, t) = \sum_{k=1}^{\infty} \eta_k(t) U_{3k}(\theta)$ and, thus, modal control effects can be evaluated independently. Studies of prototype models are evaluated next.

3. Case Studies

Scaled demonstration models of the micro-space structures were presented in Figures 2 and 3. Piezoelectric bimorphs or piezoelectric laminates were used as the parabolic structural struts for precision control of the final micro-shell structures. These two configurations are analyzed as case studies and theoretical solutions are compared with numerical results.

3. 1. CASE STUDY 1 - PARABOLIC BIMORPH STRUTS

For this case a parabolic piezoelectric bimorph strut is made of two piezoelectric (PVDF) layers laminated together with opposite polarity. The beam is 401.6 mm long, 3.3 mm wide and 2×0.65 mm thick. All geometric and material properties are provided in Table 1. The beam is modeled as an assemblage of 280 identical triangular elements, 140 for each layer and 70 meshes along the beam length (see [6] and [7]). The left end is clamped and the right is free (see Figure 4). A unit voltage (one Volt) is applied across the thickness. The objective of this model is to evaluate the precision actuation capability and to verify the converse piezoelectric effect by applying a constant voltage on the parabolic bimorph struts. Two fundamental techniques, namely, theoretical and finite element methods, are used in the analysis of the parabolic piezoelectric bimorph beam struts of the micro-space structures. Table 2 summarizes the theoretical and finite element results of the constant-radius beam struts with various angles (0-90°) and compares the analytical solutions and numerical result.

TABLE 1. Geometric and material properties.

Length, L (m)	0.4016
Width, b (m)	3.3×10^{-3}
Thickness, h (m)	0.5×10^{-4}
Density, ρ (Kgm^{-3})	1.8×10^{3}
Young's modulus, Y (Pa)	2.0×10^{9}
Poisson's constant, μ	0.29
Piezostrain constant, d_{31} (mV^{-1})	0.22×10^{-10}
Electric permitivity, ε_{11} (Fm^{-1})	1.062×10^{-10}

The static deformation of the free end of the parabolic bimorph beam struts has two deformation components: the transverse deformation and the horizontal deformation. It is observed that the transverse displacement component increases and the horizontal displacement component decreases with the increment of central angle. And the total static deformation decreases with the increment of the central angle. The analytical and numerical results are compared very well and the error range is within 5%. Also, the precision actuation capability is excellent and suitable in micro-shell actuation and control.

TABLE 2. Comparisons of the displacement by theoretical and finite element results of parabolic beam struts (free end).

Angle	0°	15°	30°	45°	60°	75°	90°
Theory	3.2931	3.2862	3.2674	3.2364	3.1933	3.1387	3.0729
FEM	3.1493	3.1467	3.1325	3.1060	3.0682	3.0189	2.9589
Error (%)	4.36	4.24	4.13	4.03	3.91	3.82	3.71

Unit: (m)$\times 10^{-6}$

3.2. CASE STUDY 2 - LAMINATED PARABOLIC STRUTS

The parabolic beam studied here is laminated with two pieces of PVDFs on the top and bottom surfaces of a flexible beam strut. It has parabolic curvature radius in the longitudinal direction, and infinite curvature radius in the width direction as shown in Figure 8 (beam B). (Note that beams A and C are constant radius beam, in which the minor radius is the radius of beam A and the major radius is the radius of beam C.) It is assumed that the inner piezoelectric layer serves as a distributed sensor, the outer layer serves as a distributed actuator. All geometric and material properties are provided in Table 3.

The beam is clamped at the left end and the right end is free. A point force with magnitude of 0.01 N is applied at the tip end. In FE analysis, the beam strut is modeled with 300 triangular shell elements, 100 for each layer, and 50 meshes along the strut length.

TABLE 3. Geometric and material properties

Properties	PVDF	Steel
Length, L (m)	0.4016	0.4016
Width, b (m)	3.3×10^{-3}	3.3×10^{-3}
Thickness, h (m)	0.5×10^{-4}	1.3×10^{-3}
Density, ρ (Kgm^{-3})	1.8×10^{3}	7.8×10^{3}
Young's modulus, Y (Pa)	2.0×10^{9}	2.1×10^{11}
Poission's constant, μ	0.29	0.30
Piezostrain constant, d_{31} (mV^{-1})	0.22×10^{-10}	
Electric permittivity, ε_{11} (Fm^{-1})	1.062×10^{-10}	

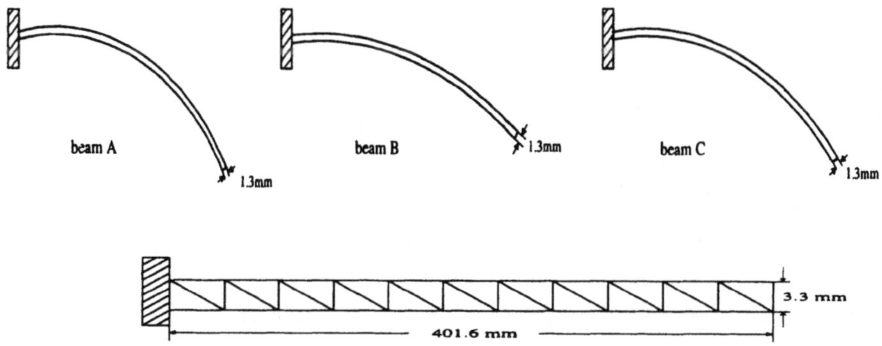

Figure. 8. Circular struts (A & C) and parabolic strut (B).

3.2.1. *Static Actuation*

The minimum and maximum curvature radii of the parabolic beam B are 0.52686 m and 0.9679 m, respectively. Two additional constant-radius beam struts (A and C), i.e., beam A has a constant curvature radius 0.52686m and beam C has a constant curvature radius 0.9679m, are studied. Based on the analytical derivations presented previously, the transverse static deformation (W) of the tip end of the parabolic beam B should be smaller than that of beam A and larger than that of beam C. The horizontal static deformation (U) should be larger than that of beam A and smaller than that of beam C. The static deformations of the tip end of these three ring shells are provided in Table 4; comparisons are outlined in Table 5.

TABLE 4. Static deformations of the tip end of the beams.

	Beam A	Beam B	Beam C
W	0.151349D-02	0.160175D-02	0.164002D-02
U	0.444910D-03	0.321288D-03	0.260831D-03
S	0.1577528D-2	0.1633655D-2	0.1660632D-2

TABLE 5. Transverse deformations of tip end of beams.

	Beam A	Beam C
Analytical solution	1.5146×10^{-3}	1.6437×10^{-3}
Numerical solution	0.151349D-02	0.164002D-02
Percentage error	0.073	0.22

3.2.2. *Dynamic Analysis and Control*

In dynamic analysis, an initial displacement is imposed at the free end. The initial damping ratio is assumed to be 0.2% and the total time with a time step $\Delta t = 2.5\times10^{-3}$ second is set to be 5 seconds. The first natural frequency is 6.7488 Hz; the free displacement response of the tip end is shown in Figure 9. Controlled responses under the negative velocity control are shown in Figures 10.

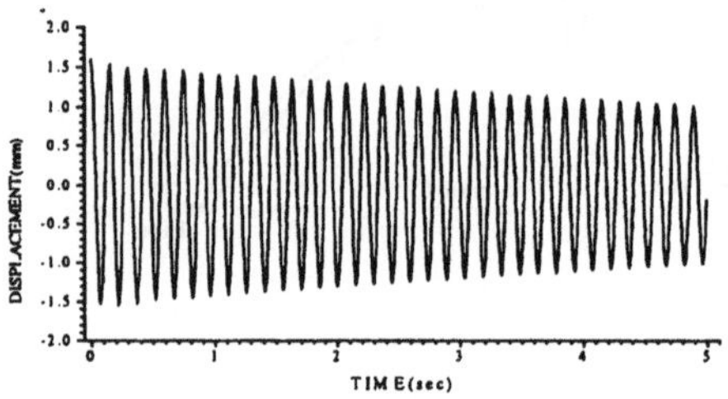

Figure. 9. Free displacement response of the tip end

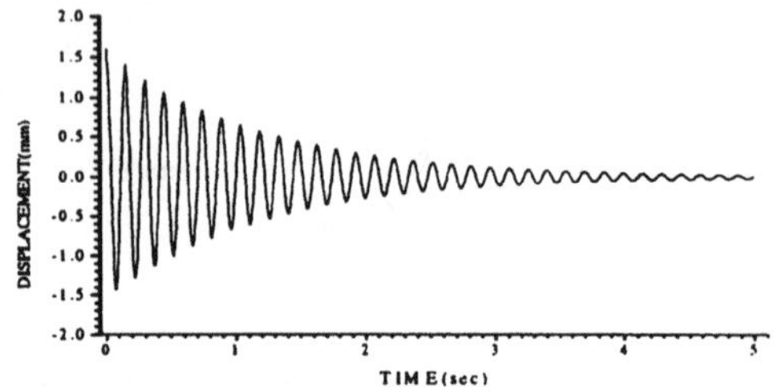

Figure. 10. Controlled response of the tip end (gain=3.0).

4. Conclusions

Precision actuation and control of micro-shell space structures made of smart materials such as those evaluated in this paper enable a new class of space mechanisms. The actuation and control capabilities of laminated parabolic struts were analyzed and the analytical solutions show very good agreement and correlation with finite element results. This study demonstrates the feasibility of precision actuation and control capabilities of deployable shell-type micro-space structures.

Acknowledgement

This research is supported by the Deformable Thin Shell Nano-Laminate Mirror Task at Jet Propulsion Laboratory, California Institute of Technology under contract with the National Aeronautics and Space Administration. Contributors to this work included: Yasser Al-Saffar, Andrew Clem, Ron Couch, Jim Jackson, Craig Moseley, Chris Kuhn, Jason Clark, Kathy Hardesty, and Casey McIntosh in the Department of Mechanical Engineering at the University of Kentucky.

5. References

1. Tzou, H.S. and Anderson, G.L. (Editors) (1982), Intelligent Structural Systems, Kluwer Academic Publishers, Dordrecht/Boston/London.
2. Tzou, H.S. (1998) "Multi-field Transducers, Devices, Mechatronic Systems and Structronic Systems with Smart Materials," Shock and Vibration Digest, Vol.30, No.4, pp.282-294
3. Tzou, H.S., (1989) "Development of a Light-weight Robot End-effector using Polymeric Piezoelectric Bimorph", Proceeding of the 1989 IEEE international Conference on Robotics and Automation, pp.1704-1709.
4. Boresi, P. (1978) "Advanced Mechanics of Materials", 3rd ed., John Wiley and Sons, New York.
5. Tzou, H.S. (1993) Piezoelectric Shells (Distributed Sensing and Control of Continua), Kluwer Academic Publishers, Dordrecht/Boston/London.
6. Tzou, H.S. and Ye, R. (1996) "Analysis of Piezoelastic Structures with Laminated Piezoelectric Triangle Shell Element", AIAA Journal, Vol. 34, No. 1, pp.110-115
7. Hom, C.L. and Shankar, N. (1996) "A finite element method for electrostrictive ceramic devices", Int. J. Solids & Structures, Vol. 33, pp.1757-1779.

EXPERIMENTAL STUDIES ON SOFT CORE SANDWICH PLATES WITH A BUILT-IN ADAPTIVE LAYER

H. ABRAMOVICH*, H.-R. MEYER-PIENING*
**Faculty of Aerospace Engineering, Technion, I.I.T., 32000 Haifa, Israel.*
Tel: +972-4-8293199, Fax: +972-4-8231848
E-Mail: haim@aerodyne.technion.ac.il
**Institut für Leichtbau und Seilbahntechnik ETH Zurich, CH-8092, Switzerland.*

1. Introduction

A common form of an adaptive structure is a thin type structure equipped with piezoelectric laminae. These laminae are made of piezoelectric materials such as Polyvinylidene Fluoride (PVDF) - a piezoelectric copolymer film [1] or Lead Zirconia Titanate (PZT)[2] - a piezoceramic based material available at present in relatively small rectangular patches. The figure of merit of such laminae is their capability of transducing electric fields into mechanical strains, and mechanical strains into electrical charges. These "active" laminae are used either to actuate the hosting structure by inducing strains in the non-piezoelectric, "passive" laminae, or to sense deflections of the hosting structure by measuring the local strain fields. The active laminae, the actuators, can be continuous over the entire domain of the structure, as in the PVDF case, or discontinuous as in the case of piezoceramic (PZT) patches. A survey on piezoelectricity and its use can be found in reference [3].

A vast number of studies deal with the use of piezoceramic patches to control the vibrations of flexible structures like beams and plates and their deflections. Such are the studies of Cudney et al. [4], Clark et al. [5], Akella et al. [6] and Batra and Ghosh [7] to quote only a few. Other researchers, like Main et al. [2], Kim et al. [8], Chandra and Chopra [9] and Pletner and Abramovich [10], try to model composite structures, which include PZT actuators and to evaluate the influence of the actuators induced strains on the overall behavior of the structures.

The subject of noise attenuation using active beside passive means has been presented in numerous studies in the literature(see Refs.11-18). Those references represent only a few references in which theoretical and experimental efforts were made to apply active acoustical control to reduce noise levels. In some of the works [13-18], use was made of PZT patches acting to induce strains and to control and then attenuate interior noise.

While the control part of the smart-adaptive structures problem is well advanced it seems that the development of accurate models for induced strain actuation and sensing

U. Gabbert and H.S. Tzou (eds.), IUTAM Symposium on Smart Structures and Structronic Systems, 223–230.

local strains baked by many experimental studies are essential for the correct and efficient design of smart/adaptive structures and their use for vibration suppression, noise attenuation, shape control and vibration steering.

At present, the majority of the smart structures do not involve sandwich structures. A few studies [19-22], do concentrate on sandwich type structures, using the PZT actuators either on the outer faces [19] or in the thickness-shear mode [20-21]. Recently, Abramovich & Pletner [22] presented a sandwich beam having a flexible inner aluminum beam equipped with PZTs. The sandwich beam had no faces and the actuation and sensing was done using the inner flexible beam, yielding very promising results.

Sandwich structures consisting of stiff facing sheets and a relatively soft lightweight core such as rigid foam or honeycomb are highly efficient in bending. An excellent review on computational models of sandwich structures can be found in reference [23]. Providing the sandwich structure with means of sensing the vibration motions induced by external noise and means of actuation to reduce the amplitudes of those motions will yield a new advanced structure having the efficiency of a sandwich one together with the ability of active noise attenuation.

This new structure will be constructed from five main layers: a thin metal plate equipped with PZT patches for sensing and actuation sandwiched in between the outer faces and foam core. This new emerging sandwich construction offers many advantages over the conventional surface-mounted constructions. For example, a far better protection of the piezoceramic patches and its electrical leads and wires are obtained. Moreover, the sensing and the actuation is done independently without the need to mount and change the host structure, and at relatively low bending stresses which will not be detrimental to the brittle piezoceramic patches. One should note the built-in adaptive layer is located at the middle of the cross section of the sandwich construction. It was **not** designed to excite the whole structure, as it is located on the neutral plane. The actuators/sensors located on the built-in adaptive layer are aimed to excite/sense only the adaptive layer itself, which vibrates due to the outer noise induced vibrations. If these noise induced vibrations could be sensed clearly by the built-in adaptive layer, and the actuators can excite it at its present position, then many other configurations can be tailored to yield an optimal location for noise attenuation. This will be done in a follow-up research, which was already initiated.

The experiments carried out and presented in this paper were aimed at trying to provide answers to the following general imposed questions:

1. Can the structure induce enough strain to reduce the noise-induced vibrations radiated through the structure into an enclosure. This question deals with the converse effect of the piezoelectric phenomena, namely actuation. Its application might be as smart walls for active control of sound radiated by harmonically excited thin walled structures, to be found in the aircraft and automotive industries as well as in other areas where noise levels reduction is a must.

2. Can the structure "feel" external disturbances, like impact (short, sharp type signal) which would involve a propagating wave phenomena, leakage (long continuous type signal) or other imposed loads, which would lead to the dynamic response of the plate. This question deals with the direct effect of the piezoelectric phenomena, namely

sensing. Its possible applications might be identified with smart sandwich walls for the marine and space structures by using their enhanced capability to sense either impacts or leakage by both pinpointing the location and its induced stress levels.

2. Specimen and Test Set-up

Two soft-core sandwich plates were designed and manufactured. The first one, having the dimensions of 50 x 50 cm^2, has a nominal thickness of 41.1 mm (see Fig. 1). As it can be seen, it consists of a common sandwich configuration with a built-in flexible plate made of aluminum and carrying the PZT transducers. The dimensions and the material properties used to manufacture the sandwich plate are given in Table 1.

The locations of the 5 pairs of PZTs on the flexible aluminum plate are given in Ref. [24]. The location of the various PZTs stemmed from the desire to control the first bending vibration modes of the plate, thus placing the transducers at the various peeks of the modes (where one would expect maximum bending deflection) being constrained by the overall number of available PZTs (5).

Note that the PZTs forming the pairs #1 and #2 have a thickness of 0.5 mm for each transducer, while the other three pairs, #3, #4, #5 have a thickness of .25 mm for each transducer. This was done to check the influence of the PZT thickness on its output (both sensing and actuating). It turns out ([24]) that the sensing and actuating capabilities are doubled if the thickness of the PZT is increased by two.

The second sandwich plate, aimed for full scale experiments, has the dimensions of 100 x100 cm^2, and a thickness similar to the previous one. It has nine pairs of PZTs, symmetrically located to control the first bending modes of the plate. The electrical connections are as the small prototype soft-core sandwich plate.

Following the experience gained with the sandwich beam [24], a 50x50 cm^2 sandwich plate was then manufactured. Five pairs of PZTs, with soldered wires, were bonded to the middle flexible carrying aluminum plate, using the method developed earlier for a sandwich beam. Note that due to the electrical connections (out-of-phase voltage) the PZTs faces touching the flexible aluminum plate are electrically connected due to the conductivity of the aluminum plate. Then the two foam plates were glued simultaneously to both sides of the host plate, using Epoxy adhesive. The aluminum faces of the sandwich plate were glued, one by one, a day after.

3. Results

Typical results will be presented herein. A detailed description can be found in detail in reference [24]. The experiments were aimed to obtain answers to the two questions imposed in the Introduction and deal with the sensing and the actuation abilities of the soft core sandwich plate with a built-in adaptive layer. First, some of the results obtained from experiments with the prototype sandwich plate will be presented followed by the full-scale experimental results.

3.1. PROTOTYPE 50x50 CM^2 SOFT CORE SANDWICH PLATE

First the actuation capabilities of the sandwich plate were tested. The plate was placed on a table in free air. The various PZTs excited the built-in adaptive layer (Al+PZT in Table 2) using a power supply fed in by a wave generator, at different frequencies.

Note that only a pair of PZTs performed the excitation at a time. The generated noise levels were measured with a microphone, placed at a height of 30 cm. above the center of the plate. Its output was read on an oscilloscope. The results are presented in Table 2, for various configurations of the sandwich plate. The results show that the adaptive layer can be activated, and its output can be sensed easily. The damping features of the soft-core do not interfere and the PZTs are capable to produce enough power to excite the adaptive layer.

Next the PZT sensing capabilities of the piezolaminated sandwich plate under loudspeaker excitation were tested. The loudspeaker produced noise with a constant output of: 3 V (peak-to-peak)

Two microphones were employed. One microphone was inside the box, and the other one outside the box. The level of noise generated inside the box was 10 times higher than those measured outside. Various sequential arbitrary waveforms were fed through the wave generator and the amplifier to the loudspeaker. As before it turns out that the adaptive plate inside the sandwich core is sensitive enough to detect the noise induced vibrations and to produce a relatively high output, over the background noise.

The sensing capabilities of the piezolaminated sandwich plate under a light impact were also tested. The results, as measured by PZT #1 (in mV p-p), are presented in Table 3. As the impact could not be identical for all the cases, the comparison should be made only relatively, namely to evaluate the output of the PZT on Side A as compared to that of the PZT on Side B of the plate, for a particular transducer only.

Other possible applications, like sensing a leakage through a hole induced in the sandwich plate were also tried, and it was concluded that provided the leakage through the hope produces noise, it can be sensed by the PZT patches on the built-in adaptive layer [24].

3.2. FULL SCALE 100x100 CM^2 SOFT CORE SANDWICH PLATE

The full-scale soft-core sandwich plate with a built-in adaptive layer was placed in a 2x2x2 m^3 anechoic chamber. First the plate was excited by only one PZT located in the middle of the adaptive layer. The level of the excitation was 12.5 V (p-p= peak to peak) with a sine wave and a varying frequency. All other PZTs served as sensors yielding a high detectable output having the range of volts. The output reached a relatively high peak at 750 Hz, one of the first bending natural frequencies of the plate. This showed that even with one actuator, the plate could be activated without any special problems. Using all the nine pairs of actuators would bring to a very powerful actuation and thus enabling vibrations at any desired frequency aimed to cancel out any unwanted noised induced vibrations.

A load speaker at various frequencies then exposed the full-scale sandwich plate to external non-attached disturbances using the noise produced.

The sensors attached to the middle adaptive layer (the nine pairs of PZTs) detected the noise-induced vibrations. For an excitation level of 10 V (p-p) produced by the loud speaker, the various sensors detected an output in the range of 30 mV (p-p) which is a relatively high one (more details see Ref.[24]).

4. Conclusions and Recommendations

Two soft-core sandwich plates with a built-in adaptive layer were designed and manufactured.

Based on the large number of experiments performed on the sandwich plates, it turns out the built-in adaptive layer is capable of inducing enough strain to reduce the noise induced vibrations radiated through the structure into an enclosure. Its application might be as smart walls for active control of sound radiated by harmonically excited thin walled structures, to be found in the aircraft and automotive industries as well as in other areas where noise levels reduction is a must. An effective control method must be used to successfully attenuate the noise levels. Further experiments are on their way to prove this new concept accompanied with theoretical calculations using the finite element method. Their results will be reported in due time.

Parallel to the actuation issue, it seems that the adaptive built-in layer is capable of "feeling" external disturbances, like impact, noise-induced vibrations or leakage. Its possible applications might be identified with smart sandwich walls for the marine and space structures by using their enhanced capability to sense either impacts or leakage by both pinpointing the location and its induced stress levels. Experimental studies on locating the impact point based on the PZT sensing output are currently carried out.

It looks that the new sandwich configuration, based on a built-in adaptive layer, has many real-life applications. Its introduction in full-scale structures must be accompanied by well conducted experiments to back the vast amount of theoretical studies found in the literature.

5. References

1. Miller, S. E., Abramovich, H. and Oshman, Y., Active distributed vibration control of anisotropic piezoelectric laminated plates, *J. of Sound and Vibration* **183**(1995), 797-817.
2. Main, J. A., Garcia, E. and Howards D. , Optimal placement and sizing of paired piezoactuators in beams and plates, *Smart Mater. Struct.* **3**(1994), 373-381.
3. Rao, S.S. and Sunar, M., Piezoelectricity and its use in disturbance sensing and control of flexible structures: A survey, *Appl. Mech. Rev.,* **47**(1994),113-123.
4. Cudney, H.H., Alberts, T.E. and Colvin, J.A. (1992). A classical approach to structural control with piezoelectrics, AIAA-92-2462-CP , 33rd AIAA/ASME/ASCE/AHS/ASC Structures, Structural Dynamics, and Materials Conference, Dallas, TX, USA, pp. 2118-2126.
5. Clark, R.L., Flemming, M.R. and Fuller, C.R., Piezoelectric actuators for distributed vibration excitation of thin plates: A comparison between theory and experiment, *J. of Vibration and Acoustics, Trans. of the ASME,* **115**(1993), 332-339.
6. Akella, P., Chen, X., Cheng, W., Hughes, D. and Wen, J.T., Modeling and control of smart structures with bonded piezoelectric sensors and actuators, *Smart Mater. Struct.*, **3**(1994),344-353.

7. Batra, R.C. and Ghosh, K., Deflection control during dynamics deformations of a rectangular plate using piezoceramic elements, *AIAA J.*, **33**(1995),1547-1548.
8. Kim, J., Varadan, V.V., Varadan, V.K. and Bao, X. Q., Finite-element modeling of a smart cantilever plate and comparison with experiments, *Smart Mater.Struct.*, **5**(1996),165-170.
9. Chandra, R. and Chopra, I., Structural modeling of composite beams with induced-strain actuators, *AIAA J.*, **3**(1993), 1692-1701.
10. Pletner, B. and Abramovich, H., A consistent methodology for the modeling of piezolaminated shells, *AIAA J.*,**35**(1997), 1316-1326.
11. Crane, S.P., Cunefare, K.A., Englestad, S.P. and Powell, E.A. (1996). A comparison of optimization formulations for design minimization of aircraft interior noise, AIAA-96-11480-CP, 37th AIAA/ASME/ASCE/AHS/ASC Structures, Structural Dynamics, and Materials Conference, Salt Lake City, UT, USA, pp. 1504-1514.
12. Dungan, M.R., Mollo, C., Vlahapoulos, N. and Anderson, W.J. (1996). Acoustic vibration model of a composite shell with sound absorption material, AIAA-96-1348-CP, 37th AIAAA/ASME/ASCE/AHS/ASC Structures, Structural Dynamics, and Materials Conference, Salt Lake City, UT, USA, pp. 267-273.
13. Wang, B. T. , Active control of far-field sound radiation by a beam with piezoelectric control transducers: physical system analysis, *Smart Mater. Struct.*, **3**(1994), 476-484.
14. Kim, J., Varadan, V.V. and Varadan, V.K., Finite-element optimization methods for the active control of radiated sound from a platestructure, *Smart Mater. Struct.*, **4**(1995), 318-326.
15. Bao, X., Varadan, V.V. and Varadan, V.K., Active control of sound transmission through a plate using a piezoelectric actuator and sensor, *Smart Mater. Struct.*, **4**(1995), 231-239.
16. Ko, B. and Tongue, B.H., Acoustic control using a self-sensing actuator, *J. of Sound and Vibration*, **187**(1995), 145-165.
17. Balachandran, B., Sampath, A. and Park, J., Active control of interior noise in a three-dimensional enclosure , *Smart Mater. Struct.*, **5**(1996), 89-97.
18. Pletner, B, Abramovich, H. and Idan, M. (1996). A new methodology for active control of structure-radiated noise. Proc. of the 7th Int. Conference on Adaptive Structures, Rome, Italy.
19. Birman, V. and Simonyan, A., Theory and applications of cylindrical sandwich shells with piezoelectric sensors and actuators, *Smart Mater. Struct.*, **3**(1994), 391-396.
20. Sun, C. T. and Zhang, X. D., Use of thickness-shear mode in adaptive sandwich structure, *Smart Mater. Struct.*, **4**(1995), 202-206.
21. Zhang, X. D. and Sun, C. T., Formulation of an adaptive sandwich beam, *Smart Mater. Struct.*, **5**(1996), 814-823.
22. Abramovich, H. and Pletner, B., Actuation and sensing of piezolaminated sandwich type structures, *Composite Structures*, **38**(1997), 17-27.
23. Noor, A. K., Burton, W. S. and Bert, C. W., Computational models for sandwich panels and shells, *Appl. Mech. Rev.*, **19**(1996), 155-199.
24. Abramovich, H.,(1997). Experimental Studies on Soft Core Sandwich Structures with a Built-in Adaptive Layer, TM 278, Inst. für Leichtbau und Seilbahntechnik, ETH Zurich, Switzerland.

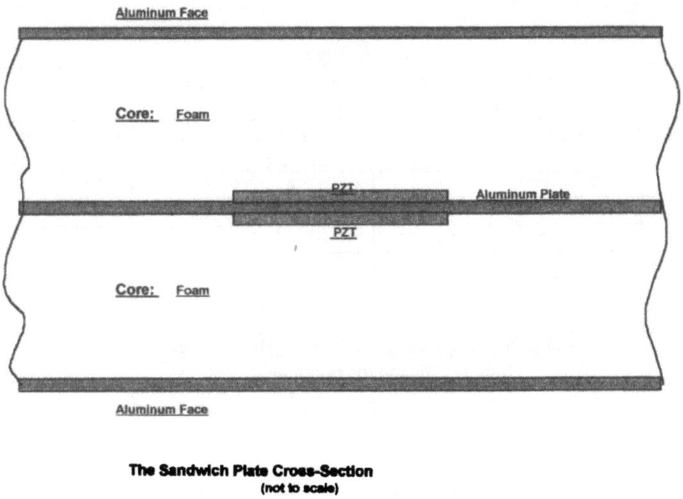

Fig.1 The cross-section of the sandwich plate

TABLE 1 Dimensions and Material Properties

	PZT Patch	Aluminum Plates	Foam Plates
Producer and Index type	PI CERAMIC GmbH, PIC-151	Anticorodel-110 730.64 or Alusuisse, 6110	AIREX, KAPEX® C50
Thickness (mm)	0.25 or 0.50	0.5 or 0.3	20
Width (mm)	25	500 or 1000	500 or 1000
Density, ρ(kg/m^3)	7800	2700	60
Young's Mod.(Pa)	$E_p= 6.67*10^{10}$	E_s(nominal)=$7*10^{10}$	E_f(comp.)= $2.5*10^7$ E_f(tensile)= $1.0*10^7$
Length (mm)	l=75	500 or 1000	500 or 1000
Piezoelectric Const.,d_{31}(m/V)	$-210*10^{-12}$	--	--

Note: The total sandwich thickness (nominal) is: 41.1 mm.

TABLE 2 Noise Levels Generated by Exciting a Pair of PZTs (Ambient noise level: 2mV, p-p)

Excited at PZT pair*:					
Sandwich configuration: Foam,Al+PZTs,Foam; Side B is facing the microphone.					
Frequency Hz	#1 (2 PZTs) mV(p-p)	#2 (2 PZTs) mV(p-p)	#3 (2 PZTs) mV(p-p)	#4 (1 PZT)† mV(p-p)	#5 (2 PZTs) mV(p-p)
200	3	2	4	2	2
300	-	-	6	5	-
400	6	10	-	-	-
500	12	14	-	-	-
600	>18	>18	10	6	6
1000	-	-	16	6	14
Sandwich configuration: Al,Foam,Al+PZTs,Foam; Side B is facing the microphone.					
200	2	2	3	2	2.5
300	-	-	-	-	-
400	3	2.5	4	3	6
450	>16	-	-	-	-
500	-	>16	-	-	-
600	-	-	6	2	6
800	-	-	-	2	
1000	-	-	7.5	3	9
Sandwich configuration: Al,Foam,Al+PZTs,Foam; Side A is facing the microphone.					
200	2	2	3	2	4
300	-	-	-	-	-
400	3	2.5	6	3	6
450	>16	-	-	-	-
500	-	10	-	-	-
600	-	>16	8	4	10
800	-	-	-	4	
1000	-	-	7.5	2.5	16

*Note that the thickness of the PZTs forming pairs #1 and #2 is twice the thickness of the other PZTs (0.5 mm vs. 0.25 mm for pairs #3,#4,and #5).

† Note that for pair #4, only one PZT is active (glued on face B), as the other one glued on face A was found short-circuited after the assembly of the sandwich construction.

TABLE 3 Sensing Capabilities of the Piezolaminated Sandwich Plate under a Light Impact as Measured by PZT #1(in mV p-p) †

Applied at the location of PZT #:	1	2	3	4	5
Side A	60	50	50	50	50
Side B	45	35	35	35	35

† The impact was applied on one face of the sandwich (at which Side A of the aluminum plate equipped with PZTs is pointing out) while the other face was resting on a 60 x 60 x 3 cm^3 chipboard wooden plate.

SIMULATION OF SMART COMPOSITE MATERIALS OF THE TYPE OF MEM BY USING NEURAL NETWORK CONTROL

V.D. KOSHUR
Professor, Head of Chair of Computational Experiment,
Krasnoyarsk State Technical University,
Leading Researcher, Institute of Computational Modelling of RAS
Institute address:
Institute of Computational Modelling of RAS
660036, Krasnoyarsk, Academgorodok, RUSSIA
E-mail: koshur@icm.krasn.ru
Copy to: koshur@fivt.krasn.ru

1. Introduction

Solving complex engineering problems such as adaptive control of miscellaneous systems in real time, the design of smart matter, intelligent composite materials and structures, demand new approaches in information representation, storage and its transformation. In particularly it requires the development of high-performance computers with novel computer architectures, massively parallel computing and so on. At the present time, one of the prospective ways for solving such problems is an application of neural computing based on various artificial neural networks and their realisation in different neurochip architectures.

The presented results concern the computer modelling of the proposed animated composite materials as a complex mechanical, neural network and electronic system which has been called Matrix Electronic Material (MEM) [1-4]. The MEM can transform physical signal and/or information, which are distributed in space and time, so that the MEM has an intelligent physical reaction to external actions. The new type of MEM can be defined as a combined system with neural network control and a supervisor computer system. These novel composite materials use distributed neural controls provided by piezoelectric sensors and actuators. They use active and high-speed signal transformation in parallel analog and digital analog form.

2. General System of Simulation

The MEM can be defined as a complex system with neural network control and a supervisor computer system in the form

$$S_{MEM} = \{[[[[M], E], NN], G], SV\},$$

where:

- M is a structural composite material with embedded sensors and actuators which possess piezoelectric properties and/or ferromagnetic properties,
- E is a matrix electronic circuit for producing the interactions between sensors, the neural network, actuators, the energy block and the supervisor system,

U. Gabbert and H.S. Tzou (eds.), IUTAM Symposium on Smart Structures and Structronic Systems, 231–238.

- NN is an artificial neural network to realise the required control of the mechatronic system,
- G is an energy block or a power station for active energetic intervention,
- SV is a supervisor computer system with traditional or/and neurochip architecture,
- [...] - the brackets denote the different levels of connections between subsystems.

3. Modelling and Objective Functions

The computer model of the reduced system S^0_{MEM} = {[[M], NN],G} for laminated metal-ceramics composites is developed on the basis of the Discrete Variation Method (DVM) [5-7]. The DVM, like the FEM (Finite Element Method), is used to implement the computer modelling. The DVM employs structural discretization, it constructs discrete constitutive relations for active and passive components and accommodates discontinuities due to embedded sensors and actuators. The ceramic layers are analysed by using linear piezoelectricity theory [8]. The input signals for neural network are the voltages in sensors piezoelectric layers and the output signals are the voltages for actuators piezoelectric layers, as shown in Figure 1.

The solution of the problem of adjusting the weight parameters for a three-layered neural network has been presented [9] as minimisation of functional J_1 (for the first problem) and J_2 (for the second problem) in the form

$$J_1 = (1/T) \int_0^T (v_s)^2 dt, \tag{1}$$

$$J_2 = (1/T)(\int_0^T (u_s - u_s^{opt})^2 dt)^{1/2}, \tag{2}$$

where

- T is the time of double running elastic perturbations along thickness of the laminated composite plate,
- v_s is the velocity of vibration of the plate back surface, u_s is the displacement of the back surface of the composite plate,
- $u_s^{opt}=f(t,\omega^{opt})$ is the defined form of the back surface displacement with the defined frequency ω^{opt}.

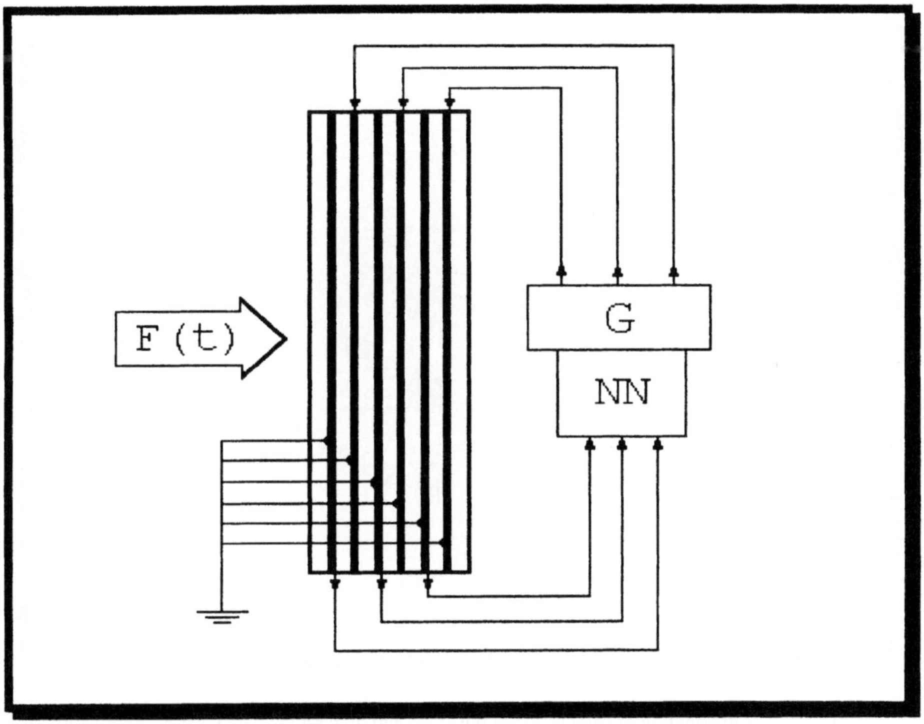

Figure 1. The scheme of connections for the neural network control.

4. Numerical Results

The numerical results are presented for the elastic wave transformations along the thickness for the laminated aluminium-ceramic plate with the structure

$$F(t) \rightarrow [A+C_1^0+A+C_2+A+C_3^0+A+C_4+A+C_5^0+A+C_6+A]$$

here A is the layer of aluminium, C_i^0 for i=1,3,5 are the sensors layers and C_i for i=2,4,6 are the actuators layers, $F(t) = F_0\sin(\omega t)$ is the external pressure. By integrating the dynamic system [4,5] with respect time t in the interval [0, T] we compute the values of the functional $J_1(\mathbf{W})$ and $J_2(\mathbf{W})$. For minimisation of the functional $J_1(\mathbf{W})$ and $J_2(\mathbf{W})$ we used the Hook-Jives algorithm since it does not require the cost function gradients [10].

Figures 2 – 6 show the numerical results. As the base, we use deformation process without control when every ceramic layer is passive. For passive process of dynamic deformation the distribution of the velocity v(z, t) along thickness of the plate and time t as the surface VVpas is shown in Figure 2a. The nearest vertical cross-section in Figure 2a shows the co-ordinate on the back surface with respect to time t in microseconds increasing right to left. The level zero corresponds to the unperturbed state of the laminated metal-ceramic plate during the dynamic process of deformation. Figure 2b shows the electric field distribution.

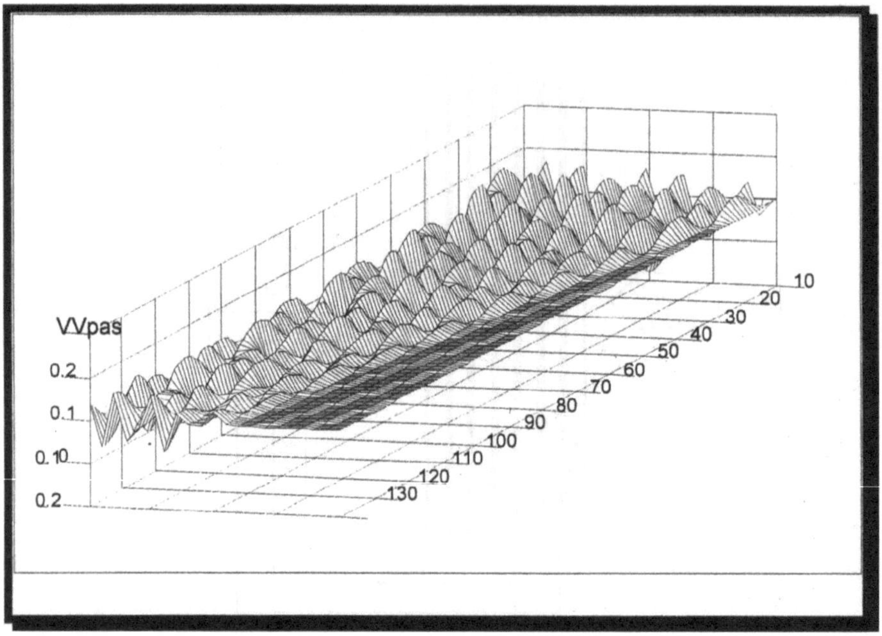

Figure 2a. The distribution of velocity along thickness and time for passive dynamic deformation.

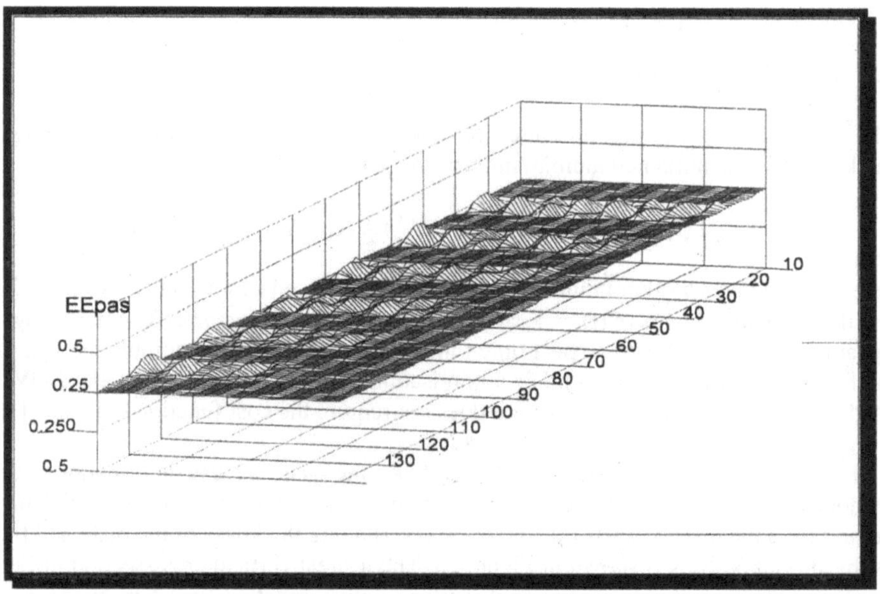

Figure 2b. The distribution of electric field E along thickness and time for passive dynamic deformation.

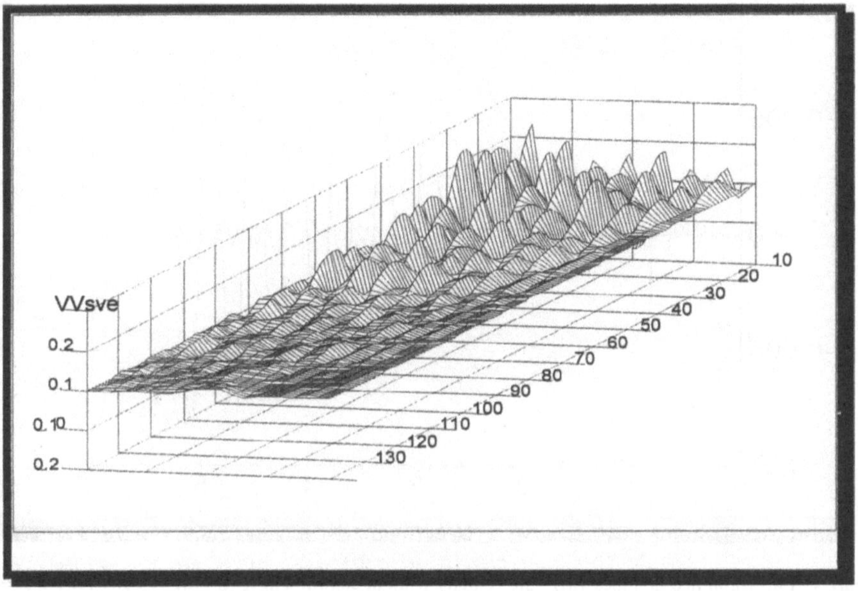

Figure 3a. The distribution of velocity along thickness and time for the obtained neural network control with $J_1 \rightarrow \min$.

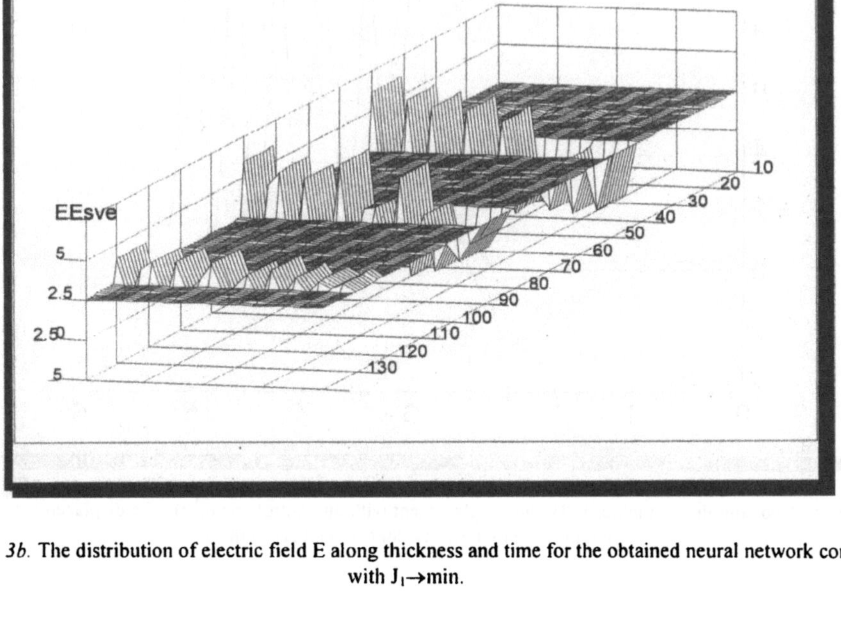

Figure 3b. The distribution of electric field E along thickness and time for the obtained neural network control with $J_1 \rightarrow \min$.

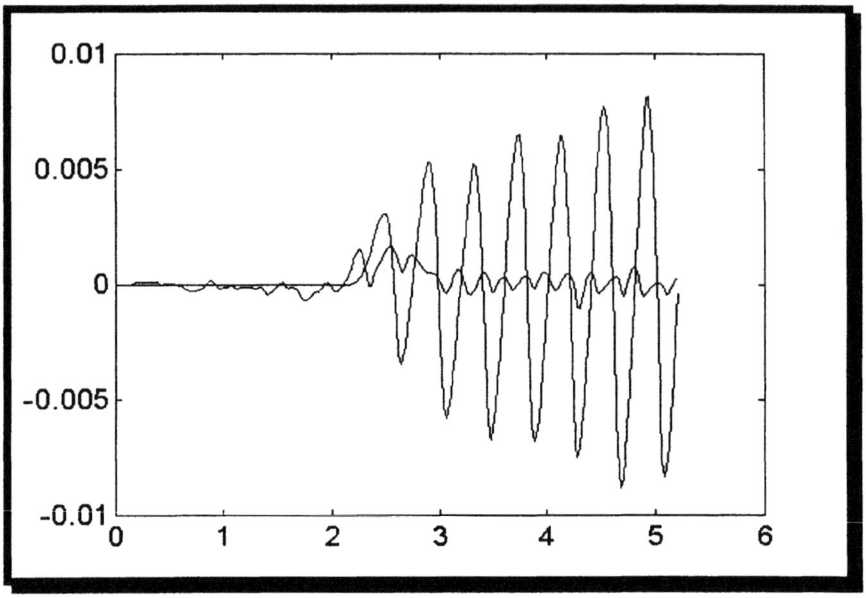

Figure 4. Two combined graphics $v_s(t)$ - the velocity without control and $v_s^*(t)$ - the velocity for the obtained neural network control with $J_1 \rightarrow \min$, horizontal axis is time t (μs).

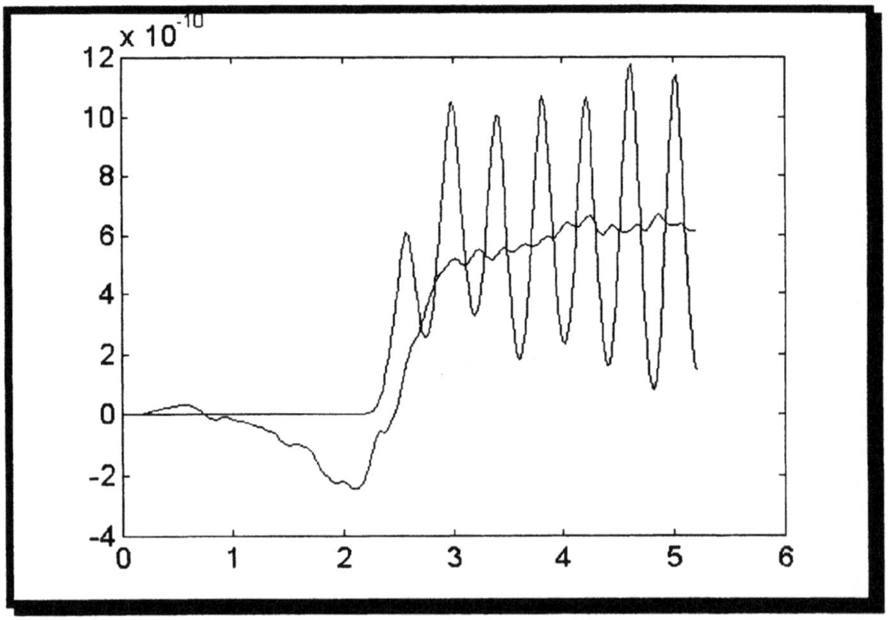

Figure 5. Two combined graphics $u_s(t)$ - the displacement without control and $u_s^*(t)$ - the displacement for the obtained neural network control with $J_1 \rightarrow \min$.

Figure 3a shows the distribution of the velocity v(z, t) along the thickness of the plate and time t as the surface VVsve for the obtained active neural network control with $J_1 \rightarrow min$; Figure 3b shows the demanded electric field distribution. During minimisation (J_1) the values of the amplitude of velocity v_s and displacement u_s have been decreased by about ten times (shown in Figure 4 and Figure 5).

For the solution of the second problem of transformation, the frequency of the displacement on the back surface of the plate for cost functional J_2 is presented in Figure 6. In this case for chosen class of neural network controls we have not achieved the high degree of accuracy needed to transform the frequency to the given value.

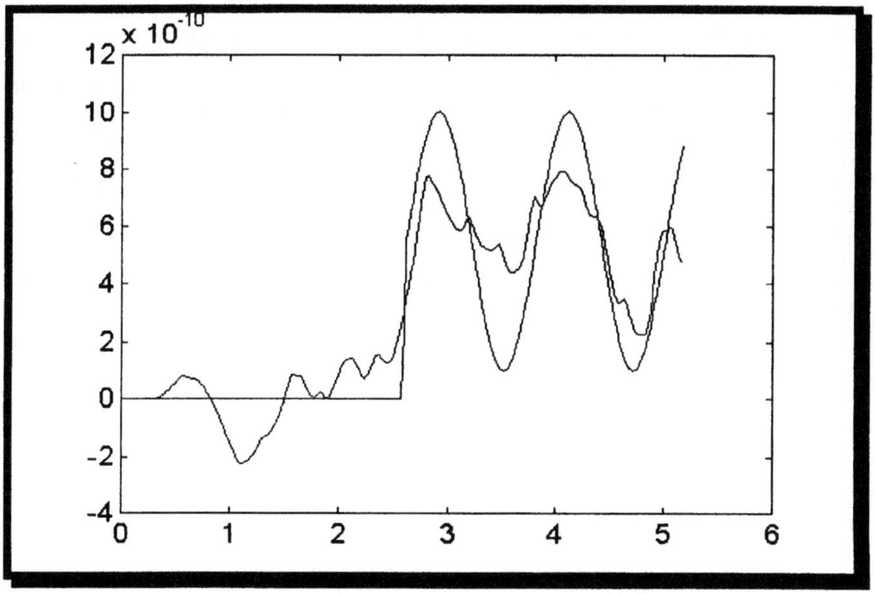

Figure 6. The graphics $u_s^{opt}=f(t,\omega^{opt})$ - the defined optimal displacement ($\omega^{opt} = \omega/3$) and $u_s^{**}(t)$ - the displacement for the obtained neural network control with $J_2 \rightarrow min$, horizontal axis is time t (μs).

5. Conclusion

Computer modelling shows that MEM with neural network can control the wave transformation with respect to amplitudes and frequency of oscillations. In applications it may be used for active suppression of noise and digital-analog wave conversion.

6. References

1. Koshur V.D. Design and computer simulation of new animated composite materials, media, systems and processes with adaptive dynamic control of deformation shock-wave processes. *International Conference AMCA-95, Advanced Mathematics, Computations and Applications, Abstracts,* NCC Publisher, Novosibirsk, 1995, pp. 185-186

2. Koshur V.D. Modelling of control process of elastic waves transformation in laminated metal-ceramic composites, concept of the Matrix Electronics Materials. *Report of Academy of Science. Russian Academy of Science*, Moscow, Nauka, 1998, Vol. 363, No.2, pp. 181-183
3. Koshur V.D. Neural Networks Control of Wave Deformational Transformation in the Laminated Piezoelectric Composites. *Fifth All-Union Seminar on Neuroinformatic and its Applications, 3-5 October 1997, Krasnoyarsk, Abstracts*, Published by Krasnoyarsk State Technical University, Krasnoyarsk, 1997, pp.107-109
4. Koshur V.D. Active and passive neural network control of shock waves transformation in laminated metal-ceramic composites, concept of Matrix Electronic Materials (MEM). *Proceedings of the EUROMECH 373 Colloquium Modelling and Control of Adaptive Mechanical Structures. Otto-von-Guericke University of Magdeburg, 11-13 March 1998*, Published by University of Magdeburg, 1998, pp. 311-316.
5. Koshur V.D. and Nemirovsky Yu.V. *Continual and Discrete Models of Dynamic Deformation of Elements of Structures.* - Novosibirsk: Nauka, 1990, 198 P.
6. Koshur V.D. and Nemirovsky Yu.V. *Discrete Structural Models of Non-linear Dynamic Processes of Deformation and Fracture in Composite Materials.* AMSE Transactions, Scientific Siberian, A, Vol.12, AMSE Press. Tassin, France, 1994, 301 P.
7. Koshur V.D. and Bykovskih A.M. The combination of discrete and structural approach for simulation dynamic behaviour of composites subjected to impulse loading and impact. *Proceedings of the 10th International Conference on Composite Materials, Whistler, Canada, August 14-18, 1995, Volume 5*, Whistler, 1995, pp. 195-202.
8. Maugin G.A. *Continuum Mechanics of Electromagnetic Solids.* Elsevier Science Publisher B. V., New York-Oxford-Tokyo, 1988. 560 P.
9. Gorban A.N. and Rossiev D.A. *Neural Networks on Personal Computer.* Nauka, Novosibirsk, 1996, 276 P.
10. Gill P.E., Murray W. and Wright M.H. *Practical Optimisation.* Academic Press, London-New York-Toronto, 1981, 509 P.

DAMAGE DETECTION IN STRUCTURES BY ELECTRICAL IMPEDANCE AND OPTIMIZATION TECHNIQUE

V. LOPES, Jr.
Department of Mechanical Engineering – UNESP
15385-000 Ilha Solteira SP, Brazil

H. H. MÜLLER-SLANY, F. BRUNZEL
University of Duisburg, Institute of Mechanics
47048 Duisburg, Germany

D. J. INMAN
Center for Intelligent Material Systems and Structures
Virginia Polytechnic Institute and State University
Blacksburg, VA – USA, 24061-0261

1. Introduction

The current interest in research regarding predictive maintenance is the result of several factors, involving economic reasons, recent developments in sensor and actuator technology and the elimination of dangerous failures involving risk of human life. Vibration-based methods have been investigated for a long time, and new approaches are continuously proposed. However, the interpretation of vibration signals to identify damage is not an easy and straightforward task. In this context, smart material technology has become an area of increasing interest, and the electrical impedance technique has been accepted as an effective method for structural health monitoring because its easy implementation and simple structural evaluation. In this paper a new methodology is proposed, which combines impedance technique and the application of a diagnostic model based on vibration measurements and generated by optimization procedures.

The impedance-based technique, which utilizes the electromechanical coupling property of piezoelectric materials, is based on high frequency structural excitation. This technique is very sensitive to minor changes in the near field of piezoelectric patches, and it will be used to locate the damage areas. Possible damaged parameters are identified by using a diagnostic model.

The methodology of combining smart materials and diagnostic models can be used as a tool to detect and quantify damage. The first step of the methodology uses an impedance-based technique to give information about the areas of the damage positions. By using a diagnostic model, candidates of structure elements are found with identified variations of system parameters due to damage. The accuracy of the model based procedure is a function of the degree of condensation of the model. An example based on experimental measurements shows the results of the proposed methodology.

U. Gabbert and H.S. Tzou (eds.), IUTAM Symposium on Smart Structures and Structronic Systems, 239–246.

2. Damage Detection in Structures by Electrical Impedance Technique

There are several distinct requirements for bonded or embedded piezoceramic material to be reliably applied as self-excitation and self-sensing devices for detection and characterization of damage. The partial differential equations describing the dynamics of a beam with surface bonded piezoceramic patches is determined by Banks et al. [1].

The electrical impedance is defined as the ratio of the input voltage to the resulting current. The impedance-based health monitoring method that utilizes piezoceramic path for both actuator and sensor can be written as

$$Y(\omega) = i\omega \frac{w_A l_A}{h_A}\left(\varepsilon_{33}^T (1 - i\delta) - \frac{Z_S(\omega)}{Z_S(\omega) + Z_A(\omega)} (d_{33})^2 \bar{Y}_{22}^E \right) \tag{1}$$

where ε_{33}^T, δ, d_{33}, and $\overline{Y}_{22}^E$ are PZT's material constants, and w_A, l_A, and h_A are PZT's dimensions [2]. Z_A and Z_S are the PZT and the structural mechanical impedance, respectively. The first term of the above equation is the capacitance admittance of the free PZT and the second term is the result of the electromechanical interaction of the PZT with the structure. Therefore, any variation of the structural impedance can be directly related with the electrical impedance of the PZT bonded on the structure.

The damage metric, defined as the sum of the squared differences of the real and imaginary impedance changes at each frequency step, is used to simplify the interpretation of the impedance variations and provides a summary of the information obtained from the impedance response signals [3]. The monitoring of each PZT impedance signal is done separately, i.e., each PZT is continuously monitored in a specific frequency range, and the threshold level, above which is an indication of fault in that position, must be defined. Operational conditions and temperature variations can change the modal characteristics of the structure in the high frequency range. Hence, these values must be defined by trial and error or by prior knowledge of the structural dynamics of the system. More details about the application of the damage metric chart can be found in Lopes Jr. et al [4].

3. Model based Damage Detection by Optimization Technique

3.1 MODEL BASED DAMAGE DETECTION PROCEDURE

The problem of identifying variations of system parameters from measured system behavior does not have a unique solution. In general, in model based damage detection we have to deal with the inverse vibration problem, which is known as an ill-conditioned mathematical problem. The first step in model based damage detection is the generation of a suitable simulation model which describes the dynamical behavior of the undamaged elastomechanical system in a correct way. The second step is to identify the system parameters that are damaged. Both steps are solved by optimization techniques (see figure 1).

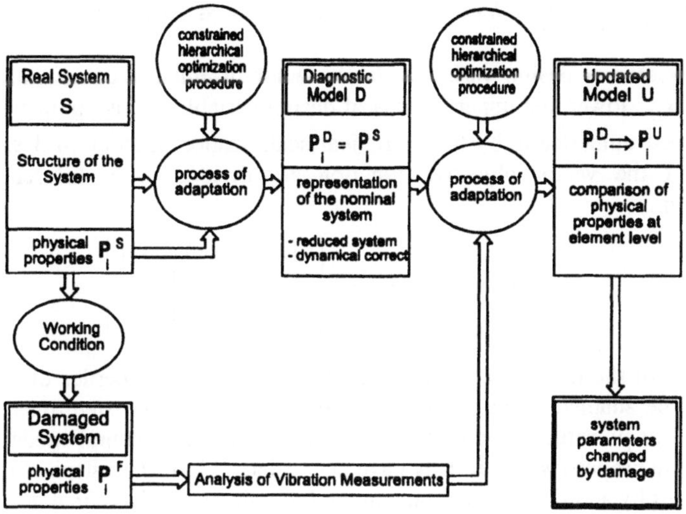

Figure 1. Model based damage detection procedure

3.2 GENERATION OF THE DIAGNOSTIC MODEL

The diagnostic model has the geometrical structure of the real system, using super-elements for its construction which are described by design variables $\boldsymbol{x} = [x_1, \ldots, x_n]^T$ [5, 6]. In a vector-optimization process this model is adapted to the important dynamical properties of the real system. In the case of using beam-elements this model has the following types of design parameters: coordinates of p beam nodes: x_i, y_i, z_i, $i = 1, \ldots, p$; masses of m beam elements: m_i, $i = 1, \ldots, m$; stiffness parameters of m beam elements: GI_{xi}, EI_{yi}, EI_{zi}, $i = 1, \ldots, m$. The most important dynamical properties of the real elasto-mechanical system under consideration are:

- the total mass m and the position of center of mass $\boldsymbol{r}_C(\boldsymbol{x})$,
- the elements of the tensor of inertia $\Theta_C(\boldsymbol{x})$ and
- the eigenfrequencies and eigenmodes under consideration: $\alpha_i(\boldsymbol{x})$, $\boldsymbol{q}_i(\boldsymbol{x})$, $i = 1, \ldots, r$.

All dynamical properties are functions of the design variables $\boldsymbol{x}$. The process of adaptation can be written as a multicriteria optimization procedure in which a vector objective function $\boldsymbol{f}(\boldsymbol{x})$ has to be minimized. The components of $\boldsymbol{f}(\boldsymbol{x})$ are error expressions $\varepsilon(\boldsymbol{x})$ in which measured physical properties $P(\boldsymbol{x})$ of the original system S and simulated values of the diagnostic model D are compared:

$$\boldsymbol{f}(\boldsymbol{x}) = \begin{bmatrix} \varepsilon_1(\boldsymbol{x}) \\ \varepsilon_2(\boldsymbol{x}) \\ \varepsilon_3(\boldsymbol{x}) \\ \varepsilon_4(\boldsymbol{x}) \\ \varepsilon_5(\boldsymbol{x}) \end{bmatrix} = \begin{bmatrix} \textit{error expression of : complete mass} \\ \textit{.... position of centre of mass} \\ \textit{.... elements of inertia tensor} \\ \textit{.... considered natural frequencies} \\ \textit{.... considered natural modes} \end{bmatrix}, \tag{2}$$

$$\min_{x \in \Sigma} \left\{ f\left[\varepsilon(x)\right] \mid h(x) = 0 \right\}, \quad \Sigma := \left\{ x \in \Re^n \middle| x_L \le x \le x_U \right\} , \tag{3}$$

where: x: vector of design variables, $h(x)$: vector of equality constraints, Σ: feasible range, x_L , x_U : lower and upper bounds of design variables. This optimization problem can be solved numerically by a hierarchical scalarization strategy in 3 steps [7]. The elements of the vector objective function are divided in 3 different groups by scalarization:

$$f(x) \Rightarrow s\{f[\varepsilon(x)]\} = \begin{bmatrix} s_1[\varepsilon_1(x),\varepsilon_2(x),\varepsilon_3(x)] \\ s_2[\varepsilon_4(x)] \\ s_3[\varepsilon_5(x)] \end{bmatrix} . \tag{4}$$

In each step of the hierarchical optimization one group of properties of the diagnostic model will be adapted to the properties of the real system: *1st step*: mass geometrical values, *2nd step*: natural frequencies of interest, *3rd step*: natural modes of interest. Those system properties that already have been adapted in former steps of the procedure must be fixed by additional constraints in the following steps. The final design vector x^D creates the diagnostic model. For the solution of the highly nonlinear optimization problem a SQP-algorithm from the NAG-library was used [8].

3.2 IDENTIFICATION OF DAMAGE

The electrical impedance technique will indicate that there are changes of physical parameters of a system. The identification of the damaged system parameters can be done by an optimization procedure in which an error function $f(x^U)$ will be minimized and the design vector x^U of the damage model U will be generated:

$$\min_{x^U \in \Sigma} \left\{ f\left[\varepsilon\left(x^U\right)\right] \middle| h(x^U) = 0 \right\}, \quad \Sigma := \left\{ x^U \in \Re^n \middle| x^U{}_L \le x^U \le x^U{}_U \right\} . \tag{5}$$

If we assume that damage means a reduction of stiffness, in the design vector x^U only the stiffness parameters of the system are free to change and the mass-geometrical system properties will be fixed by constraints. The variations of system parameters due to damage can be found by comparing the design vectors x^D and x^U of the diagnostic model D and of the damage model U, respectively.

4. Experimental Application

4.1 TEST STRUCTURE

The application of the combined methodology will be demonstratet on an elasto-mechanical structure shown in figure 2. The damage consists of two cuts near points ③ and ⑧ (see figures 2 and 3). Both cuts have a width of 1 mm. The first cut has a depth of 7.5 mm and the second one a depth of 2 mm. Four PZT-elements are fixed on the structure as shown in figure 2. For vibration measurements 3D-accelerometers (type: PCB) are used on the structure points 1 ... 14.

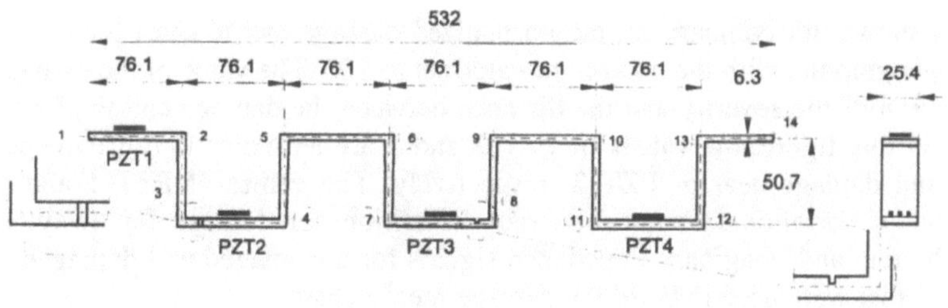

Figure 2. Elasto-mechanical test structure with PZT-elements and 2 damages (①... structure points)

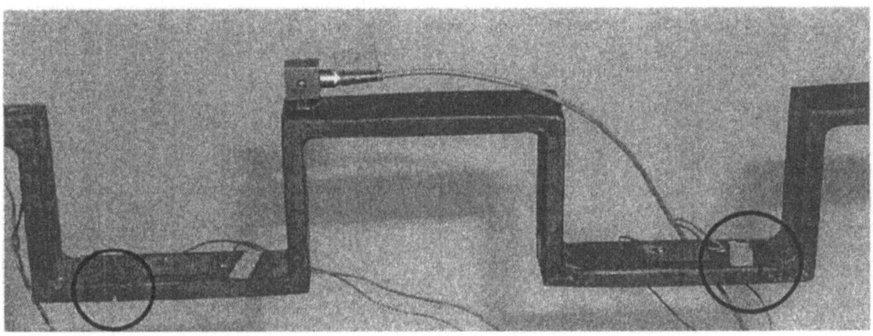

Figure 3. Test structure, position of 2 damages ○

4.2 RESULTS OF ELECTRICAL IMPEDANCE TECHNIQUE

The impedance signals were monitored in the frequency ranges of 20 to 40 kHz, 40 to 60 kHz, and 45 to 55 kHz. Each PZT was monitored independently in order to identify the damage in its sensing area. The results shown below were obtained for the frequency range of 45 to 55 kHz.

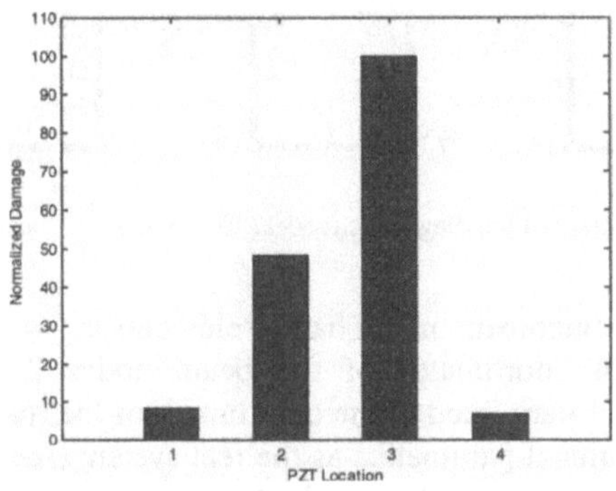

Figure 4. Damage metric chart from PZT 1 to PZT 4

Figure 4 shows, for comparison, the normalized damage metric chart for all four PZTs in a simgle graphic, with the maximum value set to 100. The value of the damage metric is a function of the severity and the distance between the damage and the PZT. As can be seen in this figure, the values of PZT 2 and 3 are an indication of damage near of PZT 2 and damage near of PZT 3, respectively. The values of PZT 1 and 4 are an indication of variation in some boundary conditions or damage far away. Figure 5 shows the real and imaginary impedance signals for undamaged and damaged structure for PZT 1 that were used to build the damage metric chart.

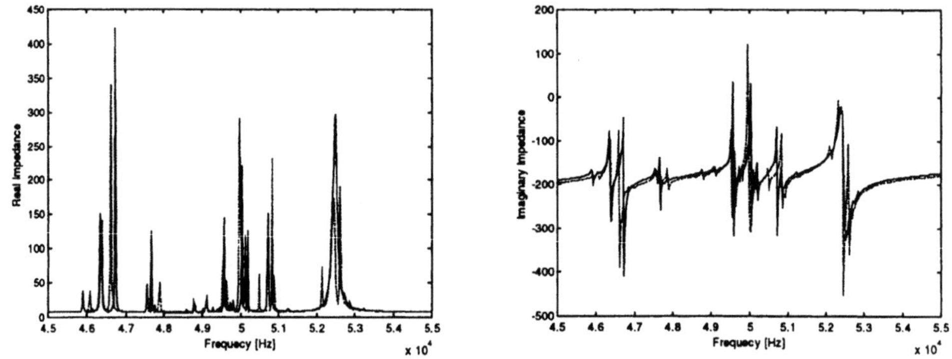

Figure 5. Impedance signal from PZT 1; intact (——), damaged (··········)
a) real part of impedance signal; b) imaginary part of impedance signal

4.3 IDENTIFICATION OF DAMAGED PARAMETERS

In the first step of the identification of damaged parameters by a model based technique a substantially condensed but dynamical highly correct diagnostic model has to be generated for the real structure. This model is built of 13 beam elements (see figure 6).

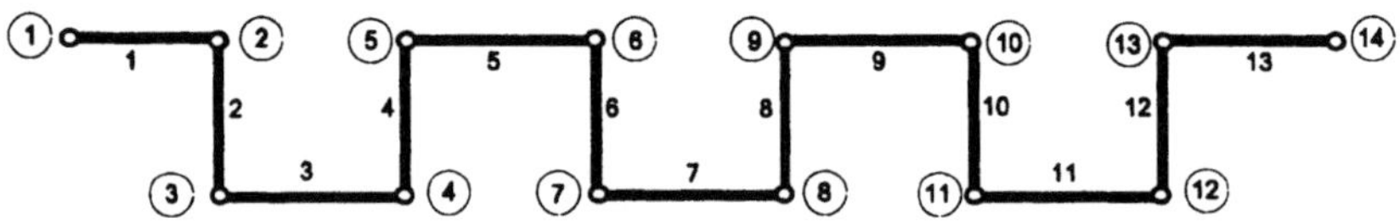

Figure 6. Structure of the diagnostic model (① ... nodes, 1 ... beam elements)

The design vector of the diagnostic model has 76 elements: $\boldsymbol{x}^D = [x_1, \ldots x_{76}]^T$. These are: 13 mass parameters, 24 coordinates of the beam nodes 2...13, and 39 stiffness parameters. Nodes 1 and 14 are fixed to the coordinates of the real structure. This model shall have the same dynamical parameters as the real system (see table 1). The variation of the natural frequencies due to the damage by two cuts is also shown in table 1.

TABLE 1. Dynamical properties of the original structure and of the damaged structure

mass-geometrical parameters			original structure: natural vibration			damaged structure: natural vibration	
			mode		f [Hz]	f [Hz]	diff.%
Mass	m [kg]	1.002	1	1. bending in plane of projection	70.48	69.70	1.1
centre of mass	x_S [m]	0.267	2	1. bending normal to plane of proj.	137.31	136.90	0.3
	y_S [m]	0.028	3	2. bending in plane of projection	181.05	180.08	0.5
moment of inertia	J_{Sx}[kg m²]	0.00125	4	1. longitudinal vibration	230.22	226.34	1.7
	J_{Sy}[kg m²]	0.0926	5	1. torsion	284.94	283.39	0.5
	J_{Sz}[kg m²]	0.0938	6	3. bending in plane of projection	314.84	308.32	2.1
	J_{Sxy}[kg m²]	- 0.00741					

The numerical solution of the optimization problem (3) is the design vector x^D of the diagnostic model. Table 2 shows that it is possible to generate a simple beam structure as diagnostic model, whose dynamical properties are very close to those of the real structure. The further process of damage identification will be possible if this model is sensitive enough to detect the local damage cut into the structure.

TABLE 2. Errors of the diagnostic model D and of the damage model U compared with measurements

dynamical property	relative error		natural modes	MAC-value	
	diagnostic model D	damage model U		diagnostic model D	damage model U
total mass	1.0E-5	1.0E-5	mode 1	0.981	0.980
centre of mass	< 1.0E-5	< 1.0E-5	mode 2	0.993	0.990
moments of inertia	< 1.4E-4	< 1.4E-4	mode 3	0.967	0.978
natural frequencies $f_1, \dots, f_6$	< 2.2E-4	< 5.0E-4	mode 4	0.987	0.981
			mode 5	0.995	0.945
			mode 6	0.940	0.971

The next step (see figure 1) is the adaptation of the diagnostic model to the measured dynamical behaviour of the real damaged structure by solving the optimization problem (5). The objective function $f(x^U)$ is created from the errors in the first 6 natural frequencies and natural mode shapes, respectively:

$$f(x^U) = \begin{bmatrix} \varepsilon_4(x^U) \\ \varepsilon_5(x^U) \end{bmatrix} = \begin{bmatrix} \textit{error expression of: considered natural frequencies } f_1(x^U) \dots f_6(x^U) \\ \dots \textit{ considered natural modes } q_1(x^U) \dots q_6(x^U) \end{bmatrix}. \quad (6)$$

The solution of (5) is the design vector x^U of the damage model U (for the dynamical properties see table 2). This model is a representation of the real structure which is damaged by two cuts. As both models, the diagnostic model *D* and the damage model *U*, respectively, have the same layout, the differences of their design vectors are indicators for the real damaged parameters (see table 3).

The identified damaged parameters are reasonable indicators for real damage of the system, with one exception: The identification of damage in element 12 is wrong. Obviously, the sensitivity of the diagnostic model was not high enough for a clear identification of the real damage, which has only a slight influence on the dynamical system behaviour (see tab. 1). As the results of the impedance technique exclude a damage in element 12, the combined methodology gave a correct description of the damage in the elements 3 and 7.

TABLE 3. Differences of design variables: Damage model x^U ./. diagnostic model x^D. Tabled are values of differences > 0.5 %

number of beam element	stiffness-parameter	difference: damage model./. diagnostic model [%]	number of beam element	stiffness-parameter	difference: damage model./. diagnostic model [%]
3	I_z	- 6.8	7	I_{Px}	- 29.9
7	I_y	- 2.4	12	I_z	- 6.2
7	I_z	- 10.6			

These identification results belong to a design vector x^U of the damage model U in which *all* model stiffness parameters were free to vary in the adaptation process (5). We get a clearer result from the identification procedure if we use the information of the impedance-based procedure, *which* system elements are candidates for damage. With a reduced design vector $x^U = [I_{2Px}, I_{2y}, I_{2z}, ..., I_{10Px}, I_{10y}, I_{10z}]^T$ in which only the stiffness parameters of the elements 2 ...10 (see figure 6) are free for variations, the results of identification are nearly the same as in table 3, but without element 12.

5. Conclusion

A combined damage detection methodology of electrical impedance technique and model generation by optimization procedures is reported. The solution of both procedures gives indicators for damage in a structure. The impedance technique gives clear information *where* damages occur. The model based damage detection gives the information: *which* parameters of *which* elements of the highly condensed structure are candidates for the description of a damage. The amount of damage is described by values of parameter variation on the model design level. If both procedures corresponds in their statements the damaged parameters have been found with high probability.

6. References

1. Banks, H.T., Inman, D.J., Leo, D.J., and Wang, Y.: An Experimentally Validated Damage Detection Theory in Smart Structures, *Journal of Sound and Vibrations*, **191** (1996) 859-880.
2. Sun, F.: *Piezoelectric Active Sensor and Electric Impedance Approach for Structural Dynamic Measurement*, Master Thesis, Virginia Polytechnic Institute and State University – CIMSS, 1996.
3. Park,G.: Assessing Structural Integrity using Mechatronic Impedance Transducers with Applications in Extreme Environments, Phd Thesis, Virginia Polytechnic Institute and State University – CIMSS, April/2000.
4. Lopes Jr., V., Park, G., Cudney, H.H., and Inman, D.J.: Structural Integrity Identification Based on Smart Material and Neural Network. *XVIII IMAC – International Modal Analysis Conference, San Antonio, Texas, Feb/2000*, pp. 510-515.
5. Müller-Slany, H.H.; Pereira, J.A.; Weber, H.I.; Brunzel, F.: Schadensdiagnose für elastomechanische Strukturen auf der Basis adaptierter Diagnosemodelle und FRF-Daten. *VDI-Berichte 1466 (1999)*, pp. 323-340.
6. Müller-Slany, H.H.; Brunzel, F.: Generierung und Anwendung realitätsnaher, dynamisch hochgenauer Rechenmodelle in der modellgestützten Schadensdiagnose. *VDI-Berichte 1550 (2000)*, pp. 631-650.
7. Müller-Slany, H.H.(1993) A Hierarchical Scalarization Strategy in Multicriteria Optimization Problems, in B. Brosowski (ed.), *Multicriteria Decision*, Verlag Peter Lang, Frankfurt am Main: pp. 69-79.
8. NAG Library Mark 17, NAG Ltd, Oxford, UK, 1997.

OPTIMAL PLACEMENT OF PIEZOELECTRIC ACTUATORS TO INTERIOR NOISE CONTROL

I. HAGIWARA, Q. Z. SHI, D. W. WANG AND Z. S. RAO
Dept. of Mechanical Engineering and Science, Tokyo Institute of Technology
2-12-1 Ookayama, Meguro, Tokyo, Japan 152-8552

1. Introduction

An important problem to systems such as automobiles, ships and aircrafts is the control of interior noise which is being generated by the vibration of an elastic structure. In the present study, active control of interior noise is of primary interest In the past decade, closely related to 'sound quality' and spectrum shaping, the research on noise control has been paid more and more attention in attenuating low frequencies and reshaping the acoustical modes to obtain a satisfactory acoustical environment [1]. Active control with smart material becomes a topic in the recent development of material science and has been gaining a rapid development in real application [2].

Active interior noise control using piezoelectric material involves the modeling techniques which includes coupling analysis of structure, acoustic and electric system. The coupling problem between structural and acoustical system is one of the challenging topics in vibration and acoustics and has been studied by many researchers [3],[4]. Modal synthesis method (MSM) is a useful method to model the structurally and acoustically coupled system for low frequency problems as booming noise, ride comfort, etc. In this paper, the analytical model is developed for active minimization of interior noise inside a three-dimensional car cabin cavity using distributed piezoelectric actuators using the modal synthesis technique. Researches showed that optimal control effect is greatly dependent on the number of piezoelectric actuators, their thickness, sizes and placement locations [5], [6]. The optimization technique should deal with the discrete design variables such as the number of piezoelectric actuators and continuous design variables such as the locations of piezoelectric [7]. Furthermore, multi-optima always appear for the kind of optimum problems. A new response surface methodology MPOD (most probable optimal design) proposed by the authors is applied to solve the global optimal locations of piezoelectric patches. Numerical investigations to the effect of optimization of piezoelectric actuators' locations on SPLs are presented.

2. System Modeling by Modal Synthesis Method

A model is demonstrated in Fig. 1 to describe the mechanical behavior of a

U. Gabbert and H.S. Tzou (eds.), IUTAM Symposium on Smart Structures and Structronic Systems, 247–254.

three-dimensional cavity. The cavity has four rigid wall boundaries and the dimensions are specified by l_x, l_y, and l_z respectively. Two panels, denoted by a and b, are located respectively at $z = l_z$ and $z = 0$ and form the flexible boundaries of the enclosure. These panels are assumed to be thin and isotropic. Piezoelectric ceramic (PZT) patches are bonded symmetrically to the top and bottom of the panel a to form the actuators, and all the patches are assumed to have the same geometrical and material characteristics. The external disturbing point forces are exerted perpendicularly to the bottom of the panel b.

The acoustical field of cavity can be described in homogeneous wave equation by

$$\nabla^2 p - \frac{1}{c_0^2}\frac{\partial^2 p}{\partial t^2} = -\rho_0 \frac{\partial q}{\partial t} \tag{1}$$

where $p(r)$ is the sound pressure, q is the distribution of sound source volume velocity per unit volume. Assume that the normal structural displacement to be positive outward, then we have,

$$q = -\frac{\partial w_a}{\partial t}\delta(\xi - \xi_a) - \frac{\partial w_b}{\partial t}\delta(\xi - \xi_b) \tag{2}$$

where w_a and w_b are displacements of panels a and b respectively, $\delta(\cdot)$ is one-dimensional Dirac delta function of a coordinate normal to the surface.

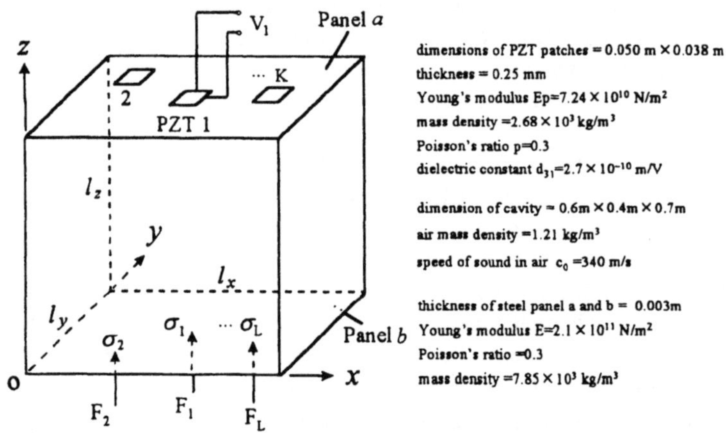

FIG. 1. Acoustic cavity model with two elastic panels

Expending p in terms of the time-dependent sound pressure modal coordinates $p_n(t)$ and the acoustic pressure normal modes of the fluid volume with rigid boundaries, $\psi_n(r)$,

$$p = \sum_n p_n \psi_n \tag{3}$$

Expending panels displacements in terms of their displacement modal coordinates $q_r^a(t)$, $q_s^b(t)$ and the in-vacuo normal modes as

$$w_a = \sum_r q_r^a \varphi_r^a, \; w_b = \sum_s q_s^b \varphi_s^b \tag{4}$$

where φ_r^a, φ_s^b are mode shapes corresponding to the panel a and panel b natural frequencies respectively.

The internal dissipation mechanisms in panels and the fluid in enclosure can be taken into account by including viscous damping terms in the modal coordinate. Hence, we can obtain a complete set of equations for the coupled system,

$$\ddot{p}_n + 2\zeta_n\omega_n\dot{p}_n + \omega_n^2 p_n = -\frac{\rho_0 c_0^2 A_a}{M_n}\sum_r C_{nr}^a \ddot{q}_r^a - \frac{\rho_0 c_0^2 A_b}{M_n}\sum_s C_{ns}^b \ddot{q}_s^b \tag{5}$$

$$\ddot{q}_s^b + 2\zeta_s^b\omega_s^b\dot{q}_s^b + \omega_s^{b2} q_s^b = \frac{A_b}{M_s^b}\sum_n p_n C_{ns}^b - \frac{1}{M_s^b}\sum_L \theta_{sL} F_L \tag{6}$$

$$\ddot{q}_r^a + 2\zeta_r^a\omega_r^a\dot{q}_r^a + \omega_r^{a2} q_r^a = \frac{A_a}{M_r^a}\sum_n p_n C_{nr}^a - \frac{1}{M_r^a}\sum_K \varepsilon_{rK} V_K \tag{7}$$

where ζ_n, ζ_r^a and ζ_s^b are modal-damping ratios of fluid and panels respectively, which are usually determined based upon the empirical data, $\theta_{sL} = \varphi_s^b(\sigma_L)$, C_{nr}^a and C_{ns}^b are dimensionless modal coupling coefficients integral over the coupling areas A_a and A_b,

$$C_{nr}^a = \frac{1}{A_a}\int_{A_a} \psi_n \varphi_r^a dA, C_{ns}^b = \frac{1}{A_b}\int_{A_b} \psi_n \varphi_s^b dA \tag{8}$$

ε_{rK} is the coupling coefficient between the structure and piezo patch.

$$\varepsilon_{rK} = \frac{(h_a + h_p)E_p d_{31}}{(1-\nu_p)}\int_{A_a} \varphi_r^a (\nabla^2 \chi_K) dA \tag{9}$$

The amplitudes of mode pressure without and with piezoelectric patches can be expressed in form of matrices and time-independent complex amplitude vectors explicitly,

$$\{P_F\}_{(n\times 1)} = [G]^{-1}[Z^b][H^F]\{F\} \quad , \quad \{P\} = \{P_F\} + [B]\{V_{con}\} \tag{10}$$

3. The Minimization of Enclosure Sound Pressure

3.1 OBJECTIVE FUNCTION

So far in the literature, several kinds of objective functions for active noise feedforward control are described. These functions mainly involve: (1) The spatial-average mean-square sound pressure; (2) The mean-square sound pressure for specified discrete positions in the enclosure. In fact, the first one is equivalent in the meaning of 'global control', and the second one, in the meaning of 'local control', is often used because only a limit number of sensors are needed in its implementation. Here, the two objective functions will be discussed.

Scheme I The temporal and spatial average of the sound pressure squared is given by

$$\langle \overline{P}^2 \rangle_{gl} = \frac{1}{2V} \int_V pp^* dV \tag{11}$$

where the subscript *gl* denotes the global control and the asterisk denotes the complex conjunction. Hence, the optimal values for the control voltage amplitude and the mean-square pressure are given by

$$\{V_{con}\} = -[a_1]^{-1}[b_1] \quad , \quad \langle \overline{P}^2 \rangle_{gl}^{opt} = [c_1] - [b_1]^H [a_1]^{-1} [b_1] \tag{12}$$

where,

$$[a_1]_{(K\times K)} = [B]^H [\Delta][B],$$

$$[b_1]_{(K\times 1)} = [B]^H [\Delta]\{P_F\}, [c_1]_{(1\times 1)} = \{P_F\}^H [\Delta]\{P_F\},$$

Scheme II In order to implement a global sound pressure control, an infinite number of microphones are needed inside the enclosure. This is obviously impractical. Hence, for practical purpose, an alternate is using N number of sensors to perform a local minimization. Thus, we define the local optimal objective function as

$$\langle \overline{P}^2 \rangle_{loc} = \frac{1}{2N} \{P\}^H [\Psi_N][\Psi_N]^T \{P\} \tag{13}$$

where the subscript T denotes the transpose, $[\Psi_N]$ is $(n \times N)$ matrix of n cavity mode shapes at the N sensor locations and the subscript *loc* denotes the local control. We can see that when N tends to infinite, this control will tend to a result of the global control. In order to obtain an optimal control input, we assume that $N \geq K$, and the optimal values for the control voltage amplitude and the mean-square pressure are found to be

$$\{V_{con}\} = -[a_2]^{-1}[b_2] \quad , \quad \langle \overline{P}^2 \rangle_{loc}^{opt} = [c_2] - [b_2]^H [a_2]^{-1} [b_2] \tag{14}$$

where,

$$[a_2]_{(K\times K)} = (1/2N)[B]^H [\Psi_N][\Psi_N]^T [B] \quad ,$$

$$[b_2]_{(K\times 1)} = (1/2N)[B]^H [\Psi_N][\Psi_N]^T \{P_F\} \quad ,$$

$$[c_2]_{(1\times 1)} = (1/2N)\{P_F\}^H [\Psi_N][\Psi_N]^T \{P_F\}.$$

3.2 RESULTS AND DISCUSSION

For the numerical simulation predictions, we assume that two panels are simply supported along their four sides. The sound-pressure level in the cavity is characterized by

$$SPL = 20\log_{10}\left(\frac{\sqrt{\langle \bar{P}^2 \rangle}}{20\times 10^{-6}}\right) \tag{15}$$

where the reference pressure is taken as 20 μPa. The dimensions of the enclosure are $l_x \times l_y \times l_z$ = 0.6m × 0.4m × 0.7m. PZT patches are bonded symmetrically to the top and bottom of the panel a. Three PZT patches, two point forces and three sensors are considered in the analysis. The centers of PZT pairs 1~3 are located on panel *a* respectively at σ_1 = (0.300, 0.105) m, σ_2 = (0.145, 0.320) m, and σ_3 = (0.455, 0.200) m. The locations of point forces 1~2 on panel *b* are $\hat{\sigma}_1$ = (0.300, 0.200) m (i.e., the center of the panel) and $\hat{\sigma}_2$ = (0.125, 0.110) m respectively, and the locations of sensors 1~3 in enclosure are s_1 = (0.300, 0.200, 0.485) m, s_2 = (0.485, 0.345, 0.625) m and s_3 = (0.115, 0.055, 0.075) m respectively

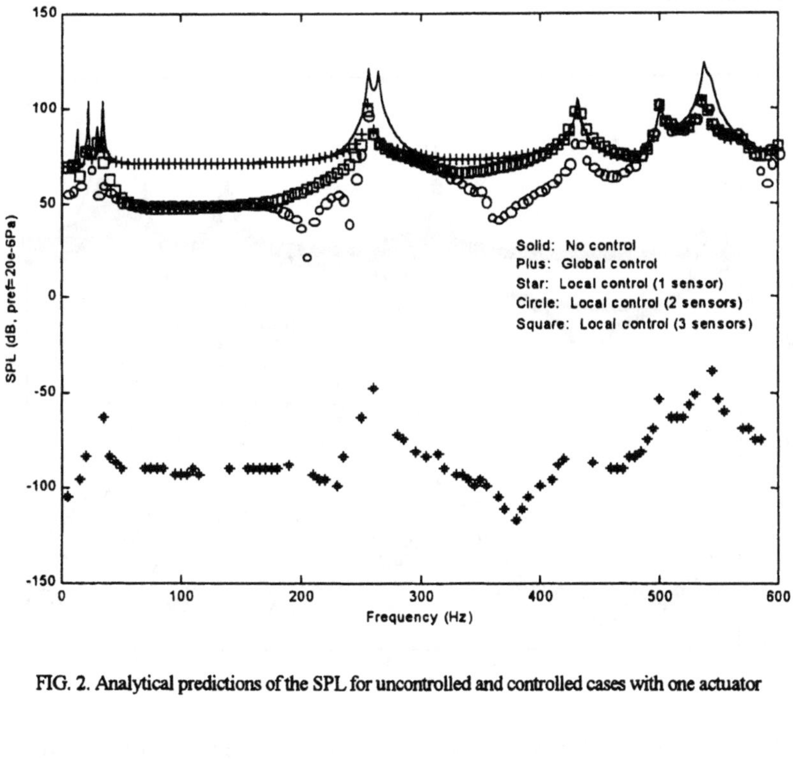

FIG. 2. Analytical predictions of the SPL for uncontrolled and controlled cases with one actuator

In order to improve the precision of modal coupling analysis, the number of enclosure modes and panel modes should be taken in such way that the highest modal frequencies of both enclosure modes and panel modes within the interested frequency range are equal or close. Here, they are taken to be 40 and 47 respectively and the corresponding highest modal frequencies are 1060.33 Hz and 1061.42 Hz respectively. Only one point force located at $\hat{\sigma}_1$ with amplitude of 0.01 N is used in generating the analytical predictions, unless otherwise stated.

Figure 2 shows the analytical predictions of the SPL for the uncontrolled case and four controlled cases over a range of excitation frequencies, with one actuator located at σ_1. The controlled case with the legend 'Global control' is based on Scheme I and the cases with the legend 'Local control' is based on Scheme II. It demonstrates that, for global control case, a maximum reduction of about 20 dB is achieved at some peak SPL frequencies. For local control cases, as expected, when the number of sensors is increased, the SPL predictions tend to the results of the global control.To investigate the effect of the number of actuators on the SPL inside the enclosure, the analytical predictions of the SPL using different number of actuators are demonstrated in figure 3, in which three sensors are used for all local control cases. It is observed that, with the number of actuators increasing, the SPLs decrease obviously for local control, but almost keep unchanged for global control cases. It indicates that, in the viewpoint of global control, the increase of a limited number of actuators does not necessarily give rise to the reduced values of SPL.

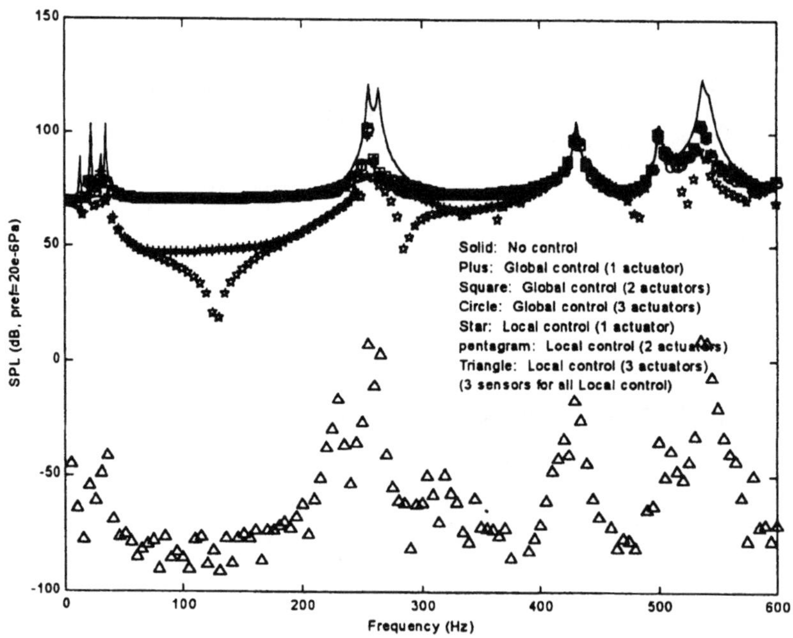

FIG. 3. Analytical of the SPL for uncontrolled and controlled cases with different number of actuators

4. Optimization of Piezoelectric Location

The dimensions and materials of piezoelectric materials and their placement location bonded on host structures, etc. will greatly influence the control results. The optimization of piezoelectric materials generally involves different problems which should be considered in geometry, voltages, dimensions and integral numbers. The optimization technique should deal with the problems which have both continuous and discrete design variables, multi-optima, etc. The most probable optimal design (MPOD) technique proposed by authors in previously study is employed to the optimization of piezoelectric material actuators, [8]. As the first stage, the study take the dimensions and materials, the number of piezoelectric patches fixed, only the placement locations of piezoelectric patch are variable. Since it is significant that the control effect depends on the location of actuators.

The optimization example is taken the dimensions and materials, the number of piezoelectric patches fixed, only the placement location of piezoelectric patch is the design variable. The formulation of this optimization problem may be state as follows,

minimize $J(X)$
subject to,

$|V_i| \leq 100$ V voltage limitation;

$X_l < X < X_u$ side limitation constraint.

where, $J(X)$ is the sound pressure level defined in Eq.(15), X is the design variable which is the central location of piezoelectric patch, X_l and X_u are the lower and upper limit of design variable respectively. The design variable range is X_l=(0.06, 0.04)m, X_u=(0.54,0.36)m. In addition, the constraint condition for piezoelectric actuator without overlapping should also be taken into account. The frequency range for optimization is set to 275 to 300Hz, the optimal location of sound pressure level when one actuator employed is solved by MPOD algorithm. The reduction of sound pressure level is shown in Fig 4. From the figure, it is clear that the sound pressure level before optimization is 36.85dB, and it decreases to 18.70dB after optimization of piezoelectric patch's location.

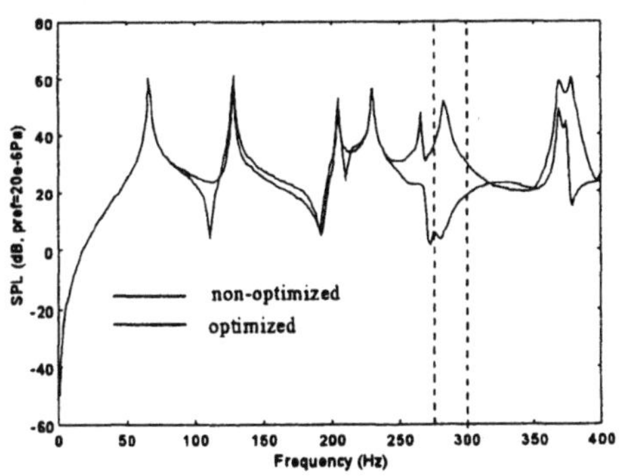

FIG.4. Comparison between the non-optimized and optimized sound pressure level

5. Conclusions

A mechanical model is developed for the active minimization of the interior cavity noise using piezoelectric actuators. The governing motion equations of fully coupled acoustics-structure-piezoelectric patch system are established by using modal synthesis method (MSM). The MSM has the great advantages that the control model can be made by mode parameters of structure, acoustic and electric-strain property of piezoelectric materials before coupling, and the modeling technique can be performed in personal computer facility even for complex shaped structural-acoustical system. Optimization with this modeling technique can be easily performed. The discussion on the placement of actuator locations indicates that, in the viewpoint of practical implementation, in order to realize the optimal noise control, an optimal placement of actuators will give a very great control effect. The results obtained in this work should be relevant for optimal interior noise control in enclosures.

References

[1] Eriksson, L. J., 1994, Recent Trend in the Development of Active Sound and Vibration Control Systems, *Proceedings of NOISE-CON 94*, Ft. Lauderdale, Florida, pp. 271-278.

[2] Bao, X. Q., et al., 1995, Active Control of Sound Transmission through a Plate Using a Piezoelectric Actuator and Sensor, *Smart Materials and Structures*, No. 4, pp. 231-239.

[3] Dowell, E. H., Gorman, G. F., and Smith, D. A., 1977, Acoustoelasticity: General Theory and Force Response to Sinusoidal Excitation, Including Comparison with Experiments, *J. of Sound and Vibration*, Vol.52, No.4, pp.519-542.

[4] Wolf, J., 1977, Modal Synthesis for Combined Acoustical-Structural Systems, *AIAA Journal*, pp.743-747.

[5] Pan, J. and Hansen, C. H., 1991, Active Control of Noise Transmission through a Panel into a Cavity. III: Effect of the Actuator Location, *Journal of the Acoustical Society of America*, Vol. 90, No. 3, pp. 1493-1501.

[6] Nam, C. et al., 1996, Optimal Sizing and Placement of Piezo-actuators for Active Flutter Suppression, *Smart Materials and Structures*, No. 5, pp. 216-224.

[7] Yang, S. M. and Lee, Y. J., 1993, Optimization of Non-collocated Sensor/Actuator Location and Feedback Gain in Control Systems, *Smart Material; s and Structures*, No. 2, pp. 96-102.

[8] Q. Shi, I. Hagiwara, Structural Optimization Based on Holographic Neural Network and its Extrapolation, *Technical Paper AIAA 98-4975, presented at 7th AIAA/USAF/NASA/ISSMO Symposium on Multidisciplinary Analysis and Optimization*, Sept.2-4,1998, pp.2124-2132.

SIMULTANEOUS OPTIMIZATION OF ACTUATOR PLACEMENT AND STRUCTURAL PARAMETERS BY MATHEMATICAL AND GENETIC OPTIMIZATION ALGORITHMS

G. LOCATELLI, H. LANGER, M. MÜLLER, H. BAIER
Institute of Lightweight Structures
Aerospace Department, Technische Universität München
85747 Garching, Germany

1. Introduction

Because of their potential for significant performance improvement, smart structures gain increasing consideration for aerospace or precision system applications. Nevertheless the performance increase can only be realized by proper if not optimal selection of structural and control parameters. The related design problem can often be formulated as a nonlinear multicriteria optimization problem. Objective functions are mostly related to performance (e.g. amplitudes and settling time), mass, required power etc., while constraints are to be satisfied e.g. with respect to eigenvalues, control force amplitudes etc. This increases the complexity of the structural design, where the design variables are not only geometry, mass and stiffness distribution but also actuator/sensor positions and control parameters. In this paper, the optimal design of adaptive structures for active damping is addressed.

2. Optimization problem

The objectives are to maximize active damping of different modes on the one side and also to minimize the mass on the other side. This is therefore a multi-objective nonlinear optimization problem. The constraints are on the position of the actuators that must stay on the structure and cannot overlap and as well as on the structure stiffness and eigenfrequencies. The design parameters considered in this paper are the actuator and sensor position, control parameters, stiffness and mass distribution. It has to be noted that the influence of these parameters are coupled and as already shown in previous publications of the authors [1], control parameters have to be considered as design variables to determine the optimal position of actuators and sensors.

Actuator and sensor position on finite element structure models are discrete variables. This makes an additional effort for interpolation necessary with continuous optimization algorithms and thus the use of Genetic Algorithms more attractive. The different algorithms were tested on typical cases for beam, plate

U. Gabbert and H.S. Tzou (eds.), IUTAM Symposium on Smart Structures and Structronic Systems, 255–263.

and truss structures with defined geometry but with variable stiffness properties. These three benchmarks are described more in the detail in section 5.

3. Model description

3.1. ACTUATORS/SENSORS AND CONTROL ALGORITHMS

To improve system stability, only collocated actuators and sensor pairs are considered on the three types of structure mentioned above. These consist of PZT plates (actuator and sensor), proof-mass actuator and accelerometer as well as force sensors and linear actuators integrated in active bars (see Fig. 1). Actuator forces are proportional to the applied voltage and are modeled as concentrated forces and moments. As shown in figure 1, these moments are distributed over the nodes lying on the edges in the case of piezo-plates.

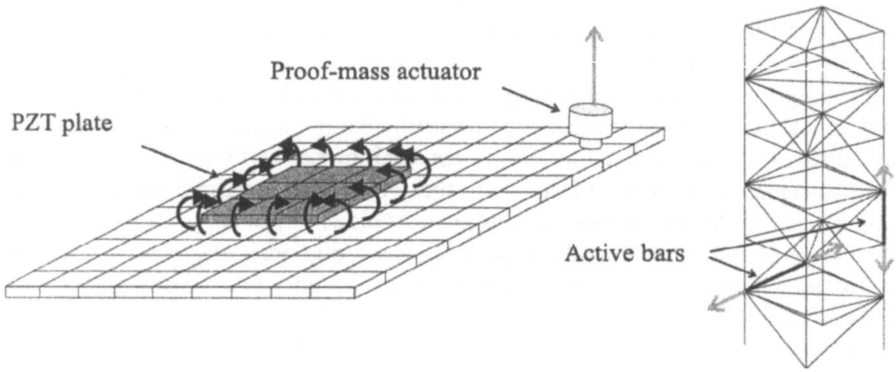

Figure 1: Actuators and simplified force models

The control algorithms considered are Direct Velocity Feedback [2] and Positive Position Feedback [3] for PZT plates as well as for proof-mass actuators and accelerometers. In the case of truss structures and active bars, the control algorithm is Integral Force Feedback [4] .

3.2. MODAL REDUCTION

A modal reduction of the system has been performed. However, it has been shown that the number of modes that have to be taken into account depends very much on the type of actuator and sensor pair considered. The number of modes needed to reach the same accuracy in damping prediction is much higher for bending actuator-sensor pairs like PZT plates than for transverse actuator-sensor pairs like proof-mass actuators and accelerometers. This can be observed in Fig. 2 where the normalized reachable damping values are plotted with respect to the number of modes considered in the case of a simply supported beam with bending and transverse actuator-sensor pairs respectively. The residuals are plotted on the same figure. They indicate the error done when truncating the transfer function modal expansion. Their variation with respect to the number of modes shows to be strictly related to the variation of the reachable damping.

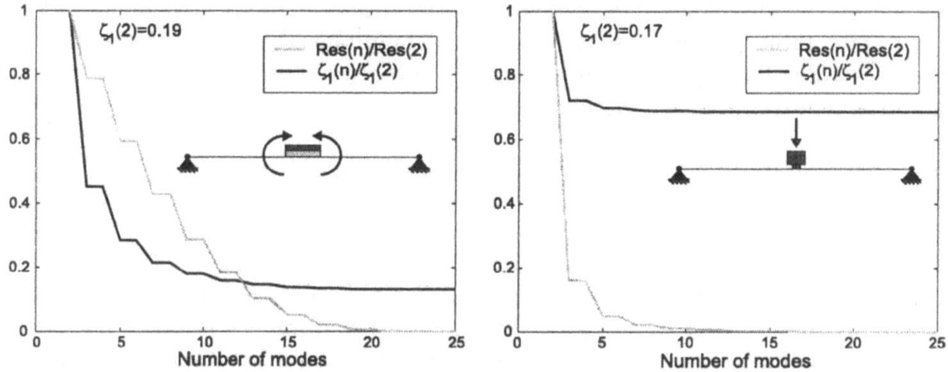

Figure 2: Reachable damping and residuals with respect to the number of modes for bending and transversal actuator/sensor pairs

It can be seen in Fig. 2 that the convergence of reachable damping and residual to their asymptotic value is much faster for the transverse actuator-sensor pair than for the bending actuator-sensor pair. More over, as expected, the damping value obtained with transverse actuator and sensor is higher than with PZT plates, although the initial value considered $\zeta_1(2)$ (damping of mode 1 calculated with 2 modes) is lower (17% vs. 19%). The quality of the model is related to the ability to approximate the static stiffness between actuator and sensor with a modal expansion.

The error caused by modal reduction can be compensated taking into account a feedthrough component in the state space representation of the controlled system. Physically, this is due to the collocation of actuator and sensor and represents a direct proportionality between input and output. This feedthrough or static-correction of the transfer function allows to use a much lower number of modes to calculate accurate modal damping. However, a structure controlled with Direct Velocity Feedback and feedthrough has a transfer function which is not strictly proper and its roots cannot be calculated anymore if no compensator is used. In our investigations a low-pass filter is used as a compensator.

3.3. SOFTWARE TOOLS

The numerical investigations and the optimization are performed with the program MATLAB. This is also used as an application manager to start Finite Elements calculations and set design variables of existing NASTRAN or ANSYS models. The simulation of the controlled system and its optimization is performed in MATLAB. More details on the optimization methods are presented in the next section.

4. Optimization methods

As mentioned above the optimization problems are multiobjective (damping ratios of several eigenmodes, mass) and include characteristics like discrete design variables (actuator placement on a FEM mesh) which pose special challenges for the optimizer. In this paper three potential optimization methods are evaluated for their applicability to this class of problems: A multiobjective genetic algorithm, a gradient based method and finally a mixed approach of both. The main elements and characteristics of each method is discussed in this section.

4.1. MULTIOBJECTIVE GENETIC ALGORITHM (MOGA)

Genetic Algorithms (GA) are known to have noticeable skills in solving problems with discrete variables or combinatorial problems, so they are a natural choice concerning the problems with discrete parameters treated in this paper. But besides this they can be advantageously applied to multiobjective problems, too. GAs work with a group of design points (individuals) simultaneously, the so called population, so the idea to persue several objectives simultaneously is near by. However, GAs basically use a scalar measure of quality (fitness in GA terms) like gradient based methods. Conventionally the different objectives in a multiobjective optimization are combined in a scalar measure by certain methods, the most popular of these is the weighted sum approach. In this case the decision about the relative importance of each objective is taken a priori without any knowledge about the tradeoff between the different objectives. Further more, only a single solution is attained for each run. But generally the nature of a multiobjective optimization is to minimize all objectives simultaneously. There is usually no unique solution, practical problems are often characterized by contradicting and incompatible objectives. But there is a set of non dominated alternative solutions known as the pareto optimal set or the set of best compromises. The goal of a multiobjective optimization is always to attain these pareto-optimal solutions, the pareto frontier.

The basic idea behind the MOGA used in this paper is to employ this concept of non domination and pareto-optimality for fitness assignment, as a measure of quality. But first let us recall the definitions for pareto-dominance and pareto-opimality.

Dominance: Given a vector of n objectives $\boldsymbol{f} = [f_1\ f_2 \dots f_n]$, two result vectors $\boldsymbol{r} = [f_1(\boldsymbol{x}_r)\ f_2(\boldsymbol{x}_r) \dots f_n(\boldsymbol{x}_r)]$ and $\boldsymbol{s} = [f_1(\boldsymbol{x}_s)\ f_2(\boldsymbol{x}_s) \dots f_n(\boldsymbol{x}_s)]$ ($\boldsymbol{x}_r$, $\boldsymbol{x}_s$ design variable vectors), then $\boldsymbol{r}$ dominates $\boldsymbol{s}$, if and only if $\boldsymbol{r}$ is partially less than $\boldsymbol{s}$, $(\boldsymbol{r}\ _p < \boldsymbol{s}\ )$: $\forall i \in [1, \dots, n], r_i \leq s_i \cap \exists i \in [1, \dots, n], r_i < s_i$

Pareto optimality: A solution $\boldsymbol{x}_r$ is pareto optimal, if and only if there is no $\boldsymbol{x}_s$ in the designspace for which $\boldsymbol{s} = \boldsymbol{f}(\boldsymbol{x}_s) = [f_1(\boldsymbol{x}_s)\ f_2(\boldsymbol{x}_s) \dots f_n(\boldsymbol{x}_s)]$ dominates $\boldsymbol{r} = \boldsymbol{f}(\boldsymbol{x}_r) = [f_1(\boldsymbol{x}_r)\ f_2(\boldsymbol{x}_r) \dots f_n(\boldsymbol{x}_r)]$.

The basis to assign fitness to an individual is to determine the number of individuals of the population that dominate it. Accordingly for all currently best individuals, the non dominated individuals, this number is zero and therefore they are assigned the highest fitness. The advantage of this strategy is that all members of the current non dominated frontier are assigned the same fitness, so that the population should finally converge to the pareto optimal frontier. In the scientific literature several fitness assignment strategies based on this fundamental idea can be found [5] [6]. For the MOGA used here the fitness assignment strategy proposed by Fonseca and Fleming [5] has been implemented. Beside multiple objective their proposal also encompasses the handling of multiple constraints and different priority levels among the objectives. Details can be read in the referenced paper [5].

One major drawback of GA in general is the high number of function evaluations. In the multiobjective case this disadvantage is attenuated somewhat because in a single run now the (near) pareto-optimal frontier is attained. With a gradient based method in combination with the weighted sum approach numerous runs with different weighting coefficients would be necessary, as shown in section 5.2.

4.2. GRADIENT BASED METHOD

Gradient based methods are developed for the optimization of continuous functions. Compared to evolutionary algorithms one major advantage is that not only the objective function values are used, but also the gradients. Because of this additional information an optimization can be performed with only very few iterations and function evaluations. The optimization problem presented in section 2. is by nature a continuous problem and only because of the use of FEM models the geometry becomes discrete. To recover continuous functions for the optimizer an interpolation for parameters between nodal positions is required. It turned out, that a good compromise between effort and smoothness is a simplified linear interpolation:

$$f(\boldsymbol{x}) = f(\boldsymbol{x}_0 + \Delta\boldsymbol{x}) = f(\boldsymbol{x}_0) + \frac{1}{\alpha} \left. \frac{\partial f(\boldsymbol{x})}{\partial \boldsymbol{x}} \right|_{\boldsymbol{x}_0} \Delta\boldsymbol{x}$$

f : objective function
$\boldsymbol{x} = \boldsymbol{x}_0 + \Delta\boldsymbol{x}$: selected position
$\boldsymbol{x}_0$: closest nodal position
$\Delta\boldsymbol{x}$: distance from closest nodal position
α : smoothing factor

4.3. MIXED APPROACH

In order to combine the specific strengths of both strategies, GAs and gradient methods, a mixed approach is also investigated. The idea is to employ each method only where it is especially strong, e.g. the GA to deal with discrete parameters and multiple objectives and the gradient method for continuous parameters and problems with high efficiency required. Because it was possible to divide parameter and objectives into two more or less physically independent groups a two level combination was applied to the truss structure problem (see flowchart in Fig. 3).

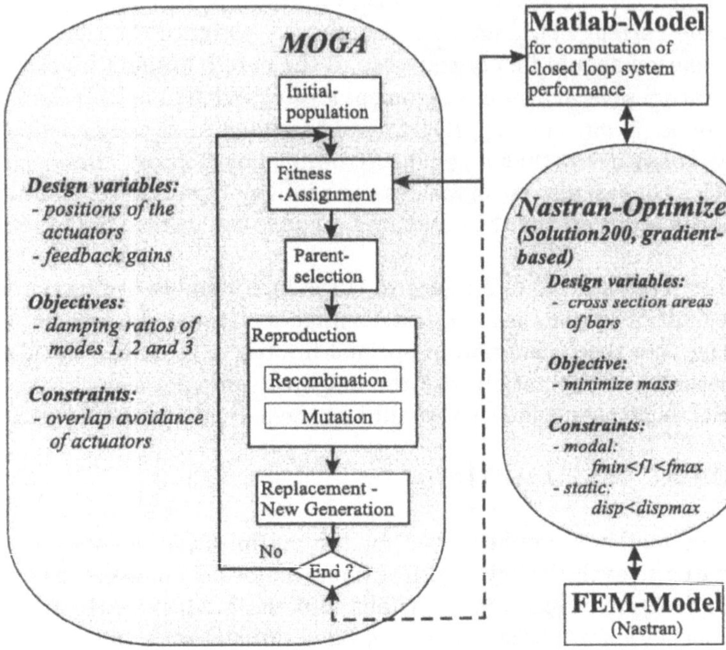

Figure 3: Two-level optimization of an active damped truss structure

5. Results

5.1. BENCHMARK PROBLEMS

The discussed modeling, control and optimization methods are applied to three benchmark problems: a simply supported beam, a clamped plate and a truss structure. The detailed information about each model can be seen in Table 1.

Table 1: Benchmark problems - detailed description

example	**beam**	**plate**	**truss structure**
actuators	3 pzt-plates	3 pzt-plates	3 active bars
design variables	act. pos., act. size, feedback gains	act. pos., feedback gains	act. pos., feedback gains, bar cross section
constraints	max. act. size, max. overall act. size , overlap avoidance	overlap avoidance	overlap avoidance, max. deflection under static load, min. first eigenfrequency
objectives	damping of first 3 modes	damping of first 3 modes	damping of first 3 modes, mass
optimization method	MOGA, *fmincon* (MATLAB)	MOGA, *fmincon* (MATLAB)	mixed approach

5.2. OPTIMIZATION RESULTS FOR BEAM AND PLATE

For the beam and the plate both the gradient method as well as the MOGA are used. The experiences for the plate proved to be similar to the beam, so that only the beam results are presented in detail. For the gradient based method the *fmincon* function provided by MATLAB is used. To attain a scalar measure of quality the weighted sum approach is used. Five different weighting vectors were applied:

$$f_1 = \zeta_1, \quad f_2 = \zeta_2, \quad f_3 = \zeta_3, \quad f_4 = \frac{\zeta_1 + \zeta_2}{2}, \quad f_5 = \frac{\zeta_1}{2} + \frac{\zeta_2 + \zeta_3}{4}$$

To see the influence of the initial condition for each objective function 32 different starting vectors were used. For the the results presented below a population of 200 individuals has been chosen and the algorithm has been run for 50 generations.

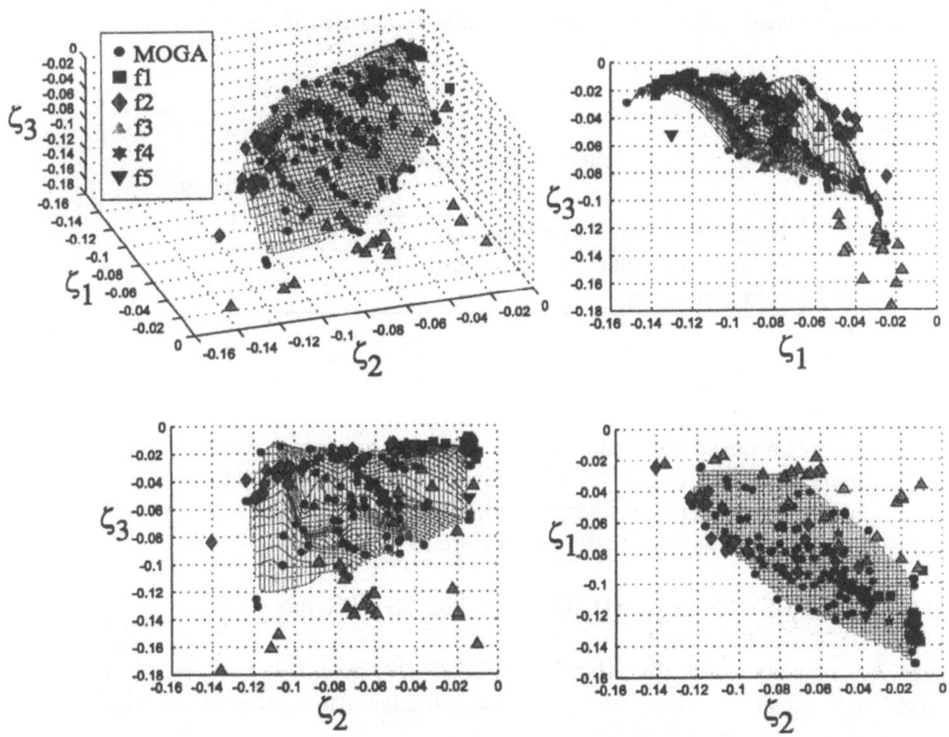

Figure 4: Optimization results: damping ratios achieved with the MOGA and *fmincon* for five different objective functions (damping increases in the negative axis direction)

In Fig. 4 the results of the different optimization runs are presented, the achieved damping ratios for the target modes 1, 2 and 3 are plotted for the MOGA as well as for the five different objective functions for *fmincon*. For MOGA the objectives values for the final population are plotted (filled circles). The mesh is only an optical help to visualize the final distribution of non dominated solutions (because there is no proof how close it is to the real pareto optimal surface, the expression non dominated surface is preferred). It can be seen that the final solutions cover a sufficient area to provide a good basis for trade-off decisions between the target modes. For *fmincon* the results for 32 starting positions for each objective function are plotted. The dependence of the final optimum on the initial position can be clearly seen and confirms the assumption of several local optima. Concerning the quality of the results *fmincon* is able to find better solutions for single target modes than MOGA (e.g. mode 3 : 17.7% (*fmincon*), 13.2% (MOGA)). But for a combined objective function like $f_2 = (\zeta_1 + \zeta_2)/2$ MOGA achieves solutions that clearly dominate the results of *fmincon*.

The optimization costs concerning time and the number of function evaluations are as expected much higher for MOGA than *fmincon* with respect to a single run (see Table 2). However with one single MOGA run 200 useful solutions (near pareto-optimal frontier useful for trade-off studies) are obtained. So the computation costs per solution are even less than for *fmincon*.

Table 2: Optimization costs

	fmincon	MOGA
Number of function evaluation	75	10100
Number of function evaluation / solution point	75	51

It has to be noted that the cost for *fmincon* are increased here compared to an ordinary optimization problem because of the interpolation of the objective function due to the discrete design variables. These costs are even higher when more discrete variables have to be considered.

5.3. OPTIMIZATION RESULTS FOR TRUSS STRUCTURE

For the truss-structure the 2-level optimization approach is used (see Fig. 3). All together there are 81 design variables. An reliable optimization of this high dimensional design space with MOGA alone would have required a considerable large population and thus high computational costs. In order to minimize these costs only the discrete actuator positions and the related feedback gains are handled by MOGA. The population size has been chosen to be 50 and the algorithm has again been run for 50 generations, which lead to satisfactory results and proved again the abilities of GAs to deal with discrete variables. The optimization of the cross section areas of the bars is done on the second level of the optimization with the NASTRAN Solution 200. This approach proved to be very efficient.

6. Conclusions

In this paper the problem of optimal design of adaptive structures was addressed. Methods to optimize actuator and sensor position, control and structure parameters were presented in the case of active damping problems. The optimization methods considered are a combination of gradient based methods and Multi Objective Genetic Algorithms (MOGA) developed by the authors. Some results obtained on typical beam, plate and truss structure have been presented. This work is an important step towards an automatic design of complex adaptive structures.

References

[1] H. Baier, G. Locatelli, "Optimization of actuator placement and structural parameters in smart structures," in *Mathematics and Control in Smart Structures*, Vasundara V. Varadan, ed., *Proc. SPIE* **3667**, pp. 267–276, 1999.

[2] M. J. Balas, "Direct velocity feedback control of large space structures," *AIAA Journal of Guidance* **2**(3), pp. 252–253, 1979.

[3] J. L. Fanson,T. K. Caughey, "Positive position feedback control for large space structures," *AIAA Journal* **28**, pp. 717–724, April 1990.

[4] A. Preumont, Jean Paul Dufour, Christian Malékian, "Active Damping by a Local Force Feedback with Piezoelectric Actuators," *AIAA Journal of Guidance* **15**, pp. 390–395, March-April 1992.

[5] P. J. Fleming ,C. M. Fonseca, "Multiobjective Optimization and Multiple Constraint Handling with Evolutionary Algorithms-Part I and II: A Unified Formulation," *IEEE Transactions on System Man and Cybernetics, Part. A: Systems and Humans* **28**, pp. 26–37, 1998.

[6] J. Horn, N. Nafpliotis, D. E. Goldberg, "A niched Pareto genetic algorithm for multiobjective optimization," in *First IEEE Conference on Evolutionary Computation (ICEC 94), IEEE World Congress on Computational Intelligence* **1**, pp. 82–87, 1994.

SUITABLE ALGORITHMS FOR MODEL UPDATING AND THEIR DEPLOYMENT FOR SMART STRUCTURES

M.W.ZEHN, O.MARTIN
Otto-von-Guericke-University Magdeburg,
Institute of Mechanics, D-39106 Magdeburg, Germany

1. Introduction

Smart structures can sense and actuate in a controlled manner in response to variable ambient stimuli combined with load carrying functions. If we are using finite element analysis up-front to design parts of our smart structures it is beset with problems like uncertainties from the manufacturing and in material properties. Attempts to improve the model with more detailed, sophisticated, refined discretised models are limited in either way. Hence, uncertainties not can be eliminated by shear mesh refinement or more detailed modelling in general. Moreover, model reduction techniques might be necessary to make a dynamical simulation feasible. Model updating aims to correct or at least alleviate invalid assumptions, omissions, and uncertainties as well as model reduction errors by processing vibration test results so *that the theoretical model is closer to reality*. Yet, model updating is limited in its applicability as well by the choice of the right parameters and weighting matrices. It also requires substantial computational effort because of the inverse character of the mathematical problem it involves; successful application requires a number of preparatory steps. Several methods for validation and error localisation should be applied to the FE model and a good selection of correction parameters (either sensitive or representative - a big problem). Today a wide variety of different model updating methods exist. We will confine ourselves to iterative updating methods. The updating of FE model parameters is based on a minimisation of a cost or penalty function at each iteration. These parameter estimation methods depend on the proper choice of the weighting matrices used ensure a good initial in the numerical process, and in the results. The weighting matrices usually chosen by engineering judgement. A more natural approach for coping with the uncertainties in parameters and results is a statistical one. A still unanswered question involves the updating of electrical parameter and equipment like filters, amplifiers, etc. in smart structure design. In the paper we illustrate approaches to the model updating of smart structures; we do not provide an exhaustive description of all suitable updating algorithms; there is seldom a single *best* method.

2. Model Updating and Statistics

Modal Analysis (FEM): After FE discretisation the linear dynamic vibration problem is

U. Gabbert and H.S. Tzou (eds.), IUTAM Symposium on Smart Structures and Structronic Systems, 265–274.

represented by the discrete differential equation for the structure

$$\mathbf{M}\ddot{\mathbf{q}} + \mathbf{C}\dot{\mathbf{q}} + \mathbf{K}\mathbf{q} = \mathbf{f} \tag{1}$$

and the matrix eigenvalue problem

$$(\mathbf{K} - \lambda_{a_i}\mathbf{M})\Phi_{a_i} = 0$$

$$\Phi_a = [\Phi_{a_1}, \ldots, \Phi_{a_i}, \ldots, \Phi_{a_m}] \in \Re^{n \times m} \text{ (modal matrix)}, \; \Lambda = \text{diag}(\lambda_{a_i}) \in \Re^{m \times m} \text{ (spectral matrix)} \tag{2}$$

$$\text{with} \quad \Phi^T\mathbf{M}\Phi = \mathbf{E} = \text{diag}(\mathbf{1}) \quad \text{and} \quad \Phi^T\mathbf{K}\Phi = \Lambda = \text{diag}(\lambda_{a_i})$$

Problems in the simulation with the FE model are caused by model projection error (3D,shell, plate, beam, etc.), simplifications, omissions, uncertainties (manufacturing, material, loading, constrains, multi-physics, etc.), discretization, truncation of modal space, model reduction, numerical errors, systematically error of methods, subjective errors (model preparation, application of methods, etc.) and so on.

Modal Testing (EMA): Experimental Modal Analysis provides us with the necessary measured modal data

$$\Phi = [\Phi_{e_1}, \ldots, \Phi_{e_i}, \ldots, \Phi_{e_m}] \quad \text{and} \quad \Lambda = diag(\lambda_{e_i}) \tag{3}$$

Problems arise from noise corruption of data, subjective errors, equipment errors and limitations, product and measurement variability, matching of data and so forth. The modal data from vibration testing and FE analyses are common ways to characterise the linear dynamics of a structure. The comparison of modal results is vital for assessing model and data quality. For any model improvement, besides a sensible starting model, we need model parameters suitable for reducing the model error and sufficiently sensitive to produce numerical results. To get better model parameters, we use as estimator the following cost or penalty function (Extended Weighted Least Squares Estimation (EWLS))

$$P(\theta) = (z_{ex} - z(\theta))^T \cdot W_{ex} \cdot (z_{ex} - z(\theta)) + (\theta - \theta^{(0)})^T \cdot W_\theta \cdot (\theta - \theta^{(0)}) \Rightarrow \text{Min.} \tag{4}$$

In many publications it is stressed that the proper choice of weighting matrices is crucial for the resulting parameters $\theta^{opt)}$ and the numerical solutions. The experiences of the authors comply with that. Whereas for the measurement error some assumptions could be found to construct a matrix W_{ex}, for W_θ, in most cases only knowledge and experience about the character of parameter uncertainties can help. Equation (4) is expanded in a truncated (after the second term) Taylor series

$$P(\theta) = P(\theta^{(0)}) + \frac{\partial P(\theta^{(0)})}{\partial \theta}(\theta - \theta^{(0)}) + \frac{1}{2}(\theta - \theta^{(0)})^T \cdot H(\theta^{(0)}) \cdot (\theta - \theta^{(0)}) \tag{5}$$

with the Hessian matrix $\mathbf{H}(\theta^{(0)})$. That leads to the following iteration scheme to compute updated parameters $\theta^{(j+1)}$

$$\theta^{(j+1)} = \theta^{(j)} + (W_\theta + S(\theta^{(j)})^T \cdot W_{ex} \cdot S(\theta^{(j)}))^{-1} \cdot [S(\theta^{(j)})^T \cdot W_{ex} \cdot (z_{ex} - z(\theta^{(j)}) + W_\theta \cdot (\theta^{(0)} - \theta^{(j)})] \tag{6}$$

where $S(\theta^{(j)})\ni\Re^{m,mp}$ is the sensitivity matrix,

$$S(\theta^{(j)}) = \frac{\partial z(\theta^{(j)})}{\partial \theta^{(j)}} \tag{7}$$

For a controllability of numerical stable behaviour the rank of $S(\theta^{(j)})$ must not be less than the number of parameters. If we take the parameters and measured results with their inherent statistics, the *a priori* distribution can be regarded in most cases as normally distributed. Thus, the mean for the parameters is given by $E(\theta)=\theta^{(0)}$, where $E(\Delta\theta)=0$ ($E(\;)$ is the expected value) and the co-variance matrix is $Cov(\theta)=Cov(\Delta\theta)=E(\Delta\theta_i,\Delta\theta_k)=V^{(0)}$ (or short $\theta \sim N(\theta^{0}, V^{(0)})$). For the output values the mean is given by $E(\Delta z)=E(\varepsilon)=0$ (ε measurement error) and the co-variance matrix can be written as $Cov(\Delta z)=E(\varepsilon_i,\varepsilon_k)=\Sigma$. For $\theta^{opt)}$ the calculated output values become optimal as well $z(\theta^{opt)})=z^{(opt)}$, so that $z_{ex}= z^{(opt)} + \varepsilon$. Furthermore, it will be assumed that there is no correlation between the measurement error and the estimated parameter changes, that means $E(\Delta\theta_i,\varepsilon_k)=0$. In reality there is a correlation because the measurements are used in the updating process. The conditional density function for z_{ex} (θ given) is $N(z(\theta),\Sigma)$ and for θ with a given z_{ex} is the so-called *a posteriori* density function, which can be expressed by

$$f(\theta|z_{ex}) = \frac{1}{C}exp[-\frac{1}{2}(z(\theta)-z_{ex})^T \cdot \Sigma^{-1} \cdot (z(\theta)-z_{ex}) - \frac{1}{2}(\theta-\theta^{(0)})^T \cdot V_0^{-1} \cdot (\theta-\theta^{(0)})] \tag{8}$$

where C is a scaling constant. According to a maximum likelihood the maximum of the density function can be obtained by vectors θ that minimises

$$P(\theta) = (z(\theta)-z_{ex})^T \cdot \Sigma^{-1} \cdot (z(\theta)-z_{ex}) + (\theta-\theta^{(0)})^T \cdot V_0^{-1} \cdot (\theta-\theta^{(0)}) \Rightarrow \text{Min.} \tag{9}$$

Equation (9) is similar to equation (5); the weighting matrices in (5) have been replaced by the inverse co-variance matrices in (9). Such an estimator is then called a Bayes estimator. For iterative application, equation (6) is also valid for the determination of parameter changes. There is no guarantee that the iteration will not converge to a local minimum. The best way to avoid this consists in a good parameter choice, error localisation, and a good starting model. Candidates for ***response quantities*** z could be eigenfrequencies, eigenvectors, MAC and COMAC values, modal displacements, mass properties of the rigid body, etc. and for ***model parameter*** θ, e.g., mass and stiffness parameters (discrete or distributed), material properties, boundary and joint properties, geometrical parameter (thickness, shape, position, ...), and so on. If a substructure technique is deployed (reduced model) often mass or stiffness substructure parameter are used

$$\mathbf{M} = \mathbf{M}_0 + \sum_{j=1}^{n_\theta} \mathbf{M}_j = \mathbf{M}_0 + \sum_{j=1}^{n_\theta} \theta_{M_j}\tilde{\mathbf{M}}_j \qquad \mathbf{K} = \mathbf{K}_0 + \sum_{j=1}^{n_\theta} \mathbf{K}_j = \mathbf{K}_0 + \sum_{j=1}^{n_\theta} \theta_{M_j}\tilde{\mathbf{K}}_j \tag{10}$$

where $\mathbf{M}_0$ and $\mathbf{K}_0$ are the substructures, which are not taken into a correction and θ_{Mj} and θ_{Kj} are the mass and stiffness substructure parameter, respectively.

At the first glance we can use different quantities $z(\theta)$ and different parameter types in θ to enrich the information content in measurement and calculation. If we apply different sensitivities, equation (7) can be written as

$$\mathbf{S}(\theta^{(j)}) = [c_{S_\lambda} \cdot \mathbf{S}_\lambda,\ c_{S_\phi} \cdot \mathbf{S}_\phi,\ c_{S_{MAC}} \cdot \mathbf{S}_{MAC},\ c_{S_{CMD}} \cdot \mathbf{S}_{CMD},\ c_{S_{Mass}} \cdot \mathbf{S}_{Mass},\ \ldots]^T \quad (11)$$

The scaling factors $c_{s\lambda}$, $c_{s\phi}$, c_{sMAC}, and c_{sMass} are introduced to overcome ill-conditioning of the total sensitivity matrix. These factors are difficult to determine because the parameters θ are more or less sensitive to the response quantities $z(\theta)$. The attendant numerical difficulties out weigh the benefits. Also the derivatives in equation (11) needed to obtain the sensitivities require computational effort out of all proportion to possible improvements. We will confine ourselves to natural frequencies and eigenvectors on the one hand, and stiffness, mass and thickness parameters on the other. For the eigenfrequencies as response parameter $z(\theta)$ the sensitivity matrix can be obtained from the eigenvalue problem in equations (2)

$$\frac{\partial \lambda_i}{\partial \theta_j} = \phi_i^T \left(\frac{\partial \mathbf{K}}{\partial \theta_j} - \lambda_i \frac{\partial \mathbf{M}}{\partial \theta_j}\right)\phi_i = \mathbf{S}_{\lambda_{ij}} \quad (12)$$

For the substructure parameter in equation (10) this yields

$$\mathbf{S}_{\lambda_i \theta_{\mathbf{M}_j}} = \frac{\partial \lambda_i}{\partial \theta_{\mathbf{M}_j}} = -\lambda_i \phi_i^{(j)^T} \tilde{\mathbf{M}}_j \phi_i^{(j)} \qquad \mathbf{S}_{\lambda_i \theta_{\mathbf{K}_j}} = \frac{\partial \lambda_i}{\partial \theta_{\mathbf{K}_j}} = \phi_i^{(j)^T} \tilde{\mathbf{K}}_j \phi_i^{(j)} \quad (13)$$

The calculation of eigenvector sensitivities is much more complicated due to singularities. One possible approach is Nelson's method [9]. The idea is to calculate the sensitivities of the mass normalised eigenvectors ϕ_i indirectly. First we have to determine the sensitivities of the eigenvectors ϕ^*_i that are obtained, when each associated eigenvector ϕ_i is divided by its largest valued co-ordinate ϕ_i^{max}, we have

$$\frac{\partial \phi_i}{\partial \theta_j} = \phi_i^{max} \frac{\partial \phi_i^*}{\partial \theta_j} + \frac{\partial \phi_i^{max}}{\partial \theta_j} \phi_i^* \quad (14)$$

The derivative of ϕ^*_i is calculated by solving

$$(\mathbf{K} - \lambda_i \mathbf{M})^* \frac{\partial \phi_i^*}{\partial \theta_j} = \left[-\left(\frac{\partial \mathbf{K}}{\partial \theta_j} - \frac{\partial \lambda_i}{\partial \theta_j}\mathbf{M} - \lambda_i \frac{\partial \mathbf{M}}{\partial \theta_j}\right)\phi_i^*\right]^* \quad (15)$$

in which ( )* and [ ]* denote that in the matrices the rows and columns associated with ϕ_i^{max} are removed. The derivative of ϕ_i^{max} can be calculated with the derivation of the ortho-normalisation equation of the eigenvectors (see equation (2)) and becomes

$$\frac{\partial \phi_i^{max}}{\partial \theta_j} = -\phi_i^{max} \left(\frac{1}{2} \phi_i^T \frac{\partial \mathbf{M}}{\partial \theta_j} \phi_i + \phi_i^{max} \phi_i^T \mathbf{M} \frac{\partial \phi_i^*}{\partial \theta_j}\right) \quad (16)$$

Fox and Kapoor [6] proposed for the eigenvector differentiation a linear combination of the eigenvectors

$$\mathbf{S}_{\phi_{ij}} = \frac{\partial \phi_i}{\partial \theta_j} = \sum_{l=1}^{n} \alpha_{ijl} \phi_l \quad (17)$$

The coefficients α_{ijl} can be determined by the derivation of the eigenvalue problem and the orthogonality conditions in equations (2) with regard to the parameters. The advantage compared to Nelson's method is that there is no matrix equation to be solved, but for an exact solution all n eigenvectors are necessary; these are not available from a FE calculation. A truncation of the row in equation (17) leads often to considerable errors. Lim et al. [7] and Wang [8] have proposed an improved modal method that corrects or at least alleviates those errors induced by the higher modes. But the additional numerical effort prevents it from being competitive to Nelson's method. Nelson's procedure is so far the most efficient way to calculate eigenvector sensitivities. However, for updating of large FE-models our experience shows that the inclusion of eigenvector sensitivity in iterative updating algorithms vastly increases computer time per iteration step. Alternatively, the equation error of the eigenvalue problem can be used in the updating process instead of direct inclusion of the eigenvectors (see Natke [9]). The error in the eigenvalue equation is determined by replacing the computed modal properties in equation (2) with the measured ones

$$\mathbf{R}_i = \begin{bmatrix} \left(\mathbf{K} - \lambda_{iex}\mathbf{M}\right)\phi_{iex} \\ \phi_{iex}^T\mathbf{M}\phi_{iex} - 1 \end{bmatrix} \tag{18}$$

3. Minimum Variance (MVE) and other Estimators

Another iterative method is the *Minimum Variance Estimator (MVE)*, see [3]. In that we consider, from step $k \geq 1$ on, that there is a correlation between the measurement error and the estimated parameter changes with $Cov(\theta,\varepsilon)=E(\theta.\varepsilon^T)=-\mathbf{D}_k$ (where $\mathbf{D}_0=0$). The joint distribution of (z_{ex},θ) is a normal distribution with

$$E\begin{pmatrix} \mathbf{z}_{ex} \\ \theta \end{pmatrix} = \begin{pmatrix} \mathbf{z}(\theta_k) \\ \theta_k \end{pmatrix}$$

$$Cov\begin{pmatrix} \mathbf{z}_{ex} \\ \theta \end{pmatrix} = \begin{pmatrix} \Sigma - \mathbf{S}(\theta_k)\mathbf{D}_k - \mathbf{D}_k^T\mathbf{S}^T(\theta_k) + \mathbf{S}(\theta_k)\mathbf{V}_k\mathbf{S}^T(\theta_k) & -\mathbf{D}_k^T + \mathbf{S}(\theta_k)\mathbf{V}_k \\ -\mathbf{D}_k + \mathbf{V}_k\mathbf{S}^T(\theta_k) & \mathbf{V}_k \end{pmatrix} =: \begin{pmatrix} \mathbf{A}_k & \mathbf{C}_k^T \\ \mathbf{C}_k & \mathbf{V}_k \end{pmatrix} \tag{19}$$

Hence, the Bayesian estimator $\hat{\theta}$, i.e. the estimator with the minimum mean squared error, is given by the expected value of θ (given z_{ex}). For the iteration we therefore have

$$\begin{aligned} \theta_{k+1} &= \hat{\theta} = E(\theta \mid \mathbf{z}_{ex}) = \theta_k + \mathbf{C}_k\mathbf{A}_k^{-1}(\mathbf{z}_{ex} - \mathbf{z}(\theta_k)) \\ \mathbf{V}_{k+1} &= Cov(\theta \mid \mathbf{z}_{ex}) = \mathbf{V}_k - \mathbf{C}_k\mathbf{A}_k^{-1}\mathbf{C}_k^T \\ \mathbf{D}_{k+1} &= -Cov(\theta,\varepsilon \mid \mathbf{z}_{ex}) = -E(\theta\varepsilon^T \mid \mathbf{z}_{ex}) = E((\hat{\theta} - \theta)\varepsilon^T) = \mathbf{D}_k - \mathbf{C}_k\mathbf{A}_k^{-1}(\mathbf{S}(\theta_k)\mathbf{D}_k - \Sigma) \end{aligned} \tag{20}$$

Σ is not updated by $cov(\varepsilon|z_{ex})$! A similar procedure is ECHHK (Estimator after **C**ollins, **H**art, **H**aselman, and **K**ennedy, see [3]). It simply takes the *posterior* distribution as the new *prior*, that means $\mathbf{D}_k=0$ for all k, and we get

$$\begin{aligned} \theta_{k+1} &= \theta_k + \mathbf{V}_k\mathbf{S}^T(\theta_k)(\Sigma + \mathbf{S}(\theta_k)\mathbf{V}_k\mathbf{S}^T(\theta_k))^{-1}(\mathbf{z}_{ex} - \mathbf{z}(\theta_k)) \\ \mathbf{V}_{k+1} &= \mathbf{V}_k - \mathbf{V}_k\mathbf{S}^T(\theta_k)(\Sigma + \mathbf{S}(\theta_k)\mathbf{V}_k\mathbf{S}^T(\theta_k))^{-1}\mathbf{S}(\theta_k)\mathbf{V}_k \end{aligned} \tag{21}$$

Both MVE or ECHHK need more operations and processing time than Equation (6). It has been shown in [1] that MVE has stability problems with regard to standard derivation of measurement errors. WLS (weighted least square estimator) and EWLS (extended weighted least square estimator) show in general a good convergence rate. In our experience convergence rates for MVE and ECHHK do not show remarkable improvement.

4. Spatial Statistical Generation of Thickness Data According to Manufacturing Uncertainties

Parameters due to uncertainties in manufacturing or other reasons possess a statistical character; in most cases this statistic is a special for it is spatially distributed. In [5], Zehn and Saitov present a method for tackling such problems, using the example of thickness variations in composite plates. For a statistical description of the varying

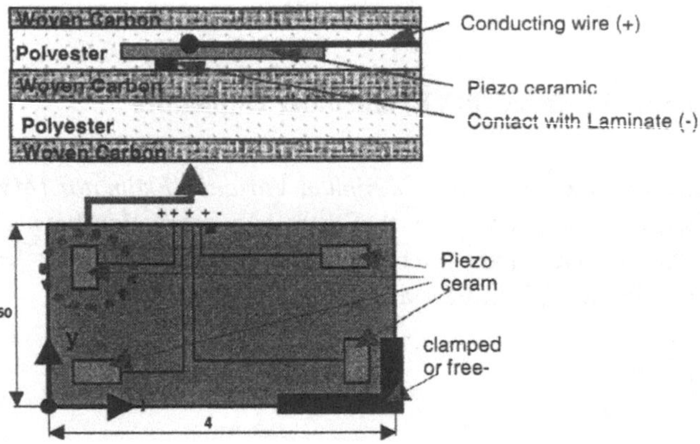

Figure 1. structure of carbon fibre-reinforced test plates

thickness d of a plate, see figure 1, d is regarded as a random number with probability density $f(d)$. Assuming that the thicknesses of the plate at different locations in the mid-surface $\boldsymbol{D}$ of the undeformed plate are statistically independent, we can construct an

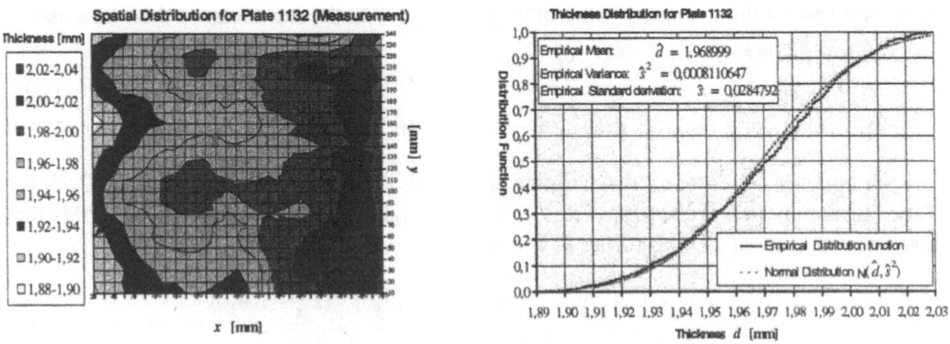

Figure 2. Measured Thickness and Empirical Distribution Function

empirical distribution function for the thickness (random test function) from a representative area of the sample. Corresponding to such an empirical determined distribution, a thickness distribution can be generated in the FE model by means of pseudo-random numbers. Figure 3 depicts the resulting distribution, if we assign these numbers

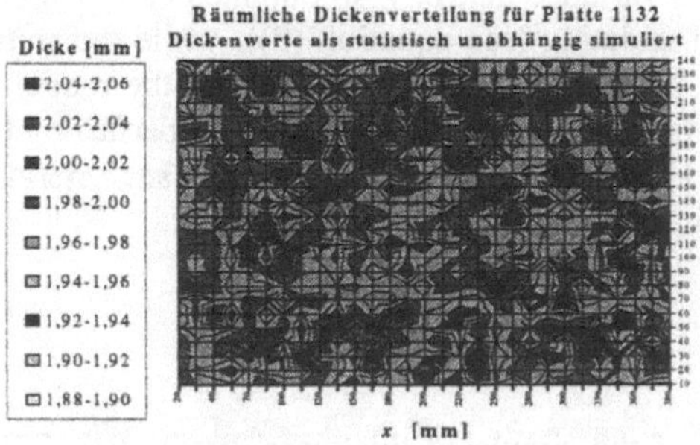

Figure 3. Generation of Uncorrelated Normal Distribution Thickness

to the FE mesh knot or element wisely, without any respect to the spatial distribution and correlation. Such a distribution has nothing in common with the real distribution of manufacturing errors. For that purpose, on the basis of the theory pioneered by Cressie [4], a procedure and algorithm for the calculation of a spatial-correlated probabilistic thickness distribution has been developed [5].

Provided that the description of the spatial distribution for the random value d (thickness) over the mid-surface of the plate depends on the (deterministic) co-ordinates of the mid-surface $\boldsymbol{D}$, then d can be regarded as a stochastic process $\{d(\boldsymbol{v}): \boldsymbol{v} \in \boldsymbol{D} \subset \Re^2\}$. It refers to the undeformed mid-surface $\boldsymbol{D}$, where $\boldsymbol{v}=(x, y)^T$ is the vector of the co-ordinates. Supposing that $\{d(\boldsymbol{v}): \boldsymbol{v} \in \boldsymbol{D}\}$ is a stationary Gaussian process $N(\theta_0 \boldsymbol{1}_n, \boldsymbol{V})$, where $\theta_0=E[d(\boldsymbol{v})]$ is the mean, $\boldsymbol{1}_n$ a unit vector, and $\boldsymbol{V}$ the co-variance matrix $([\boldsymbol{V}]_{ij}=Cov[d(v_i),d(v_j)],\ i,j=1,\dots,n)$, then θ_0 is constant for all $\boldsymbol{v} \in \boldsymbol{D}$. The co-variance matrix can then be described by a function $Cov[d(v_i),d(v_j)]= C(v_i - v_j)$ for all $v_i, v_j \in \boldsymbol{D}$. And with that function $C(\boldsymbol{h})$, $\boldsymbol{h} = v_i - v_j$, the so-called co-variogram function or by

$$
\begin{aligned}
C(0) - C(\boldsymbol{h}) &= \gamma(\boldsymbol{h}), \qquad \boldsymbol{h} = \boldsymbol{v}_i - \boldsymbol{v}_j \\
Var[d(\boldsymbol{v}_i), d(\boldsymbol{v}_j)] &= \gamma(\boldsymbol{v}_i - \boldsymbol{v}_j) \equiv \gamma(\boldsymbol{h}) \qquad \text{for all } \boldsymbol{v}_i, \boldsymbol{v}_j \in D
\end{aligned}
\tag{22}
$$

the semi-variogram function $\gamma(\boldsymbol{h})$ has been defined.

This assumption implies that the co-variance matrix $\boldsymbol{V}$ can be parameterised and be determined following a parametric concept

$$
\mathbf{V}_{ij} = C(\boldsymbol{h}) = C(0) - \gamma(\boldsymbol{h}), \quad \boldsymbol{h} = \|\boldsymbol{v}_i - \boldsymbol{v}_j\|, \quad i,j = 1,\dots,n \tag{23}
$$

An admissible parametric assumption has to guarantee that the co-variance matrix based on this assumption is positive definite. A number of parametric variogram assumptions are proposed in the literature [4]. The distribution parameter of the assumed Gaussian process for the thickness have to be estimated on the basis of the measured thickness values (empirical (non-parametric) estimation of the semi-variogram taken form a representative area). With the help of the empirical variogram (and its graphic representation), it is possible to select a type for the parametrical variogram assumption. Then, with statistical correlated description of the varying of the plate thickness on hand, its distribution can be simulated in a more natural way with regard to the manufacturing errors. This can be use in Monte Carlo and other computations of the dynamical behaviour of the structure. The co-variance matrix V so obtained can then be employed in the Bayes estimator for model updating (see equation (9)), while the statistical description used in model updating is kept unchanged. A next step in the model updating is to attempt to obtain a further improvement of the statistical properties on the basis of vibration measurement and the FE model results. The theory of such iterative procedure has been described firstly in [5].

5. Examples

Figure 4 shows the result of the simulated thickness distribution following the theory explained in the previous chapter.

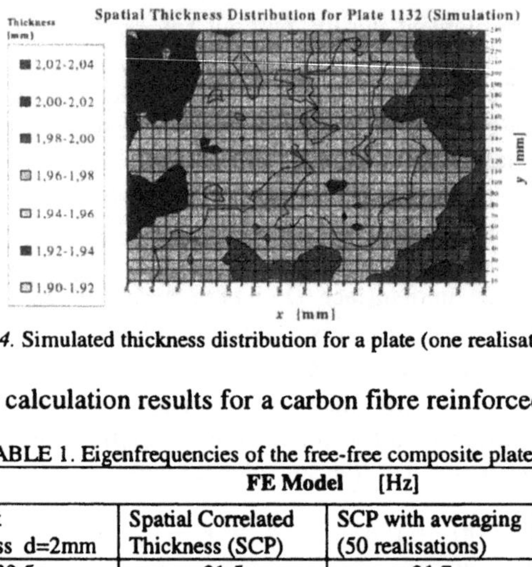

Figure 4. Simulated thickness distribution for a plate (one realisation only)

Table 1 shows further calculation results for a carbon fibre reinforced plate.

TABLE 1. Eigenfrequencies of the free-free composite plate

Measurement [Hz]	FE Model [Hz]			
	Constant Thickness d=2mm	Spatial Correlated Thickness (SCP)	SCP with averaging (50 realisations)	SCP with updating, cons. co-var. matrix
28.7	33.5	31.5	31.7	31.9
61.4	63.5	59.67	60.1	60.43
85.8	93.3	88.19	88.3	89.3
155.7	163.1	153.8	154.0	155.8
164.8	171.2	161.8	162.1	163.9
172.1	180.7	170.2	171.1	173.0
193.6	202.7	192.1	192.1	194.6
214.3	223.7	212.0	211.8	214.5

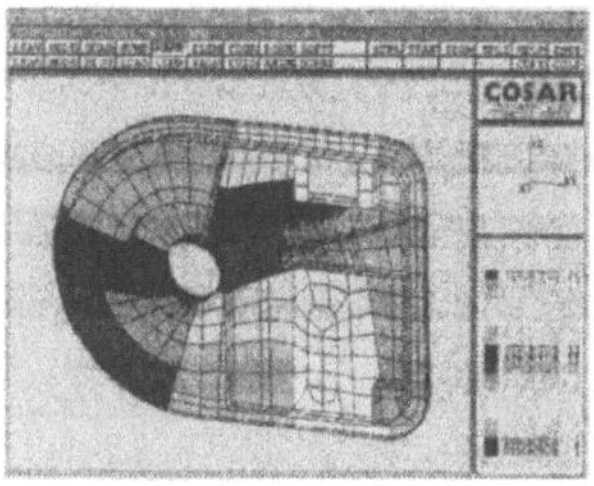

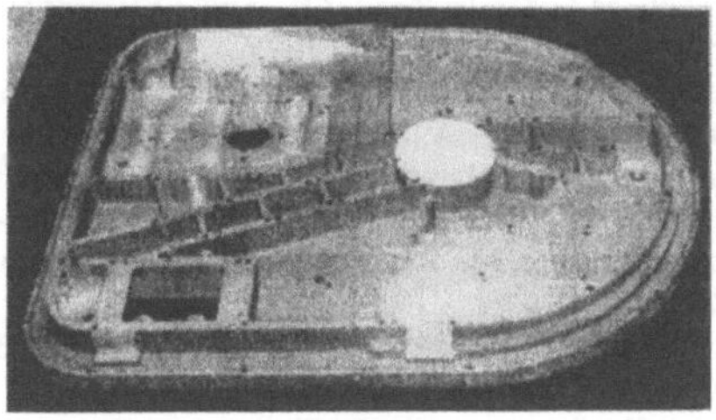

Figure 5. Base plate

The FE-model of the base plate, shown in figure 5, is supposed to illustrate the influence of the estimation method on the model updating process. The structure is modelled with 706 semiloof shell elements using superelement technique with the FE system COSAR and its pre-processor COSMESH.

TABLE 2. Eigenfrequencies of the plate using different estimators

	1.	2.	3.	4.	5.	6.
Measured	223.5	396.3	590.7	668.9	750.9	1094.6
Initial	208.4	364.0	534.6	633.6	730.4	1031.6
EWLS (mass update, 7 It.)	212.0	375.2	566.2	676.0	752.1	1094.0
EWLS (stiffness update, 7 It.)	221.3	370.6	552.8	644.6	734.3	1037.3
ECHHK (mass update, 7 It.)	211.9	380.0	566.9	675.6	753.9	1093.5
MVE (mass update, 7 It.)	211.6	377.0	564.4	675.4	747.5	1085.1

6. Conclusions

Convergence behaviour of EWLS, MVE, and ECHHK depends heavily on the chosen *a priori* weighting or co-variance matrices. We found no significant difference in terms of convergence speed between EWLS, MVE, and ECHHK. The choice of updating parameters can influence convergence behaviour tremendously. In order to assess the potential advantages of a smart structure, the entire structure has to be modelled and simulated, and the FE Model must be augmented, e.g. with electrical parts. A better statistical description for spatial distributed uncertainties can be obtained for simulations and model updating procedures. Although, the procedure for the spatial parameter distribution has been illustrated for thickness distribution, the idea can readily be employed to other parameter uncertainties of statistical correlated distributed nature. Difficulties could arise in the determination of acceptable measurements in a representative area.

7. References

1. Zehn, M.W., Martin, O., Offinger R.: Influence of Parameter Estimation Procedures on the Updating Process of Large Finite Element Models, 2nd International Conference on *"Identification in Engineering Systems"* (1999), University of Wales Swansea, pp. 240 –250.
2. Zehn, M.W., Schmidt,G.: FE-Modellierungen und Modellverbesserungen von CFK-Laminaten miteingebetteten Piezokeramiken auf der Grundlage experimenteller Untersuchungen, 4. Magdeburger Maschinenbautage *"Entwicklungsmethoden und Entwicklungsprozesse im Maschinenbau"* (1999), Magdeburg, pp. 361-368.

3. Friswell, M.I., Mottershead, J.E.: *Finite Element Model Updating in Structural Dynamics*, Kluwer Academic Publishers, Dordrecht, 1995.
4. Cressie, N.A.C.: *Statistics for spatial data*, John Wiley & Sons, New York, 1993.
5. Zehn, M.W., Saitov,A.: Determination of spatially distributed probability density functions for parameter estimation in modal updating procedures, Proceedings *"ISMA25 - 2000 International Conference on Noise and Vibration Engineering"*, P.Sas (ed.), 13.-15. September 2000, Leuven/Belgium, pp. 155-162.
6. Fox, R., Kapoor, M.: Rates of Change of Eigenvalues and Eigenvectors, *AIAA Journal* 6(1968), pp. 2426-2429.
7. Lim,K., Junkins, J., Wang, B.: Re-examination of Eigenvector Derivatives, *AIAA Journal of Guidance, Control and Dynamics*, 10(1987), pp. 581-587.
8. Natke, H.: *Einführung in die Theorie und Praxis der Zeitreihen- und Modalanalyse*, Vieweg-Verlag, Braunschweig, Wiesbaden, 1992.
9. Nelson, R.B.: Simplified Calculation of eigenvector derivatives, *AIAA Journal*, 9(1976), pp. 1201-1205.

BENDING ANALYSIS OF PIEZOELECTRIC LAMINATES

M.H. ZHAO, C.F. QIAN, S. W. R. LEE, P. TONG, T.Y. ZHANG *

Department of Mechanical Engineering,
Hong Kong University of Science and Technology,
Clear Water Bay, Kowloon, Hong Kong, China

1. Introduction

Due to excellent characteristics of piezoelectricity, actuators and sensors made of piezoelectric materials are widely used in smart structures and systems. One kind of smart structures and systems is of multilayer plates which take advantages of the integrated properties of each layer [1,2]. Therefore, analysis of a piezoelectric laminated plate is of theoretical significance and engineering importance and thus attracts many researchers [3-8].

Though there are a number of two-dimensional plate bending models for piezoelectric/piezoelectric laminated plates [3-8], it is sometimes difficulty to obtain exactly analytic solutions. Therefore, developing a simplified model is still desirable, wherein the Infinite-Plane-Capacitor (IPC) assumption is frequently used for piezoelectric layers whose two surfaces are fully electroded [3, 8]. The IPC assumption means that the electric field strength inside a piezoelectric lamina under a given applied voltage is constant. Thus, it reduces problems for piezoelectric laminated plates to the ones for conventional plates and hence substantially simplifies the solution procedure. The simplified solutions based on the IPC assumption predict, to a large extend, the flexural-extensional vibration behavior and the stress field induced by the applied electric voltage [3, 8]. However, the IPC assumption may introduce significant errors for stresses in the piezoelectric layer when mechanical loads are applied. Clearly, stress analysis is essential in designs with consideration of reliability. That is the motivation for us to develop a simple model to improve the IPC model for piezoelectric laminated plates. This note briefly introduces the new model and the detailed derivation and discussion will be published soon [9].

2. Analysis

2.1 DISPLACEMENTS AND STRAINS IN A LAMINATED PLATE

* Corresponding author, E-mail: mezhangt@ust.hk, Tel. (852) 2358-7192, Fax: (852) 2358-1543.

U. Gabbert and H.S. Tzou (eds.), IUTAM Symposium on Smart Structures and Structronic Systems, 275–282.

Figure 1 shows schematically a general (N+M)-layered piezoelectric laminated plate with M piezoelectric laminae and N non-piezoelectric laminae. The piezoelectric laminae are transversely isotropic, with all the planes of isotropy being parallel with the surfaces of the laminae, and their poling direction is along the thickness direction. The two surfaces of each PZT lamina are fully electroded. The Cartesian coordinate system

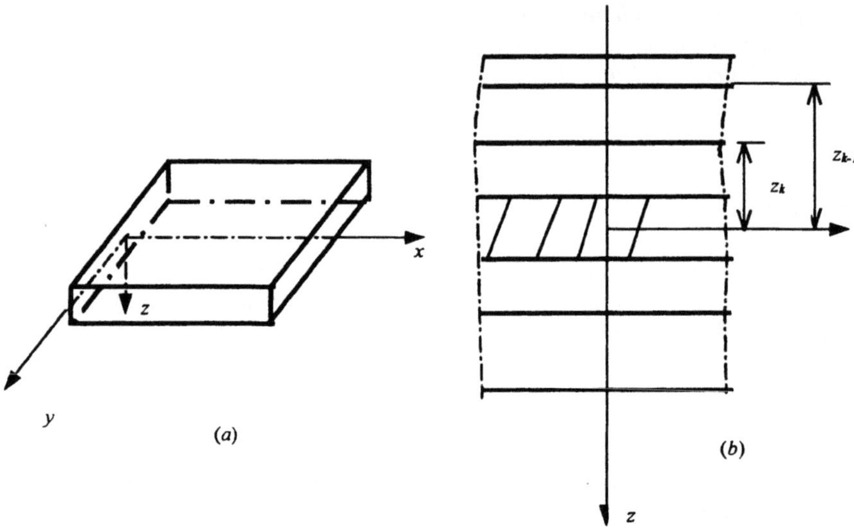

Figure 1. (N+M)-layered piezoelectric laminated plate of N non-piezoelectric laminae and M piezoelectric lamina

is set up such that the z-axis is along the thickness direction of the plate. We use z_k and z_{k-1} to denote the two surfaces of the kth layer, as shown in Figure 1.

In this paper, the following assumptions are made:
(1) Perfect bonding at all interfaces;
(2) A zero value of the stress component σ_z ;
(3) Validity of the Kirchhoff hypothesis of normal-remain-normal;
(4) No in-plane electric components of E_x and E_y; and
(5) No electric fields in the non-piezoelectric laminae.

According to the Kirchhoff hypothesis, the displacements of the plate can be expressed as follows [10]

$$u(x,y,z) = u^0(x,y) - z\frac{\partial w}{\partial x},$$
$$v(x,y,z) = v^0(x,y) - z\frac{\partial w}{\partial y}, \tag{1}$$
$$w(x,y,z) = w(x,y),$$

where $u^0(x,y)$ and $v^0(x,y)$ are the in-plane displacements and $w(x,y)$ is the out-plane deflection of the middle-surface. From equation (1), we have the strain fields

$$\varepsilon_x = \varepsilon_x^0 + zk_x, \quad \varepsilon_y = \varepsilon_y^0 + zk_y, \quad \gamma_{xy} = \gamma_{xy}^0 + zk_{xy}, \quad \gamma_{xz} = 0, \quad \gamma_{yz} = 0, \tag{2a}$$

where

$$\begin{aligned} &\varepsilon_x^0 = \frac{\partial u^0}{\partial x}, \quad \varepsilon_y^0 = \frac{\partial v^0}{\partial y}, \quad \gamma_{xy}^0 = \frac{\partial u^0}{\partial y} + \frac{\partial v^0}{\partial x}, \\ &k_x = -\frac{\partial^2 w}{\partial x^2}, \quad k_y = -\frac{\partial^2 w}{\partial y^2}, \quad k_{xy} = -2\frac{\partial^2 w}{\partial x \partial y}. \end{aligned} \tag{2b}$$

2.2 LAMINA CONSTITUTIVE RELATIONSHIP

Following assumption (2) and from the constitute equations for a PZT medium, we have the reduced constitutive equations [5]

$$\begin{bmatrix} \sigma_x \\ \sigma_y \\ \tau_{xy} \end{bmatrix} = \begin{bmatrix} Q_{11}^p & Q_{12}^p & 0 \\ Q_{12}^p & Q_{11}^p & 0 \\ 0 & 0 & Q_{66}^p \end{bmatrix} \begin{bmatrix} \varepsilon_x \\ \varepsilon_y \\ \gamma_{xy} \end{bmatrix} - \begin{bmatrix} 0 & 0 & e_{31}^p \\ 0 & 0 & e_{31}^p \\ 0 & 0 & 0 \end{bmatrix} \begin{bmatrix} E_x \\ E_y \\ E_z \end{bmatrix} \equiv \left[Q^p\right] \begin{bmatrix} \varepsilon_x \\ \varepsilon_y \\ \gamma_{xy} \end{bmatrix} - \left[e^p\right] \begin{bmatrix} E_x \\ E_y \\ E_z \end{bmatrix}, \tag{3}$$

$$\begin{bmatrix} D_x \\ D_y \\ D_z \end{bmatrix} = \left[e^p\right]^T \begin{bmatrix} \varepsilon_x \\ \varepsilon_y \\ \gamma_{xy} \end{bmatrix} + \begin{bmatrix} \varepsilon_{11} & 0 & 0 \\ 0 & \varepsilon_{11} & 0 \\ 0 & 0 & \varepsilon_{33}^p \end{bmatrix} \begin{bmatrix} E_x \\ E_y \\ E_z \end{bmatrix}, \tag{4}$$

where the reduced constants are given by the elastic, piezoelectric and dielectric constants

$$Q_{ij}^p = c_{ij}^p - \frac{c_{i3}^p c_{j3}^p}{c_{33}^p}, \quad i,j = 1,2,6, \quad e_{31}^p = e_{31} - \frac{c_{13}^p}{c_{33}^p} e_{33}, \quad \varepsilon_{33}^p = \varepsilon_{33} + \frac{e_{33}^2}{c_{33}^p}. \tag{5}$$

The IPC hypothesis assumes that the electric field is the same as that in an infinite plane capacitor, which results in the uniform electric field in the PZT layer

$$E_x = 0, \quad E_y = 0, \quad E_z = E_z^0 = -\frac{V}{t_p}, \tag{6}$$

where E_z^0 is the apparent electric field strength, and V and t_p are the applied electric voltage between the two electrodes and the thickness of the PZT layer, respectively. Due to the piezoelectric effect, an external mechanical load, in addition to the loading of electric voltage, induces an electric field, which is generally nonuniform. In return, the electric field induced by both electric and mechanical loading will produce a mechanical field, which will be essential and important to reliability designs. To account for this coupling effect, we introduce assumption (4) [11] to replace the IPC assumption and accordingly derive the electric field from equation (4). Using the dielectric governing equation

$$D_{i,i} = 0, \tag{7}$$

and assumption (4), we have

$$e_{31}^{p}(k_x + k_y) - \varepsilon_{33}^{p}\frac{\partial^2 \varphi}{\partial z^2} = 0. \tag{8}$$

The general solution of equation (8) is

$$\varphi = \frac{e_{31}^{p}}{\varepsilon_{33}^{p}}\frac{z^2}{2}(k_x + k_y) + z\varphi_1 + \varphi_2, \tag{9}$$

where φ_1 and φ_2 are two undetermined functions of (x, y). If an electric potential voltage V is applied between the two electrodes, φ_1 and φ_2 can be determined and equation (9) has the form of

$$\varphi = -E_z^0 z + \frac{1}{2}\frac{e_{31}^{p}}{\varepsilon_{33}^{p}}(k_x + k_y)(z - z_u)(z - z_l), \tag{10}$$

where z_u and z_l are the z-coordinates of the two surfaces of the corresponding piezoelectric lamina. Note that, similar results were obtained in piezoelectric beam [12] and thin shell analyses [5]. Consequently, differentiating equation (10) with respect to z leads to the electric field strength

$$E_z = E_z^0 - \frac{e_{31}^{p}}{\varepsilon_{33}^{p}}(k_x + k_y)(z - z_0), \tag{11}$$

where $z_0 = (z_u + z_l)/2$. As indicated in equation (11), the electric field varies linearly through the thickness of the PZT lamina. Substituting equation (11) into the constitutive equations, i.e., equations (3) and (4) and considering assumption (4), we have the modified reduced constitutive equations for the piezoelectric lamina

$$\begin{bmatrix} \sigma_x \\ \sigma_y \\ \tau_{xy} \end{bmatrix} = \left[Q^p\right]\begin{bmatrix} \varepsilon_x^0 \\ \varepsilon_y^0 \\ \gamma_{xy}^0 \end{bmatrix} + \left(z\left[Q^p\right] + (z - z_0)\left[Q^{pp}\right]\right)\begin{bmatrix} k_x \\ k_y \\ k_{zy} \end{bmatrix} - \begin{bmatrix} e_{31}^{p}E_z^0 \\ e_{31}^{p}E_z^0 \\ 0 \end{bmatrix}, \tag{12}$$

$$D_z = \varepsilon_{33}^{p}E_z^0 + e_{31}^{p}(\varepsilon_x^0 + \varepsilon_y^0) + e_{31}^{p}(k_x + k_y), \tag{13}$$

where

$$\left[Q^{pp}\right] = \frac{(e_{31}^{p})^2}{\varepsilon_{33}^{p}}\begin{bmatrix} 1 & 1 & 0 \\ 1 & 1 & 0 \\ 0 & 0 & 0 \end{bmatrix} \tag{14}$$

is the modified reduced stiffness matrix. As can be seen in equations (12) and (13), the variables of strains and electric field strengths have been specifically expressed in terms of the middle-surface strains, the curvatures and the apparent electric field strength. Under a given applied electric voltage, the apparent electric field strength is a constant for a fixed PZT lamina thickness. In the right hand side of equation (12), the first term gives the stress field due to the middle-surface stretching, the second term is the bending stress, and the third term represents the reverse piezoelectric effect. The bending stresses are related not only to the reduced stiffness matrix, but also to the modified reduced stiffness matrix, i.e., equation (14), which will change the bending stresses significantly. Dropping out the modified reduced stiffness matrix $[Q^{pp}]$ from equation (12) reduces equation (12) to the constitutive equation based on the IPC assumption.

For each of the non-piezoelectric lamina, we assume them to be orthotropic and hence have the reduced constitute equation [10]

$$\begin{bmatrix} \sigma_x \\ \sigma_y \\ \tau_{xy} \end{bmatrix} = \begin{bmatrix} Q_{11}^e & Q_{12}^e & Q_{16}^e \\ Q_{12}^e & Q_{22}^e & Q_{26}^e \\ Q_{16}^e & Q_{26}^e & Q_{66}^e \end{bmatrix} \begin{bmatrix} \varepsilon_x \\ \varepsilon_y \\ \gamma_{xy} \end{bmatrix} \equiv \left[Q^e\right] \begin{bmatrix} \varepsilon_x \\ \varepsilon_y \\ \gamma_{xy} \end{bmatrix}. \tag{15}$$

2.3 LAMINATE CONSTITUTIVE EQUATIONS

Substituting equations (12) and (15) into the expressions of the stress resultants and moment in a plate [10] yields the constitute equations of the piezoelectric laminated plate

$$\begin{bmatrix} N_x \\ N_y \\ N_{xy} \\ M_x \\ M_y \\ M_{xy} \end{bmatrix} = \begin{bmatrix} A & B \\ C & D \end{bmatrix} \begin{bmatrix} \varepsilon_x^0 \\ \varepsilon_y^0 \\ \varepsilon_{xy}^0 \\ k_x \\ k_y \\ k_{xy} \end{bmatrix} - \begin{bmatrix} N_{x0} \\ N_{y0} \\ N_{xy0} \\ M_{x0} \\ M_{y0} \\ M_{xy0} \end{bmatrix}, \tag{16}$$

where the elements of the matrices are

$$\begin{aligned} A_{ij} &= \sum_{n=1}^{N}\left(Q_{ij}^g\right)_n (z_n - z_{n-1}) + \sum_{m=1}^{M}\left(Q_{ij}^P\right)_m (z_m - z_{m-1}), \\ B_{ij} &= \frac{1}{2}\sum_{n=1}^{N}\left(Q_{ij}^g\right)_n (z_n^2 - z_{n-1}^2) + \frac{1}{2}\sum_{m=1}^{M}\left(Q_{ij}^p\right)_m (z_m^2 - z_{m-1}^2), \\ C_{ij} &= B_{ij}, \qquad\qquad i, j = 1, 2, 6 \\ D_{ij} &= \frac{1}{3}\sum_{n=1}^{N}\left(Q_{ij}^g\right)_n (z_n^3 - z_{n-1}^3) \\ &\quad + \frac{1}{3}\sum_{m=1}^{M}\left[\left(Q_{ij}^p\right)_m (z_m^3 - z_{m-1}^3) + \frac{1}{4}\left(Q_{ij}^{pp}\right)_m (z_m - z_{m-1})^3\right] \end{aligned} \tag{17}$$

$$N_{x0} = \sum_{m=1}^{M}(e_{31}E_z^0)_m(z_m - z_{m-1}),\ N_{y0} = N_{x0},\ N_{xy0} = 0,$$
$$M_{x0} = \frac{1}{2}\sum_{m=1}^{M}(e_{31}E_z^0)_m(z_m^2 - z_{m-1}^2),\ M_{y0} = M_{x0},\ M_{xy0} = 0. \quad (18)$$

In equations (16) and (17), the A_{ij} are extensional stiffnesses, the B_{ij} or C_{ij} are bending-extension coupling stiffnesses, and the D_{ij} are bending stiffnesses.

Equations (17) and (18) show that the stiffnesses A_{ij} and B_{ij} in the present model are identical to those in the IPC model, but the bending stiffness D_{ij} are different in the two models due to the modified reduced stiffness $[Q^{pp}]$.

2.4 GOVERNING EQUATIONS OF LAMINATED PIEZOELECTRIC PLATE

Taking the displacements and the deflection as the independent variables, and using the principle of virtual work [5], we obtained the governing equations for piezoelectric laminated plate, which have the same forms as those governing equations for conventional composite laminated plates [10], and can be solved by using the available existed methods including the approximate methods, e.g., the reduced bending stiffness method [10].

3. Examples

We consider a rectangular symmetric sandwich plate of Al/PZT-5H/Al of $a \times b$ with a=0.078 m, where a and b are the plate length (dimension in the x-direction) and width (dimension in the y-direction), respectively. The simply supported boundary conditions are applied to the four edges of the plate. The thickness of the top or bottom Al lamina is t=1.00×10^{-3} m and the thickness of PZT core is t_p=0.87×10^{-3} m [9]. The material

TABLE 1. Material properties of Al and PZT laminae
(E and G in units of GPa, e in C/m^2 and ε in 10^{-9} F/m)

	E_1	E_2	E_3	G_{12}	G_{13}	ν	ε_{11}	ε_{33}	e_{33}	$-e_{31}$
Al	70.3	70.3	70.3	26.1	26.1	0.34	--	--	--	--
PZT	61	61	48	23.3	19.1	0.31	6.0	6.0	21.319	14.645

properties are listed in TABLE 1. The middle-surface of the PZT lamina is in the oxy plane. In this case, the bending-extension coupling stiffnesses B_{ij} and the stiffness components D_{16} and D_{26} are all equal to zero, which leads $u^0 = v^0 = 0$. The deflection is to be determined.

If the applied electric potential drop V=0 and the applied transverse load is uniformly distributed along the y-axis at x=$a/2$, i.e.,

$$q(x,y)=\delta(x-\frac{a}{2})q_0, \tag{19}$$

where δ is the Dirac Delta function and $q_0=1.0\times10^{-3}$ N/m, we solve the problem by the Levy's method [10]. The normalized maximum bending stress in the PZT layer at the plate center is given by

$$\beta=\sigma_x\Big/\frac{q_0 a}{t_p^2}. \tag{20}$$

Figures 2 shows β as a function of the aspect ratio b/a. The corresponding results

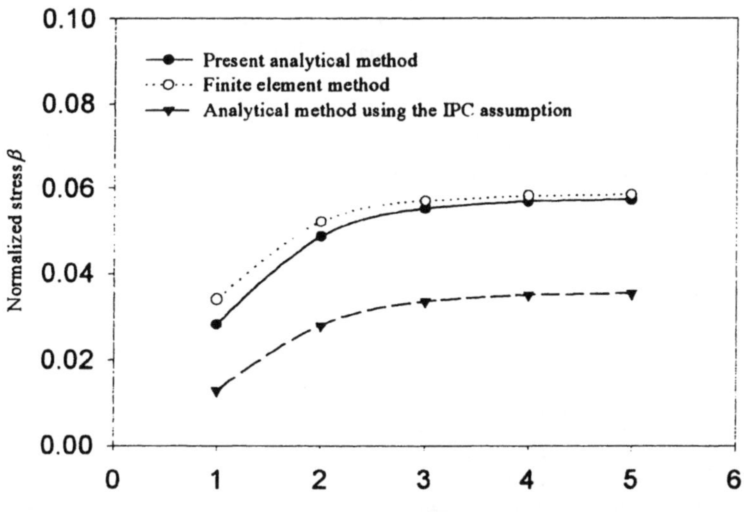

Figure 2. Normalized bending stress σ_x at the plate center

based on the IPC assumption are also shown in Figure 2 for comparison. The value of the stress at the PZT core calculated from the present model could be twice larger than that calculated from the model based on the IPC assumption, as shown in Figure 2. Finite element analysis (FEA) with the commercial software ABAQUS is resorted to verify the proposed method. 3D with 20-node solid elements are employed and there are 2 elements through the thickness of each layer of the lamina. The mesh was refined to check the accuracy and the final one comprises 1200 elements. Figure 2 indicates that the FEA results agree well with those by the present analytical method.

4. Concluding remarks

Neglecting the electric field components E_x and E_y, we derive the electric potential and expressed it in terms of the deflection in the bending piezoelectric lamina. In contrast to a constant electric field strength inside the piezoelectric lamina assumed in the IPC model, the present model yields a linear electric field strength with the piezoelectric lamina thickness. This electric field can be easily incorporated in the reduced constitutive equation of the PZT lamina. As a result, the bending stiffness of the PZT lamina contains the piezoelectric and dielectric parameters. The forms of the governing equation of the general piezoelectric laminated plates, derived by the principle of virtual work in terms of the in-plane displacements and the deflection, are identical to those for conventional composite laminated plates. Therefore, the available solution method for conventional composite laminated plates can be used for the piezoelectric laminated plates. The results of rectangular sandwich plate plates, Al/PZT/Al, with different aspect ratios of width to length, show that the linear electric field strength in the thickness direction has a great influence on the stress distribution and must be taken into account in the structural analysis of piezoelectric laminated plates.

Acknowledgement -- The work was fully supported by an RGC grant (HKUST6050/97E) from the Research Grants Council of the Hong Kong Special Administrative Region, China.

Reference

1. Tzou, H. S., Fukuda, T.: *Precision sensors, Actuators, and System*, Kluwer Academic Publishers, Dordrechet, 1992.
2. Lee, S. W. R., Li, H. L.: Development and characterization of a rotary motor driven by anisotropic piezoelectric composite laminate, *Smart Mater. Struct.*7 (1998), 327- 336.
3. Adelman, N. T., Stavsky, Y.: Flexural-extensional behavior of composite piezoelectric circular plates, *J. Acoust. Soc. Am.* **67** (1980), 819-822.
4. Mindlin, R. D.: Frequence of piezoelectrically forced vibrations of electroded, doubly rotated quartz plates, *Int. J. Solids Struct.* **20** (1984), 141-157.
5. Parton, V. Z., Kudryavtsev, B. A.: *Electromagnetoelasticity*, Gordon and Breach Science Publishers, New York, 1988
6. Mitchell, J. A., Reddy, J. N.: A refined hybrid plate theory for composite laminates with piezoelectric laminae, *Int. J. Solids Struct.* **32**(1995), 2345-2367.
7. Heyliger, P.: Exact solution for simply supported laminated piezoelectric plates, *J. Apll. Mech.* **64** (1997), 299-306.
8. Zhang, X. D., Sun, C. T.: Analysis of a sandwich plate containing a piezoelectric core, *Smart Mater. Struct.* **8** (1999), 31-40.
9. Zhao M. H., Qian C. F., Lee, S. W. R., Tong, P, and Zhang, T. Y.: Electro-elastic analysis of piezoelectric laminated plates, submitted to *Int. J. Solids Struct.*.
10. Jones, R. M.: *Mechanics of composite materials*, Taylor & Francis, Pennsylvania, 1999.
11. Tzou, H. S., Bao, Y.: A theory on anisotropic piezothermoelastic shell laminates with sensor/actuator applications, *J. Vib. Sound* **184** (1995), 453-473
12. Michael, K., Hans, I.: On the influence of the electric field on free transverse vibrations of smart beams, *Smart Mater. Struct.* **8** (1999), 401-410.

BUCKLING OF CURVED COLUMN AND TWINNING DEFORMATION EFFECT

Yuta URUSHIYAMA and David LEWINNEK
Honda R&D Co., Ltd.
4630 Shimotakanezawa, Haga-gun, Tochigi, Japan

Jinhao QIU and Junji TANI
Institute of Fluid Science, Tohoku University
2-1-1Katahira, Aoba-ku, Sendai 980-8577, Japan

1. Introduction

Intelligent materials and structures can function autonomously in response to varying environmental and operating conditions. This capability has been used to improve the vibration characteristics, increase the durability, lower the weight, and improve other performance criteria of structures, e.g.[1].

One way to achieve these improvements in structural systems is the application of shape memory alloys (hereafter abbreviated to SMA). Ti-Ni alloys have many practical applications including cell phone antennas, which utilize the superelasticity of SMA, and mixing valves and other valves, which utilize the shape memory effect, e.g.[2]. In addition to the features which are currently applied in commercial products, SMAs also have an unusual stress-strain property resulting from twinning deformation in the martensitic phase.

Buckling is crucially important in structural design. Although a great deal of research has been done on buckling of traditional materials, hardly any research can be found on the buckling characteristics of shape memory alloys.

In the process of their research, the first authors found that curved SMA columns trended to straighten before buckling when subjected to axially compressive load. The buckling load of the curved SMA columns was close to the buckling load of straight columns. This contradicts the popularly accepted results for conventional materials that the curvature of a column tends to increase under axially compressive load, and curved columns bear lower loads than straight columns. The unique buckling characteristic of SMA columns is attributed to the unique stress-strain diagram, resulting from the twinning deformation effect (hereafter abbreviated to "TD-effect"). Since straight columns can support larger axially compressive loads and usually resist buckling better than curved columns, the straightening behavior of curved SMA columns is called self shape optimization in this paper.

This paper first describes the conditions and results of self shape optimization experiments of SMA columns. Next a parameter called 'spring constant' is introduced to measure the ability of columns to resist buckling under axial compression, and the variation of the spring constant of a SMA column with the axial load is calculated. Finally, numerical analysis on the mechanism of anti-buckling effect of SMA columns

U. Gabbert and H.S. Tzou (eds.), IUTAM Symposium on Smart Structures and Structronic Systems, 283–290.

caused by the self shape optimization is presented with the unique stress-strain curve of SMA taken into consideration.

2. Experiments on Compressive Buckling

The experimental conditions and results for the straightening phenomenon of curved SMA columns (or as called self shape optimization) are described in this chapter.

2.1 CHARACTERISTICS OF SAMPLE SMA MATERIAL

$Ti\text{-}Ni_{40.8}\text{-}Cu_{9.9}$ SMA was used as test piece material in this study. The four temperatures for phase transformation are shown in Table 1. Since the A_s point of the material is 51.9℃, the alloy is in martensitic phase at room temperature.

The original shape of the test piece is a round column with a diameter of 5 mm and a length of 50 mm and they are deformed in the steps as shown in Figure1 to obtain curved columns. First the test pieces are compressed by 5.5% (20kN) under axial loading and then unloaded. The residual strain is about 3.5%. Second, the columns are bent so that the shape of the columns upon unloading is a full cosine wave from 0 to 2π with a peak-to-peak amplitude of 1.9 mm as shown in Figure 1.

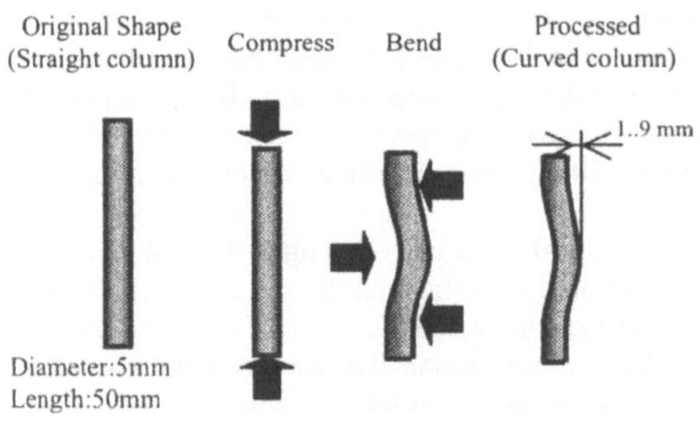

Figure 1. SMA column shapes

TABLE 1. Transformation temperatures of $Ti\text{-}Ni_{40.8}\text{-}Cu_{9.9}$

Transformation	Temperature (°C)
A_s	51.9
A_f	64.9
M_s	44.6
M_f	30.0

2.2 SELF SHAPE OPTIMIZATION AND BUCKLING LOAD

The experimental results of compressive tests of the curved SMA columns are presented in this section. In the compressive tests, a column was placed between the upper and lower fixtur es of a test machine and compressed by the downward movement of the upper fixture at a constant speed. The change of shape of the column was photographed successively and the load-deformation relationship was recorded.

Figure 2 shows the shape of a column for each 0.5 mm increment of the stroke in succession. From this figure, it can be seen that the curvature of the column begins to decrease when the stroke exceeds 1mm, and the curvature approaches zero when the stroke is 3mm. The straightening behavior of SMA columns under compressive load is completely different from the behavior of conventional engineering materials, where the curvature of columns monotonically increases with the axially compressive load and curved columns buckle at lower critical loads than straight columns.

The curved SMA column test pieces eventually buckle when the displacement of the upper fixture reaches about 3.2 mm, immediately after having fully straightened out. Figure 3 shows the force-displacement relationship of curved SMA columns compared to that of a straight SMA column. The critical buckling load of the curved SMA column is almost the same as that of the straight column, due to the self shape optimization effect of the curved column before buckling, though the deformation of curved column is smaller than that of the straight one at the buckling point.

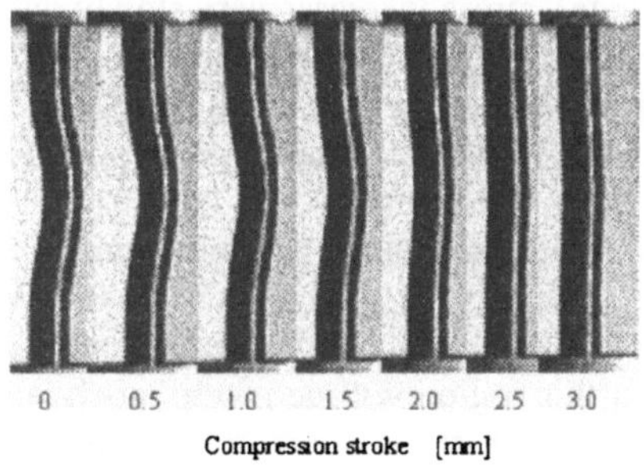

Figure 2. Changing column shape by compression

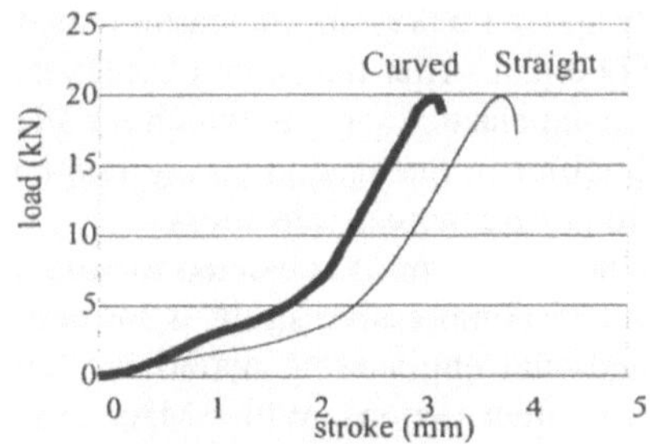

Compression velocity: 1mm/min
Material:Ti-Ni-Cu_{10}
Column size: D=ϕ 5 L=50mm

Figure 3. Stroke-load of SMA columns

Since the temperature of the test piece is affected by strain variation and the strain changes for a wide range in this buckling test, the temperature rise is not negligible. When a material is deformed, the temperature tends to rise. Hence, the temperature was measured during compression. Figure 4 shows the results of the temperature measurements. When the straightening behavior occurred, the temperature of the curved SMA columns rose by 3.5℃, and the highest temperature among the five measuring points was 25.6℃. This temperature is much lower than the A_s point of the material

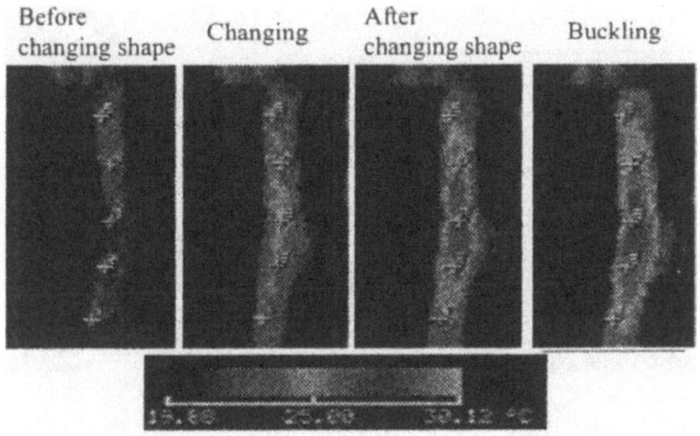

Figure 4. Temperature of curved SMA columns

(51.9℃), even without taking into account the stress-induced rise of the phase transformation temperature (e.g. [3][4]). It can be asserted that the straightening behavior of the curved SMA columns occurs in the martensitic phase.

3. Numerical Analysis

The high buckling-resisting capacity of SMA column can be attributed to the unique stress-strain property of SMAs in martensitic phase. As shown in Figure 5, SMAs have a very low yield stress (about 50MPa) and the elastic range is very small at the beginning of compression. The yield is due to the TD-effect of SMA. When the strain is below about 3.5% strain (hereafter called TD-range), stress increases very slowly due to the TD-effect. After the strain exceeds the TD-range, the stress-strain relationship enters the second elastic range, in which the stress-strain relationship is linear. Since SMAs are much stiffer in the second elastic range than in the TD-range, small increases in strain lead to drastic increases in stress.

When a column is subjected to both a compressive force and a bending moment, the maximum compressive strain is located on the inner side of the curved column. With conventional engineering materials, the stress in this region will be greater than the stress in other regions. With SMAs, the stress in this region will be much, much larger than the stress in other regions due to the non-linear stress-strain relationship. The unique stress distribution makes curved SMA columns more resistive to buckling. In this chapter, the distribution of stress and strain in the SMA column is analysed numerically and the buckling-resisting mechanism of SMA is illustrated.

3.1 NUMERICAL CALCULATION OF INTERNAL STRESS AND STRAIN

A simple numerical method is used to analyze stress and strain in the columns. As discussed above, the axis of the test piece after initial compression and bending is a full cosine wave, which is a planar curve. Hence, the stress and strain distributions in the columns after the initial deformation can be regarded as two-dimensional, with uniform

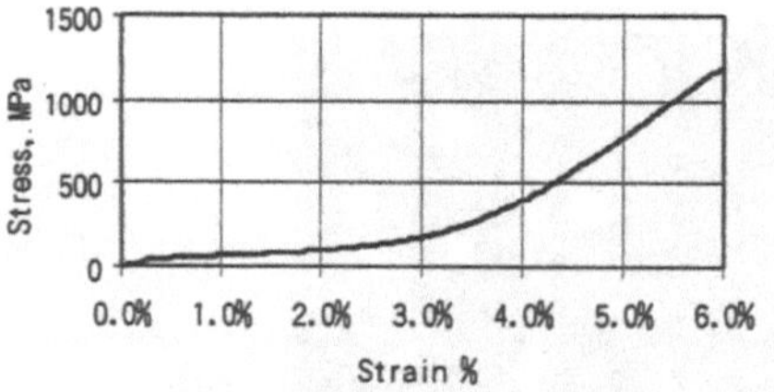

Figure 5. Stress-strain curve of SMAs

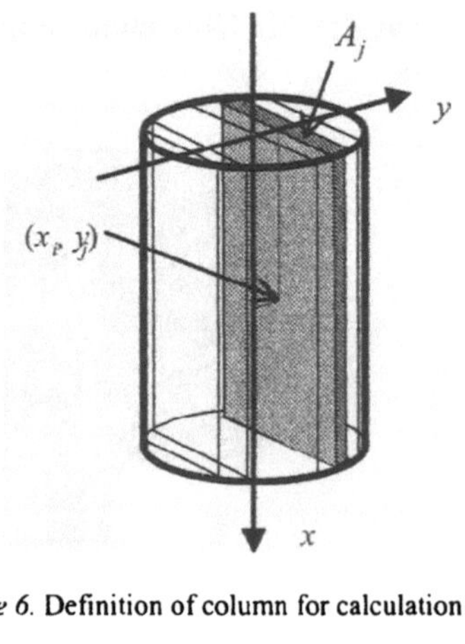

Figure 6. Definition of column for calculation

distribution in the plane perpendicular to the plane of curvature. Therefore, two coordinates x and y are used to express the position in the plane of the curve, with x lying on the axis of the column and y pointing to the inner side of the curve as shown in Figure 6, and the stress and strain can be expressed as function of x and y.

For simplicity, three assumptions were made in the analysis. First, the residual strain generated in the initial compression can be neglected. Second, for each known strain, the stress can be directly calculated from the stress-strain relationship shown in Figure 5, that is, the effect of hysteresis in the twin deformation can ignored. Third, the curvature of the initial bending does not vary with the axial load, that is, the axial compression can regarded as the superposition of axially compressive strain and the initial bending strain. In addition to these three assumptions, the boundary conditions for the two ends of the columns were assumed to be fixed-fixed.

The length and diameter of the test pieces are denoted by L and D, respectively, and lateral deflection of the columns axis is denoted by δ. For numerical calculation, the columns are divided into m elements in the x direction and n elements in the y direction. The coordinates of the nodes can be expressed in the following form:

$$x_i = \frac{L}{m} i \qquad (i = 0, 1, 2, \cdots, m, \cdots, 2m) \tag{1}$$

and

$$y_j = -\frac{D}{2} + \frac{D}{n} j \qquad (j = 0, 1, 2 \cdots n) \tag{2}$$

The curvature κ_i at x_i of the test pieces due to initial bending can be obtained by differentiating the cosine function used to express the shape of the columns as follows:

$$\kappa_i = \frac{d^2 h}{(dx)^2} = \frac{d^2}{(dx)^2}\left[\frac{1}{2}\delta - \frac{1}{2}\delta \times \cos\left(\frac{2\pi \cdot x}{L}\right)\right] = \frac{2\delta\pi^2}{L^2}\cos\left(\frac{2\pi \cdot x}{L}\right) \tag{3}$$

The strain ε^b_{ij} at (x_i, y_i) caused by the initial bending can be expressed in the following form:

$$\varepsilon^b_{ij} = \kappa_i y_j \tag{4}$$

Based on the third assumption, the strain at any point (x_i, y_i) in the test piece during the buckling test can be divided into two parts: the strain due to bending, which is expressed in Equation (6), and the strain generated by the axial force acting at the two ends. The latter is uniform throughout the cross section, that is, it is only a function of x, independent of y. Hence the strain at point (x_i, y_i) can be expressed in the following:

$$\varepsilon_{ij} = \varepsilon^b_{ij} + \Delta\varepsilon_i . \tag{5}$$

where $\Delta\varepsilon_i$ is the strain generated by the axially compressive force at the cross-section $x=x_i$. Because it is difficult to directly calculate $\Delta\varepsilon_i$,an iterative process is used.

Based on the assumption described at the beginning of this section, the stress σ_{ij} at (x_i, y_i) can be obtained from $\sigma_{ij} = \sigma(\varepsilon_{ij})$, where $\sigma(\varepsilon)$ is a function defined by the stress-strain relationship of the SMA shown in Figure 5. The resultant force F_i of the cross-section $x=x_i$ can easily be obtained using numerical integration of σ_{ij} over the whole section area. The force F_i at any cross section $x=x_i$ should equal the force F_0 acting on

TABLE 2. Buckling load of columns

Method	Type of column	Bucking load [kN]
Experiment	Curved Column	19.7
Experiment	Straight Column	19.9
Calculate	Straight Column	20.1

the two ends. The strain $\Delta\varepsilon_i$ can be obtained by solving F_i=F_0 using an iterative approach. After $\Delta\varepsilon_i$ is solved, the distribution of stress and strain is known.

3.2 COMPARISON OF BUCKLING LOADS

To judge the stability of the column under axial compression, a variable called spring constant is introduced. The spring constant is defined as the ratio $\Delta M/\Delta\delta$, where ΔM is the moment M generated by the column to resist buckling due to the increment $\Delta\delta$ of lateral deflection δ, which is small and virtually applied to the column. The column is stable when the spring constant is positive. When the spring constant becomes negative, the system will be unstable and will buckle. The spring constant varies axially throughout the column, but it can easily be found that midpoint of the column is the most unstable point and the stability of column depends on the spring constant of this cross-section. Since the curved columns have similar buckling properties to straight columns as shown in Figure 3, the stability analysis was carried out on straight columns instead of curved ones for simplicity.

After the stress distribution is solved using the method explained in the last section, spring constant can be calculated by

$$\frac{\Delta M}{\Delta\delta}=\frac{\Delta M_m-\Delta M_0}{\Delta\delta}=\frac{\sum_{j=1}^{n}\left[\left(\Delta\delta+y_j\right)\times A_j\times\Delta\sigma_{mj}\right]-\sum_{j=1}^{n}\left[y_j\times A_j\times\Delta\sigma_{0j}\right]}{\Delta\delta} \tag{6}$$

where M_m and M_0 are moments at the mid cross-section and the end of the column as shown in Figure 7, and A_j is the area of the j-th element in any cross-section.

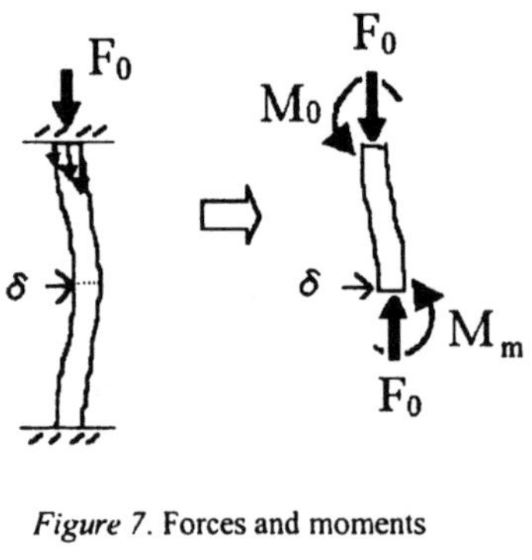

Figure 7. Forces and moments

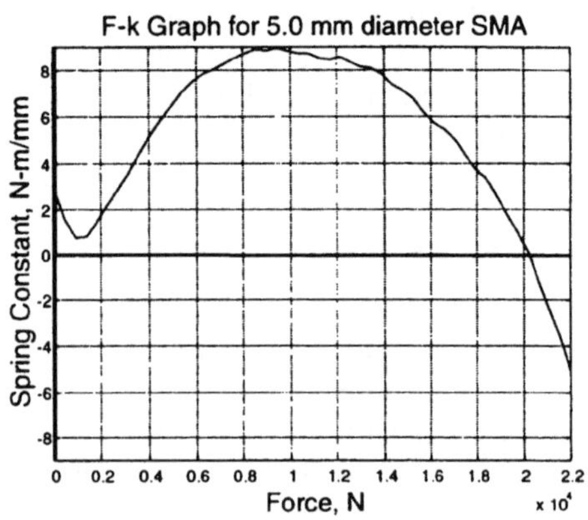

Figure 8. Spring constant vs. compressive force

In these calculations, the deflection $\Delta\delta$ was set to 0.1mm and the spring constant was solved for different axial loads. Figure 8 shows the calculation results of spring constant, which is positive up to 20.1 kN. There is a local minimum at an axial compression of about 1 kN, but the column does not buckle at this point since the spring constant is still positive. After the local minimum point, the spring constant increases, that is, the column becomes more stable, with increasing of the axial load due to the unique stress-strain relationship of SMA. The critical buckling load is about 20.1kN, where the spring constant changes its sign from positive to negative. A comparison of the numerical and experimental results of the critical buckling load is shown in Table 2.

3.3 ANALYSIS OF STRAIGHTENING BEHAVIOR

If an increment ΔF is added to axial load F, an increment of stress $\Delta\sigma$ and an increment of strain $\Delta\varepsilon$ are generated at every point in the column. Although $\Delta\sigma$ is uniform on each cross-section (or for each value of x_i), $\Delta\varepsilon$ is non-uniform. The resultant force of $\Delta\sigma$ equals ΔF and acts at the centroid of $\Delta\sigma$. The centroids of $\Delta\sigma$ in all the cross-sections constitute a continuous curve in the x-y plane, which is called the centroidal curve of stress in this paper. The centroidal curve of stress, which has a similar physical meaning to the concept of neutral fiber in a beam under pure bending, can be used to judge the direction of bending generated by the incremental force ΔF. Hence, if the centroidal curve of stress is solved for a given axial load F, we will be able to determine whether the curvature of the column will increase or decrease with increasing the axial load. The y coordinate of the centroidal curve of stress can be calculated from the following equation:

$$\bar{y}_i = \frac{\sum_{j=1}^{n} \sigma_{ij} \times A_j \times y_j}{\sum_{j=1}^{n} \sigma_{ij} \times A_j} = \frac{\sum_{j=1}^{n} E_{ij} \times A_j \times y_j}{\sum_{j=1}^{n} E_{ij} \times A_j} \tag{7}$$

where $E_{ij} = \Delta\sigma_{ij}/\Delta\varepsilon_i$ is the instantaneous Young's modulus at (x_i, y_j). After the distribution of stress and strain under a given axial load F has been solved, the

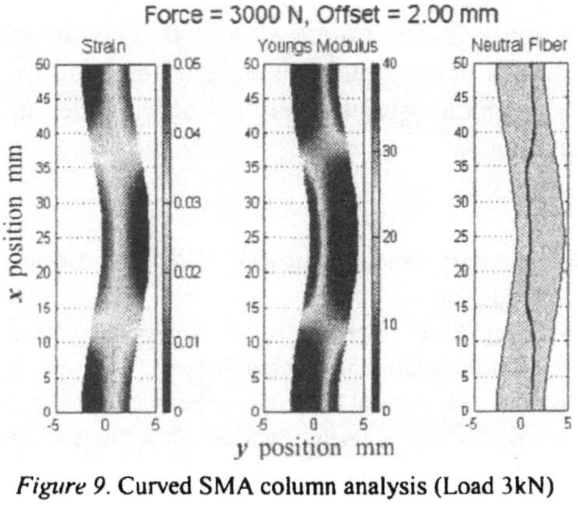

Figure 9. Curved SMA column analysis (Load 3kN)

distribution of E_{ij} can easily be obtained by differentiating the stress-strain curve $\sigma(\varepsilon)$.

From Figure 2, it can be seen that the straightening behavior is most noticable when the stroke is between 1.0 mm and 1.5 mm, which in turn corresponds to axial load between 3 kN and 4 kN as shown in Figure 3. Hence, the distributions of strain, Youngs modulus, and centroidal curve of stress were calculated for the axial loads of 3kN and 4kN. The results for 3kN are shown in Figure 9, in which the centroidal curve of stress is denoted by neutral fiber since it has a similar physical meaning to the neutral fiber in a beam under pure bending.

Figure 9 clearly shows that the centroidal curve of stress is concave to the right, while the geometric axis of the beam is bent concave to the left. This shape can be attributed to the biased distribution of Youngs modulus as shown in the same figure. Due to this shape of centroidal curve of stress, a further increase of the axial load will bend the column in the opposite direction from the initial curvature. In other words, the curved SMA column will be straightened by increasing axial load.

4. Conclusions

This research describes the straightening behavior (or so called self shape optimization behavior) of curved SMA columns under compression, which was found by the authors. Numerical analysis on the distribution of stress and strain was also performed to clarify the underlying mechanism of this behavior. From the results of experiments and numerical simulations, the following conclusions can be drawn.

1. The compression tests revealed that the curvature of initially curved SMA columns decreases under axial compression before eventually buckling. As a consequence, the critical buckling load of curved columns is virtually the same (99%) as that of straight columns. The results of temperature measurement show that straightening occurs in the Martensitic phase. The straightening behavior can be attributed to the unique stress-strain relationship of SMA, resulted from the TD-effect of the material.
2. A spring constant was introduced and numerically calculated to evaluate the stability of SMA columns under compressive load. The numerical result of the critical buckling load is in good agreement with experimentally measured ones.
3. The centroidal curve of stress, which has the similar meaning as neutral fiber of beams under pure bending, was defined and calculated numerically to exhibit the trend of bending under given compressive load. The numerical results show that the curved SMA columns under compression tend to bend in the direction opposite to initial curvature, which qualitatively explains the underling mechanism of straightening behavior.

5. References

(1) Tani, Furuya, Eda, Morishita, Nattori, Higuchi: "Intelligent Composite Materials and Intelligent Structures" (1996), Yokendo, Ltd..

(2) Committee for Developing Shape Memory Alloy Applications: "Shape Memory Alloys and Their Applications" (1987), Nikkan Kkogyo Shimbunsha, Ltd..

(3) Tanaka, Tobushi, Miyazaki: "Mechanical Properties of Shape Memory Alloys," (1993), Yokendo Ltd..

(4) Tae Hyun Nam, Toshio Saburi, Kenichi Shimizu: "Cu-Content Dependence of Shape Memory Characteristics in Ti-Ni-Cu Alloys," Material Transactions, JIM, Vol.31, No.11, 1990, pp.959-967.

ELECTRONIC CIRCUIT MODELING AND ANALYSIS OF DISTRIBUTED STRUCTRONIC SYSTEMS

H. S. TZOU, J. H. DING
Department of Mechanical Engineering, StrucTronics Lab
University of Kentucky, Lexington, KY 40506-0108 USA

1. Introduction

Smart structures and structronic (structure + electronic) systems are recognized as one of the essential technologies of the 21st century [1]. Conventional techniques used in modeling and analysis of structronic systems involves 1) theoretical analysis, 2) finite element analysis, and 3) laboratory experiments. This research is to investigate the fourth modeling and analysis technique based on the electrical analogy, i.e., using electronic circuits and components to model distributed structronic systems. A generic distributed structronic control system is shown in Figure 1.

Components of discrete mechanical systems and structures modeled by ordinary differential equations (ODE) can usually find their counterparts in electrical systems, and thus, its equivalent electronic circuits, e.g., analog computers, can emulate the system responses.

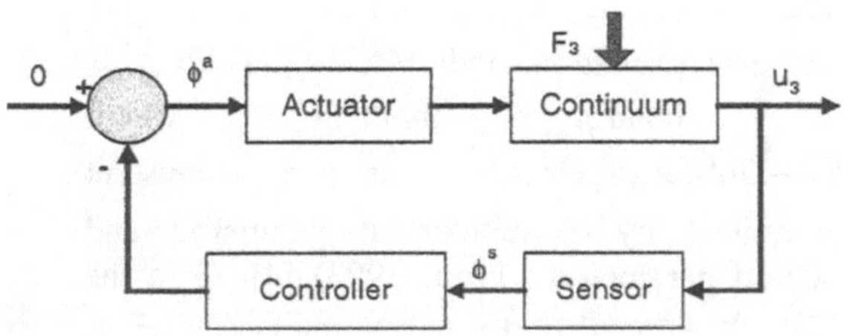

For elastic continua - distributed parameter systems (DPS) - modelled by partial differential equations (PDE), generic electronic components (i.e., resistors, capacitors, inductors, and transformers) can also be used to represent standard elastic components of discretized system PDE (Measurement, 1966) [2]. For multi-field distributed (parameter) structronic systems, Shah, et. al (1994) applied the finite element method (FEM) to discretize a piezoelectric laminated beam and then proposed to use very large scale integration (VLSI) chips to simulate the beam dynamic and control responses [3]. However, for complicated distributed structronic systems, their electronic analogies are not so easily derived. A new technique to improve electronic modeling and analysis of distributed structronic systems is reported in this paper. Circuit models (soft) and electronic circuits (hard) of the distributed structronic (beam/sensor/actuator/control) system are established to validate the new technique. Control effectiveness of the diaplcement and velocity feedbacks is evaluated. System responses (electrical signals)

U. Gabbert and H.S. Tzou (eds.), IUTAM Symposium on Smart Structures and Structronic Systems, 291–298.

of the equivalent circuit are compared with analytical, simulation and experimental (physical model) results.

2. Modeling of Distributed (Beam/Sensor/Actuator) Structronic System

Electronic modeling of a distributed structronic (structure/sensor/actuator/control) system based on the finite difference discretization is presented. Procedures of the electronic circuit design are demonstrated in this section.

2.1. PDE MODEL OF STRUCTRONIC BEAM SYSTEM

Consider a generic cantilever Euler-Bernoulli beam sandwiched between two thin piezoelectric layers serving as distributed sensor and actuator, respectively, Figure 1, i.e., a distributed cantilever beam structronic system. Sensing signal acquired from the distributed sensor is amplified and feedback to the distributed actuator actively counteracting the beam oscillation. An elastic Euler-Bernoulli beam structronic system is modeled by a fourth-order PDE [4]:

$$YI\frac{\partial^4 u_3}{\partial x^4} + \rho A\ddot{u}_3 - \frac{b\partial^2 (M_{11}^a)}{\partial x^2} = bF_3, \qquad (1)$$

where Y is Young's modulus; u_3 is the transverse displacement; ρ is the mass density; I $=bh^3/12$ is the area moment of inertia; A = bh is the cross-section area; b is the beam width; h is the beam thickness; F_3 is the transverse excitation; and $M_{11}^a = \tilde{\xi}[\partial u_3/\partial x]_0^L$ is the control moment where $\tilde{\xi} = -Gh^s r_1^a d_{31} Y_p r_1^s h_{31}/L$ is a constant determined by a number of geometric and material parameters (Tzou, 1993) [4]. G is the feedback gain; h^s is the sensor thickness; r_1^a is the actuator distance; d_{31} and h_{31} are the piezoelectric constants; Y_p is Young's modulus of piezoelectric layers; r_1^s is the sensor distance; and L is the beam length. It is assumed that the resistance of the surface electrodes is neglected so that the voltage is uniformly distributed. Since the cantilever boundary conditions are fixed at x=0 and free at x=L and the actuator layer is fully distributed, the distributed control action becomes an equivalent boundary control action, i.e., a control moment acting at the free end (Tzou, Johnson, and Liu, 1999) [5].

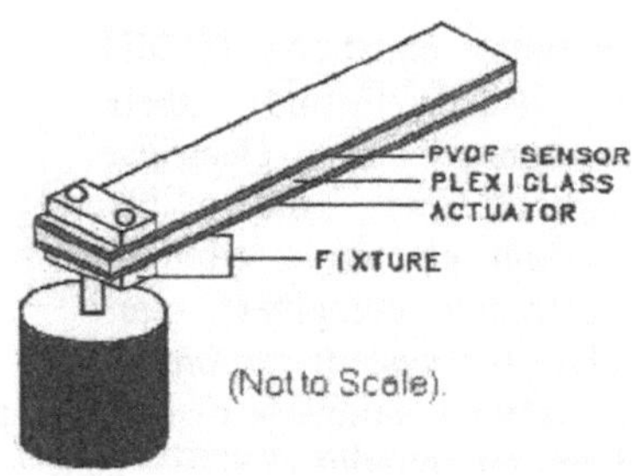

Fig.2 A structronic beam system.

2.2. DISCRETIZATION OF PDE AND CIRCUIT PARAMETER SELECTION

As discussed previously, equivalent circuit model and boundary consitions are established based on the finite difference discretization. In this way, a PDE model is discretized into m-ODE system equations determined by the number of differences; Nominal values of electronic components are determined accordlingly. Fundamentally, there are three difference techniques: backward differences, forward differences and central differences. The central difference usually leads to better accuracy and it is defined as $\frac{du}{dx} = \lim_{\Delta x \to 0} \frac{2\Delta u}{2\Delta x} = \lim_{\Delta x \to 0} \frac{u(x+\Delta x) - u(x-\Delta x)}{2\Delta x}$ (Wang, 1966) [6]. Higher-order differences can be derived based on the first order difference equation. Finite difference discretization of the distributed structronic systems is presented in this section. Their equivalant circuits are defined in the next section.

For the distributed beam structronic system, one usually takes a uniform difference Δx along the beam length L, i.e. $\Delta x = L/m$ where m is the total number of differences. Assume u_n is the displacement of the n-th node, thus, $\Delta u = u_{n+1} - u_n$. A smaller finite difference Δx would ceonceptually yield more accurate results in the circuit modeling and analysis. However, directly dividing YI by a smaller $(\Delta x)^4$ would yield a very large amplification ratio and components parameters that are difficult to realize in real circuit design. Thus, the original difference equation needs modification to avoid the signal divergence in circuit design and implementation. The fourth order differences of the beam PDE can be derived and the generic finite difference equation and the modified finite difference equation are respectively expressed at the n-th node,

$$YI\,[\frac{u_{n+2} - 4u_{n+1} + 6u_n - 4u_{n-1} + u_{n-2}}{\Delta x^4}] + \rho A \frac{\partial^2 u_n}{\partial t^2} = bF_3\ , \qquad (2a)$$

or

$$YI\,[\frac{u_{n+2} - 4u_{n+1} + 6u_n - 4u_{n-1} + u_{n-2}}{\Delta x^3}] + \Delta x \rho A \frac{\partial^2 u_n}{\partial t^2} = \Delta x b F_3 . \qquad (2b)$$

Accordingly, these n-th nodal displacments can be calculated as shown in Figure 3. Detailed active circuit will be defined in the Circuit Design section.

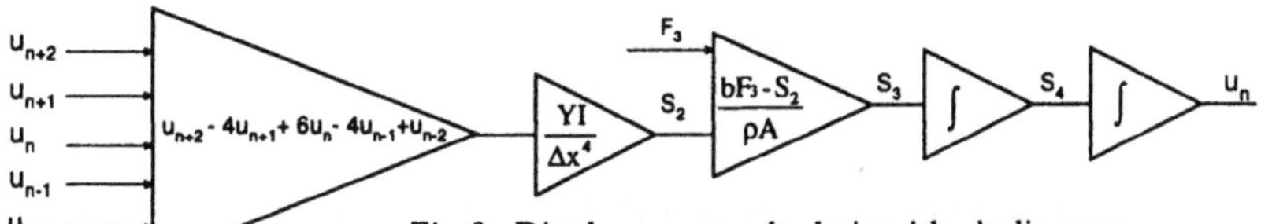

Fig.3 Displacement calculation block diagram.

2.3. BOUNDARY CONDITIONS AND EQUIVALENT BOUNDARY CONTROL

A cantilever beam coupled with distributed piezoelectric sensor/actuator layers, i.e., the distributed beam structronic system, was defined above. As discussed previously, since the distributed sensor and actuator are both fully distributed over the entire beam length, the resulting control action, i.e., the counteracting control moment, congregates at the free end (Tzou, 1993) [4]. Thus, the original distributed control problem becomes a boundary control problem. For the distributed controlled beam, boundary conditions at the fixed-end (x=0 or n=0) are the same as the original elastic beam; however, boundary conditions at the free-end (x=L or n=m) are different, corresponding to the control algorithms. (Note that the beam is divided into m differences and n denotes the node number.) Original elastic and derived control boundary conditions are presented in this section. Note that the fixed-end boundary conditions of all cases are identical. Thus, only the free-end boundary conditions are defined for the control cases. The complete feedback control circuit schematic of the distributed beam structronic system is defined afterwards.

2.3.1 Elastic Boundary Conditions (without control):

- Fixed end: (x=0, n=0): Displacement: $u_0=0$; Slope: $\partial u_0/\partial x = 0$, then $u_{-1} = u_1$.
- Free end: (x=0, n=m, m is number of elements divided, u_m is the transverse displacement at the free end): Moment: $\partial^2 u_m/\partial x^2 = 0$, then $u_{m+1} = 2u_m - u_{m-1}$; Shear force: $\partial^3 u_m/\partial x^3 = 0$, then $u_{m+2} = 2u_{m+1} - 2u_{m-1} + u_{m-2}$.

2.3.2 Displacement Feedback Control:

Assume that the control voltage is proportional to the sensing signal, i.e., $\phi^a = G\phi^s$, where ϕ^a is the actuator signal; ϕ^s is the sensing signal; and G is the gain factor. Boundary conditions at the free end (x=L) are defined as follows.

- Moment: $M^*_{11}(L) = LbM^a_{11}(L)$ where $M^a_{11}(L) = \tilde{\xi}\partial u_3(L)/\partial x = \tilde{\xi}\partial u_m/\partial x$ and $\tilde{\xi} = -Gh^s r_1^a d_{31} Y_p r_1^s h_{31}/L$ [4,5]. Thus, using the moment definition yields $-YI\partial^2 u_m/\partial x^2 = LbM^a_{11}(L) = Lb\tilde{\xi}\partial u_m/\partial x$. Since the difference equation of the double derivative is $\dfrac{\partial^2 u_m}{\partial x^2} = \dfrac{u_{m+1} - 2u_m + u_{m-1}}{\Delta x^2} = -Lb\dfrac{M^a_{11}(L)}{YI}$, thus, the displacement $u_{m+1} = -Lb(M^a_{11}(L)\Delta x^2)/(YI) + 2u_m - u_{m-1} = u_c + 2u_m - u_{m-1}$ and the control signal is $u_c = -Lb(M^a_{11}(L)\Delta x^2)/(YI)$.
- Shear force: $\partial^3 u_m/\partial x^3 = 0$, then $u_{m+2} = 2u_{m+1} - 2u_{m-1} + u_{m-2}$.

2.3.3 Velocity Feedback Control:

The control signal is now defined as $\phi^a = G\partial\phi^s/\partial t$. Free-end boundary conditions are now defined by:

- Moment: $M^*_{11}(L) = LbM^a_{11}(L)$ and $M^a_{11}(L) = r_1^a d_{31} Y_p \phi^a$ or $M^a_{11}(L) = -\tilde{\xi}\dfrac{\partial}{\partial t}[(\dfrac{\partial u_3}{\partial x})_0^L] = -\tilde{\xi}\dfrac{\partial}{\partial t}[\dfrac{\partial u_3(L)}{\partial x}] = -\tilde{\xi}\dfrac{\partial}{\partial t}(\dfrac{\partial u_m}{\partial x})$. Thus, $-YI\partial^2 u_m/\partial x^2 = LbM^a_{11}(L) = Lb\tilde{\xi}\partial(\partial u_m/\partial x)/\partial t$ and the displacement is defined by $u_{m+1} = Lb(M^a_{11}(L)\Delta x^2)/(YI) + 2u_m - u_{m-1} = u_c + 2u_m - u_{m-1}$.
- Shear force: $\partial^3 u_m/\partial x^3 = 0$, then $u_{m+2} = 2u_{m+1} - 2u_{m-1} + u_{m-2}$.

Equivalent circuits of the structronic systems, boundary conditions and boundary control are defined next.

3. Design of Electronic Circuits

To assure dynamic and signal stability over the frequency range, operational amplifiers (op-amp's) (Irvine, 1987) [7] are selected to implement the equivalent electronic circuit of the distributed structronic system. Based on the discretized PDE and boundary conditions, nominal values of resistors and capacitors can be determined by specified material, geometry, and control parameters. Figure 4 illustrates the final circuit schematic consisting of a number of operational amplifiers and resistors. Boundary conditions, boundary control, and their equivalent electronic circuits derived from the finite difference discretization are also designed. A boundary moment control circuit is given in Figure 5 where u_c is determined by the control algorithms. A circuit diagram representing the shear force is illustrated in Figure 6. The complete feedback control circuit schematic of the distributed beam structronic system is illustrated in Figure 7. Note that u_{c1} and u_{c2} correspond to the displacement and the velocity feedback respectively. The whole structronic beam system discretized into eight differences is now realized as an electronic circuit, Figure 8, including all system characteristics and distributed control effects. Besides, a numerical model of the system circuit is also setup using SIMULINK, such that the circuit performance can be validated by the numerical simulation results.

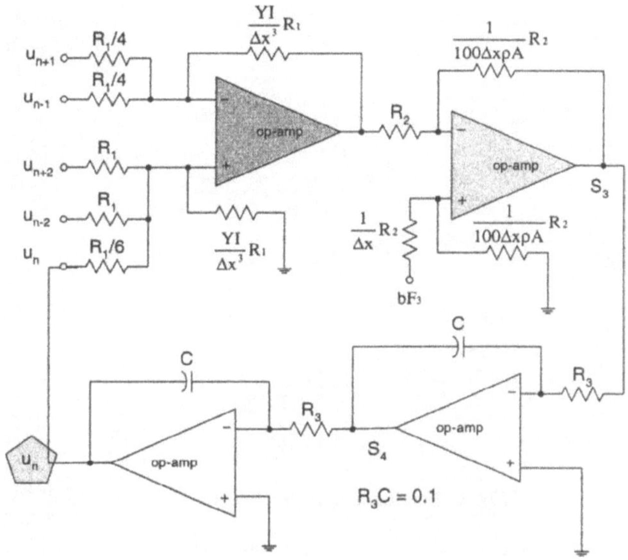

Fig.4 Circuit schematic for the n-th nodal displacement.

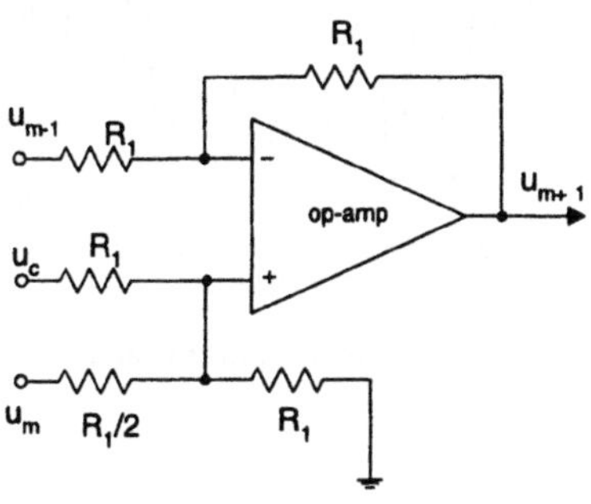

Fig.5 Boundary control moment circuit.

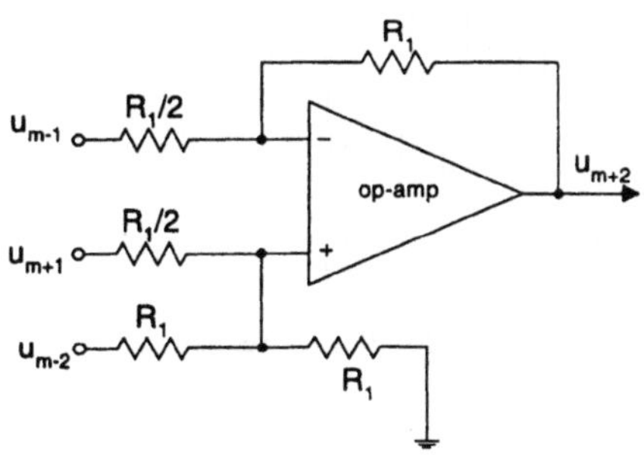

Fig.6 Boundary shear force circuit.

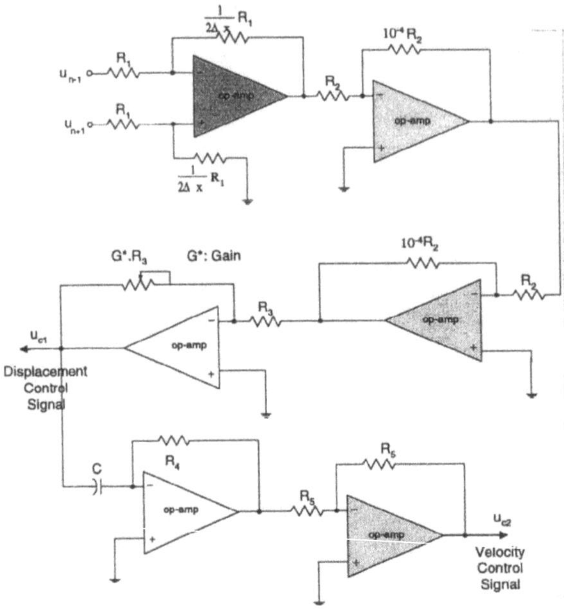

Fig.7 Feedback control circuit.

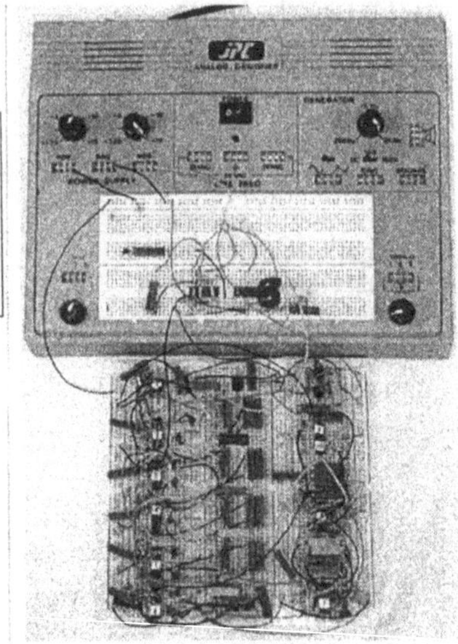

Fig.8 An active circuit for the distributed structronic beam control system.

4. Results and Discussion

Free response (Figure 9), displacement feedback responses (Figures 10a,b,c with gains 100, 500, 1000 respectively) and velocity feedback responses (Figures 11a,b,c,d with gains 50, 100, 500, 1000 respectively) with different control gains are studied. Their control characteristics are evaluated and compared with previous studies using theoretical, finite element, and experimental techniques [5].

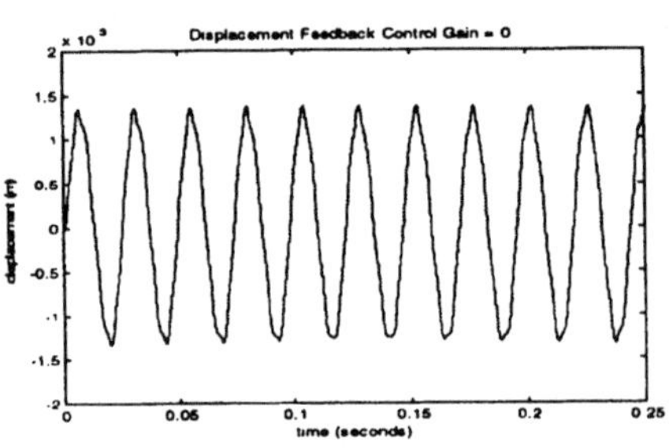

Fig.9 Free response.

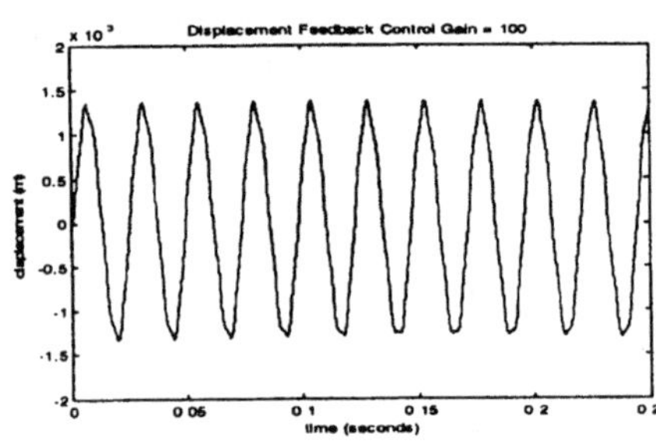

Fig.10a Displacement feedback control (gain=100).

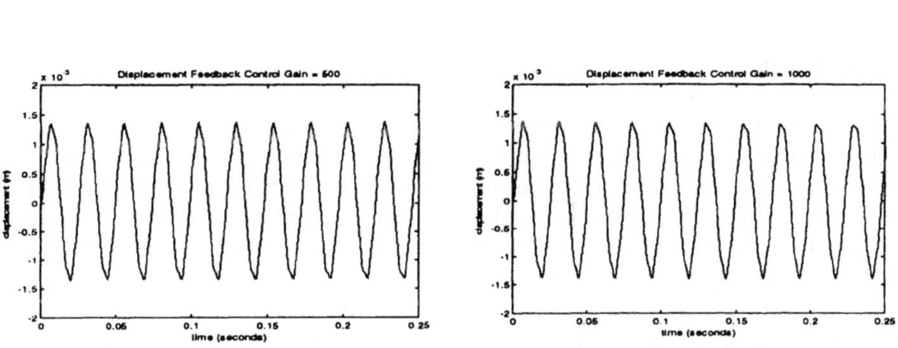

Fig.10b,c Displacement feedback control (gain=500,1000).

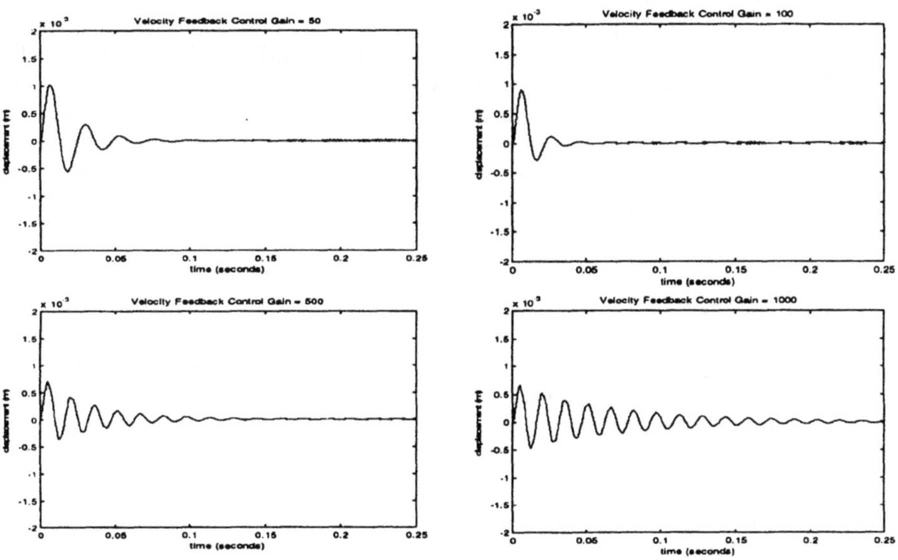

Fig.11a-d Velocity feedback control (gain=50, 100, 500, 1000).

These time histories suggest that the displacement feedback changes the frequency characteristic and the velocity feedback changes the system damping behavior. The influence to frequency variation is small; however, the influence to damping is rather significant (Tzou, 1993) [4]. It is observed that the damping is enhanced at low control gains and degraded at high control gains. Recall that the resultant control effect of a fully distributed actuator on a cantilever continuum is a counteracting control moment at the free end. The original "spatially" distributed control becomes an equivalent "boundary " control effect to the cantilever distributed system. Counteracting control moment at the free end constraints the rotational motion at high control gains, and this rotational constrain becomes a new boundary condition at high control gains (Tzou, Johnson, and Liu, 1999) [5].

5. Conclusions

Distributed smart structures and structronic systems would likely revolutionize the design and development of next-generation high-performance aerospace structures and precision systems. Modeling and analysis of distributed structronic systems has been traditionally carried out using analytical, finite element, and experimental techniques. This paper is to present a new electrical modeling technique for the distributed structronic control systems modelled by partial differential equations.

To establish the electronic equivalence of the structronic system, the original system PDE was discretized using the finite difference technique. However, directly discretizing the PDE based on a small difference, due to resolution requirement, would yield non-standard, usually very large, component parameters that are either unavailable or undesirable in circuit design. These issues were addressed and resolved to derive "reasonable" and/or "manageable" electronic component parameters. An equivalent active electronic circuit was fabricated and a numerical SIMULINK model was also established. Electric signals acquired from the active circuit and the numerical model were compared favourably with the analytical solutions, finite element results, and experimental data. Thus, the proposed new electronic modeling technique, including the numerical model and the electronic circuit, serves well and provides a viable alternative technique in modeling and analysis of distributed structronic control systems.

6. Acknowledgement

This research is supported, in part, by a grant (F49620-98-1-0467) from the Air Force Office of Scientific Research (Project managers: Dan Segalman and Brian Sanders). This support is gratefully acknowledged.

7. References

1. Tzou, H.S. and Gabbert, U., 1997, "STRUCTRONICS- A New Discipline and Its Challenging Issues," Fortschritt-Berichte VDI, *Smart Mechanical Systems – Adaptronics*, Reihe 11: Schwingungstechnik Nr.244, pp.245-250.
2. Measurement Analysis Corporation, 1966, "Electrical Analogies and the Vibration of Linear Mechanical Systems", National Aeronautics and Space Administration, Washington D.C.
3. Shah, A., Lenning, L., Bibyk, S., Ozguner, U., 1994, "Flexible Beam Modeling with Analog VLSI Circuits", ASME, Adaptive Structure and Composite Materials: Analysis and Application, AD-Vol.45/MD-Vol.54, pp.261-265.
4. Tzou, H.S., 1993, "*Piezoelectric Shells: Distributed Sensing and Control of Continua*", Kluwer Academic Publishers, Boston/Dordrecht, Chapter 6, pp.187-226.
5. Tzou, H.S., Johnson, D.V., and Liu, K.J., 1999, "Damping Behavior of Cantilever Structronic Systems with Boundary Control," *ASME Transactions, Journal of Vibration & Acoustics*, pp.402-407.
6. Irvine, R.G., 1987, "*Operational Amplifier Characteristics and Applications*", 2nd edition, Prentice-Hall, Inc., Englewood Cliffs, New Jersey, pp. 156-157.
7. Wang, P.-C., 1966, "*Numerical and Matrix Methods in Structural Mechanics-With Applications to Computers*", John Wiley & Sons, Inc., New York, pp.1-9. (PrCircuit2Fx.Iutam2000.Conf00)

AN OPERATOR-BASED CONTROLLER CONCEPT FOR SMART PIEZO-ELECTRIC STACK ACTUATORS

K. KUHNEN, H. JANOCHA
Saarland University, Laboratory for Process Automation (LPA)
Im Stadtwald, Building 13, D-66041 Saarbrücken, Germany

1. Introduction

Solid-state actuators based on piezoelectric materials are characterized by forces reaching the kilonewton range and reaction times on the order of microseconds. However, the displacements are low because the maximum strain amounts to 1.5 .. 2 ‰. If they are used in micropositioning drives, the piezoelectric stack actuators are mainly driven with high voltages $v(t)$ to achieve the largest possible displacements $s(t)$. In large-signal operation the actuator characteristic shows strong hysteresis and creep effects which can be regarded as undesired internal disturbances for the micropositioning process. Normally piezoelectric actuators are subsystems in an overlying mechanical structure. During the micropositioning process the mechanical structure reacts with a force $f(t)$ against the piezoelectric transducer. This reaction force has a strong influence on the displacement and can therefore be regarded as an undesired external disturbance for the micropositioning process. Under normal operating conditions the displacement of the piezoelectric stack transducer can be seperated, at least in a first order approximation, into a creep and hysteretic voltage-dependent part described here by a scalar operator Γ_a and a linear force-dependent part characterised by the small-signal elasticity S

$$s(t) = \Gamma_a[v](t) + S \cdot f(t) \ . \tag{1}$$

Equation (1) is the so-called operator-based actuator model of the piezoelectric stack transducer [6].

Additionally, the piezoelectric transducer has an inherent sensory capacity which originates from the same internal microphysical process as the actuator capacity. In sensor operation of the transducer the voltage-dependent part of the electrical charge signal $q(t)$ can be regarded as an external disturbance whereas the force-dependent part of the electrical charge signal contains the measurement information about the force acting on the transducer. Under normal operating conditions the electrical charge signal can be separated, at least in a first order approximation, into a creep and hysteretic voltage-dependent part described here by a scalar operator Γ_e and a linear force-dependent part characterized by the so-called small-signal piezoelectric constant d

U. Gabbert and H.S. Tzou (eds.), IUTAM Symposium on Smart Structures and Structronic Systems, 299–306.

$$q(t) = \Gamma_e[v](t) + d \cdot f(t). \tag{2}$$

Equation (2) is the so-called operator-based sensor model of the piezoelectric stack transducer [6].

2. Operator-based Controller Concept

The internal hysteretic and creep disturbances and the external force-generated disturbance lead to ambiguities in the characteristic of piezoelectric actuators and thus to a considerable reduction of the repeatability attainable in open-loop control. A possible solution of this problem is to compensate the hysteresis, creep and force-dependence simultaneously using the inverse feed-forward compensator realized by

$$v(t) = \Gamma_a^{-1}[s_c - S \cdot f](t). \tag{3}$$

In this equation $s_c(t)$ is the desired diplacement. The force $f(t)$ acting on the actuator has to be measured by an external force sensor. Γ_a^{-1} is the inverse operator of Γ_a.

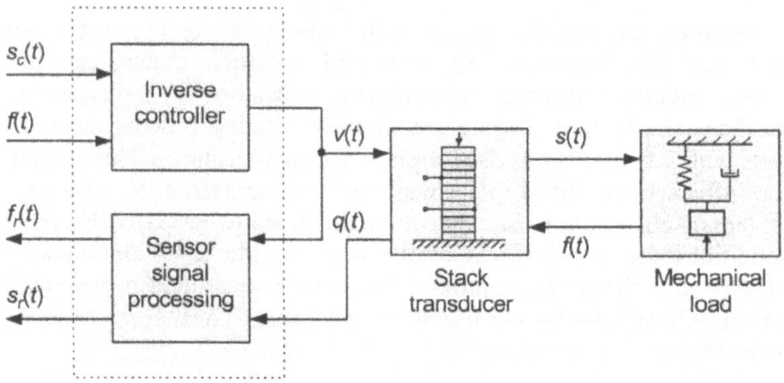

Figure 1. Inverse control of a smart piezoelectric stack actuator

Due to the multi-functional property of the piezoelectric stack transducer the displacement $s(t)$ of the actuator and the force $f(t)$ acting on it can be reconstructed during operation using a signal processing unit, see Figure 1. The signal processing unit bases on the equation

$$f_r(t) = \frac{1}{d} \cdot (q(t) - \Gamma_e[v](t)) \tag{4}$$

to compute the reconstructed force $f_r(t)$ and on the equation

$$s_r(t) = \Gamma_a[v](t) + \frac{S}{d} \cdot (q(t) - \Gamma_e[v](t)), \tag{5}$$

to compute the reconstructed displacement $s_r(t)$ of the transducer. Solid-state transducers which are used simultaneously as both actuators and sensors are frequently called smart actuators. In addition to the inverse controller the controller concept shown in Figure 1 contains a signal processing unit in order to realize such a smart piezoelectric stack actuator for the large-signal operating range. The basis of the feed-forward controller and the signal processing unit is comprised of so-called elementary creep and hysteresis operators which are mathematically simple and which reflect the qualitative properties of the transfer characteristic of the transducer.

3. Operator-based Hysteresis and Creep Modelling

In the mathematical literature the notation of 'hysteretic nonlinearity' will be equated with the notation 'rate independent memory effect' [1]. This means that the output signal of a system with hysteresis depends not only on the present value of the input signal but also on the order of their amplitudes, especially their extremum values, but not on their rate in the past. Because of its phenomenological character the concept of hysteresis operators developed by Krasnosel'skii and Pokrovskii in the 1970's allows a very general and precise modelling of hysteretic system characteristics [3].

The basic idea consists of the modelling of the real hysteretic transfer characteristic by the weighted superposition of many elementary hysteresis operators, which differ in terms of one or more parameters depending on the type of the elementary operator. One type of such an elementary hysteresis operator is the so-called play operator

$$z_r(t) = H_r[v, z_{r0}](t) \tag{6}$$

which is defined by the recursive equations

$$z_r(t) = \max\{v(t) - r, \min\{v(t) + r, z_r(t_i)\}\} \tag{7}$$

with

$$z_r(t_0) = \max\{v(t_0) - r, \min\{v(t_0) + r, z_{r0}\}\} \tag{8}$$

for piecewise monotonous input signals with a monotonicity partition $t_0 \leq t_1 \leq .. \leq t_i \leq t \leq t_{i+1} .. \leq t_N$. The operator is characterized by its threshhold parameter r, $z_r(t)$ is the operator output and z_{r0} its initial value. Figure 2a shows the rate-independent output-input trajectory of this simple hysteresis operator.

For the precise modelling of real hysteresis phenomena n play operators with different threshold values r_i can be multiplicated with weights b_i and then superimposed. This parallel connection of elementary hysteresis operators leads to the complex hysteresis operator

$$H[v](t) = \sum_{i=1}^{n} b_i \cdot H_{r_i}[v, z_{r_i0}](t). \tag{9}$$

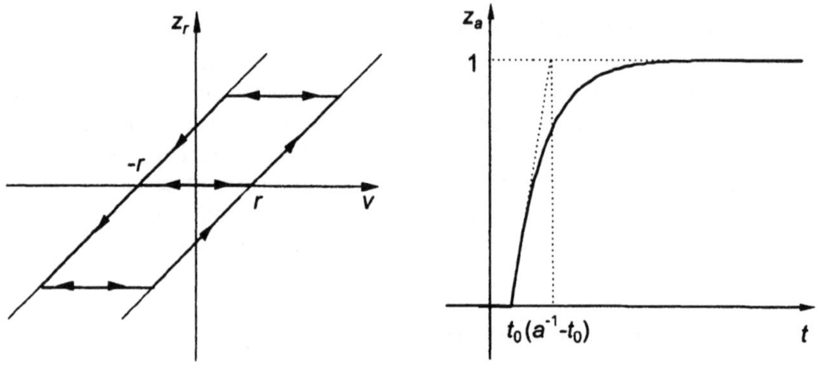

a) b)

Figure 2. a) Rate-independent transfer characteristic of the play operator
b) Step response of a rate-dependent, elementary creep operator

The notion of creep originates from the field of solid-mechanics and describes the time-variant deformation behaviour of a body due to a sudden mechanical load [2]. It is a strongly damped, rate-dependent phenomenon, which can be found in a similar form in the field of ferromagnetism and ferroelectricity. Like hysteresis phenomena electrically induced creep effects have a considerable influence on the large-signal transfer characteristic of a piezoelectric transducer [5]. In a first order approximation this creep phenomena is presumed linear. As a consequence it can be described, analogously to the hysteresis modelling process, by a complex linear creep operator

$$L[v](t)=\sum\nolimits_{j=1}^{m} c_j \cdot L_{a_j}[v,z_{a_j 0}](t), \tag{10}$$

given by a weighted superposition of m elementary linear creep operators with different creep eigenvalues a. c_j are the weights of the elementary operators. In this case the elementary linear creep operator is the solution operator

$$L_a[v,z_{a0}](t)=\mathrm{e}^{-a(t-t_0)}\cdot z_{a0}+a\cdot\int_{t_0}^{t}\mathrm{e}^{a(\tau-t)}\cdot v(\tau)\,\mathrm{d}\tau \tag{11}$$

of a linear, first-order differential equation with an initial value z_{a0}. Figure 2b shows the step response of the elementary linear creep operator, which has the same qualitative features as the step response of the creep phenomena in the real system.

In the following we derive a first-order approximation model of the actuator characteristic Γ_a and the electrical characteristic Γ_e by the linear superposition of a weighted reversible part, a rate-independent irreversible part described by the complex hysteresis operator H and a rate-dependent part described by the complex linear creep operator L. From this follows

$$\Gamma_a[v](t)=d\cdot v(t)+H_a[v](t)+L_a[v](t) \tag{12}$$

and

$$\Gamma_e[v](t) = C \cdot v(t) + H_e[v](t) + L_e[v](t). \tag{13}$$

In electrical small signal operation (12) and (13) can be reduced to the reversible part with the piezoelectric constant d and the small-signal capacity C. From this follows that the operator-based approach is a logical extension of small-signal modelling to the large-signal range.

The problem to find the inverse control value $v(t)$ for a given control value $s_c(t)$ and thus to get the inverse operator Γ_a^{-1} is equivalent to the solution of the operator equation

$$v(t) = P_a^{-1}[s_c - L_a[v]](t) \tag{14}$$

with the hysteresis operator

$$P_a[v](t) := d \cdot v(t) + H_a[v](t). \tag{15}$$

The solution of the operator equation (14) and thus the inverse operator Γ_a^{-1} exists and is unique under the natural inequality constraints

$$0 < d < \infty\ , \quad 0 \le b_i < \infty\ ; \quad i = 1\,..\,n, \tag{16}$$

and

$$0 \le c_j < \infty\ ; \quad j = 1\,..\,m\ , \quad 0 < a_j < \infty\ ; \quad j = 1\,..\,m \tag{17}$$

[4]. An efficient numerical procedure for the solution of the operator equation (14) in real-time is also given in [4].

4. Results and Discussion

The inverse controller for the compensation of hysteresis, creep and force-dependence of the piezoelectric transducer and the measurement signal processing for the reconstruction of the displacement of the transducer and the force on it was realized on a digital signal processor (DSP) with a sampling rate up to 10 kHz.

To verify the performance of the compensation concept the inverse compensator was driven with the desired displacement signal $s_c(t)$ shown in Figure 3a. Figure 3c shows the characteristic of the conventional linear controller as a gray line. It is an ideal linear rate-independent characteristic typical for conventional voltage-amplifiers. As a consequence the characteristic of the serial combination conventional controller-transducer, shown in Figure 3d as a gray line, shows the hysteresis and creep effects of the transducer. The characteristic of the operator-based inverse compensator, shown in Figure 3c as a black line, is obviously inverse to the characteristic of the transducer. As a consequence the characteristic of the serial combination inverse compensator-transducer, shown in Figure 3d as a black line, is almost completely free of hysteresis and creep effects and the displacement error caused by creep and hysteresis effects is reduced from 2.47 μm using the conventional controller to 0.25 μm using the inverse compensator. This is an improvement of one order of magnitude.

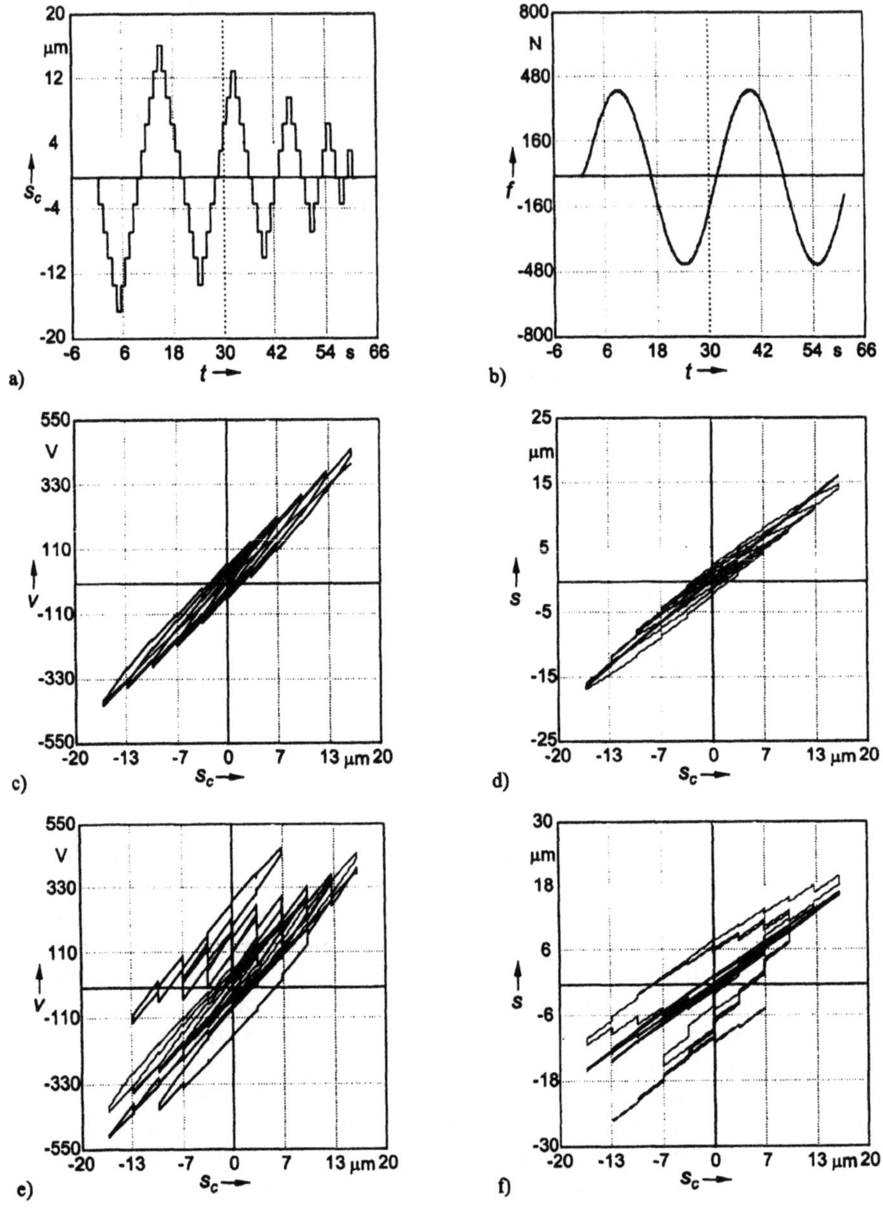

Figure 3. Hysteresis, creep and force-dependence compensation results

The gray line in Figure 3f shows the strongly disturbed characteristic of the serial combination inverse compensator-transducer. The disturbance is generated by the additional external force signal shown in Figure 3b. In this case the force-dependent part of the displacement is not compensated by the inverse controller, see the gray line in Figure 3e. This force-dependence effect leads to a displacement error of 11.8 μm. The black line in Figure 3e shows the strongly disturbed characteristic of the inverse controller. The disturbance of the inverse hysteretic and creep characteristic is caused by an additional compensation of displacement generated by the external force signal. As a consequence the force generated disturbances in the characteristic of the serial combination inverse compensator-transducer is strongly reduced, see the black line in Figure 3f. As the main result the displacement error caused by the external force is reduced to 1.55 μm. This is also an improvement of nearly one order of magnitude.

To verify the performance of the reconstruction concept the transducer was driven with the measured voltage signal $v(t)$ shown in Figure 4a. At the same time the force signal $f(t)$ shown in Figure 4c as a black curve acts on the transducer. The voltage- und force-generated displacement and charge signal of the transducer, $s(t)$ resp. $q(t)$, are shown in Figure 4d and Figure 4b as black curves.

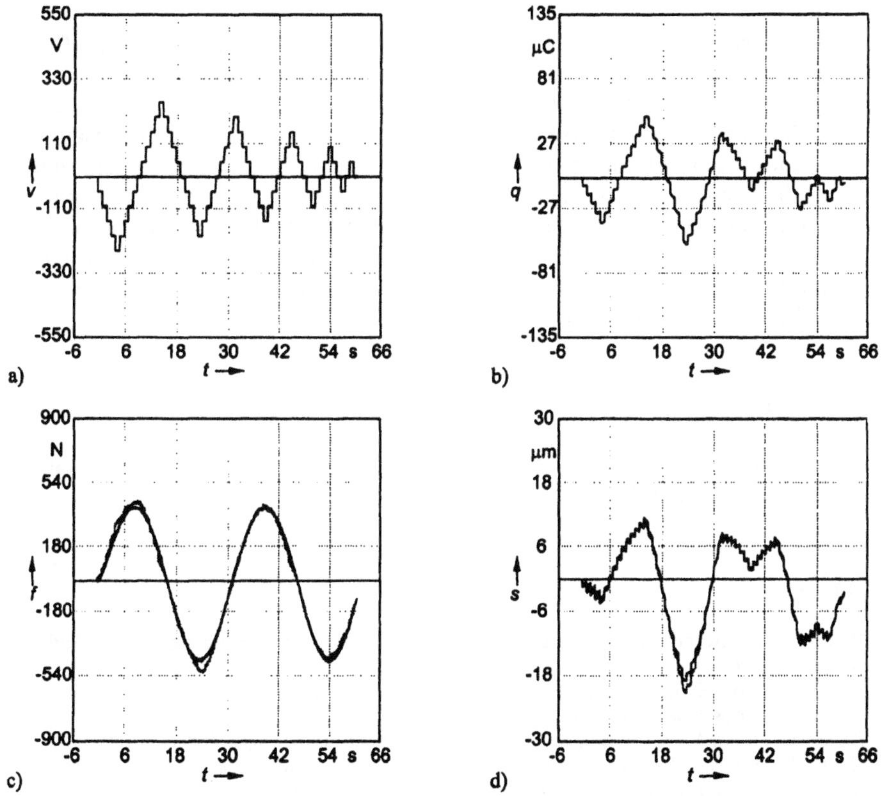

Figure 4. Reconstruction results

The gray curve in Figure 4c is the force signal reconstructed by operator equation (4). The maximum value of the relative deviation between the real force signal and the reconstructed force signal amounts to 15 %. The gray curve in Figure 4d shows the displacement signal reconstructed by operator equation (5). The maximum value of the relative deviation between the real displacement signal and the reconstructed displacement signal amounts to 10 %.

If we neglect the hysteresis and creep operators in (12) and (13) we get simplified linear reconstruction models for (4) and (5) which follow directly from the small-signal constitutive relations of the piezoelectric transducer. But due to the hysteresis and creep operators in (12) and (13) the deviation between the measured and calculated voltage-dependent part of the charge signal and the displacement signal in (4) an (5) is reduced by about one order of magnitude [6]. Therefore the reconstruction error of the force signal increases up to 150 % and the reconstruction error of the displacement signal increases up to 100 % if we use the linear reconstruction modells instead of the operator-based versions.

5. Summary

This paper has shown that complex creep and hysteresis operators offer an efficient method to model the hysteresis and creep in the electrical and actuator characteristic of a piezoelectric stack transducer if it is driven electrically in the large-signal range. Based on this method an inverse controller for the simultaneous compensation of hysteresis and creep effects in real-time and a sensor signal processing unit to reconstruct the displacement of the transducer and the force acting on it was presented. In addition to hysteresis and creep phenomena which can be regarded as intrinsic disturbances, the influence of the force on the displacement signal and the control voltage on the charge signal, the external disturbances of the system, are also considered by the new controller concept.

Acknowledgements

The authors thank the Deutsche Forschungsgemeinschaft (DFG, German National Research Council) for the financial support of this work.

References

1. Brokate, M., Sprekels, J.: *Hysteresis and Phase Transitions*, Springer-Verlag, Berlin Heidelberg NewYork, 1996.
2. Kortendieck, H.: *Entwicklung und Erprobung von Modellen zur Kriech- und Hysteresiskorrektur*, VDI Verlag, Düsseldorf, 1993.
3. Krasnosel'skii, M. A., Pokrovskii, A. V.: *Systems with Hysteresis*, Springer-Verlag, Berlin, 1989.
4. Krejci, P., Kuhnen, K.: Inverse Control of Systems with Hysteresis and Creep, submitted for publication in *IEE Proceedings of Control Theory and Applications*.
5. Kuhnen, K., Janocha H.: Compensation of the Creep and Hysteresis Effects of Piezoelectric Actuators with Inverse Systems, *Proc. of the 6th Int. Conf. on New Actuators* (1998), Messe Bremen GmbH, pp. 309-312.
6. Kuhnen, K., Janocha H.: Nutzung der inhärenten sensorischen Eigenschaften von piezoelektrischen Aktoren, *tm-Technisches Messen* **66** (1999), pp. 132-138.

EXPERIMENTS WITH FEEDBACK CONTROL OF AN ER VIBRATION DAMPER

N.D. SIMS, R. STANWAY, A.R. JOHNSON
Department of Mechanical Engineering
The University of Sheffield
Sheffield S1 3JD
United Kingdom

1. Introduction

It is now becoming well-established that there is enormous potential for the application of smart fluids to a variety of problems in vibration control. In the development of materials and devices emphasis has been placed upon electro-rheological (ER) fluids which consist of a suspension of micron-sized, semiconducting particles in a dielectric base liquid. The application of an electric field of sufficient intensity causes a rapid (and reversible) increase in the ER fluid's resistance to flow. A 1996 survey paper [1] explained the fundamental operating principles of ER fluids and described the possible modes of operation. However the main purpose of the paper was to review applications of ER fluids to vibration control.

Practically all of the applications described in reference [1] were at the laboratory or prototype stage. However, recent developments with an alternative form of smart fluid – magnetorheological (MR) fluid – have quickly led to commercial exploitation and mass-produced vibration control devices [2]. MR fluids employ magnetisable particles in suspension and as a result require the application of a magnetic (or electromagnetic field) to bring about an increase in resistance to flow. As a result much greater force levels are available from MR fluids but, perhaps more importantly, excitation can be provided by a low-voltage source. There is little doubt that the need to supply up to several kilovolts to excite an ER fluid has deterred many potential users.

One feature which both ER and MR fluids share is the form of the force-velocity characteristic. In both cases this is highly non-linear and is often likened to the constitutive relationship associated with a Bingham plastic [1]. As a result, most attempts to apply feedback control make use of non-linear control algorithms. The present authors took a different view, arguing that it might be possible to linearise the force-velocity characteristics of a smart damper using linear feedback techniques. Numerical studies followed by a comprehensive experimental validation showed that this was indeed the case. Furthermore the investigation led to questions concerning the possibility of using feedback techniques to determine the shape of the force-velocity characteristic.

U. Gabbert and H.S. Tzou (eds.), IUTAM Symposium on Smart Structures and Structronic Systems, 307–314.

In the present paper, the authors summarise progress which has been made towards shaping the force-velocity characteristics of an industrial-scale ER vibration damper. In the body of the paper, emphasis is placed on a description of the experimental facility and a discussion of the mathematical model which is used as the basis of the control algorithms. A representative series of numerical and experimental results is included to show how, with feedback control, the ER vibration damper can provide the force-velocity characteristics associated Coulomb friction, viscous damping or quadratic damping. Some suggestions for further work are also included.

2. Experimental Facility

The operation of the experimental facility is best described with reference to the schematic, Figure 1. Essentially an ER flow control valve acts as a bypass across a hydraulic piston/cylinder arrangement. ER fluid is delivered to the piston/cylinder (i.e. the damper) and the ER valve by a pump which draws fluid from a reservoir with a nitrogen-pressurised accumulator.

Referring to Figure 1, as the piston moves from left to right so the volume A decreases and the non-return valve B remains closed, thus forcing fluid through the ER valve. Meanwhile the volume C increases and the non-return valve D opens, relieving this side of the main piston to the bladder pressure. The fluid circulation valve is left open to ensure that fluid is not drawn through the pump and filter during operation of the device. The auxiliary pistons ensure that the pressures at the ends of the cylinder are always the same as the bladder pressure, thus enabling the use of bellows which overcomes the problems associated with sealing a slurry-like liquid. Although not shown on Figure 1, arrangements are in place for controlling the temperature of the ER fluid.

The ER valve itself consists of annular concentric electrodes of opposing electrical polarity. Application of an electric field results in an increased resistance to flow across the piston head and hence a pressure drop across the valve. The force on the main piston is therefore given by the differential pressure multiplied by the effective piston area.

The piston is driven using a variable speed electric motor, via a reduction gearbox and cam arrangement. The gearbox has four reduction settings, offering a range of output shaft speeds from 150 to 1500 rev/min for an input speed of 1500 rev/min. This enables the piston frequency to be chosen as any value up to 25 Hz. The maximum piston displacement is 16 mm and the piston stroke is adjustable, off-line, by moving the connecting rod along the crank arm.

The electric field applied to the ER valve is supplied from a commercially available high-voltage generator with a remote input. With this arrangement, electric field strengths of up to 8 kV/mm are available. The key measurement variables are cylinder pressure of each side of the main piston, cooling water temperature in the ER valve and baths, piston displacement and trigger signal and the electric field strength applied to the ER valve.

For fuller details of the experimental facility, the interested reader is referred to references [3] and [4].

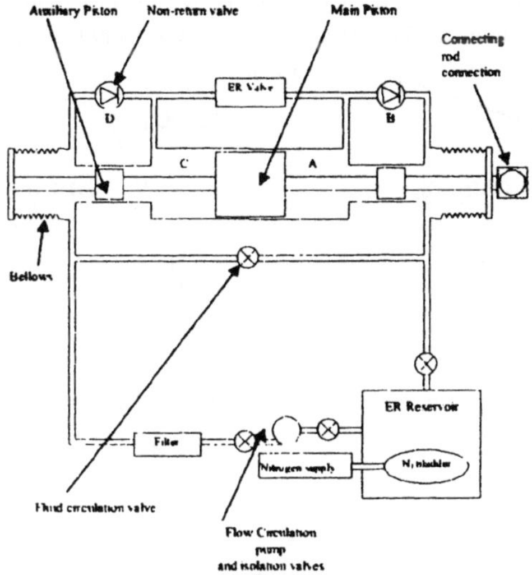

Figure 1: Schematic representation of the experimental facility

3. Mathematical Model of ER Vibration Damper

A detailed derivation of the mathematical model of the ER vibration damper is given in reference [3]. Here, space allows only a brief summary.

The basis of the modelling technique is a non-dimensional, quasi-steady analysis of flow through an ER valve [5]. A Bingham plastic is a material which combines the yield properties of solids with the Newtonian flow properties of fluids and, to a first approximation, represents the macroscopic behaviour of a smart fluid. By assuming that in a smart fluid a yield stress is generated by the application of an electric or magnetic field it is possible to formulate a cubic equation. One particular solution of this equation yields a numerical value for the pressure drop across the valve. In the basic quasi-steady analysis it is further assumed that the pressure drop across the valve is the same as that across the main piston. Thus the corresponding piston force is given by the product of the pressure drop and the effective area of the main piston.

Assuming that the piston and valve geometry together with the smart fluid properties are constant for the duration of an experiment then the piston force, F_d, is effectively a function of the piston velocity, i.e. $\dot{x}$, and the applied electric field strength, E. Mathematically

$$F_d = \chi(\dot{x}, E) \tag{1}$$

At frequencies higher than (say) 1 Hz the quasi-steady analysis fails to account for observed behaviour and dynamic elements, principally the smart fluids compressibility and inertia, need to be included. A relatively simple dynamic model which has proved effective is shown in Figure 2. The development of the model is described in reference [3]. Here we will simply state the main features.

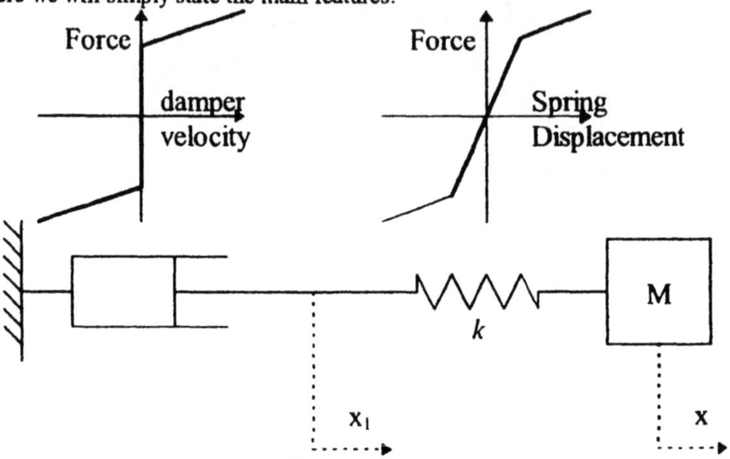

Figure 2: Lumped parameter model.

Referring to Figure 2, the damping characteristics of the device are represented by the Bingham plastic model mentioned earlier. Note that since the dynamic model involves two degrees of freedom, the velocity across the damper is now denoted $\dot{x}_1$ rather than $\dot{x}$, as in equation (1). The compressibility of the smart fluid is represented by a lumped non-linear spring element. The force, F_s, developed by the spring is given by:

$$F_s = \kappa(y, E) = \left\{ \begin{array}{ll} 2ky & \text{if } |y| \le \frac{F_y}{2k} \\ ky - F_y/2 & \text{if } y < -\frac{F_y}{2k} \\ ky + F_y/2 & \text{if } y > \frac{F_y}{2k} \end{array} \right\}, \qquad (2)$$

where y is the applied displacement across the spring and F_s is the resulting spring force. The force F_y corresponds to the yield stress developed within the smart fluid when an electric or magnetic field is applied.

Combining equation (2) with the quasi-steady damper function in equation (1) and lumping the fluid inertia into a single mass m, we can write the equations of motion:

$$\left. \begin{array}{ll} -\chi(\dot{x}_1, E) + \kappa(x_2 - x_1, E) = 0 & \text{(a)} \\ F - \kappa(x_2 - x_1, E) = m\ddot{x}_2 & \text{(b)} \end{array} \right\} \qquad (3)$$

where the co-ordinates x_1 and x_2 are shown in Figure 2.

The solution of the Equations (3) requires care, primarily because of the discontinuity in the Bingham plastic force-velocity plot when the velocity is zero. Novel methods of approaching the solution are described in reference [3].

Obviously before the model, Equations (3), could be used as the basis for control system design, a comprehensive validation procedure was performed. For full details see references [3] and [4]. As an example, a typical result is shown in Figure 3. Here the force-velocity and force/displacement traces are shown for a mechanical excitation frequency of 2 Hz and electric field strength of 2 kV/mm. It can be seen that the model predictions well capture the essential features of the experimental responses.

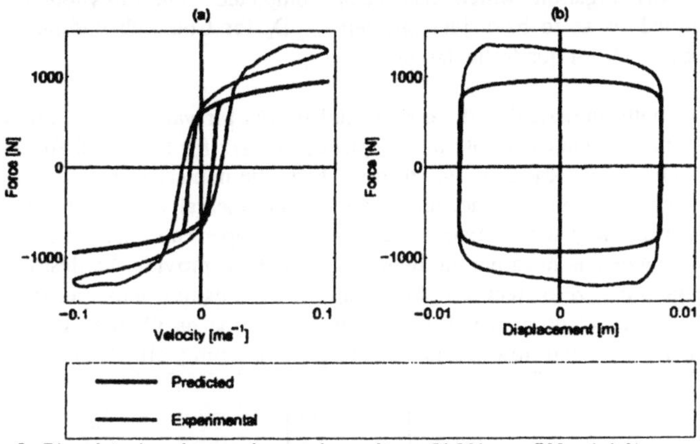

Figure 3: Simulated and experimental results at 2kV/mm, 2Hz. (a) Force - velocity response, (b) force - displacement response.

4. Experimental Shaping of the Force-velocity Response

Following validation of the mathematical model, a control strategy was designed in a series of numerical experiments using Simulink [5]. Particular attention had to be paid to generating a piston velocity signal from a measurement of the piston displacement. After trying various numerical differentiating/filtering techniques (with limited success) the velocity signal was generated artificially. Essentially a sine wave was generated based upon the operating frequency and displacement amplitude of the mechanical excitation signal. After suitable scaling, this technique produced a velocity signal which was not degraded by time delay or signal noise, both of which are inherent in numerical differentiation.

In earlier descriptions [4], attention was focussed upon linearising the force-velocity characteristics of the ER vibration damper. This required specification of a set point which defined the effective damping coefficient. Based upon this coefficient, feedback of the force generated by the ER damper was used to manipulate the applied electric field such that the damper force was proportional to the damper velocity. A natural extension of this philosophy is to investigate the effectiveness of using arbitrary polynomial functions to define the shape of the force-velocity response. As an example, a second-order polynomial shaping function may be written

$$r = D_0 + D_1\dot{x} + D_2\dot{x}.\mathrm{sgn}(\dot{x}) \tag{4}$$

where r is the set point, $\dot{x}$ is the damper velocity and D_n, n=0, 1, 2 is the nth-order set point gain. Note that the quadratic term is multiplied by $\mathrm{sgn}(\dot{x})$ so that it becomes an odd function, producing a negative set point for negative velocities.

Using this strategy we will now present a small sample of results from the comprehensive investigation which has been completed. The polynomial shaping function is limited so as to have just one term – D_0 for Coulomb damping, D_1 for viscous damping or D_2 for quadratic damping.

Figure 4 shows both numerical and experimental results for values of D_0 ranging from zero to 2 kN. The mechanical excitation frequency is 2 Hz for these and for the results which follow. There are various points to note. First the numerical results, Figure 4(a) show that the model predictions account for the experimental behaviour, Figure 4(b). Second, because like any semi-active device the ER damper can only dissipate energy. Thus when the velocity is negative the damper is unable to provide the desired positive force and resorts to the 'zero electric field' condition. For positive velocities the response attempts to track the set point which it reaches after some oscillation. Third, there is significant hysteresis around flow reversal due to compressibility effects in the ER fluid.

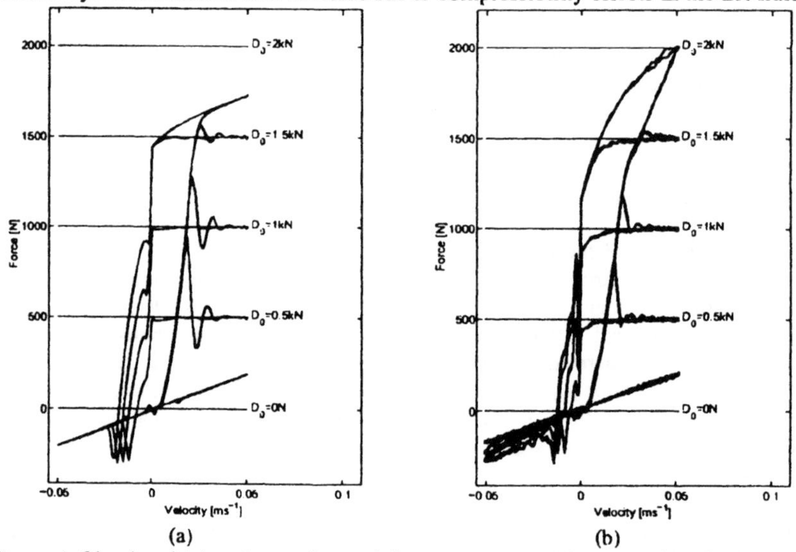

Figure 4: Simulated (a) and experimental (b) response to with zero-order (Coulomb) gain, D_0, frequency 2 Hz

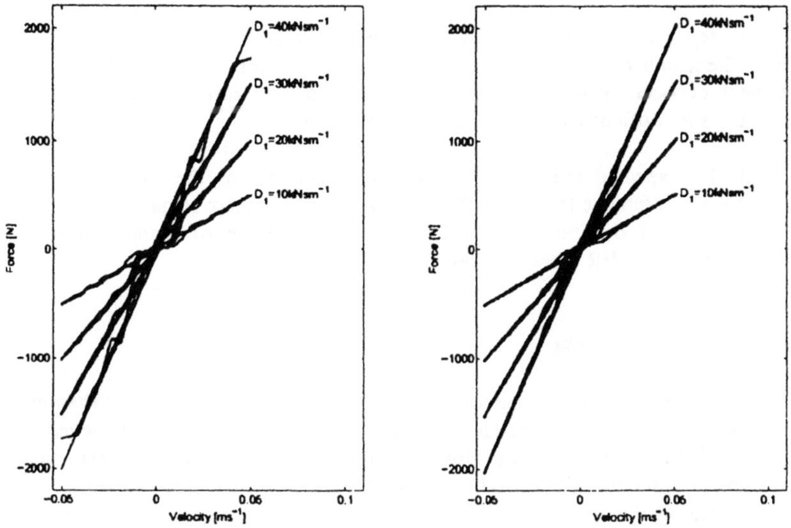

(a) (b)

Figure 5: Simulated (a) and experimental (b) response to a 1st order (D_1) gain, frequency 2 Hz

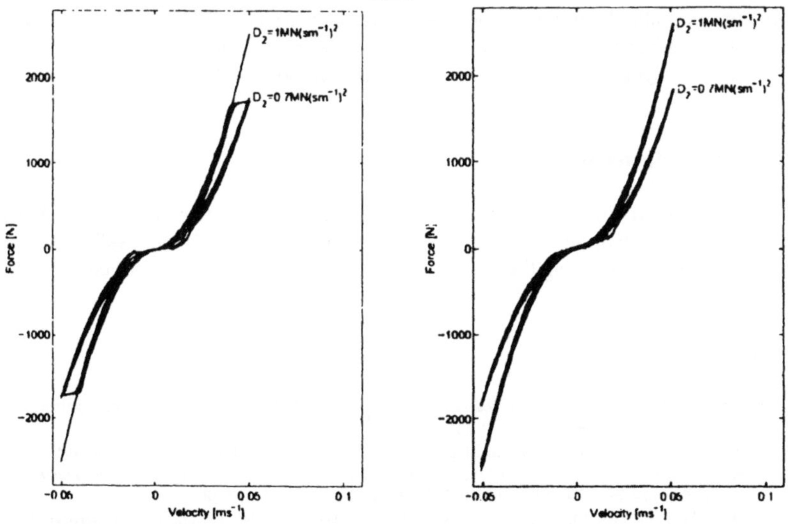

(a) (b)

Figure 6: Simulated (a) and experimental (b) response to a 2nd order (D_2) gain, frequency 2 Hz

Linearisation of the force-velocity response through setting D_1 is shown in Figure 5. In both numerical and experiments, which show excellent agreement, the effect of increasing D_1 from 10 kNsm^{-1} to 40 kNsm^{-1} is to produce a viscous characteristic with the desired slope.

The final step is to generate a quadratic damping characteristic by specifying D_2. Figure 6 shows the results of setting D_2 to two values, 0.70 $MN(sm^{-1})^2$ and 1.0 $MN(sm^{-1})$. Again agreement between model predictions and experimental results is excellent and the response has no difficulty in following the reference input.

As might be expected the results do degrade somewhat as the excitation frequency increases – see reference [6] for results at 8 Hz. However, there remains enormous scope for the design of dynamic controllers which are capable of compensating for the characteristics of the ER vibration damper.

5. Concluding Remarks

In this paper the authors have summarised recent work aimed at arbitrary shaping of the force-velocity characteristics of an ER vibration damper. Shaping is achieved through feedback of force and velocity from the main piston and using these signals to manipulate the strength of electric field applied to an ER flow-control valve.
After describing the experimental facility and the modelling procedure results were presented to illustrate the validation of the model. The model was then used as the basis of the feedback control design. By specifying the reference input to the feedback control system in the form of a second-order polynomial, it has been shown that Coulomb, viscous or quadratic damping characteristics can be generated. Future work will focus on refining the mathematical model to account for behaviour at higher frequency and an investigation of dynamic compensators in the control loop.

6. Acknowledgements

The authors would like to thank the European Research Office of the US Army for funding laboratory equipment under research control N68171-98-M-5388. The help of Mrs N A Parkes in preparing the manuscript is also gratefully acknowledged.

7. References

1. Stanway, R, Sproston, J L and El-Wahed, A K : Applications of ER fluids in vibration control: a survey, *Smart Materials and Structures* 5 (1996), 464-482.
2. Carlson, J D, Catanzarite, D M and St Clair, K A (1996) Commercial Magneto-rheological devices, in W A Bullough (ed), *Proceedings of the 5th International Conference on ER fluids, MR suspensions and associated technology*, World Scientific Publishing, Singapore, pp 20-28.
3. Sims, N D, Peel, D J, Stanway, R, Johnson, A R and Bullough, W A : The ER long-stroke damper: a new modelling technique with experimental validation, *Journal of Sound and Vibration* **229** (2000), 207-227.
4. Sims, N D, Stanway, R, Peel, D J, Bullough, W A and Johnson A R : Controllable viscous damping: an experimental study of an ER long-stroke damper under proportional feedback control, *Smart Materials and Structures* **8** (1999), 601-615.
5. Simulink (1998) The Mathworks, Inc, Nattick, MA, USA.
6. Sims, N D, Stanway, R, Johnson, A R, Peel, D J and Bullough, W A : Smart fluid damping: shaping the force-velocity response through feedback control, *Journal of Intelligent Material Systems and Structures*, submitted.

COLLOCATIVE CONTROL OF BEAM VIBRATIONS WITH PIEZOELECTRIC SELF-SENSING LAYERS

H. IRSCHIK, M. KROMMER, U. PICHLER
Johannes Kepler University of Linz, Division of Technical Mechanics
Altenbergerstr. 69, A-4040 Linz, Austria

1. Introduction

In the first part of the paper, we present a short summary of a simple but accurate electromechanically coupled theory for plane flexural vibrations of slender smart beams. The beams under consideration are assumed to be composed of electroded piezoelastic layers perfectly bonded to substrate layers. For a detailed derivation, see Krommer and Irschik [1], Irschik et. al. [2] and Krommer and Irschik [3]. In the present paper, spatially distributed self-sensing layers with an axially varying intensity of piezoelectric activity are considered within the theory of Refs. [1] - [3]. Self-sensing piezoelectric layers are single piezoelectric layers applicable for both, actuator and sensor applications. As an amazing fact from the point of control theory, perfect collocation between sensors and actuators is automatically provided by self-sensing piezoelectric layers. For details of the self-sensing sensor/actuator concept, see for example Dosch and Inman [4], Tzou and Hollkamp [5], Vipperman and Clark [6] and Oshima et. al. [7]. The main purpose of our derivations is the solution of a dynamic shape control problem, namely to find shape functions for a piezoelectric self-sensing layer such that vibrations due to known external forces can be exactly annihilated by piezoelectric actuation. It is shown that shape functions corresponding to the quasi-static bending moment distributions due to these external forces do represent solutions of this shape control problem. Previous investigations concerning this problem in the context of an electromechanically decoupled, not self-sensing theory have been presented in Irschik et. al., Refs. [8] - [10]. For Finite Element calculations in the context of the coupled theory without reference to self-sensing, see [11]. In the context of self-sensing layers, the shaped actuator can be also used as a collocated sensor, which is discussed in detail in the present paper. Exact dynamic shape control using self-sensing layers with the bending moment-type shape functions described above can be achieved only in case the span-wise distribution of the external loading is known in advance. If the span-wise distribution is not known, it may be useful to design self-sensing layers with the purpose of controlling the deflection at specific locations of the beam axis using shape functions with a particularly simple form. Such a concept is also presented in our paper. Furthermore, we make reference to so-called nilpotent shape functions, which are neither able to produce actuating effects nor to measure deflections of the beam. Nilpotent shape functions may be used to simplify the above derived solutions of the shape control problem. The presented theoretical findings are demonstrated for one-span beams with various boundary conditions.

U. Gabbert and H.S. Tzou (eds.), IUTAM Symposium on Smart Structures and Structronic Systems, 315–322.

2. Vibrations of Smart Beams

In the first part of this section, a short summary of a simple but accurate electromechanically coupled theory for plane flexural vibrations of smart beams is presented. For a detailed derivation, see Refs. [1] and [2]. The concept of self-sensing piezoelectric layers then is put into practice in the context of this beam theory.

A smart composite beam consisting of electroded piezoelectric layers perfectly bonded to substrate layers is studied. The beam is composed of a total of N linear elastic layers, with or without piezoelectric behavior. Flexural vibrations are considered, remaining within the range of the geometrically linearized theory. The axial coordinate of the straight beam is denoted by x, and z is the transverse coordinate. The deflection of the beam is assumed to take place in the (x,z)-plane. w_0 stands for the transverse deflection of the beam axis. Neglecting the influence of shear, the kinematic assumptions of Bernoulli and Euler, see e.g. Ziegler [12], are utilized. The dynamic equilibrium conditions, together with the boundary conditions at the ends of the beam, x=(0,l), are written as:

$$M_{,xx} = \rho^{(0)}\,\ddot{w}_0 - p_z\,, \qquad x=(0,l):\ w_{0,x} = \hat{\varphi}\ or\ M = \hat{M},\ \ w_0 = \hat{w}\ or\ M_{,x} = \hat{Q}\,. \tag{1}$$

In Eq. (1) p_z is the external force loading in transverse direction. The constitutive equations for the bending moment M read:

$$M = -D^{eff}\,w_{0,xx} - \bar{M}^e\,, \tag{2}$$

where the effective bending stiffness is given by the following cross-sectional integral

$$D^{eff} = \sum_{k=1}^{N} \int_{z^{k-1}}^{z^k} Y^k\,((1+\frac{d^k e^k}{\in^k})\,z^2 - \frac{d^k e^k}{\in^k}\,\frac{1}{2}\,(z^k + z^{k-1})\,z)\,dA(z)\,. \tag{3}$$

Young's modulus, effective piezoelectric coefficient, permittivity and piezoelectric modulus are denoted by Y^k, d^k, $\in^k$ and e^k, respectively. The extension of the k-th layer in thickness direction is given by z^k, z^{k-1}. Note that the influence of the direct piezoelectric effect is incorporated by means of effective stiffness parameters in Eq. (3). The piezoelectric moment $\bar{M}^e$, characterizing the piezoelectric actuation, is

$$\bar{M}^e = \sum_{k=1}^{N} S^k(x) \int_{z^{k-1}}^{z^k} Y^k\,d^k\,\frac{V^k}{h^k}\,z\,dz\,, \tag{4}$$

where the shape function $S^k(x) = b^{ek}(x)\,P^k(x)$ is given by the product of the polarization profile and the width of the electrodes, see Lee [13] for some practical realizations of shaped piezoelectric layers. The thickness of the layer is denoted by h^k, and V^k denotes the electric potential difference. Extensions of Eqs. (1) - (4) with respect to the influence of shear upon the deflection have been presented in Krommer and Irschik [3].

2.1. SELF-SENSING LAYERS

In the following, spatially distributed self-sensing layers with an axially varying time-dependent intensity of piezoelectric activity are considered. Self-sensing piezoelectric layers are single piezoelectric layers applicable for both, actuator and sensor purposes. Perfect collocation between sensors and actuators is automatically provided for self-sensing piezoelectric layers. For details of the self-sensing sensor/actuator concept, see for example Dosch and Inman [4], Tzou and Hollkamp [5], Vipperman and Clark [6] and Oshima et. al. [7].

The sensor behavior of self-sensing layers is discussed first, where the equations are formulated in the Laplace-domain. Utilizing the Gauss law, see Miu [14], the total electric charge at the upper electrode is calculated:

$$Q^k(s) = V^k(s)\, C^k - \frac{1}{h^k} \int_0^l \int_{z^{k-1}}^{z^k} S^k(x)\, e^k\, \varepsilon_{xx}(x, s)\, dz\, dx\ , \tag{5}$$

with the shape-function $S^k(x)$ of Eq. (4). For a detailed derivation of Eq. (5) see Irschik et. al. [15]. Next the actuating behavior is discussed. The actuating effect is described by means of dynamic Maysel's formula generalized to the case of piezoelastic problems, see Irschik et. al [16] and Irschik and Ziegler [17],

$$w_{0\,(\varepsilon)}(\xi, s) = -\int_0^l \bar{M}^e(x, s)\, w_{0\,(F_z),xx}(x, \xi, s)\, dx\ , \tag{6}$$

where $w_{0(Fz)}(x, \xi, s)$ is the deflection due to a single impulsive dummy force $F_z(s) = 1$ applied in the point ξ in transverse direction. In the following self-sensing layers are realized by applying an electric potential difference $V^k(s)$ at the electrodes and measuring the total charge $Q^k(s)$ at the electrodes, see Tzou and Hollkamp [5] for a proper electric circuit. From Eq. (5) it can be seen that the sensor signal is composed of two different parts. One part is due to the axial strain and the other is produced by the applied potential difference characterizing the piezoelectric actuation. In order to obtain a sensor signal that is proportional to the mechanical strain only, the influence of the actuating electric voltage has to be annihilated in the sensor signal of Eq. (5). In practice, this can be achieved by means of a differential amplifier, compare Tzou and Hollkamp [5].

3. Dynamic shape control

In order to make the idea of our proposed solution of the shaping control problem more clear, we particularly refer to a single self-sensing layer in the following. The case of several self-sensing layers can be treated by attributing different portions of the force loading to each piezoelastic layer.

3.1. FORMULATION OF THE DYNAMIC SHAPE CONTROL PROBLEM

Consider a piezoelectric layer of the laminate. Let the extension of the layer in thickness direction be defined by $z_1 \leq z \leq z_2$. Reformulating the sensor equation and actuator equation for the self-sensing layer gives, see Eqs. (5) and (6),

$$y(s) = Q(s) = e\,\frac{1}{2}\,(z_1 + z_2) \int_0^l S(x)\, w_{0,xx}(x, s)\, dx\,, \tag{7}$$

$$w_{0\,(\varepsilon)}(\xi, s) = -\,V(s)\,Y d\,\frac{1}{2}\,(z_1 + z_2) \int_0^l S(x)\, w_{0\,(F_z),xx}(x, \xi, s)\, dx\,. \tag{8}$$

It is the scope of this section to find solutions of the following dynamic shape control problem for the self-sensing layer:

Find a shape function for a piezoelectric self-sensing layer, such that flexural vibrations $w_{0(pz)}(\xi, s)$ *due to external forces are exactly annihilated by the piezoelectric actuation, i.e. find* $S(x)$ *such that*

$$w_0\,(\xi, s) = w_{0\,(\varepsilon)}(\xi, s) + w_{0\,(p_z)}(\xi, s) = 0\,. \tag{9}$$

In the context of self-sensing layers, an actuator can be also used as a collocated sensor. In a self-sensing layer the sensor signal can only be trivial in case the deflection of the beam axis vanishes identically, i.e., if the deflections are exactly annihilated by the piezoelectric actuation according to the requirements of Eq. (9). Previous investigations concerning the problem of Eq. (9) in the context of an electromechanically decoupled, not self-sensing theory have been presented in Irschik et. al., Refs. [8] - [10]. For Finite Element calculations in the context of the coupled theory without reference to self-sensing, see [11].

3.2. SOLUTIONS OF THE DYNAMIC SHAPE CONTROL PROBLEM

In order to find solutions of the dynamic shape control problem, Eq. (9), the dynamic Principle of Virtual Forces is utilized. Restricting to the case of pure bending without axial force, vibrations due to distributed external forces in transverse direction $p_z(x, s)$ follow from this Principle in the form

$$w_{0\,(p_z)}(\xi, s) = -\int_0^l M^q_{(p_z)}(x, s)\, w_{0\,(F_z),xx}(x, \xi, s)\, dx\,, \tag{10}$$

where $M^q{}_{(pz)}(x, s)$ is the quasi-static bending moment due to the external force loading. A comparison of Eq. (10) to Eq. (8) leads to a solution of the dynamic shape control problem as follows. If the piezoelectric actuating moment of the self-sensing layer is chosen proportional to the quasi-static bending moment due to an external distributed transverse force loading, $[V(s)\,Y\,d\,(z_1+z_2)/2]\;S(x) = -\,M^q{}_{(pz)}(x, s)$, vibrations imposed by this external force loading can be exactly annihilated. It can be seen from this relation that exact annihilation is possible, if the quasi-static bending moment $M^q{}_{(pz)}(x, s)$ is separable in space and time with a given space-wise distribution and an arbitrary time-evolution of the intensity, $M^q{}_{(pz)}(x, s) = M^q{}_{(pz)}(x)\,f(s)$, which is the case, if the external force loading is also separable, $p_z(x, s) = P(x)\,f(s)$. The static part of the quasi-static bending moment therefore is governed by the ordinary differential equation,

$M^q{}_{(pz)}(x)_{,xx} = - P(x)$. The solution of the dynamic shape control problem under consideration finally reads:

If on the one hand the shape function of the self-sensing layer is chosen proportional to the negative static part of the quasi-static bending moment $M^q{}_{(pz)}(x)$, *and on the other hand the time-evolution of the applied electric potential difference is chosen as the time-evolution of the external force loading* $f(s)$,

$$S(x) = -\left[Y d \frac{1}{2} (z_1 + z_2)\right]^{-1} M^q_{(p_z)}(x) \quad and \quad V(s) = f(s)\ , \tag{11}$$

vibrations due to the external force loading are exactly annihilated by the piezoelectric actuation.

As a structural example, beams with different types of support conditions loaded by a span-wise constant distributed transverse force, $p_z(x, s) = p_0 f(s)$ (p_0 = const.), with an arbitrary time-evolution $f(s)$ are considered.

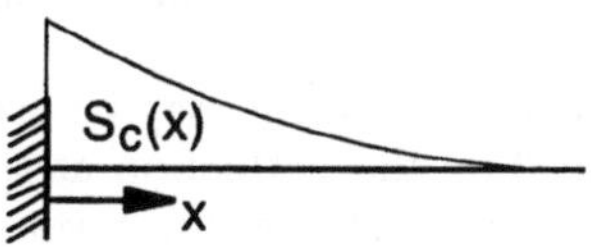

$$S_c(x) \mathrel{\hat{=}} -\frac{1}{2} p_0 (l-x)^2$$

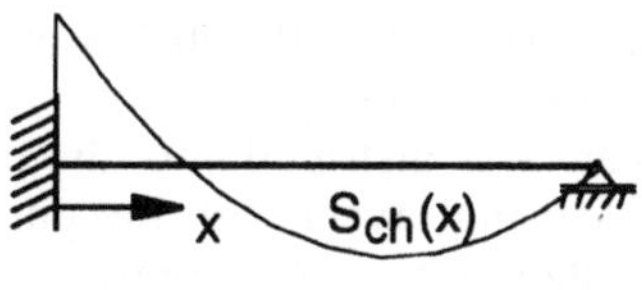

$$S_{ch}(x) \mathrel{\hat{=}} -\frac{1}{2} p_0 (l-x)^2 + \frac{3}{8} p_0\, l\, (l-x)$$

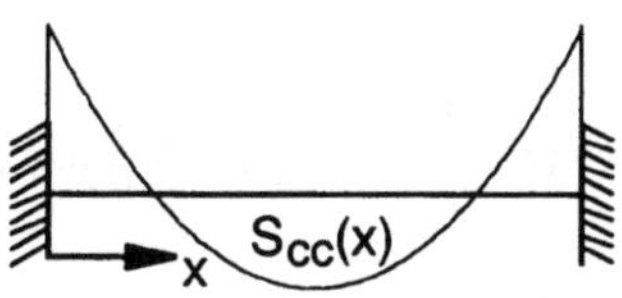

$$S_{cc}(x) \mathrel{\hat{=}} -\frac{1}{2} p_0 (l-x)^2 + \frac{1}{12} p_0\, l\, (5\, l - 6\, x)$$

Figure 1. Shape functions for controlling the vibrations due to a span-wise constant transverse force loading

In Figure 1 corresponding shape functions are presented for three different types of support conditions. The constant prefactor in Eq. (11) is understood for $S(x)$ in Figure 1. It is seen from Figure 1 that the shape function for the clamped-hinged beam is composed of a parabolic part and a span-wise triangular part. The shape function for the clamped-clamped beam is composed of a parabolic part, a span-wise constant part and a span-wise triangular part. The span-wise linear parts correspond to quasi-static bending moment distributions due to thermal loadings, which have been shown to represent nilpotent shape functions by Irschik et. al [2]. Nilpotent shape functions are shape functions not capable of measuring and of inducing any deflection. Such shape functions therefore can be added to any other shape function of the self-sensing layer without changing the behavior of the self-sensing layer. It therefore follows that the shape function of a self-sensing layer designed for the control of vibrations induced by a span-wise constant transverse force loading should be chosen as

$$S(x) \,\widehat{=}\, -\frac{1}{2}\, p_0\, (l-x)^2 \tag{12}$$

for all three of the beams shown in Figure 1.

Next the sensor signal is calculated. The sensor signal is found in the form,

$$y(s) = e\,\frac{1}{4}\,(z_1+z_2) \int_0^l p_0\, w_0(x,s)\, dx\,, \tag{13}$$

whence sensor and actuator are collocated. The sensor signal only becomes trivial, if the total deflection vanishes. The proposed design of the self-sensing layer corresponding to Eq. (12) thus can be used for the sake of controlling the vibrations of the three beams of Figure 1 due to a span-wise constant transverse force loading. Sensor and actuator are automatically perfectly collocated, a property, which is not lost by utilizing nilpotent shape functions in the design.

If the span-wise distribution of the force loading is not known, it may be useful to design self-sensing layers with the purpose of controlling the deflection at specific locations of the beam axis. Therefore it is suggested that a self-sensing layer with a sensor signal proportional to the deflection at the specified location should have an actuating behavior similar to the one of a single force acting at the specified location, say ξ, with an arbitrary time-evolution. From Eq. (11), it is seen that a shape function proportional to the static part of the quasi-static bending moment distribution due to this single dummy transverse force,

$$S(x) = -\left[Y d\,\frac{1}{2}\,(z_1+z_2)\right]^{-1} M^q_{(F_z)}(x)\,, \tag{14}$$

will exactly annihilate deflections due to a single force $F_z(x, s) = \delta(x-\xi)\, f(s)$ throughout the beam. Moreover, choosing $V(s)$ properly, any value of the deflection $w_{0(\varepsilon)}(\xi)$ can be reached in principle at the location ξ by means of a shape function of the form of Eq. (14). A suitable solution for the dynamic shape control problem of finding shape functions for controlling the deflection of the beam axis at a specific location therefore has been found in the form of the static part of the quasi-static bending moment distribution due to the unit transverse force with arbitrary time-evolution applied at this location, c. f. Eq. (14). The sensor measures the deflection at the specific location and the actuator acts like a single force applied at this location, sensor and actuator therefore are collocated.

In the example the same types of beams as in the previous example are taken into account. Solutions are shown in Figure 2, where $H(x)$ denotes the Heaviside step function. As in the previous example, the shape functions in Figure 2 include nilpotent shape functions. Therefore the shape function

$$S(x) \mathrel{\hat{=}} H(\xi - x)\,(x - \xi) \tag{15}$$

suffices to control the vibrations of the beam axis at location ξ for all three of the beams shown in Figure 2.

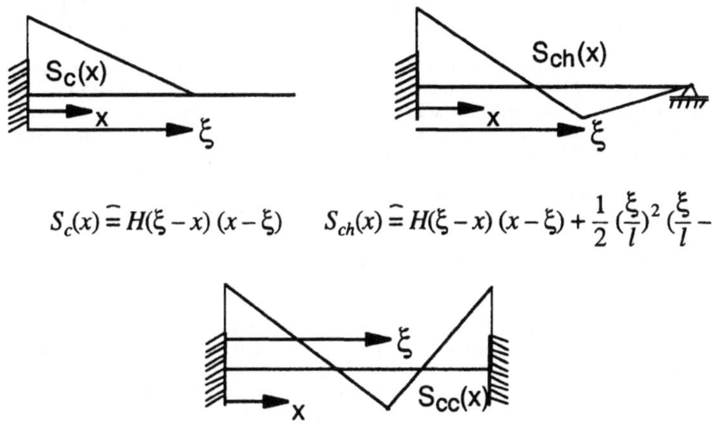

$$S_c(x) \mathrel{\hat{=}} H(\xi - x)\,(x - \xi) \qquad S_{ch}(x) \mathrel{\hat{=}} H(\xi - x)\,(x - \xi) + \frac{1}{2}\left(\frac{\xi}{l}\right)^2 \left(\frac{\xi}{l} - 3\right)(x - l)$$

$$S_{cc}(x) \mathrel{\hat{=}} H(\xi - x)\,(x - \xi) + \left(\frac{\xi}{l}\right)^2 \left(x\left(2\frac{\xi}{l} - 3\right) + 2\,l - \xi\right)$$

Figure 2. Shape functions for controlling the vibrations at a specific location ξ of the beam axis

Finally the sensor signal is calculated and it is found that the sensor signal becomes proportional to the deflection at the specified location in all three cases,

$$y(s) = -\,e\,\frac{1}{2}\,(z_1 + z_2)\,w_0(\xi, s)\ . \tag{16}$$

Designing a self-sensing layer with the shape function of Eq. (14), the sensor measures the deflection of the beam at location ξ of the beam axis.

4. Acknowledgement

Support of the authors H. Irschik, M. Krommer and U. Pichler by a grant of the Austrian National Science Fund (FWF), Contract P11993-TEC, is gratefully acknowledged.

5. References

1. Krommer, M. and Irschik, H.: Influence of coupling terms on the bending of piezothermoelastic beams, *Short Communications in Mathematics and Mechanics, GAMM 98 - Annual Meeting, University of Bremen, Germany, April 6-9, 1998, ZAMM* **79**, *Suppl.* 2 (1999), S421-S422.

2. Irschik, H., Krommer, M., Belyaev, A. K. and Schlacher, K.: Shaping of Piezoelectric Sensors/Actuators for Vibrations of Slender Beams: Coupled Theory and Inappropriate Shape Functions, *International Journal of Intelligent Material Systems and Structures* **9** (1999), 546-554.
3. Krommer, M. and Irschik, H.: On the Influence of the Electric Field on Free Transverse Vibrations of Smart Beams, *Journal of Smart Materials and Structures* **8** (1999), 401-410.
4. Dosch, J. J. and Inman, D. J.: A Self-Sensing Piezoelectric Actuator for Collocated Control, *International Journal of Intelligent Material Systems and Structures* **3** (1992), 166-185.
5. Tzou, H. S. and Hollkamp, J. J.: Collocated independent modal control with self-sensing orthogonal piezoelectric actuators (theory and experiment), *Journal of Smart Materials and Structures* **3** (1994), 277-284.
6. Vipperman, J. S. and Clark, R. L.: Hybrid Modal-Insensitive Control Using a Piezoelectric Sensoriactuator, *International Journal of Intelligent Material Systems and Structures* **7** (1996), 689-695.
7. Oshima, K., Takigami, T. and Hayakawa, Y.: Robust Vibration Control of a Cantilever Beam Using Self-Sensing Actuator, *JSME International Journal*, Series C **40**(4) (1997), 681-687.
8. Irschik, H., Belyaev, A. K. and Schlacher, K. (1994), Eigenstrain analysis of smart beam-type structures, in M. J. Acer and E. Penny (eds.), *Mechatronics: The Basis for New Industrial Denvelopments*, Comp. Mechanics Publ., Southampton pp. 487-492.
9. Irschik, H., Heuer, R., Adam, Ch. and Ziegler, F. (1998), Exact Solutions for Satic Shape Control by Piezoelectric Actuation, in Y.A. and G. J. Dvorak (eds.), *Proc. of IUTAM Symposium on Transformation Problems in Composite and Active Materials, Cairo, Egypt 1997*, Kluwer, Dordrecht, pp. 247 - 258.
10. Hagenauer, K., Irschik, H. and Ziegler, F. (1997), An Exact Solution for Structural Shape Control by Piezoelectric Actuation, in U. Gabbert (ed.), *VDI-Fortschrittsberichte, Smart Mechanical Systems - Adaptronics, Reihe 11: Schwingungstechnik Nr.244*, VDI-Verlag, pp.93-98.
11. Irschik, H., Krommer, M. and Pichler, U. (1999), Annihilation of beam vibrations by shaped piezoelectric actuators: Coupled theory, In: *CD-Rom Proceedings of the Joint Meeting: 137th regular meeting of the Acoustical Society of America, 2nd convention of the EAA: Forum Acusticum – integrating the 25th German Acoustics DAGA Conference*, Berlin, Germany, March 14-19, 1999.
12. Ziegler, F.: *Mechanics of Solids and Fluids*, 2nd ed., Springer, New York, 1995.
13. Lee, C. -K. (1992), Piezoelectric Laminates: Theory and Experiments for Distributed Sensors and Actuators, in H. S. Tzou and G. L. Anderson (eds.), *Intelligent Structural Systems*, Kluwer, Dordrecht, pp. 75 - 167.
14. Miu, D. K.: *Mechatronics: Electromechanics and Contromechanics*, Springer, New York, 1993.
15. Irschik, H., Krommer, M. and Pichler, U. (1999), Shaping of distributed piezoelectric sensors for flexural vibrations of smart beams, in V. V. Vradan (ed.), *Proc. of SPIE's 6th Annual International Symposium on Smart Structures and Materials, March 1-5, 1999, Newport Beach: Mathematics and Control in Smart Structures*, SPIE Vol. 3667, pp. 418-426.
16. Irschik, H., Krommer, M. and Ziegler, F. (1997), Dynamic Green's Function Method Applied to Vibrations of Piezoelectric Shells, in *Proc. of Int. Conf. 'Control of Oscillations and Chaos' (COC'97)*, August 27-29, 1997, St.Petersburg, pp. 381-388.
17. Irschik, H. and Ziegler, F.: Maysel's formula generalized for piezoelectric vibrations: Application to thin shells of revolution, *AIAA-Journal* **34** (1996), 2402-2405.

EFFICIENT APPROACH FOR DYNAMIC PARAMETER IDENTIFICATION AND CONTROL DESIGN OF STRUCTRONIC SYSTEMS

P. K. KIRIAZOV
Institute of Mechanics, Bulgarian Academy of Sciences
"Acad.G.Bonchev" Str., bl.4, BG-1113 Sofia, Bulgaria

1. Introduction

The proposed study addresses various-type structronic systems (SS) like robots having elastic joints/links or engineering structures with vibration/shape control. SS can be considered as functionally directed compositions of mutually influencing subsystems: control, actuator, structural, and sensor subsystems. Actuators may be of various-type, e.g., electrical motors, electro-hydraulic cylinders, piezo-electric, electro-magnetic actuators. SS may have, therefore, highly complex dynamics and the parameters describing them are often very difficult or impossible to estimate with the required accuracy. To model and control such complex mechanical systems is a challenging problem and one that has been addressed by many researchers, [5-7], [10], [15].

There are continuously increasing demands to SS for faster and stable response, more precise motion and reduced energy loss. To meet these complicated performance requirements, a conceptual framework is needed that considers at the same time the problems of full dynamic modelling, accurate parameter identification, and optimal robust control. In practice, one always has to seek for appropriate trade-off solutions of such interconnected problems.

Before setting the problems of parameter identification and control design of a SS, we have to define the controls, the controlled outputs, and the structure of the input-output relations between its subsystems [13]. Our intention is to formulate and develop control design concepts for SS having decentralised controllers at the lowest (servo) level. In practice, such controllers are preferred, as they are simpler, faster and more reliable than centralised ones.

As for the parameter identification, stochastic methods are often applied, which use gradient-based, least-square techniques to minimise the difference between the outputs of the system and its nominal dynamic model. Typically, some measured (exciting) trajectories are generated and this implies identifying all the model parameters simultaneously. Large-scale systems of equations are to be solved and the identification accuracy depends, on not only the choice of exciting trajectories and the mismatch between the model and the real system, but also on how close the solutions are to the global minimisers of the output error.

U. Gabbert and H.S. Tzou (eds.), IUTAM Symposium on Smart Structures and Structronic Systems, 323–330.

In this study, a deterministic approach for systematic identification of the dynamic parameters in multibody system (MBS) models of SS is proposed that has the potential to overcome all the above problems. It is an extension of the approach developed in [4] and has the following features:

- Decomposition of the identification task into a sequence of several easy-to-solve identification problems;
- Deterministic identification with simple estimation equations; possibility to evaluate the errors of identification for the purposes of robust control design;

In the current literature there are a few control-design methods which consider the robustness of the control performance in the design, for example, the μ-synthesis and the Quantitative Feedback Theory, [9]. The latter theory is based on classical frequency domain methods and all loops of multi-input multi-output systems need to be shaped individually. As verified in [1], the μ-synthesis, as the other well-known control design techniques for linear time-invariant systems, can not guarantee optimality and even feasibility in case of parameter inaccuracies/external disturbances (which is a rather realistic assumption for SS). In our opinion, that common drawback is due to the lack of trade-off design relations between the disturbance bounds and control force limits.

For MBS having decentralised control subsystems, we can overcome the above problems applying the control design approach proposed in [3]. It is based on a generalised diagonal dominant (GDD) condition on the control transfer matrix (CTM) which has been proven to be necessary and sufficient for such controllers to be robust against arbitrary-in-time disturbances. With CTM being GDD, we can derive optimal trade-off relations between bounds of disturbances and control function magnitudes. In this way, sliding-mode controllers or other bounded-input controllers with maximum degree of robust decentralised controllability can be designed.

The main points in the above-proposed approaches for identification and control design will be illustrated considering an example of a dynamic model of a platform (car body) with active suspensions.

2. Mathematical Modelling

2.1. DEFINITION OF PERFORMANCE VARIABLES AND CONTROLS

Assume that the mechanical subsystem of the SS (to be designed) is already determined. We define first a set of variables q (mechanical displacements) which will be controlled for the purpose of satisfying the requirements on the dynamic performance of the SS. They (like generalised co-ordinates) have to be independent of each other and their number should not be greater than the number of all actuators that can be assigned for controlling the SS-motion. Our concept is that, for a SS to be fully controllable in a decentralised manner, especially in the case of external disturbances, it must not be under-controlled.

Second, to enable the SS to fulfil the required motion tasks, controls u (driving inputs) are to be properly defined. Their number and points of application are closely related to the number and placement of the actuators. In principle, an appropriate controllability measure is needed in order to make optimal decisions about such and other design parameters. A design index, defined over all parameters responsible for the SS controllability, will be given in Section.4. By using it in the initial design stage regarding the number and placement of actuators, and sensors as well, we can give the SS maximum capability for robust decentralised controllability.

2.2. DYNAMIC MODELS

We need nominal dynamic models, relating the controls u and the controlled outputs q of SS, to be used for both purposes: accurate parameter identification and robust control design. SS are compositions of, basically, four mutually influencing subsystems: control, actuator, mechanical, and sensor subsystems, and, based on physical principles, one can find, in general, structures of full dynamic models.

Typically, control designers can efficiently work employing reduced-order dynamic models. To find such models, we can apply the so-called multibody system (MBS) approach [11]: the mechanical subsystem of a SS, though actually flexible, can be approximated by a system of rigid bodies connected by real or fictitious joints, springs, and dampers. Applying the Lagrange formalism, the dynamic behaviour of such a system can be described by the following vector differential equation

$$M(q_0)\ddot{q} = BF - v(q,\dot{q}) - s(q - q_0) - g(q_0) \tag{1}$$

where the main terms are the inertia $M(q)\ddot{q}$, driving BF, velocity $v(q,\dot{q})$, stiffness $s(q-q_0)$, and gravity $g(q)$ forces; F is the control force (torque) vector; B is the matrix representing the control force locations and, in our considerations; it is a square matrix as the number of degrees of freedom of the MBS is the same as the number of the controls; q_0 is an initial reference configuration.

The lumped parameter model (1) is accomplished with an appropriate mathematical model of the relationship between the control input and the control force $F = F(u)$. In practice, the following linearized model is often employed

$$F_i = -m_i\ddot{q}_i - c_i\dot{q}_i + k_i u_i \ , \tag{2}$$

with coefficients specifying the physical characteristics of the drive system corresponding to the i-th degree of freedom. In some cases (for example D.C. motors), these parameters can easily be estimated from the usually available data. The acceleration and velocity terms of (2) can be included in the corresponding terms of (1).

In this way, we can have a properly simplified dynamic model of the given SS, which will be used in Section.4 to present our control design approach. In the next section, we will propose a deterministic approach for identification of parameters in such models.

For simulating the identification and control of the SS, we should use much more accurate mathematical models. Such models for three-dimensional flexible SS including actuator/sensor dynamics are FEM-models developed in [6].

3. Parameter Identification

We have found that the identification process can be done much more precisely and efficiently if we estimate the parameters step-by-step in accordance with the order of input-output responses in the SS: control input, driving force, acceleration, velocity, and displacement. Our principle is, starting always from standstill $(q_0, 0)$, to chose as turning points in the identification tests to be those time-moments when the controlled outputs first change sensibly their values. When parameters to be identified depend on the SS configuration, we have to perform the identification procedure with different values of q_0.

The overall identification task can be decomposed into a sequence of several simpler identification problems for the groups of dynamic parameters describing the main dynamic terms:

1) actuator dynamics;
2) gravity forces;
3) (inverse) inertia matrix;
4) damping forces;
5) stiffness forces;

The corresponding procedures use the parameters determined in the preceding steps and only one control input is applied at each identification test. In this way, the nominal models are greatly simplified and the errors of identification can be reduced to a maximum degree. With such simplified nominal models, the upper limits for the identification errors can easily be estimated as functions of the model mismatch [4], and such estimates are indispensable for designing controllers with pre-specified robustness.

The approach above proposed enables sufficiently accurate identification of all the main dynamic parameters and they can be used in considering important analysis and control design issues. A central role in solving control design problems plays the CTM $M^{-1}B$, that is the product of the inverse inertia matrix and the matrix of control locations. Moreover, it can also be used to define explicit design relations for efficiently solving the problem of integrated control-structure design [3].

4. Control Design

Depending on the functionality of the SS, controlled actuators are assigned to carry out different motion tasks, e.g., achieve a target configuration (shape), produce some specified vibration (oscillation) motion, or damp unnecessary vibrations. In any of these motion tasks, we have to deal with the following control design problem: with known bounds of disturbances and desired system output accuracy, design a system of decentralised controllers that need minimum actuator forces to robustly stabilise the motion. To do that, the following natural decentralised controllability condition is assumed: the signs of any control input and the corresponding output (the acceleration) are the same, at least when the control input is at its maximum absolute value. In the face of bounded, but arbitrary-in-time disturbances, this condition can be fulfilled, as shown below, if only the CTM is GDD.

By feedforward control, we can compensate, to some extent, for the velocity, stiffness, and gravitation forces. Then, the following reduced model for the error dynamics can be used for the purpose of feedback control design

$$\ddot{e} = A(q)u + d \tag{3}$$

where $e = q - q^{ref}$, $A = M(q)^{-1}B$ is the CTM; the vector d stands for uncompensated terms and model inaccuracies, as well as for measurement and environment noises.

As a measure of tracking precision, we take the absolute value of $s = \dot{e} + \lambda e$, $\lambda > 0$. We will use decentralised controllers, which means that during motion the stabilising control u_i of each actuator depends solely on the corresponding controlled output s_i.

Definition: A decentralised controller is robust against random disturbances d with known upper bounds d^+ if it gets the local subsystem state $(\dot{q}, q)$ at each joint (degree of freedom) to track the desired state $(\dot{q}^{ref}, q^{ref})$ with maximum allowable absolute values s^+ of errors s.

Basic result [2]: A necessary and sufficient condition for a dynamic system (3) to be robustly controlled by a decentralised controller is that matrix A be GDD.

When matrix A is GDD, the non-negative matrix theory (see, e.g., [8]) states that there always exists a positive vector u^+ solving the following system of equations

$$A_{ii}u_i^+ - \sum_{j \neq i} |A_{ij}| u_j^+ = d_i^+, \quad i = 1, \ldots, n \tag{4}$$

Eqs. (4) present optimal trade-off relations between the bounds of model uncertainties and the control force limits. The greater the determinant Δ of this system of linear equations, the less control effort is required to overcome the disturbances. In other words, Δ quantifies the capability of MBS to be robustly controlled in a decentralised manner. For these reasons, Δ can be taken as a relevant design index for the subsystems whose parameters enter the CTM. The linearity of (4) makes it possible a decomposition of the integrated structure-control design problem into design problems for the SS's components [3]. For example, we can find optimal actuator sizes and optimal actuator/sensor placement.

For the feedback stabilisation, we can use sliding-mode controllers or other bounded-input controllers, and the magnitudes of the control functions are estimated from (4). To avoid chattering, the following continuous control functions of saturation type can be used [12]

$$\begin{aligned} &u_i(s_i) = u_i^+ \mathtt{sat}(s_i / \delta_i), \; i = 1, \ldots, n \\ &\mathtt{sat}(y) = y \quad \text{for} \quad |y| < 1 \quad \text{and} \quad \mathtt{sat}(y) = \mathtt{sgn}(y) \quad \text{for} \quad |y| \geq 1 \end{aligned} \tag{5}$$

Besides the control magnitudes u_i^+, the other control parameters λ_i and δ_i can be a subject to design if we need to further optimise the SS performance.

5. Case Study: active body control (ABC)

To illustrate the main points in the proposed approaches for identification and control of SS, we take into consideration a simple example a car body model with active suspensions at the four wheels. A similar example could be a platform or a box structure supported by four piezoelectric rods for active vibration isolation.

We have to deal with the problems of parameter identification, and control design for the three-degree-of-freedom motion (pitch/roll/bounce) of a rigid body with four actuators. In this case-study, a mathematical model similar to that given in [7] is used. For such a statically indeterminate mechanical system, it is reasonable to consider the two motions, pitch/bounce and roll/bounce, in separate [14]. Those motions are weakly coupled when the displacements are small. Moreover, with two actuators/sensors assigned for each of them, Fig. 1, such two-degree-of-freedom controlled dynamic systems satisfies the GDD-condition and our control design approach can be applied. Before that, we will consider some details regarding the identification of the main dynamic parameters.

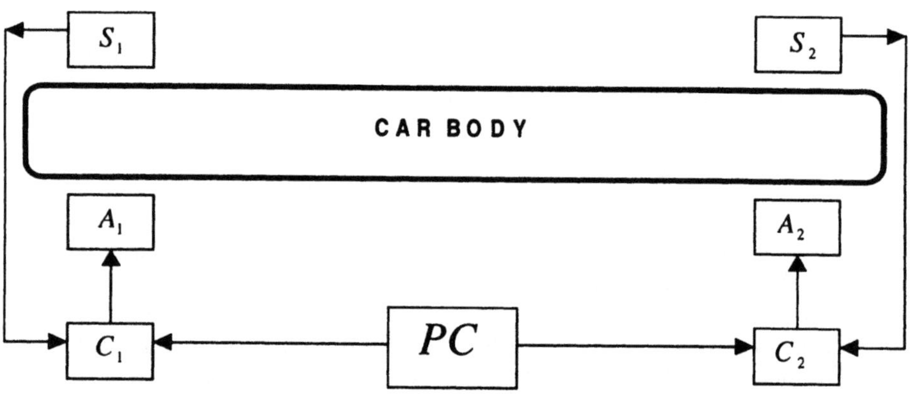

Figure 1: Block diagram of controlled pitch/bounce motion

where: *PC* – main computer for the ABC; *C* is a controller; *A* - actuator (electro-hydraulic cylinder, or other type of actuator, which may work in parallel with a spring-damper system); and *S* is displacement (velocity, or acceleration) sensor, measuring these variables with respect to a body or real space co-ordinate system.

1) To identify the gain coefficient k_i in (2) of any electro-hydraulic cylinder, we can apply a step control input and measure the generated force, just before indicating any movement of the mechanical part attached to the output link of the actuator.

2) The gravitation forces $g(q_0)$ at the suspension points can be determined finding those control inputs which provide a static balance for the suspended body.
3) The coefficients of the inverse inertia matrix A can be estimated from one-input identification tests where the outputs (accelerations) are measured at the very beginning of body motion, when we can neglect velocity and stiffness terms in (1).
4) The coefficients of the viscous friction are identified from tests similar to those in step 3) but the turning points are when the velocities first achieve substantial values and when we still can neglect the stiffness terms.
5) And the turning points for estimation of the spring constants are next time-moments when the stiffness terms acquire substantial values.

In the numerical considerations, random disturbances with non-zero mean value were introduced whose amplitudes were allowed to take values up 90% of the nominal accelerations (produced by the nominal control forces). Sliding-mode controllers (5) were designed according to the optimal trade-off relations (4). The robust transient behaviour, starting from a non-zero initial state is shown in Fig. 2.

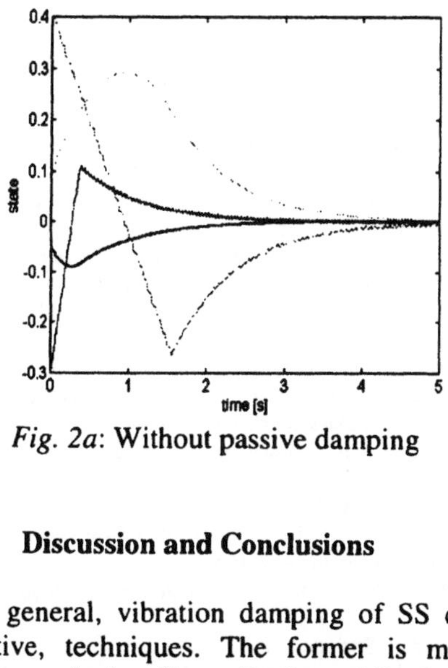

Fig. 2a: Without passive damping

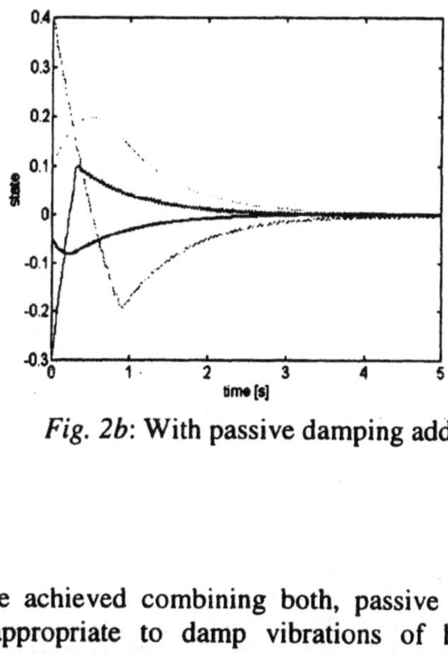

Fig. 2b: With passive damping added

6. Discussion and Conclusions

In general, vibration damping of SS can be achieved combining both, passive and active, techniques. The former is more appropriate to damp vibrations of high frequencies/small amplitudes, while the latter – for vibrations of low frequencies/large amplitudes. Therefore, our concept to use MBS models for designing active control is reasonable. The corresponding model structures can be described by a relatively small number of parameters. A deterministic approach for their parameter identification has been proposed which has the following advantages

- Decomposition of the identification task into a sequence of several easy-to-solve identification problems
- Accurate identification with simple estimation equations
- Minimum time/cost for the identification work

Moreover, the approach makes it possible finding bounds on the identification errors, which is very important for the robust control design. The main features of the proposed control design approach are

- It is based on the GDD-condition of the control transfer matrix - necessary and sufficient for robust decentralised controllability
- Explicit design relations defined over all the parameters that influence the dynamic performance - optimal trade-off relations between the bounds of model uncertainties and the control force limits
- For decentralised control, maximum degree of robustness can be achieved

Developing the proposed approaches for more complex SS, we can get an efficient methodology for solving the interconnected problems of their parameter identification and control design.

Acknowledgement: The financial support from the German Research Foundation *DFG* is gratefully acknowledged.

7. References

1. Keel, L.H. and S.P. Bhattacharyya (1997) Robust, fragile, or optimal? *IEEE Transactions on Automatic Control*, Vol.42, No.8, 1098-1105.
2. Kiriazov, P (1994) Necessary and sufficient condition for robust decentralized controllability of robotic manipulators. In: *Proc. American Control Conference*, Baltimore MA, 2285-2287.
3. Kiriazov, P. (1996) Robust integrated design of controlled multibody systems, in *Solid Mechanics and its Applications*, Vol.43, Eds. W. Schiehlen and D.Bestle, Kluwer Publ., 155-162
4. Kiriazov, P. (1996) On identification of controlled multibody systems, *Proc. of Int. Conf. on Identification in Mechanical Systems*, Swansea, U.K., Eds. M. Friswell & J. Mottershead, 127-135.
5. Kiriazov, P. (1999) On robust control of robots with elastic joints, *Proc. of the 4th Mashinenbau-Tage*, Magdeburg, Germany, Eds. Kasper, R. *et al.*, Logos-Verlag, 121-128.
6. Köppe, H., U. Gabbert, and H.-S. Tzou (1998) On three-dimensional layered shell elements for the simulation of adaptive structures, in U. Gabbert (Ed.), *Modelling and Control of Adaptive Mechanical Structures*, VDI Forschritt-Berichte, Reihe 11, Nr. 268, VDI Verlag Düsseldorf, 103-114.
7. Krtolica, R. and Hrovat, H. (1992) Optimal active suspension control based on a half-car model: an analytical solution, *IEEE Trans. On Automatic Control*, No. 4, 528-532.
8. Lunze, J. (1992) *Feedback Control of Large-Scale Systems*, Prentice Hall, UK.
9. Nwokah, O. D. I. and Yau, C.-H. (1993) Quantitative feedback design of decentralized control systems, *ASME Journal of Dynamic Systems, Measurement and Control*, 115 , pp. 452-466.
10. Pfeiffer, F. (1989) A feedforward decoupling concept for the control of elastic robots, *Journal of Robotic Systems*, 6(4), 407-416.
11. Schiehlen, W. (Ed.) (1990) *Multibody Systems Handbook*. Berlin: Springer.
12. Slotine, J.-J. and S. S. Shastry (1983) Tracking control of nonlinear systems using sliding surfaces with application to robotic manipulators, *Int. Journal of Control*, No.2, 465-492.
13. Skogestad, S. and Postlethwaithe I. (1996) *Multivariable Feedback Control: Analysis and Design*, Wiley.
14. Takezono, S., H. Minamoto, and K. Tao (1999) Two-dimensional motion of four-wheel vehicles, *Vehicle System Dynamics*, Vol. 32, No. 6, 441-458.
15. Tzou, H.-S., Guran, A. (Eds.) (1998) *Structronic Systems: Smart Structures, Devices and Systems*, World Scientific.

AN INTEGRAL EQUATION APPROACH FOR VELOCITY FEEDBACK CONTROL USING PIEZOELECTRIC PATCHES

J.M. SLOSS[1], J.C. BRUCH, JR[1,2]., S. ADALI[3], AND I.S. SADEK[4]
[1]*Dept. of Math. Univ. of Cal. , Santa Barbara, CA 93106, USA*
[2]*Dept. of Mech. & Envr. Engr., Univ. of Cal., Santa Barbara, CA 93106, USA*
[3]*School of Mech. Engr., Univ. of Natal, Durban, South Africa*
[4]*Dept. of Comp. Sci., Math., and Stat., Amer. Univ. of Sharjah, Sharjah, United Arab Emirates*

1. Introduction

Piezoelectric actuators and sensors are being used increasingly in applications involving vibration/position/shape control devices. Examples where they have been effectively employed include acoustics for noise cancellation [1-3] with applications to reduce interior noise in aircraft, aerodynamics to adjust wing surfaces [4] and electronics where they are used in the reading heads in video cassette recorders and compact discs as positioning devices [5].

Another example of the use of piezo materials in adaptive structures is shape control by piezo-actuation [6-8]. Adaptive materials and structures are presently being used in a variety of applications involving static control such as robotic end effectors and space structures [5, 9]. One of the important issues in the use of piezo actuators is their optimal deployment to minimize their weight and improve performance. In the present study the effectiveness of piezoelectric actuators, whose input is obtained by time differentiation of the sensor output, is studied by means of an integral equation formulation for varying actuator and sensor location. This type of control introduces a damping component.

Herein, the distributed vibration control of an elastic beam is considered. The beam is assumed to have a laminated patch piezoelectric sensor on the bottom of the beam and a laminated piezoelectric patch actuator on the top of the beam. The control will be a closed-loop velocity feedback control. The signal from the patch sensor is time differentiated and amplified and sent to the patch actuator. The formulation of this problem in terms of a differential equation is given in Tzou [10] and Banks *et al* [11].

The differential equation formulation for the problem is equivalent to an integral equation formulation in which the kernel is given explicitly. This will simplify finding a solution to the problem since the differential equation approach is non-standard, the equation itself contains distributions, whereas the integral equation approach is standard. Consequently the integral equation formulation facilitates the study of the effectiveness of the control. Eigensolutions of the integral equation are eigensolutions of the differential equation. In the event that the actuator patch and sensor patch are aligned,

U. Gabbert and H.S. Tzou (eds.), IUTAM Symposium on Smart Structures and Structronic Systems, 331–338.

the eigensolutions of the differential equation are eigensolutions of the adjoint integral equation. In [12] a similar integral equation approach was used for a displacement feedback control problem using patches.

2. Formulation

Assume the beam is uniform and of length ℓ with rectangular cross section and has a single sensor at $[x_{s_1}, x_{s_2}]$ and a single actuator at $[x_{a_1}, x_{a_2}]$ with $0 \le x_{s_1} < x_{s_2} \le \ell$ and $0 \le x_{a_1} < x_{a_2} \le \ell$. Let W(x, t) denote the transverse deflection of the beam, then the output signal $\varphi^s(t)$ of the distributed sensor is assumed to be [10]

$$\varphi^s = -\frac{h_s}{x_{s_2} - x_{s_1}} h_{31} r_1^s \left[\frac{\partial W}{\partial x}\right]_{x = x_{s_1}}^{x = x_{s_2}} \tag{1}$$

where h_s is the sensor thickness, h_{31} is the sensor piezoelectric constant and r_1^s is the distance from the central axis to the mid-plane of the sensor layer.

In the feedback control, the voltage φ^a applied to the actuator is proportional to the time differentiated output signal $\left(\dot{\varphi}^s = \frac{\partial \varphi^s}{\partial t}\right)$, namely

$$\varphi^a = G\, \dot{\varphi}^s$$

where G is the gain. The feedback voltage creates an externally applied distributed control moment M^a. In the absence of mechanical excitation, the equation of motion with externally applied control moment becomes

$$\rho A \frac{\partial^2 W}{\partial t^2} + YI \frac{\partial^4 W}{\partial x^4} - b \frac{\partial^2 M^a}{\partial x^2} = 0 \tag{2}$$

with A = bh, b = beam width, h = beam thickness, Y = Young's modulus, and I $= bh^3/12$, the moment of inertia. The closed loop feedback gives

$$M^a = r_1^a d_{31} Y_p \varphi^a = \overline{\xi} \frac{\partial}{\partial t}\left[\frac{\partial W}{\partial x}(x, t)\right]_{x = x_{s_1}}^{x = x_{s_2}} \left[H\left(x - x_{a_1}\right) - H\left(x - x_{a_2}\right)\right] \tag{3}$$

in which r_1^a is the effective moment arm, d_{31} is the actuator piezoelectric constant, Y_p is Young's modulus of the actuator, H(x) is the Heaviside function and

$$\bar{\xi} = -\frac{G\, h_s\, r_1^a\, d_{31}\, Y_p\, r_1^s\, h_{31}}{x_{s_2} - x_{s_1}} \tag{4}$$

Substituting the expression for M^a into equation (2) yields the equation of motion

$$\rho A\, W_{tt} + L[W] = 0 \qquad 0 < x < \ell \tag{5}$$

where

$$L[W] = L_o[W] + \alpha \frac{\partial}{\partial t} L_1[W] \tag{6}$$

with

$$L_o[W] = YI\, W_{xxxx} \text{ and } L_1[W] = -b\bar{\xi}\left[\frac{\partial W}{\partial x}\right]_{x = x_{s_1}}^{x = x_{s_2}} \left[H''\left(x - x_{a_1}\right) - H''\left(x - x_{a_2}\right)\right]$$

in which subscripts x and t refer to differentiation with respect to these variables and the primes refer to derivatives with respect to x. In equation (6) α is a parameter that has the value either 0 or 1. Its value is 0 if the actuator is extended to the entire length of the beam. Otherwise, its value is 1.

Case 1: A cantilever beam with $x_{s_1} = x_{a_1} = 0$, $x_{s_2} = x_{a_2} = \ell$. In equation (6) set $\alpha = 0$ and obtain

$$\rho A\, W_{tt} + YI\, W_{xxxx} = 0 \qquad \text{on} \qquad 0 < x < \ell \tag{7}$$

with boundary condition clamped at x = 0 and feedback moment proportional to the beam slope at $x = \ell$. Then

$$B_1[W] = W(0,t) = 0\,, \quad B_2[W] = W_x(0,t) = 0 \tag{8a}$$

and with $\beta = b\bar{\xi}/YI$

$$B_3[W] = W_{xx}(\ell,t) + \beta W_{xt}(\ell,t) = 0 \quad , \quad B_4[W] = W_{xxx}(\ell,t) = 0 \tag{8b}$$

and the control becomes a boundary control.

Case 2: $0 < x_{s_1} < x_{s_2} < \ell$ and $0 < x_{a_1} < x_{a_2} < \ell$. In the equation of motion set $\alpha = 1$ and the linear boundary conditions can be quite general. In this case we denote them by

$$B_1[W] = 0 \;,\; B_2[W] = 0 \;\text{ at } x = 0 \text{ and } B_3[W] = 0 \;,\; B_4[W] = 0 \;\text{ at } x = \ell\,. \tag{9}$$

3. Free Vibration Analysis

Setting

$$W(x,t) = \psi(x)\, e^{i\omega t} = \psi(x)\, e^{i(\omega_1 + i\omega_2)t}$$

then equation (6) becomes when $\omega \neq 0$

$$\sigma(\omega)\{L_o[\psi] + i\omega\alpha L_1[\psi]\} = \psi \qquad 0 < x < \ell \tag{10}$$

where

$$\sigma(\omega) = \left(\omega^2 \rho A\right)^{-1} \tag{11}$$

with

$$\begin{gathered} 0 \leq x_{s_1} < x_{s_2} \leq \ell \quad , \quad 0 \leq x_{a_1} < x_{a_2} \leq \ell \\ B_1[\psi] = 0 \ , \ B_2[\psi] = 0 \quad \text{at } x = 0 \\ B_3[\psi] = 0 \ , \ B_4[\psi] = 0 \quad \text{at } x = \ell \end{gathered} \tag{12}$$

Case 1: Set $\alpha = 0$ in equation (10) to obtain the differential equation

$$\sigma(\omega) L_o[\psi] = \sigma(\omega) YI\, \psi_{xxxx} = \psi \tag{13}$$

and take equation (8) for the boundary conditions.

Case 2: Set $\alpha = 1$ in equation (10) to obtain the differential equation

$$\sigma(\omega)\{L_o[\psi] + i\omega L_1[\psi]\} = \psi \tag{14}$$

and take equation (12) for the general linear boundary conditions.

For case 1 the differential equation approach to finding the eigenvalues and corresponding eigenfunctions is fairly straightforward. For case 2 the situation is complicated by the fact that the differential equation contains terms that are distributions. Because of the difficulty of finding eigensolutions directly, we introduce the following integral equation.

4. An Integral Equation

Let

$$g(x,s) = g_1(x,s)\, H(s-x) + g_1(s,x)\, H(x-s) \tag{15}$$

where for

Case 1: Patch of full beam length for a cantilever beam:

$$g_1(x,s)=\left[-\frac{1}{6}x^3+\frac{1}{2}sx^2-\frac{\beta}{4(1+\beta\ell)}s^2x^2\right] \tag{16}$$

and for

Case 2: Patch of partial beam length:

g(x, s) satisfies the following conditions

(i) $$\left(\frac{1}{YI}\right)L_o\left[g(x,s)\right]=g_{xxxx}(x,s)=\delta(x-s) \tag{17}$$

(ii) $$\begin{aligned} &B_j\left[g(x,s)\right]=0 \ \text{ at } \ x=0 \ \text{ for } \ j=1,2 \\ &B_j\left[g(x,s)\right]=0 \ \text{ at } \ x=\ell \ \text{ for } \ j=3,4 \end{aligned} \tag{18}$$

and for a cantilever beam

$$g_1(x,s)=-\frac{1}{6}x^3+\frac{1}{2}sx^2 \tag{19}$$

Let

$$K(x,s,\omega)=\frac{g(x,s)}{YI}+p(x)\,q(s,\omega) \tag{20}$$

where

$$\begin{aligned} p(x)&=0 \qquad 0\le x\le x_{a_1} \\ &=p_2(x)\left[H(x-x_{a_1})-H\left(x-x_{a_2}\right)\right]+p_3(x)\,H\left(x-x_{a_2}\right) \end{aligned} \tag{21}$$

with

$$p_2(x)=\frac{1}{2}\left(x-x_{a_1}\right)^2$$

$$p_3(x)=\left(x_{a_2}-x_{a_1}\right)\left(x-x_{a_2}\right)+\frac{1}{2}\left(x_{a_2}-x_{a_1}\right)^2$$

and

$$q(s,\omega)=\frac{i\omega b\bar{\xi}\left[g_x\left(x_{s_2},s\right)-g_x\left(x_{s_1},s\right)\right]}{\left[YI-i\omega b\bar{\xi}\left(p'\left(x_{s_2}\right)-p'\left(x_{s_1}\right)\right)\right]}\frac{1}{YI} \tag{22}$$

and consider the integral equation:

$$\sigma(\omega)\psi(x)=\int_0^{\ell} K(x,s,\omega)\,\psi(s)\,ds \tag{23}$$

5. Method of Solution of the Integral Equation

The method for solving the integral equation involves the following steps.

1. Choose a complete orthonormal set of functions.
2. Expand the kernel in terms of the Fourier Series of the orthonormal functions.
3. Reduce the integral equation to an infinite set of linear equations.
4. Determine an integer n so large that the truncated Fourier Series containing N terms is adequate.
5. Consider the integral equation in which the kernel is replaced by the N-term Fourier Series expansion of the kernel.
6. This new integral equation has a degenerate kernel and its solution is reduced to solving a linear system of N equations in N unknowns.

6. Example

As an illustration of the integral equation method, numerical results for the natural frequencies of a plexiglas beam with polyvinylidene fluoride sensor and actuator layers will be considered. Some of these frequencies will be compared with those found by Tzou [10] in which material properties of the beam, sensors, and actuators were taken as follows:

Plexiglas Beam

$Y=3.1028\times10^{9}\left(N/m^{2}\right)$ (Young's modulus) $h=1.6\times10^{-3}$ (m) beam thickness)

$\rho=1190.0\left(Kg/m^{3}\right)$ (mass density) $b=0.01$ (m) beam width)

$\ell=0.1$ (m) beam length)

Polyvinylidene Flouride Sensor/Actuator

$Y_p = 2.00\times10^9 \left(N/m^2\right)$ (Young's modulus) $h_s = 40\ (\mu m)$ (sensor thickness)

$\rho_p = 1800.0 \left(Kg/m^3\right)$ (mass density) $h_a = 40\ (\mu m)$ (actuator thickness)

$d_{31} = 2.3\times10^{-11} \left(\frac{m/m}{V/m}\right)$ (piezoelectric constant)

$h_{31} = 4.32\times10^8 \left(\frac{V/m}{m/m}\right)$ (piezoelectric constant)

For this example, 3 terms in the truncated Fourier Series expression of the solution using the eigenfunctions for the uncontrolled cantilever beam as the approximating functions were adequate to obtain a high degree of accuracy.

7. Numerical Results

Numerical results for the case when the sensor patch and the actuator patch are the full length of the beam were compared with the respective analytic results in [10] and are shown in Table 1.

TABLE 1. Comparison of the integral equation (I.E.) results for the natural frequencies (Hz) with those in [10] (Tzou) for Sensor = Actuator = ℓ (note $G^* = G/\ell$)

(I.E.) Mode	G = 100		(Tzou) Mode	G* = 1000	
	ω_1	ω_2		ω_1	ω_2
1	66.25	2.33	1	66.2	2.4
2	358.77	3.19	2	358.7	2.9
3	893.95	3.59	3	885.9	3.2
	G = 10			G* = 100	
1	51.69	16.18	1	51.7	16.2
2	351.71	31.26	2	351.7	31.2
3	888.03	35.23	3	881.2	31.2

Note because of a technical reason the interval for the patch $[0,\ell]$ was taken to be $\left[10^{-8}, \left(1-10^{-8}\right)\ell\right]$. The reason for this is that we are treating the boundary control as the limit of the patch distributed control for increasing patch sizes.

In Table 2 an aligned sensor and actuator are considered and the modification of the natural frequencies due to the patch control are compared with the uncontrolled (G = 0) natural frequencies.

TABLE 2. I. E. results (ω_1, ω_2) in Hz for patches where Sensor = Actuator = [0.4ℓ, 0.6ℓ]

G	1st mode	2nd mode	3rd mode
0	(41.735, 0.0)	(261.551, 0.0)	(732.350, 0.0)
0.1	(41.736, 0.026)	(261.701, 4.102)	(732.354, 0.032)
1.0	(41.765, 0.254)	(277.521, 37.404)	(732.506, 0.123)
10.0	(43.030, 1.027)	(339.351, 11.658)	(732.602, 0.020)
100.0	(43.794, 0.158)	(340.679, 1.176)	(732.604, 0.002)

8. Conclusions

The integral equation approach introduced in this study provides a method for converting single patch velocity feedback control problems formulated as differential equations into integral equations. The eigenvalues of the integral equation can be easily and accurately approximated when the kernel of the integral equation is expanded in the eigenfunctions of the uncontrolled beam and only a few of the terms are considered. When this is done the integral equation becomes degenerate and is easily solved.

The eigensolutions of the integral equation are eigensolutions of the differential equation. When the sensor and actuator patches are aligned, the eigensolutions of the differential equation are eigensolutions of the adjoint integral equation. Using a numerical example, the effectiveness of the patch control is contrasted with full beam length sensor/actuator control.

Acknowledgement
This material is based upon work supported by the National Science Foundation under award No. INT-9906092.

9. References

1. Clark, R.L. and Fuller, C.R.: A model reference approach for implementing active structural acoustic control, *J. Acoustical Society of America* **92** (1992), 1534 - 1545.
2. Clark, R.L. and Fuller, C.R.: Experiments on active control of structurally radiated sound using multiple piezoelectric actuators, *J. Acoustical Society of America* **91** (1992), 3313 - 3320.
3. Hsu, C.Y., Lin, C.C. and Gaul, L.: Vibration and sound radiation controls of beams using layered modal sensors and actuators, *Smart Materials and Structures* **7** (1998), 446 - 455.
4. Barrett, R.: Active plate and wing research using EDAP elements, *Smart Materials and Structures* **1** (1992), 214 - 226.
5. Huber, J.E., Fleck, N.A. and Ashby, M.F.: The selection of mechanical actuators based on performance indices, *Proceedings of the Royal Society, London A* **453** (1997), 2185 - 2205.
6. Koconis, D.B., Kollar, L.P. and Springer, G.S.: Shape control of composite plates and shells with embedded actuators, I. Voltage specified, *J. Composite Materials* **28**(5) (1994), 415 - 458.
7. Koconis, D.B., Kollar, L.P. and Springer, G.S.: Shape control of composite plates and shells with embedded actuators, II. Desired shape specified, *J. Composite Materials* **28**(5) (1994), 459 - 482.
8. Lin, C.C., Hsu, C.Y. and Huang, H.N.: Finite element analysis on deflection control of plates with piezoelectric actuators, *J. Composite Structures* **35** (1996), 423 - 433.
9. Rao, S.S. and Sunar, M.: Piezoelectricity and its use in disturbance sensing and control of flexible structures: A survey, *Applied Mechanics Reviews* **47** (1994), 113 - 123.
10. Tzou, H.S.: *Piezoelectric Shells*, Kluwer Academic Publishers, Dordrecht, The Netherlands, 1993.
11. Banks, H.T., Smith, R.C. and Wang, Y.: The modeling of piezoceramic patch interactions with shells, plates and beams, *Quarterly of Applied Mathematics* **LIII**(2) (1995), 353 - 381.
12. Sloss, J.M., Bruch, Jr., J.C., Adali, S. and Sadek, I.S.: Piezoelectric patch control using an integral equation approach, *J. Thin-Walled Structures*, (in press).

DECENTRALISED MULTIVARIABLE VIBRATION CONTROL OF SMART STRUCTURES USING QFT

M. ENZMANN, C. DÖSCHNER

IFAT - Otto-von-Guericke-Universität Magdeburg
Postfach 4120, 39016 Magdeburg

{ *Marc.Enzmann*} { *Christian.Doeschner*} *@E-Technik.Uni-Magdeburg.DE*

1. Introduction

Structronic systems consist of the basic mechanical structure, the actuators, sensors, and a number of electronic components. All the components of a smart structure can be considered to be linear if they are operated appropriately. However, as all components are inherently nonlinear and possibly time-variant robustness becomes an issue in controller design.
Quantitative Feedback Theory (QFT) is a "robust" method with a strong orientation towards practical application, the mathematical background is not too complicated and the application of the theory allows direct insights in the trade-offs between conflicting design goals [2]. Furthermore, QFT design has been shown to compare favourably with other design algorithms, e.g. for the ECC flexible transmission benchmark [5, 6].
As QFT is based on classical frequency domain methods, all loops of a MIMO system have to be shaped individually. A number of MIMO design approaches which simplify the resulting complex design problem can be found in the literature (see [2, 3, 9]).
In this presentation we use a method, which shares the basic idea -sequential loop closure- with Maynes's approach [7]. Yet our design uses QFT to accomodate stability-, performance- and integrity-issues in the sequential design.

2. Basics of QFT

QFT is a theory, that was developed to design controllers for plants with large uncertainties. It is an extension of the classical frequency-domain approaches.

U. Gabbert and H.S. Tzou (eds.), IUTAM Symposium on Smart Structures and Structronic Systems, 339–346.

QFT allows to work with structured and unstructured model uncertainties. In most cases however, real parametric uncertainties are considered.
Due to the model uncertainties at a frequency Ω_k the system's response to the excitation is a set of values in the complex plane:

$$G(\jmath\Omega_k) \in \mathbb{G}(\jmath\Omega)$$

Assume now, that a controller $H(s)$ is found, which is capable of restricting the behaviour of all possible plant variations for the closed loop to a set of allowable transfer-functions at Ω_k for disturbances at the input $\mathbb{T}_D(\jmath\Omega_k)$. Then

$$\frac{\mathbb{G}(\jmath\Omega_k)}{1+\mathbb{G}(\jmath\Omega_k)H(\jmath\Omega_k)} \in \mathbb{T}_D(\jmath\Omega_k)$$

holds. Solving for $H\,(\jmath\Omega_k)$ and premultiplying with the nominal plant $G_0\,(\jmath\Omega_k)$ results in:

$$G_0(\jmath\Omega_k)H(\jmath\Omega_k) \in \frac{G_0(\jmath\Omega_k)}{\mathbb{G}(\jmath\Omega_k)}\frac{\mathbb{G}(\jmath\Omega_k)-\mathbb{T}(\jmath\Omega_k)}{\mathbb{G}(\jmath\Omega_k)} \tag{1.1}$$

The set on the right-hand side of equation 1.1 is called a *bound*, $\mathbb{T}_D(\jmath\Omega_k)$ is a *boundary* and $\mathbb{G}(\jmath\Omega_k)$ is a *template*. Bounds can be defined for all closed-loop transfer functions, e.g. the sensitivity-function $S(s)$ or the complementary sensitivity function $T(s)$.
The idea is to find templates and boundaries for a number of relevant frequencies Ω_k and to shape the nominal open loop, so that $G_0(\jmath\Omega)H(\jmath\Omega)$ is contained in all bounds (if possible).
QFT allows to design a controller using the NOMINAL plant model, while at the same time the problems of robust stability and robust performance are considered. Furthermore the application of QFT makes the trade-offs obvious, which are inherent in any controller design. Assume that we have found bounds for $S(s)$ and $T(s)$ and the intersection is empty:

$$\mathbb{S}(\jmath\Omega_k) \cap \mathbb{T}(\jmath\Omega_k) = \emptyset$$

then the problem is infeasible and we will have to reconsider our boundaries (i.e. our expectations for the closed-loop).

3. Decentralised Control for Flexible Structures

One well known property of intelligent structures is, that the system's response to any excitation can be decomposed modally. In terms of transfer-functions this can be written as follows:

$$G(s, f_{s_a}, f_{s_s}) = \sum_{k=1}^{\infty} G_{ALS}(s) \cdot \frac{V_k(f_{s_a}, f_{s_s})}{s^2 + 2\delta_k s + \omega_k^2} \tag{1.2}$$

with G_{ALS} the transfer-function which describes the electronic components and the transduction. The f_{s_a}, f_{s_s} are functions of the location, shape and type of the actuator and sensor. These functions correlate with the eigensolutions of the partial differential equation and with the eigenvectors of the finite element model (see e.g. [1]).
As all eigenfunctions have a countable finite number of zeros, the placement of an actuator or sensor near one of the zeros of an eigenfunction will result in low controllability/observability of the respective mode ($\|V_k\| \ll 1$). However, it is in many cases possible to find actuator and sensor locations that give good controllability and observability of the important modes.
A small number of actuators and sensors means that the number of electronic components (mainly filters and amplifiers) is small. This reduces the weight and cost of the structure and makes for a more compact build of the electronics package.
Taking all this into consideration, it would seem that there are only three good reasons to apply more than one sensor and actuator in a vibration control problem:

1 There is one (or more) mode(s) which is (are) not sufficiently controllable and/or observable for the given locations.

2 The energy output of the actuator(s) is (are) too low to enhance the damping sufficiently.

3 Safety-critical applications need redundancy.

It is evident, that controller design has to interact with the design of the structure. The placement of actuators and sensors, including the choice of actuator/sensor type and the number of actuators and sensors to be placed has to reflect the goals for the controller design.
One feasible approach might be to optimise the locations for a pre-determined number of actuators and sensors with respect to the observability and controllability of the modes in the frequency spectrum of interest. Starting with the best pairing a SISO controller is designed. If the result is not satisfying, we can add either only a second actuator or a second actuator/sensor pair and design a second controller. This procedure is continued, until the result is satisfactory.

4. Model Uncertainties Due to Control Interaction

To show how uncertainties arise due to control action, we consider the transfer function from the p-th control signal u_p to the q-th measurement y_q if the system

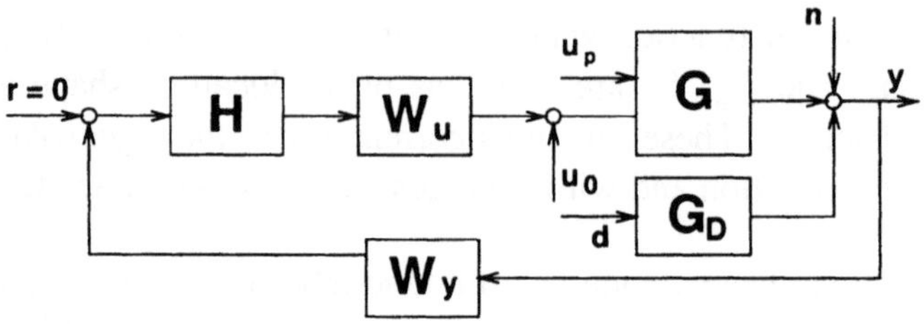

Figure 1 Uncertainties due to controller action

is given as in figure 1. In state-space notation the plant $\mathbf{G(s)}$ and the controller $\mathbf{H(s)}$ are given as

$$\dot{\mathbf{x}} = \mathbf{Ax} + \mathbf{B}\left(\mathbf{W_u u} + \mathbf{u_0}\right) + \mathbf{B_D d} \qquad \dot{\mathbf{x}}_\mathbf{H} = \mathbf{A_H x_H} + \mathbf{B_H W_y y}$$
$$\mathbf{y} = \mathbf{Cx} \qquad \mathbf{u} = -\mathbf{C_H x_H}$$

if we assume that plant and controller are strictly proper. The $\mathbf{W_u}$ and $\mathbf{W_y}$ are static diagonal matrices, with elements in the range $[0, 1]$. These matrices are a measure for the functionality of the actuators and sensors. If sensor 3 is fully functional, then $W_y[3, 3]$ is set to one, if it fails completely, the $W_y[3, 3]$ is zero. With u_p part of the vector $\mathbf{u_0}$, we can write for the closed loop

$$\begin{bmatrix} \dot{\mathbf{x}} \\ \dot{\mathbf{x}}_\mathbf{H} \end{bmatrix} = \underbrace{\begin{bmatrix} \mathbf{A} & -\mathbf{B W_u C_H} \\ \mathbf{B_H W_y C} & \mathbf{A_H} \end{bmatrix}}_{\mathbf{A_c}} \begin{bmatrix} \mathbf{x} \\ \mathbf{x_H} \end{bmatrix} + \underbrace{\begin{bmatrix} \mathbf{B} & \mathbf{B_D} \\ \mathbf{0} & \mathbf{0} \end{bmatrix}}_{\mathbf{B_c}} \begin{bmatrix} \mathbf{u_0} \\ \mathbf{d} \end{bmatrix}$$

$$\mathbf{y} = \underbrace{\begin{bmatrix} \mathbf{C} & \mathbf{0} \end{bmatrix}}_{\mathbf{C_c}} \begin{bmatrix} \mathbf{x} \\ \mathbf{x_H} \end{bmatrix}$$

The transfer-function from u_p to y_q is then given by

$$G_{qp}(s) = \mathbf{C_c}[\mathbf{q}, :] \left(\mathbf{sI} - \mathbf{A_c}\right)^{-1} \mathbf{B_c}[:, \mathbf{p}] \tag{1.3}$$

Clearly the system matrix depends on the weighting matrices $\mathbf{W_u}$ and $\mathbf{W_y}$. The resulting uncertainty can be considered in the standard QFT manner, calculating the additional uncertainties in the model (i.e. the respective templates) and including these in the calculation of the bounds.

One approach to model real parametric uncertainties in state-space models is given through affine-parameter dependent models. This type of modelling allows to describe parameter dependent matrices as linear combination of matrices:

$$\mathbf{A}(\vec{p}) = \mathbf{A}_0 + \sum_{k=1}^{N_p} p_k \cdot \mathbf{A}_k \qquad \text{with:} \quad p_k \in \left[p_k^{MIN}, p_k^{MIN}\right]$$

Obviously we can describe the matrices in the plant description and the weighting matrices as affine parameter dependent. If we consider that the product of two affine parameter dependent matrices

$$\mathbf{M}(\vec{p}) \cdot \mathbf{N}(\vec{q}) = \mathbf{M}_0\mathbf{N}_0 + \sum_{k_p=1}^{N_p} p_{k_p}\mathbf{M}_{k_p} \cdot \mathbf{N}_0 + \sum_{k_q=1}^{N_q} q_{k_q}\mathbf{M}_0 \cdot \mathbf{N}_{k_q} + \\ + \sum_{k_p=1}^{N_p}\sum_{k_p=1}^{N_p} \left(p_{k_p}q_{k_q}\right) \mathbf{M}_{k_p}\mathbf{N}_{k_q}$$

can be recast into a new affine parameter dependent matrix, then the closed loop can be considered to be affine parameter dependent, so that QFT templates can be calculated using the mapping-theorem [8].

At this point we have to introduce one distinction in the calculations. Of course we need a stable system, no matter how many actuators or sensors fail. Therefore all possible combinations of actuator and sensor failures must be considered. For the robustness of the performance however we must define allowable degradations, e.g. with one sensor or actuator failure, we still need full performance. This leads to a second uncertainty model, which is relevant for the calculation of performance bounds.

5. The Design-Scheme and its Application

Considering that the aim of the design is robust performance, the design scheme which is sketched at the end of section 3 can be elaborated as follows:
After thoroughly analysing the system:

1 Specify the overall performance goal.

2 Specify under which conditions the controller will have to deliver nominal performance, and which performance degradation is allowable.

3 Design the controller.

4 Check the performance. Test the robustness. If the specifications are not robustly met , repeat steps 3 and 4 until the performance is satisfactory or it is evident, that with the given structure the desired performance can not be obtained.

The application is now to be shown in a simulative investigation. The plant model is based on a FE model of a plate. It has four outputs and five inputs of which four

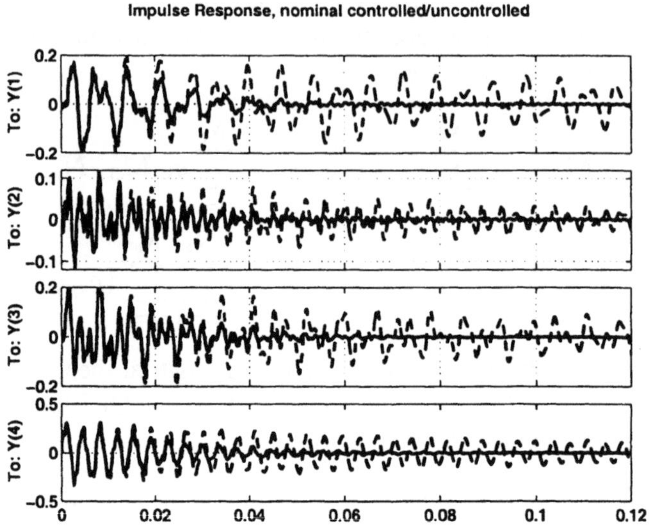

Figure 2 Comparison of Impulse-Responses Uncontrolled/Controlled (nominal case)

are available for control, one is used as a disturbance source. The model has an order of 28.
The goal of the control design is to augment the damping of the low-frequency modes, so, that disturbances of impulse-type die out within 0.1 seconds. This can be achieved, if the real-part of the all modes is ≤ 60. In the case of actuator or controller failure, the system is supposed to come to rest after 0.2 seconds. Analysis of the system shows, that it does not suffice to use a SISO controller, because there is no signal-path which allows to stabilise all relevant modes sufficiently. Therefore we have to use two controllers. We have chosen to use the first sensor and actuator for the first, the third actuator and sensor for the second controller. Both controllers were designed sequentially, starting with $H_{11}(s)$, the controller between sensor 1 and actuator 1. For the design of $H_{33}(s)$, the first loop was closed. The results of the nominal case (both loops closed) can be seen in figure 2.

To show how the uncertainty can be handled, a look at the closure of the second loop is helpful. For this simple case it will be sufficient to look at the extreme values of controller operation. Then the controller works either on the controlled or uncontrolled system. Figure 3 shows the Bode-plots of both cases. The only visible differences between the plots are around the resonances of the first and fourth mode. As the fourth mode is almost uncontrollable, we can restrict our attention to the first mode.

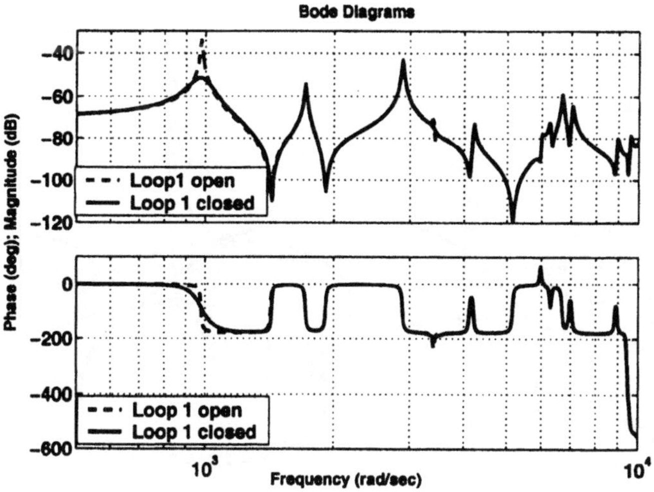

Figure 3 Comparison of Bode-plots for G(3,3) controlled/uncontrolled

After designing controller $H_{33}(s)$ (assuming controller $H_{11}(s)$ is fully functional), we use the results of Kidron and Yaniv [4] to investigate, how the controller $H_{33}(s)$ will perform on the system. Figure 4 shows, that the input disturbance attenuation varies from 0.3 to 15 dB, depending on the status of the controller H_{11}. The magnitude of the first mode for disturbances at the input will then be between -50 and -52 decibels, which roughly corresponds to a damping of 4 to 7 percent or a real part of -39 to -68, which gives us the desired performance.

6. Conclusions

We have presented a method for the design of decentralised controllers for vibration suppression based on Quantitative Feedback Theory. The method can be used to obtain robust control performance while at the same time system integrity can be considered. The method treats the interaction of the controllers as model uncertainty and uses standard QFT approaches to deal with it.
Our method will be tested in experimental investigations. Some further mathematical analysis will be necessary, e.g. the analysis of conservativity.

Acknowledgements

This work has been funded by the DEUTSCHE FORSCHUNGSGEMEINSCHAFT in the project Innovationskolleg ADAMES, project number INK 25/A1-1. This support is gratefully acknowledged.

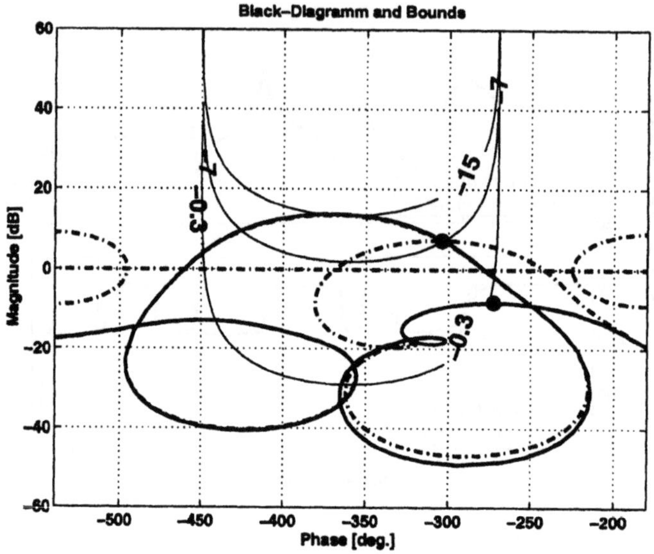

Figure 4 Black-Diagram and Bounds for the two cases

References

[1] C.R. Fuller, S.J. Elliott, and P.A. Nelson. *Active Control of Vibration*. Academic Press, London, 1996.

[2] I.M. Horowitz. *Quantitative Feedback Design Theory*. QFT Publications, Boulder, CO, 1993.

[3] C.H. Houpis and S.J. Rasmussen. *Quantitative Feedback Theory - Fundamentals and Applications*. Marcel Dekker Inc., New York, 1999.

[4] O. Kidron and O. Yaniv. Robust control of uncertain resonant systems. *European Journal of Control*, 1:104–112, 1995.

[5] I.D. Landau, D. Rey, A. Karimi, A. Voda, and A. Franco. A flexible transmission system as a benchmark for robust digital control. *European Journal of Control*, 1(2):77–96, 1995.

[6] J. Langer and A. Constantinescu. Pole-placement design using convex optimisation criteria for the flexible transmission benchmark. *European Journal of Control*, 5(2), 1999.

[7] D.Q. Mayne. The design of linear multivariable systems. *Automatica*, 9:201–207, 1973.

[8] L.H. Keel S.P. Bhattacharyya, H. Chapellat. *Robust Control: The Parametric Approach*. Prentice-Hall, Englewood-Cliffs, N.J., 1995.

[9] O. Yaniv. *Quantitative Feedback Design of Linear and Nonlinear Control Systems*. Kluwer Academic Publishers, Boston, 1999.

MULTI-OBJECTIVE CONTROLLER DESIGN FOR SMART STRUCTURES USING LINEAR MATRIX INEQUALITIES

SRIDHAR SANA, VITTAL S. RAO
Department of Electrical & Computer Engineering and
Intelligent Systems Center,
University of Missouri-Rolla,
Rolla, MO 65401, U.S.A.

1. Introduction

Smart structures have emerged as a result of integration of research in the areas of sensors, actuators, micro electronics, signal processing and controller design. In this paper, we develop an integrated controller design procedure for disturbance rejection and performance optimization in smart structures. There are two main challenges in such an integration. The first problem known as *spill over* problem (see [1]), is the degradation of the performance due to the effect of unmodeled dynamics on the closed-loop system. The second challenge is the constraint on the available actuation force which can also limit the achievable performance. Hence to design controllers for effective integration with structural systems, it is necessary that these constraints are incorporated in the controller design process, necessitating a multi-objective design approach. In recent times, Linear matrix inequalities(LMIs) (see [2]) have emerged as a powerful tool for formulating and solving such multi-objective design problems. We formulated the integration problem in smart structures as a problem of designing an output feedback robust controller in the presence of uncertainties due to unmodeled dynamics and control input limits to achieve maximum possible attenuation for a given set of finite energy disturbances. The proposed method is employed to design a controller for a smart structural test article. The controller is then implemented using dSpace system and experimental results are included.

2. Problem Formulation

The problems associated with integration of controllers with structural systems can be formulated as an output feedback robust controller design problem in the presence of uncertainty due to unmodeled dynamics and control input constraints due to limited actuation force. Actuator limits usually reflect as limits on the energy or peak of the control input. In this paper, we consider peak limits on the control input in accordance with the situation encountered in the application of piezo electric actuators. It is desired to achieve maximum possible disturbance rejection to arbitrary finite energy disturbances. These disturbances per-

U. Gabbert and H.S. Tzou (eds.), IUTAM Symposium on Smart Structures and Structronic Systems, 347–354.

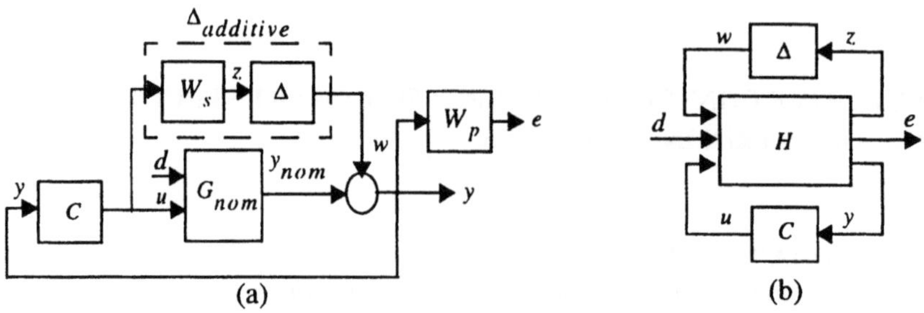

Figure 1. (a) Uncertainty formulation; (b) LFR of the generalized plant

sist only for a finite duration and represent the typical disturbances encountered by structural systems. In what follows, we formally define the specifications on the control system.

The uncertainty formulation used in the design of the robust controller for the proposed integration of controllers in smart structures is shown in Figure 1(a). In this figure, G_{nom} represents the nominal model of the structural system consisting of modes of interest. Typically, structural systems are infinite dimensional in nature and the corresponding modeling errors are represented as additive uncertainty $\Delta_{additive}$, as shown in Figure 1(a). The vector $u(t)$ is the control input, $y_{nom}(t)$ is the sensor output, $d(t)$ is the disturbance input and $w(t)$ is the contribution due to the modeling error.

The additive uncertainty $\Delta_{additive}$ is normalized using a weighting function W_s as:

$$\Delta_{additive} = \Delta W_s ; \ \|\Delta\|_{\infty} \leq 1 \tag{1}$$

The weighting function W_s is selected such that it forms the lowest upper bound of the additive uncertainty due to unmodeled dynamics.

W_p is the weighting function for the output on which the effect of disturbance needs to be minimized. In practice, it is required to achieve disturbance attenuation in a frequency range of interest. This requirement can be incorporated by appropriately shaping the weighting function W_p. The uncertainty formulation shown in Figure 1(a), can be represented in the linear fractional representation(LFR) as shown in Figure 1(b). The generalized plant is given by:

$$\begin{bmatrix} \dot{x} \\ z \\ e \\ y \end{bmatrix} = [H] \begin{bmatrix} x \\ w \\ d \\ u \end{bmatrix} = \begin{bmatrix} A & B_w & B_d & B_u \\ C_z & 0 & 0 & D_{zu} \\ C_e & D_{ew} & D_{ed} & D_{eu} \\ C_y & I & D_{yd} & D_{yu} \end{bmatrix} \begin{bmatrix} x \\ w \\ d \\ u \end{bmatrix} ; \ z = \Delta w \tag{2}$$

with C as the feed-back controller.

We consider the design of strictly proper output feedback controllers with the same order as the generalized plant and is given by,

$$\begin{bmatrix} \dot{x}_c \\ u \end{bmatrix} = \begin{bmatrix} A_c & B_c \\ C_c & D_c \end{bmatrix} \begin{bmatrix} x_c \\ y \end{bmatrix} \text{ with } D_c = 0 \tag{3}$$

The strictly proper controller simplifies the formulation and eliminates the well-posedness problems that might arise.

The corresponding closed loop system can be written as,

$$\begin{bmatrix} \dot{\tilde{x}} \\ z \\ e \end{bmatrix} = \begin{bmatrix} \tilde{A} & \tilde{B}_w & \tilde{B}_d \\ \tilde{C}_z & 0 & 0 \\ \tilde{C}_e & \tilde{D}_{ew} & \tilde{D}_{ed} \end{bmatrix} \begin{bmatrix} \tilde{x} \\ w \\ d \end{bmatrix}; \; u = \tilde{C}_u \tilde{x}; \; z = \Delta w \tag{4}$$

We define the set of finite energy disturbances utilized in our formulation as:

$$F = \left\{ d(t) \,\middle|\, \int_0^\infty d(t)^T d(t) dt < d_{max} < \infty \right\} \tag{5}$$

where d_{max} is the maximum bound on the energy any disturbance in F can have.

2.1. CONTROLLER DESIGN SPECIFICATIONS:

1. Robust stability for the normalized unmodeled dynamics given by Δ.
2. The L_2 gain from disturbance to the performance output given by $\|T_{ed}\|_\infty$ is minimized.
3. For all disturbances $d(t) \in F$, each the control inputs $u_{max,\,i}$ must satisfy,

$$|u_i(t)| \le u_{max,\,i} \text{ for all } t \ge 0 \text{ and } i = 1, 2, \ldots, n_u \tag{6}$$

3. Controller design

In this section, we give a procedure for the design of an output feedback robust controller which satisfies the given specifications. We show the matrix inequality formulation of each of the constraints, namely the robust disturbance rejection and the control input constraints. Based on these matrix inequality constraints we formulate an optimization problem involving LMI constraints for the solution of the proposed controller design.

3.1. ROBUST DISTURBANCE REJECTION

To formulate robust stability constraint for the additive uncertainty due to unmodeled dynamics, we consider the Lyapunov function as shown below,

$$V(\tilde{x}) = \tilde{x}^T \tilde{P} \tilde{x} + \int_0^t z^T z dt - \int_0^t w^T w dt \text{ where } \tilde{P} = \tilde{P}^T > 0 \tag{7}$$

Along with the robust stability, an upper bound γ on the L_2 gain of the system from d and e can be guaranteed if there exist a $\tilde{P} > 0$ such that,

$$\dot{V}(\tilde{x}) + \gamma^{-2} e^T e - d^T d < 0 \text{ with } \tilde{x}(0) = 0 \tag{8}$$

which is equivalent to finding $\tilde{P} > 0$ such that,

$$\begin{bmatrix} \tilde{A}^T \tilde{P} + \tilde{P} \tilde{A} & \tilde{P} \tilde{B}_w & \tilde{P} \tilde{B}_d & \tilde{C}_z^T & \tilde{C}_e^T \\ \tilde{B}_w^T \tilde{P} & -I & 0 & 0 & \tilde{D}_{ew}^T \\ \tilde{B}_d^T \tilde{P} & 0 & -I & 0 & \tilde{D}_{ed}^T \\ \tilde{C}_z & 0 & 0 & -I & 0 \\ \tilde{C}_e & \tilde{D}_{ew} & \tilde{D}_{ed} & 0 & -\gamma^2 I \end{bmatrix} < 0 \tag{9}$$

3.2. CONTROL INPUT CONSTRAINT

We present a matrix inequality formulation of the control input constraints given in (6) for all the disturbances defined in (5) in this subsection. For the derivation of this matrix inequality constraint, we obtain a bounding ellipsoid condition for all closed-loop state trajectories due to all the disturbances specified in F.
From the Lyapunov condition in (8), we have the ellipsoid condition,

$$\tilde{x}(t)^T \tilde{P} \tilde{x}(t) < d_{max} \text{ for any } d(t) \in F \text{ for all } t \ge 0 \tag{10}$$

Now to ensure that the control input limits in (6) are satisfied we need,

$$\tilde{x}(t)^T \tilde{C}_{u,i}^T \tilde{C}_{u,i} \tilde{x}(t) < u_{max,i}^2 ;\ \forall t \ge 0 \text{ for all } i = 1, 2, \ldots, n_u \tag{11}$$

where $\tilde{C}_{u,i}$ is the ith row of the matrix $\tilde{C}_u$.

Combining (10) and (11), the control input limit conditions can be expressed as,

$$\begin{bmatrix} \tilde{P} & \tilde{C}_{u,i}^T \\ \tilde{C}_{u,i} & \frac{u_{max,i}^2}{d_{max}} \end{bmatrix} > 0 \quad \text{for all } i = 1, 2, \ldots, n_u \tag{12}$$

3.3. OPTIMIZATION PROBLEM

The controller design problem can now be stated as an optimization problem involving $\tilde{P}$, A_c, B_c, C_c as,

minimize γ^2,

$$\text{subject to constraints (9), (12) and } \tilde{P} > 0 \tag{13}$$

The optimization problem involves the constraint (9), which is nonlinear in the unknown parameters $\tilde{P}$, A_c, B_c, C_c rendering it non-convex and hence can not be solved by the use of interior point methods (see [3]). To convert the above problem into an equivalent convex optimization we utilize the parameterization (see [4],[5]):

$$\tilde{P} = \begin{bmatrix} P & M \\ M^T & R \end{bmatrix}, \tilde{Q} = \tilde{P}^{-1} = \begin{bmatrix} Q & N \\ N^T & S \end{bmatrix} \tag{14}$$

where P and $Q \in R^{nxn}$. The matrices M and N are invertible and related as $N = (I - QP)M^{-T}$. In this case the matrices R and S are given by,

$$R = M^T(P - Q^{-1})^{-1}M \text{ and } S = N^T(Q - P^{-1})^{-1}N \tag{15}$$

Using these parameterizations the optimization problem in (13) can be restated as an optimization problem involving P, Q, A_K, B_K and C_K as,

minimize γ^2 subject to,

$$\begin{bmatrix} Q & I & C_{K,i}^T \\ I & P & 0 \\ C_{K,i} & 0 & \frac{u_{max}^2}{d_{max}} \end{bmatrix} > 0 \quad \text{for all } i = 1, 2, \ldots, n_u \tag{16}$$

$$\begin{bmatrix} \Phi_1^T + \Phi_1 & \Phi_2 & \Phi_3 & \Phi_4 & \Phi_5 \\ \Phi_2^T & -I & 0 & 0 & D_{ew}^T \\ \Phi_3^T & 0 & -I & 0 & D_{ed}^T \\ \Phi_4^T & 0 & 0 & -I & 0 \\ \Phi_5^T & D_{ew} & D_{ed} & 0 & -\gamma^2 I \end{bmatrix} < 0 \tag{17}$$

$$\text{and} \begin{bmatrix} P & I \\ I & Q \end{bmatrix} > 0 \tag{18}$$

where

$$\Phi_1 = \begin{bmatrix} AQ + B_u C_K & A \\ A_K^T & PA + B_K C_y \end{bmatrix}, \Phi_2 = \begin{bmatrix} B_w \\ PB_w + B_K D_{yw} \end{bmatrix}, \Phi_3 = \begin{bmatrix} B_d \\ PB_d + B_K D_{yd} \end{bmatrix},$$

$$\Phi_4 = \begin{bmatrix} QC_z^T + C_K^T D_{zu}^T \\ C_z^T \end{bmatrix}, \Phi_5 = \begin{bmatrix} QC_e^T + C_K^T D_{eu}^T \\ C_e^T \end{bmatrix} \tag{19}$$

and

$$A_K = PAQ + PB_u C_K + B_K C_y Q + MA_c N^T + B_K D_{yu} C_K, \; B_K = MB_c \text{ and } C_K = C_c N^T \tag{20}$$

The controller parameters A_c, B_c, C_c can be determined by solving (20) from A_K, B_K and C_K obtained by solving the optimization problem.

4. Results

To demonstrate the application of the robust controller design procedure, we consider a cantilever beam smart structure.The cantilever beam is made of aluminum with length 54.1 cm, width 1.27 cm and thickness 0.3175 cm. This beam is mounted with PZT(Lead Zirconite Titanate) patches which form the two actuators and the sensor shown in the Figure 2. The actuators are driven by power amplifiers with maximum input limits of +/-5v. The plant model used in the design uses the actuator 1 as control input. Actuator 2 is used to simulate the disturbance input to the structure. An eighth order state space model is obtained for the beam using frequency response measurements. The first four modes of the structures are at 9.53, 57.88, 161.51 and 314.81Hz and are included in the plant as well as in the disturbance models. All of the modes have a damping of 0.7%.

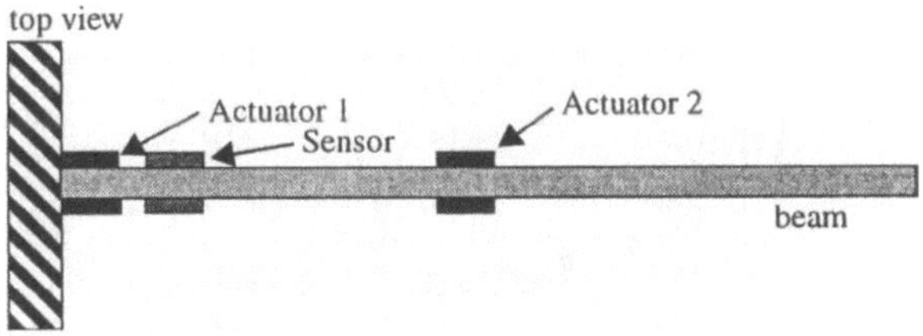

Figure 2. Cantilevered beam smart structure

In our design, we consider all disturbances with energy not exceeding 0.5 $volt^2$sec. The weighting functions used for normalizing the unmodeled dynamics and weighting the sensor output respectively are given by,

$$W_s = \frac{3.8s^2}{(s+3000)^2}\ ;\ W_p = \frac{12000s}{(s+5)(s+3500)} \tag{21}$$

The comparison of the frequency responses of the nominal plant, modeling errors with the frequency responses of the weighting functions is shown in Figure 3(a). With these weighting functions a 12th order generalized plant is obtained. Based on this generalized plant we design a 12th order controller. The controller achieved a reduction of 14.62dB, 11dB and 5.1dB in the first, second and fourth modes respectively. It is shown in Figure 3(b), that this controller also satisfies the following sufficient condition for robustness:

$$T_{uw}(\omega)\Delta_{additive}(\omega) < 1 \ \text{ for } \forall\omega > 0 \tag{22}$$

For testing the controller's experimental performance, we utilized the dSpace rapid prototyping system for implementation of the controller. The frequency responses of the closed-loop and the open-loop system are obtained and shown in Figure 4. Next, the closed -loop system is tested with a sinusoidal pulse disturbance at the first mode frequency with energy 0.5 $volt^2$sec. From the time responses (see Figure 5) it can be seen that the controller achieved good disturbance rejection while satisfying the control input limits.

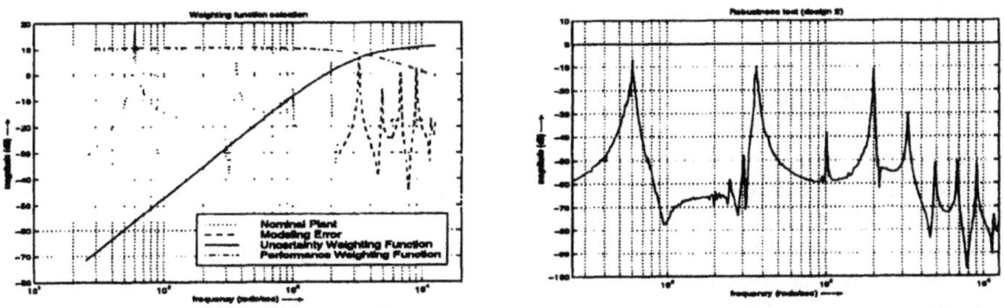

(a) (b)

Figure 3. (a) Weighting functions (b) Robustness Test

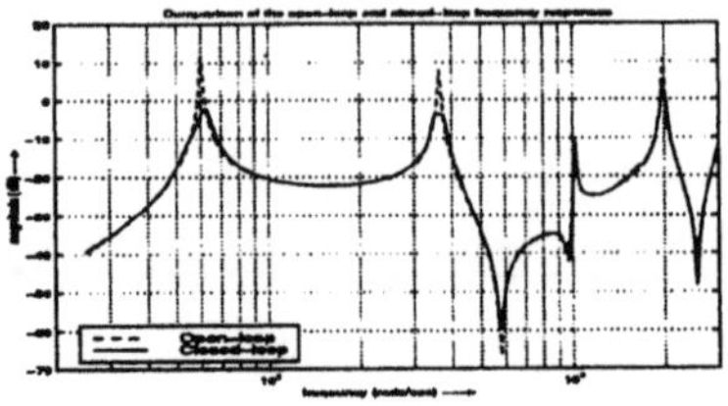

Figure 4. Experimental comparison of the open-loop and closed-loop frequency response

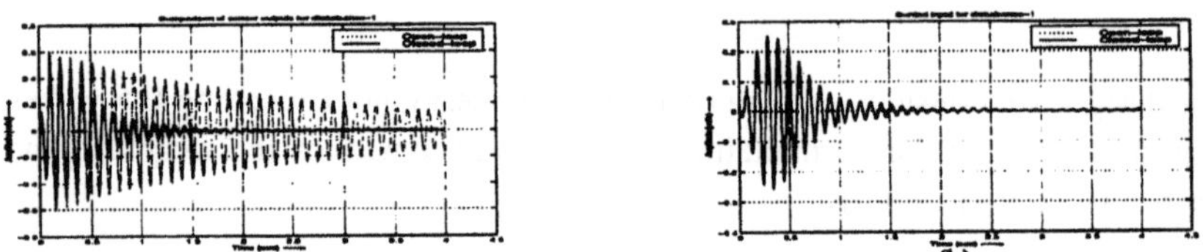

(a) (b)

Figure 5. Experimental disturbance time responses (a) Sensor (b) Control input

5. Conclusions

A multi-objective optimization procedure utilizing LMIs is formulated to design an output feedback robust disturbance rejection controller in the presence of control input limits and unmodeled dynamics for integration with smart structural systems. The design procedure is successfully tested on an experimental smart structure. Using robustness tests, it is proved that the closed loop system is unaffected by the unmodeled dynamics. Because of the use of weighting functions in the formulation of the robust controller design problem, the order of the controller is higher than that of the nominal plant. In cases, where there is a limitation on the size of the controller hardware, it desirable to have lower order controllers. We are currently investigating LMI formulations for designing lower order controllers for integration with smart structures.

Acknowledgments

This research is supported by National Science Foundation(Grant EEC - 9872392).

References

1. M. J. Balas, "Trends in large space structures control theory: Fondest hopes, wildest dreams," *IEEE Trans. Automatic Control*, vol. 27, no. 6, pp. 522-535, 1982.
2. S. Boyd et. al, *Linear matrix inequalities in Systems and control theory*, vol 15 of SIAM, Philadelphia, PA: SIAM, 1994.
3. Y. Nesterov and A. Nemirovsky., *Interior point polynomial methods in convex programming: theory and applications*. SIAM, 1993.
4. C. Scherer, P. Gahinet and M. Chilali, "Multi objective Output-Feedback Control via LMI Optimization," *IEEE Tr AC*, vol. 42, no. 7, pp. 896-911, 1997.
5. A. Packard, K. Zhou, P. Pandey and G. Becker, "A collection of Robust Control Problems Leading to LMI's," *Proc. IEEE Conf. on Decision and Control*, pp. 1245-1250, 1991.

Index of Authors

Mechanics

SOLID MECHANICS AND ITS APPLICATIONS

Series Editor: G.M.L. Gladwell

Aims and Scope of the Series

The fundamental questions arising in mechanics are: *Why?*, *How?*, and *How much?* The aim of this series is to provide lucid accounts written by authoritative researchers giving vision and insight in answering these questions on the subject of mechanics as it relates to solids. The scope of the series covers the entire spectrum of solid mechanics. Thus it includes the foundation of mechanics; variational formulations; computational mechanics; statics, kinematics and dynamics of rigid and elastic bodies; vibrations of solids and structures; dynamical systems and chaos; the theories of elasticity, plasticity and viscoelasticity; composite materials; rods, beams, shells and membranes; structural control and stability; soils, rocks and geomechanics; fracture; tribology; experimental mechanics; biomechanics and machine design.

1. R.T. Haftka, Z. Gürdal and M.P. Kamat: *Elements of Structural Optimization.* 2nd rev.ed., 1990 ISBN 0-7923-0608-2
2. J.J. Kalker: *Three-Dimensional Elastic Bodies in Rolling Contact.* 1990 ISBN 0-7923-0712-7
3. P. Karasudhi: *Foundations of Solid Mechanics.* 1991 ISBN 0-7923-0772-0
4. *Not published*
5. *Not published.*
6. J.F. Doyle: *Static and Dynamic Analysis of Structures.* With an Emphasis on Mechanics and Computer Matrix Methods. 1991 ISBN 0-7923-1124-8; Pb 0-7923-1208-2
7. O.O. Ochoa and J.N. Reddy: *Finite Element Analysis of Composite Laminates.* ISBN 0-7923-1125-6
8. M.H. Aliabadi and D.P. Rooke: *Numerical Fracture Mechanics.* ISBN 0-7923-1175-2
9. J. Angeles and C.S. López-Cajún: *Optimization of Cam Mechanisms.* 1991 ISBN 0-7923-1355-0
10. D.E. Grierson, A. Franchi and P. Riva (eds.): *Progress in Structural Engineering.* 1991 ISBN 0-7923-1396-8
11. R.T. Haftka and Z. Gürdal: *Elements of Structural Optimization.* 3rd rev. and exp. ed. 1992 ISBN 0-7923-1504-9; Pb 0-7923-1505-7
12. J.R. Barber: *Elasticity.* 1992 ISBN 0-7923-1609-6; Pb 0-7923-1610-X
13. H.S. Tzou and G.L. Anderson (eds.): *Intelligent Structural Systems.* 1992 ISBN 0-7923-1920-6
14. E.E. Gdoutos: *Fracture Mechanics.* An Introduction. 1993 ISBN 0-7923-1932-X
15. J.P. Ward: *Solid Mechanics.* An Introduction. 1992 ISBN 0-7923-1949-4
16. M. Farshad: *Design and Analysis of Shell Structures.* 1992 ISBN 0-7923-1950-8
17. H.S. Tzou and T. Fukuda (eds.): *Precision Sensors, Actuators and Systems.* 1992 ISBN 0-7923-2015-8
18. J.R. Vinson: *The Behavior of Shells Composed of Isotropic and Composite Materials.* 1993 ISBN 0-7923-2113-8
19. H.S. Tzou: *Piezoelectric Shells.* Distributed Sensing and Control of Continua. 1993 ISBN 0-7923-2186-3
20. W. Schiehlen (ed.): *Advanced Multibody System Dynamics.* Simulation and Software Tools. 1993 ISBN 0-7923-2192-8
21. C.-W. Lee: *Vibration Analysis of Rotors.* 1993 ISBN 0-7923-2300-9
22. D.R. Smith: *An Introduction to Continuum Mechanics.* 1993 ISBN 0-7923-2454-4
23. G.M.L. Gladwell: *Inverse Problems in Scattering.* An Introduction. 1993 ISBN 0-7923-2478-1

Mechanics

SOLID MECHANICS AND ITS APPLICATIONS

Series Editor: G.M.L. Gladwell

24. G. Prathap: *The Finite Element Method in Structural Mechanics.* 1993 ISBN 0-7923-2492-7
25. J. Herskovits (ed.): *Advances in Structural Optimization.* 1995 ISBN 0-7923-2510-9
26. M.A. González-Palacios and J. Angeles: *Cam Synthesis.* 1993 ISBN 0-7923-2536-2
27. W.S. Hall: *The Boundary Element Method.* 1993 ISBN 0-7923-2580-X
28. J. Angeles, G. Hommel and P. Kovács (eds.): *Computational Kinematics.* 1993 ISBN 0-7923-2585-0
29. A. Curnier: *Computational Methods in Solid Mechanics.* 1994 ISBN 0-7923-2761-6
30. D.A. Hills and D. Nowell: *Mechanics of Fretting Fatigue.* 1994 ISBN 0-7923-2866-3
31. B. Tabarrok and F.P.J. Rimrott: *Variational Methods and Complementary Formulations in Dynamics.* 1994 ISBN 0-7923-2923-6
32. E.H. Dowell (ed.), E.F. Crawley, H.C. Curtiss Jr., D.A. Peters, R. H. Scanlan and F. Sisto: *A Modern Course in Aeroelasticity.* Third Revised and Enlarged Edition. 1995 ISBN 0-7923-2788-8; Pb: 0-7923-2789-6
33. A. Preumont: *Random Vibration and Spectral Analysis.* 1994 ISBN 0-7923-3036-6
34. J.N. Reddy (ed.): *Mechanics of Composite Materials.* Selected works of Nicholas J. Pagano. 1994 ISBN 0-7923-3041-2
35. A.P.S. Selvadurai (ed.): *Mechanics of Poroelastic Media.* 1996 ISBN 0-7923-3329-2
36. Z. Mróz, D. Weichert, S. Dorosz (eds.): *Inelastic Behaviour of Structures under Variable Loads.* 1995 ISBN 0-7923-3397-7
37. R. Pyrz (ed.): *IUTAM Symposium on Microstructure-Property Interactions in Composite Materials.* Proceedings of the IUTAM Symposium held in Aalborg, Denmark. 1995 ISBN 0-7923-3427-2
38. M.I. Friswell and J.E. Mottershead: *Finite Element Model Updating in Structural Dynamics.* 1995 ISBN 0-7923-3431-0
39. D.F. Parker and A.H. England (eds.): *IUTAM Symposium on Anisotropy, Inhomogeneity and Nonlinearity in Solid Mechanics.* Proceedings of the IUTAM Symposium held in Nottingham, U.K. 1995 ISBN 0-7923-3594-5
40. J.-P. Merlet and B. Ravani (eds.): *Computational Kinematics '95.* 1995 ISBN 0-7923-3673-9
41. L.P. Lebedev, I.I. Vorovich and G.M.L. Gladwell: *Functional Analysis.* Applications in Mechanics and Inverse Problems. 1996 ISBN 0-7923-3849-9
42. J. Menčik: *Mechanics of Components with Treated or Coated Surfaces.* 1996 ISBN 0-7923-3700-X
43. D. Bestle and W. Schiehlen (eds.): *IUTAM Symposium on Optimization of Mechanical Systems.* Proceedings of the IUTAM Symposium held in Stuttgart, Germany. 1996 ISBN 0-7923-3830-8
44. D.A. Hills, P.A. Kelly, D.N. Dai and A.M. Korsunsky: *Solution of Crack Problems.* The Distributed Dislocation Technique. 1996 ISBN 0-7923-3848-0
45. V.A. Squire, R.J. Hosking, A.D. Kerr and P.J. Langhorne: *Moving Loads on Ice Plates.* 1996 ISBN 0-7923-3953-3
46. A. Pineau and A. Zaoui (eds.): *IUTAM Symposium on Micromechanics of Plasticity and Damage of Multiphase Materials.* Proceedings of the IUTAM Symposium held in Sèvres, Paris, France. 1996 ISBN 0-7923-4188-0
47. A. Naess and S. Krenk (eds.): *IUTAM Symposium on Advances in Nonlinear Stochastic Mechanics.* Proceedings of the IUTAM Symposium held in Trondheim, Norway. 1996 ISBN 0-7923-4193-7
48. D. Iesan and A. Scalia: *Thermoelastic Deformations.* 1996 ISBN 0-7923-4230-5

Mechanics

SOLID MECHANICS AND ITS APPLICATIONS

Series Editor: G.M.L. Gladwell

49. J.R. Willis (ed.): *IUTAM Symposium on Nonlinear Analysis of Fracture*. Proceedings of the IUTAM Symposium held in Cambridge, U.K. 1997 ISBN 0-7923-4378-6
50. A. Preumont: *Vibration Control of Active Structures*. An Introduction. 1997 ISBN 0-7923-4392-1
51. G.P. Cherepanov: *Methods of Fracture Mechanics: Solid Matter Physics*. 1997 ISBN 0-7923-4408-1
52. D.H. van Campen (ed.): *IUTAM Symposium on Interaction between Dynamics and Control in Advanced Mechanical Systems*. Proceedings of the IUTAM Symposium held in Eindhoven, The Netherlands. 1997 ISBN 0-7923-4429-4
53. N.A. Fleck and A.C.F. Cocks (eds.): *IUTAM Symposium on Mechanics of Granular and Porous Materials*. Proceedings of the IUTAM Symposium held in Cambridge, U.K. 1997 ISBN 0-7923-4553-3
54. J. Roorda and N.K. Srivastava (eds.): *Trends in Structural Mechanics*. Theory, Practice, Education. 1997 ISBN 0-7923-4603-3
55. Yu.A. Mitropolskii and N. Van Dao: *Applied Asymptotic Methods in Nonlinear Oscillations*. 1997 ISBN 0-7923-4605-X
56. C. Guedes Soares (ed.): *Probabilistic Methods for Structural Design*. 1997 ISBN 0-7923-4670-X
57. D. François, A. Pineau and A. Zaoui: *Mechanical Behaviour of Materials*. Volume I: Elasticity and Plasticity. 1998 ISBN 0-7923-4894-X
58. D. François, A. Pineau and A. Zaoui: *Mechanical Behaviour of Materials*. Volume II: Viscoplasticity, Damage, Fracture and Contact Mechanics. 1998 ISBN 0-7923-4895-8
59. L.T. Tenek and J. Argyris: *Finite Element Analysis for Composite Structures*. 1998 ISBN 0-7923-4899-0
60. Y.A. Bahei-El-Din and G.J. Dvorak (eds.): *IUTAM Symposium on Transformation Problems in Composite and Active Materials*. Proceedings of the IUTAM Symposium held in Cairo, Egypt. 1998 ISBN 0-7923-5122-3
61. I.G. Goryacheva: *Contact Mechanics in Tribology*. 1998 ISBN 0-7923-5257-2
62. O.T. Bruhns and E. Stein (eds.): *IUTAM Symposium on Micro- and Macrostructural Aspects of Thermoplasticity*. Proceedings of the IUTAM Symposium held in Bochum, Germany. 1999 ISBN 0-7923-5265-3
63. F.C. Moon: *IUTAM Symposium on New Applications of Nonlinear and Chaotic Dynamics in Mechanics*. Proceedings of the IUTAM Symposium held in Ithaca, NY, USA. 1998 ISBN 0-7923-5276-9
64. R. Wang: *IUTAM Symposium on Rheology of Bodies with Defects*. Proceedings of the IUTAM Symposium held in Beijing, China. 1999 ISBN 0-7923-5297-1
65. Yu.I. Dimitrienko: *Thermomechanics of Composites under High Temperatures*. 1999 ISBN 0-7923-4899-0
66. P. Argoul, M. Frémond and Q.S. Nguyen (eds.): *IUTAM Symposium on Variations of Domains and Free-Boundary Problems in Solid Mechanics*. Proceedings of the IUTAM Symposium held in Paris, France. 1999 ISBN 0-7923-5450-8
67. F.J. Fahy and W.G. Price (eds.): *IUTAM Symposium on Statistical Energy Analysis*. Proceedings of the IUTAM Symposium held in Southampton, U.K. 1999 ISBN 0-7923-5457-5
68. H.A. Mang and F.G. Rammerstorfer (eds.): *IUTAM Symposium on Discretization Methods in Structural Mechanics*. Proceedings of the IUTAM Symposium held in Vienna, Austria. 1999 ISBN 0-7923-5591-1

Mechanics

SOLID MECHANICS AND ITS APPLICATIONS

Series Editor: G.M.L. Gladwell

69. P. Pedersen and M.P. Bendsøe (eds.): *IUTAM Symposium on Synthesis in Bio Solid Mechanics*. Proceedings of the IUTAM Symposium held in Copenhagen, Denmark. 1999 ISBN 0-7923-5615-2
70. S.K. Agrawal and B.C. Fabien: *Optimization of Dynamic Systems*. 1999 ISBN 0-7923-5681-0
71. A. Carpinteri: *Nonlinear Crack Models for Nonmetallic Materials*. 1999 ISBN 0-7923-5750-7
72. F. Pfeifer (ed.): *IUTAM Symposium on Unilateral Multibody Contacts*. Proceedings of the IUTAM Symposium held in Munich, Germany. 1999 ISBN 0-7923-6030-3
73. E. Lavendelis and M. Zakrzhevsky (eds.): *IUTAM/IFToMM Symposium on Synthesis of Nonlinear Dynamical Systems*. Proceedings of the IUTAM/IFToMM Symposium held in Riga, Latvia. 2000 ISBN 0-7923-6106-7
74. J.-P. Merlet: *Parallel Robots*. 2000 ISBN 0-7923-6308-6
75. J.T. Pindera: *Techniques of Tomographic Isodyne Stress Analysis*. 2000 ISBN 0-7923-6388-4
76. G.A. Maugin, R. Drouot and F. Sidoroff (eds.): *Continuum Thermomechanics*. The Art and Science of Modelling Material Behaviour. 2000 ISBN 0-7923-6407-4
77. N. Van Dao and E.J. Kreuzer (eds.): *IUTAM Symposium on Recent Developments in Non-linear Oscillations of Mechanical Systems*. 2000 ISBN 0-7923-6470-8
78. S.D. Akbarov and A.N. Guz: *Mechanics of Curved Composites*. 2000 ISBN 0-7923-6477-5
79. M.B. Rubin: *Cosserat Theories: Shells, Rods and Points*. 2000 ISBN 0-7923-6489-9
80. S. Pellegrino and S.D. Guest (eds.): *IUTAM-IASS Symposium on Deployable Structures: Theory and Applications*. Proceedings of the IUTAM-IASS Symposium held in Cambridge, U.K., 6–9 September 1998. 2000 ISBN 0-7923-6516-X
81. A.D. Rosato and D.L. Blackmore (eds.): *IUTAM Symposium on Segregation in Granular Flows*. Proceedings of the IUTAM Symposium held in Cape May, NJ, U.S.A., June 5–10, 1999. 2000 ISBN 0-7923-6547-X
82. A. Lagarde (ed.): *IUTAM Symposium on Advanced Optical Methods and Applications in Solid Mechanics*. Proceedings of the IUTAM Symposium held in Futuroscope, Poitiers, France, August 31–September 4, 1998. 2000 ISBN 0-7923-6604-2
83. D. Weichert and G. Maier (eds.): *Inelastic Analysis of Structures under Variable Loads*. Theory and Engineering Applications. 2000 ISBN 0-7923-6645-X
84. T.-J. Chuang and J.W. Rudnicki (eds.): *Multiscale Deformation and Fracture in Materials and Structures*. The James R. Rice 60th Anniversary Volume. 2001 ISBN 0-7923-6718-9
85. S. Narayanan and R.N. Iyengar (eds.): *IUTAM Symposium on Nonlinearity and Stochastic Structural Dynamics*. Proceedings of the IUTAM Symposium held in Madras, Chennai, India, 4–8 January 1999 ISBN 0-7923-6733-2
86. S. Murakami and N. Ohno (eds.): *IUTAM Symposium on Creep in Structures*. Proceedings of the IUTAM Symposium held in Nagoya, Japan, 3-7 April 2000. 2001 ISBN 0-7923-6737-5
87. W. Ehlers (ed.): *IUTAM Symposium on Theoretical and Numerical Methods in Continuum Mechanics of Porous Materials*. Proceedings of the IUTAM Symposium held at the University of Stuttgart, Germany, September 5-10, 1999. 2001 ISBN 0-7923-6766-9

Kluwer Academic Publishers – Dordrecht / Boston / London

ICASE/LaRC Interdisciplinary Series in Science and Engineering

1. J. Buckmaster, T.L. Jackson and A. Kumar (eds.): *Combustion in High-Speed Flows.* 1994 ISBN 0-7923-2086-X
2. M.Y. Hussaini, T.B. Gatski and T.L. Jackson (eds.): *Transition, Turbulence and Combustion.* Volume I: Transition. 1994 ISBN 0-7923-3084-6; set 0-7923-3086-2
3. M.Y. Hussaini, T.B. Gatski and T.L. Jackson (eds.): *Transition, Turbulence and Combustion.* Volume II: Turbulence and Combustion. 1994 ISBN 0-7923-3085-4; set 0-7923-3086-2
4. D.E. Keyes, A. Sameh and V. Venkatakrishnan (eds): *Parallel Numerical Algorithms.* 1997 ISBN 0-7923-4282-8
5. T.G. Campbell, R.A. Nicolaides and M.D. Salas (eds.): *Computational Electromagnetics and Its Applications.* 1997 ISBN 0-7923-4733-1
6. V. Venkatakrishnan, M.D. Salas and S.R. Chakravarthy (eds.): *Barriers and Challenges in Computational Fluid Dynamics.* 1998 ISBN 0-7923-4855-9
7. M.D. Salas, J.N. Hefner and L. Sakell (eds.): *Modeling Complex Turbulent Flows.* 1999 ISBN 0-7923-5590-3

KLUWER ACADEMIC PUBLISHERS – DORDRECHT / BOSTON / LONDON

ERCOFTAC SERIES

1. A. Gyr and F.-S. Rys (eds.): *Diffusion and Transport of Pollutants in Atmospheric Mesoscale Flow Fields*. 1995 ISBN 0-7923-3260-1
2. M. Hallbäck, D.S. Henningson, A.V. Johansson and P.H. Alfredsson (eds.): *Turbulence and Transition Modelling*. Lecture Notes from the ERCOFTAC/IUTAM Summerschool held in Stockholm. 1996 ISBN 0-7923-4060-4
3. P. Wesseling (ed.): *High Performance Computing in Fluid Dynamics*. Proceedings of the Summerschool held in Delft, The Netherlands. 1996 ISBN 0-7923-4063-9
4. Th. Dracos (ed.): *Three-Dimensional Velocity and Vorticity Measuring and Image Analysis Techniques*. Lecture Notes from the Short Course held in Zürich, Switzerland. 1996 ISBN 0-7923-4256-9
5. J.-P. Chollet, P.R. Voke and L. Kleiser (eds.): *Direct and Large-Eddy Simulation II*. Proceedings of the ERCOFTAC Workshop held in Grenoble, France. 1997 ISBN 0-7923-4687-4

KLUWER ACADEMIC PUBLISHERS – DORDRECHT / BOSTON / LONDON

MIX
Papier aus verantwortungsvollen Quellen
Paper from responsible sources
FSC® C105338

If you have any concerns about our products,
you can contact us on
ProductSafety@springernature.com

In case Publisher is established outside the EU,
the EU authorized representative is:
Springer Nature Customer Service Center GmbH
Europaplatz 3, 69115 Heidelberg, Germany

Printed by Libri Plureos GmbH
in Hamburg, Germany